Advanced Green Chemistry

Part 2: From Catalysis to Chemistry Frontiers

Series on Chemistry, Energy and the Environment

ISSN: 2529-7716

Series Editors: Karl M. Kadish *(University of Houston, USA)*
Roger Guilard *(University of Bourgogne, France)*

The aim of this series on "Chemistry, Energy and the Environment" is to bring together authoritative contributions where the multidisciplinary character of Chemistry and its relationship to energy and the environment can be illustrated. In each volume, the latest advances in bio-, energy- and environmentally-related fields are described in a way that unambiguously shows the major contributions from chemists at the top of their field who describe new and improved technologies for the future.

Published

Vol. 6   *Advanced Green Chemistry*
Part 2: From Catalysis to Chemistry Frontiers
edited by István T. Horváth and Max Malacria

Vol. 5   *Bioinspired Chemistry: From Enzymes to Synthetic Models*
edited by Marius Réglier

Vol. 4   *Prospects for Li-ion Batteries and Emerging Energy*
Electrochemical Systems
edited by Laure Monconduit and Laurence Croguennec

Vol. 3   *Advanced Green Chemistry*
Part 1: Greener Organic Reactions and Processes
edited by István T. Horváth and Max Malacria

Vol. 2   *Elaboration and Applications of Metal-Organic Frameworks*
edited by Shengqian Ma and Jason A. Perman

Vol. 1   *Perovskite Solar Cells: Principle, Materials and Devices*
edited by Eric Wei-Guang Diau and Peter Chao-Yu Chen

❧ Volume 6 ☙

Series on Chemistry, Energy and the Environment

Advanced Green Chemistry

Part 2: From Catalysis to Chemistry Frontiers

Edited by

István T Horváth
City University of Hong Kong, Hong Kong

Max Malacria
ICSN, France & Sorbonne Université UPMC, France & CNRS, France

Series Editors

Karl M. Kadish
University of Houston, USA

Roger Guilard
Université de Bourgogne, France

World Scientific

NEW JERSEY · LONDON · SINGAPORE · BEIJING · SHANGHAI · HONG KONG · TAIPEI · CHENNAI · TOKYO

Published by

World Scientific Publishing Co. Pte. Ltd.

5 Toh Tuck Link, Singapore 596224

USA office: 27 Warren Street, Suite 401-402, Hackensack, NJ 07601

UK office: 57 Shelton Street, Covent Garden, London WC2H 9HE

Library of Congress Cataloging-in-Publication Data
Names: Horváth, István T. | Malacria, Max, 1949–   editor.
Title: Advanced green chemistry / editors, István T. Horváth, City University of Hong Kong,
    Hong Kong, Max Malacria, ICSN, France & Sorbonne Université UPMC, France.
Other titles: Green chemistry
Description: New Jersey : World Scientific, 2017- | Series: Series on chemistry, energy, and
    the environment ; volume 3 | Includes bibliographical references and index.
    Contents: part 1. Greener organic reactions and processes
Identifiers: LCCN 2017037720 | ISBN 9789813228108 (hardcover : alk. paper : pt. 1)
Subjects: LCSH: Green chemistry. | Environmental chemistry.
Classification: LCC TP155.2.E58 A37 2017 | DDC 660.028/6--dc23
LC record available at https://lccn.loc.gov/2017037720

Volume 6
Part 2: From Catalysis to Chemistry Frontiers
ISBN 978-981-121-057-0 (hardcover)
ISBN 978-981-121-058-7 (ebook for institutions)
ISBN 978-981-121-059-4 (ebook for individuals)

British Library Cataloguing-in-Publication Data
A catalogue record for this book is available from the British Library.

For any available supplementary material, please visit
https://www.worldscientific.com/worldscibooks/10.1142/11559#t=suppl

Typeset by Stallion Press
Email: enquiries@stallionpress.com

Printed in Singapore

Preface

Green chemistry grew rapidly since the beginning of the third millennium and is considered as a mature, fruitful and powerful area of research both at the academic and industrial level. Tremendous development has taken place in many important areas including renewable energy and resources, reaction environments, catalysis, synthesis, chemical biology, green polymers, and facile recycling. The combination of Green Chemistry with engineering, biology, toxicology, and physics will lead to novel interdisciplinary systems, which can now lift green chemistry to the next, advanced level.

In this addition of "Advanced Green Chemistry," the editors have assembled several authors among the best specialists in this growing area of research. This collection of reviews and perspectives provides an exciting vision of the more recent developments and the current status of several aspects of Green Chemistry. It illustrates the breath of the field and its leading role to address environmental and even health issues. These volumes will serve as books of reference showing a panoramic view of the field and a preview of its future direction as well as books of inspiration for those aiming to advance the frontiers. In Volume 2 of this collection, the emphasis is on the most recent developments in green catalysis, biosourced polymers and the study of continental organic matter for a better understanding of the carbon geochemical cycle.

In the first chapter, István Horváth provides a historical context to explain the pivotal role of chemistry in many aspects and challenges of humankind. The concepts of "green chemistry" and "sustainable chemistry" are separated from each other and precise definitions are provided. Although they are fundamentally different from each other, their simultaneous application could lead to their integration and series of events that led to the current special focus on "green and sustainable chemistry" are outlined. Importantly, a key point presented relates to the relevance of an honest and truly scientific approach towards these issues. This is underlined by a critical description of the importance of reliable and precise metrics to assert talking points.

The second chapter by Elena Motti, Nicola Della Ca', Giovanni Maestri and Max Malacria on palladium catalyzed sequential reactions involving C–H bond activation demonstrates the superiority of "one pot" operations over the traditional stepwise synthetic methodologies on a "green and sustainable" standpoint. These methodologies are both highly atom economic and environmentally benign, and able to offer a very elegant strategy to the total synthesis of complex molecules of biological and industrial interests, in a "green and sustainable" fashion. However, it is important to keep in mind that the power of C–H activation has been complementary to "non-C–H activation" strategies, and their combination provide the platform of the realization of successful efficient, rational and environmentally friendly synthesis of very complex molecules.

In the last two decades *photoredox catalysis* has emerged as a powerful and safe alternative method for the formation of radicals and also offered the opportunity for new direction in "green and sustainable" chemistry. In the third chapter Christophe Lévêque, Etienne Levernier, Vincent Corcé, Louis Fensterbank, Max Malacria and Cyril Ollivier demonstrate the importance of this methodology to the chemistry community to develop much greener radical chemistry. Moreover, they present a comprehensive review on the combination of photoredox catalysis and organometallic catalysis, the so-called "dual catalysis" which is considered as a highly versatile approach for the construction of a wide range of molecular scaffolds. The overall methodology involves either only electron transfers between both catalysts or with an additional radical trapping step by the organometallic catalyst. Various transition metals including palladium, gold, copper and nickel can be used to perform cyclizations or cross-coupling reactions *via* C–H or C–X bond activation.

Guillaume Levitre and Géraldine Masson in their chapter "Brønsted acid as efficient catalyst for synthesis of biologically active natural products" provide an overview of enantioselective total or formal syntheses relying on hydrogen bonds assisted organocatalysis. Notably, the focus has been on the activation by catalyst such as bifunctional thioureas, Cinchona alkaloids, bis-imides, and BINOL derivatives that offer an exceptional entry to the pharmaceutically valuable compounds. In order to demonstrate the versatility of such processes and the diversity of the accessible motifs, this chapter is organized according to target scaffolds and reaction types. Implementation of chiral organocatalysis in the area of pharmaceutical chemistry offers the opportunity to simultaneously introduce one or several stereogenic centers present on the target structure. It has the potential to improve overall selectivity, shorten synthetic routes and increase their convergence. Chiral organocatalysts are often easy to handle compared to other conditions, involving Lewis acids for example, while permitting the use of more readily available starting materials. Examples of using of organocatalytic methods for the construction of diverse natural or synthetic active molecules are already numerous and keep on accumulating. Thereby demonstrating the adequation of these tools to solve problems met when trying to access such chiral, complex and polyfunctional structures. Now that efficiency of chiral organocatalysis at the laboratory scale seems to be established. The next challenge in our opinion is to find ways to increase the adoption rate of this class of processes in development and production contexts. Generalization of this powerful tool would potentially have a have a great benefit in terms of ease of operation as well as sustainability.

The chapter "Bio-sourced polymers: recent advances" by Henry Cramail, Boris Bizet, Océane Lamarzelle, Pierre-Luc Durand, Geoffrey Hibert and Etienne Grau is an elegant and comprehensive contribution to the current domain of bio-sourced polymers. They have fully illustrated the current development of bio-sourced polymers motivated at first by the search for performances and for new features. This is based in particular on the valorization of two main types of resources at the industrial level: oleaginous plants and polysaccharides, in connection with the corresponding agricultural productions. Consequently, the development of building-blocks stemming from the biomass is rapidly expanding, both in the academic community and industry and the development of polymers from renewable resources is inevitable and expects a colossal potential both for the academic research and the chemical industry in the coming years. This new research theme is present from now in most of the laboratories and the conferences, justified by Green and Sustainable Chemistry and the current global and local environmental challenges. Besides, the production of monomers and bio-sourced polymers permit interdisciplinary crossings of interest with the experts of the biomass or the bioprocesses, which also take all their sense in a synergic approach of the biomass valuation.

Finally, Katell Quénéa, Sylvie Derenne and Marc Benedetti in a chapter entitled "Insight into continental organic matter: a chemist view" have clearly shown the importance to achieve a complete chemical characterization of natural organic matter in order to understand the global biogeochemical cycles and their potential role in the transportation of trace elements. Thanks to numerous analytical developments, chemists nowadays have a large array of techniques at their disposal to investigate the chemical structure of the natural organic matter. They have led to conspicuous advances in the knowledge of the composition, reactivity and fate of this complex material, which in addition exhibits a large spatial and temporal heterogeneity. As they described in the comprehensive review, most of the studies on natural organic matter presently focus on carbon despite the increasing interest in the biogeochemical cycles of other nutrients such as nitrogen, sulfur and phosphorus. However, the characterization of the chemical functions comprising these elements still constitutes an analytical challenge and further developments are needed in this context.

István T. Horváth and Max Malacria
December 2019

Contents

3. Photoredox Catalysis, an Opportunity for Sustainable Radical Chemistry   49

Christophe Lévêque, Etienne Levernier, Vincent Corcé, Louis Fensterbank, Max Malacria and Cyril Ollivier

4. Brønsted Acid as Efficient Catalyst for Synthesis of Biologically Active Natural Products 123

Guillaume Levitre and Géraldine Masson

6. Insight into Continental Organic Matter: A Chemist View   **329**

Katell Quénéa, Sylvie Derenne and Marc F. Benedetti

1 Green and Sustainable Chemistry

István T. Horváth

Department of Chemistry
City University of Chemistry
Kowloon, Hong Kong
istvan.t.horvath@cityu.edu.hk

Table of Contents

List of Abbreviations

AE	atom equivalent
BHAGs	big hairy audacious goals
DNA	deoxyribonucleic acid
DDT	dichlorodiphenyltrichloroethane
E_a	activation energy

E-factor   environmental-factor
EE   ethanol equivalent
EPA   Environmental Protection Agency
IQ   intelligence quotient
NO_X nitrogen oxides
POPs   persistent organic pollutants
RNA   ribonucleic acid
SDGs   sustainable development goals
SUS_{ind} sustainability indicator
SV_{rep} sustainability value of resource replacement
SV_{waste} sustainability value of the fate of wastes
VW   Volkswagen
WHO   World Health Organization

I. Introduction

The long term physical conditions of the Earth have become the concerns of more and more people due to the local, regional, and global environmental impacts of private and commercial activities of the rapidly growing population. The results of the scattered and/or systematic responses of private citizens, governmental and non-governmental organizations as well as businesses have resulted in the evolution of novel concepts, among which the two most important are typically labeled "green" and/or "sustainable." The foundations of *green chemistry* were laid down from the 1980's to the early 1990's resulting in a carefully-worded definition and the twelve associated principles in 1998.[1] In the contrary, the term *sustainable development* was inadequately defined a decade earlier in the late 1980's,[2] triggering the formulation of newer and newer definitions[3] to overcome the shortcomings, mostly due to the unpredictable impacts of vested or conflicts of interests of the stake holders.[4] The term "sustainable" has been used very freely for a wide range of issues, which have had very little or nothing to do with sustainability. The most unfortunate consequence of the poor definition of sustainable development has been that the terms of "green" and "sustainable" have incorrectly become exchangeable, especially in the case of chemistry. It is proposed that this could be corrected by systematically applying two independent definitions:

- ***Green chemistry*** is the utilization of a set of principles that reduces or eliminates the use or generation of hazardous substances in the design, manufacture, and applications of chemical products.[1]
- ***Sustainable chemistry*** is the use of resources, including energy, at a rate at which they can be replaced naturally, and the generation of waste, which cannot be faster than the rate of their remediation.[5]

While neither **green chemistry** nor **sustainable chemistry** could ensure perfect solutions for every challenging issues, their integration to **green and sustainable chemistry** could lead to appropriate combinations to achieve successful answers, especially by setting appropriate overarching goals or "BHAGs."[6] For example, any **green and sustainable** solar energy harvesting device should not contain toxic constituents at all and produced and installed by using renewable energy only. Furthermore, the same criteria should be applied for associated energy storage systems. If such a rule would be used for Heliogen's emerging technology to concentrate the power of sunlight to generate high temperature,[7] it should first be applied to manufacture the silicon in the mirrors. Once all the mirrors are **green and sustainable**, their use for making cement, steel, or petrochemicals would be a major breakthrough to mankind.

II. Suitable Chemistry

Chemistry has a central role in evolution, especially in the slow but spectacular transition from the robust inorganic continental crust[8] to the more and more sophisticated and fragile biosphere in the aqueous environment, frequently immobilized on and within the cracks and interstices of the landscape.[9] While it is impossible to identify the exact time when a mixture of chemicals has formed the first living system,[10] chemistry certainly has been playing an enabling role in the formation, development and sustainability of biological species. Due to the position of the slowly rotating Earth in the solar system, the average surface temperature has been between 10–20°C in the last 3.5 billion years.[11] The modest global surface temperature sent the inorganic landscape to an almost standstill and only the deep inside of our planet has remained hot enough to provide appropriate local environments for new reactions without constrains leading to the

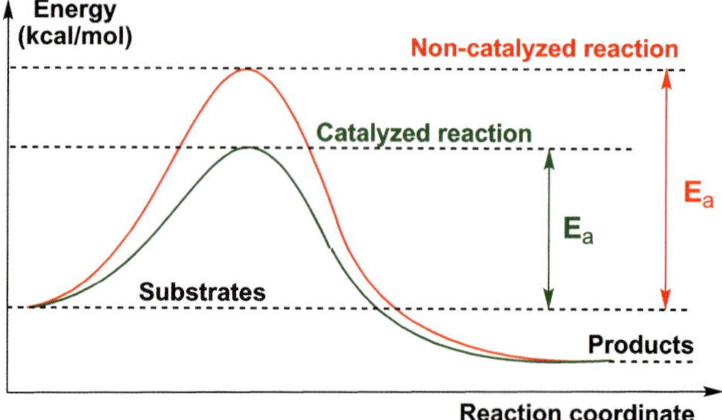

Figure 1. The activation energies (E_a) of a non-catalyzed and a catalyzed reactions.

formation of new chemicals and materials, which could survive the harsh conditions.[12] Since all chemical reactions have to overcome of their activation energy (E_a) (Figure 1),[13] the available chemical reactions on the surface of the Earth were limited by the modest temperatures for a very long time. The evolution of a wide range of catalysts has dramatically enlarged the pool of available reactions and structures with little or no constrains.

Consequently, chemistry has blossomed, many different molecules and materials such as mono-, di-, and poly-carbohydrates, aromatics and lignin, amino acids and proteins, nucleic acids and DNA/RNA were synthesized by Nature. Some of these have become building blocks of life, others have been serving Nature as simple to sophisticated reagents and catalysts, or become part of the control and information systems. In the early phase of evolution the outcomes of chemical reactions were depended only on the local environmental parameters (temperature, pressure, light, acidity, concentrations, etc.) and proceeded according to the rules of thermodynamics and kinetics. In contrast, the evolution of life has required *suitability*, e.g. *having the qualities that are right, needed, or appropriate for something*.[14] The "something" has originally been the "survival of the species," e.g. building and selecting the fittest to survive even in unexpectedly changing environments. The development of suitable biological catalysts or enzymes was an important step in adapting to the mild conditions of the Earth. It was achieved by combining the molecular level complexity through primary, secondary, tertiary and quaternary structures of proteins with an active site, frequently with ligated metal

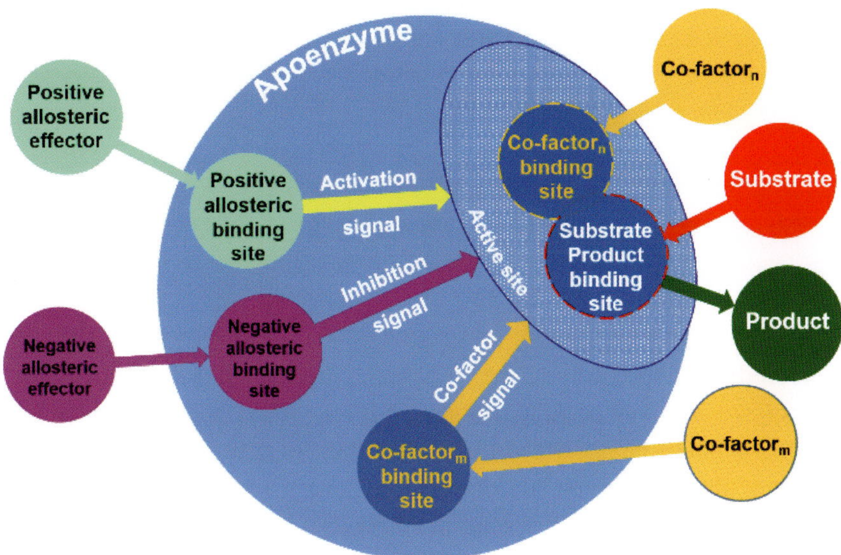

Figure 2. The structural components of enzymes.

center (s), for binding and converting the substrate and releasing the product (Figure 2). In addition, the molecular control was further increased by involving binding sites for co-factors and allosteric effectors. The net results include some level of selectivity of the desired reaction(s), which in some cases can reach absolute specificity, e.g. total control.[15]

Once humans have joined the selection processes, the objectives have been expanded beyond survival and/or achieving measurable advantages over the same or competing species,[16] though many of the new goals such as profit or recognition, all leading to personal or group power, have increased the survival rate for some and limited for others. Thus, *suitable chemistry* has emerged, which could be defined as *having reactions, chemicals and materials that are right, needed, or appropriate for something.*[17] Although suitable chemistry has been ultimately driven by the customers need resulting in financial benefits to the producers and distributors of chemicals and materials, the possible negative environmental and health effects of the used feed stocks, the processes, and the final and side products were overlooked or even concealed for centuries. Ecotoxicological impacts were generally ignored until the seminal work of Rachel Carson, entitled "Silent Spring," which was published in 1962. While DDT or

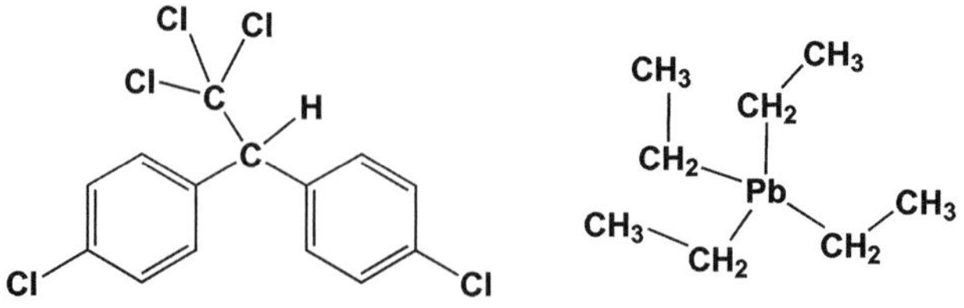

Figure 3. DDT or dichlorodiphenyltrichloroethane (a) and tetraethyl-lead or $(CH_3CH_2)_4Pb$ (b).

dichlorodiphenyltrichloroethane (Figure 3a) has been successfully used to control malaria and typhus among civilians and troops during World War II, it could cause cancer, bioaccumulate in birds, causing eggshell thinning, nesting failures and dramatic population declines of eagles and brown pelicans. Although the agricultural use of DDT in the USA was banned in 1972, indoor application in Africa countries has been supported by World Health Organization (WHO) since 2006.[18] The benefits of DDT exceed the health and environmental risks and its use has been consistent with the Stockholm Convention on Persistent Organic Pollutants (POPs), which takes a specific exemption of the use of DDT for malaria control.[19]

It should be recognized that the rapidly growing population have dramatically increased the use of natural resources and industrial production of energy, food, and chemicals resulting in pollutions and left behind serious health and environmental damages across the planet.[20] The addition of the toxic tetraethyl-lead or $(CH_3CH_2)_4Pb$ (Figure 3b) to gasoline as octane booster resulted in significant release of lead to the atmosphere along major highways, which is known to lower the IQ for regularly exposed bystanders.[21] The treatment of the morning sickness of pregnant women with the racemic form of thalidomide led to lost pregnancies or babies with missing limbs.[20] Unfortunately, respectable goals to solve major environmental and health concerns could unexpectedly lead to even greater environmental issues. Although nonflammable and nontoxic chlorofluorocarbons were introduced to replace highly toxic refrigerants such as sulfur dioxide or ammonia, their easy leakage to the atmosphere resulted in ozone depletion for decades.[22] The silent escape of gaseous methyl isocyanate or CH_3NCO from a chemical plant to the night sky of

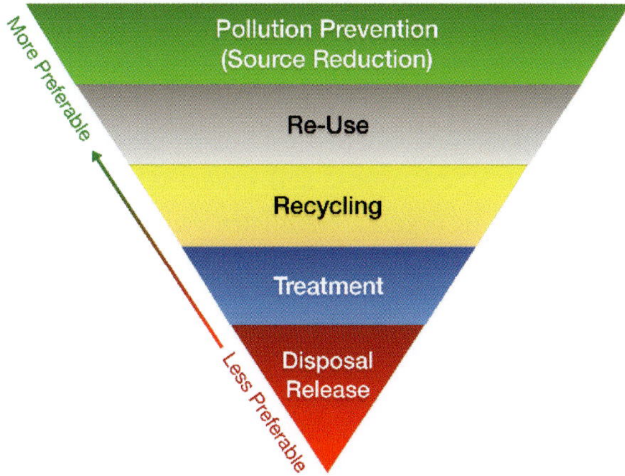

Figure 4. Environmental protection hierarchy.

Bhopal, India on December 3, 1984, caused 3500 deaths and 150,000 injuries and the impacts are still observable.[20] There have been large number of accidents involving dioxins[20] and significant environmental contaminations by crude oil,[23] melamine,[24] and ammonia[25] as well. The increasing number of environmental disasters have clearly shown that the "end-of-the-pipe" clean-up methods have not worked and a new strategy was required to address global and local environmental issues simultaneously. Scientists at the US Environmental Protection Agency (US EPA) have realized by the late 1980s that prevention of the use of hazardous chemicals, materials and processes by design is the only possible approach to eliminate the risks globally.[26] Consequently, the Pollution Prevention Act was passed by the US Congress in 1990,[27] which is described as any practice that reduces, eliminates, or prevents pollution at its source, also known as source reduction.[28] Pollution prevention has become the most preferable environmental protection method (Figure 4) resulting in the establishment of green chemistry.[1]

III. Green Chemistry

The development of green chemistry has started slow by sporadic activities of concerned scientists and engineers to prevent environmental or health issues caused by chemicals, reactions or processes, some of which were discussed in the previous section. While the oldest examples of green chemistry were not labeled green, some of the early green concepts and strategies were practiced under the name of environmentally friendly molecules and materials, reactions, and processes.[29] While water is the medium of life, its use in organic chemistry only started in 1980 for the hydrophobic acceleration of Diels–Alder reactions.[30] The fundamentals of the first aqueous process for the hydroformylation of propylene to *normal-* and *iso*-butanal was invented in 1975[31] and commercialized in 1984 (Figure 5).[32]

The introduction of atom economy by Trost in 1991[33] was an important step to understand the intrinsic environmental limitations of chemical reactions. Since substitution and elimination reactions generally have poor atom economy, combination of isomerization and addition reactions have gained dominating role in contemporary organic synthesis. The definition of the Environmental-factor (E-factor) by Sheldon in 1991 was another important milestone by providing a metrics to account for all substances involved in a procedure or a process.[34]

The first green issue of Chemical Reviews on *Environmental Chemistry* was published in 1995, with one of the earliest definitions of green chemistry: "It is no longer sufficient to make "marvelous" new molecules solely on the basis of their marketable properties. Although marketability is an appropriate goal, we, as scientists, must also be concerned with our creations' potentials for environmental impact. At the same time, we should constantly tighten our scientific standards for generating experimental data, so that any conclusions drawn from such are and will be unambiguous... ."[35]

Green chemistry was carefully defined in the book entitled *"Green Chemistry: Theory and Practice"* by Paul Anastas and John Warner in 1998.[1] The 12 principles of green chemistry were also outlined to

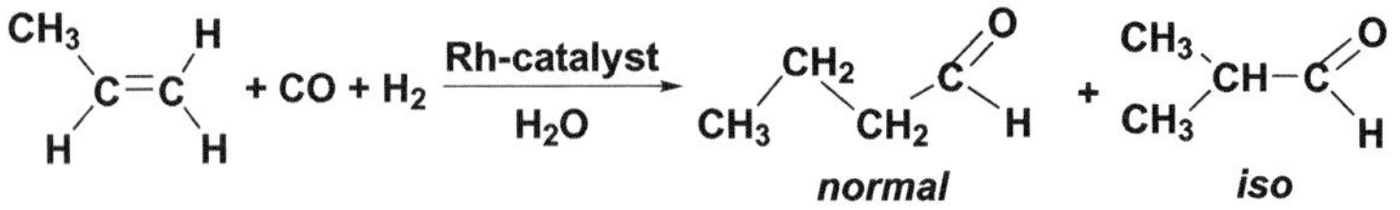

Figure 5.   Rhodiumcatalyzed hydroformylation of propylene to *n*- and *i*-butanal.

Table 1. Twelve principles of green chemistry.

1. Prevent waste
2. Atom economy
3. Less Hazardous Synthesis
4. Design benign chemicals
5. Benign solvents and auxiliaries
6. Design for energy efficiency
7. Use of renewable feedstocks
8. Reduce derivatives
9. Catalysis
10. Design for degradation
11. Real time analysis
12. Inherently safe chemistry

provide guidelines how to achieve the definition and address the most important challenges of chemistry and many other fields using chemistry (Table 1).

Books, dedicated journals, special issues in periodicals, and websites have been dedicated to green chemistry ever since. For example, the second green issue of Chemical Reviews was published in 2007, which focused on the innovative challenges in green chemistry.[36]

IV. Sustainable Chemistry

The necessity to address the globally increasing negative environmental and health impacts of the rapidly growing population resulted in the definition of *"sustainable development"* in 1987.[37] The United Nations' World Commission on Environment and Development stated that sustainable development *"should meet the needs of the present without compromising the ability of future generations to meet their own needs."* Since reliable forecasts of the needs of the future generations have been practically impossible, especially with respect to social and economic needs, the term "sustainable development" has been exploited for a wide range of issues regularly, which have had very little or nothing to do with sustainability.

Although many national and international meetings have been dedicated to different sustainability issues, the definition of "sustainable development" has remained ambiguous. Consequently, many stake holders with vested or conflict of interests have used "sustainability" to label "suitability" to make profit, secure financial support, or gain political power. For example, Volkswagen (VW) sold 11 million cheating diesel cars worldwide, including 600,000 in the United States.[38] The cheating VW diesel vehicles emitted 10 to 40 times the NO_X emissions permitted by the EPA, which increased the rates of low birth weight and acute asthma attacks among children.

The United Nations acknowledged the problems by issuing 17 sustainable development goals (SDGs) (Table 2) in 2015.[39] The progress of SDGs should be monitored by Potential and Indicative SDG Indicators and Complementary National Indicators, some of which were not even defined.[40]

Table 2. The United Nations' 17 sustainable development goals.

1. No poverty
2. Zero hunger
3. Good health and well-being
4. Quality education
5. Gender equality
6. Clean water and sanitation
7. Affordable and clean energy
8. Decent work and economic growth
9. Industry, innovation and infrastructure
10. Reduced inequalities
11. Sustainable cities and communities
12 Responsible consumption and production
13. Climate change
14. Life below water
15. Life of land
16. Peace, justice and strong institutions
17. Partnership for the goals

The SDGs may serve as a "roadmap to happiness" but not as a definition to develop simple metrics to set deliverables and achieve accountability and credibility.

We have recently tried to define an alternative definition of sustainability, which is totally independent from economical and societal aspects: *Nature's resources, including energy, should be used at a rate at which they can be replaced naturally, and the generation of waste cannot be faster than the rate of their remediation.*[4] Of course, repairing, reconditioning, remanufacturing or recycling methods, processes, and technologies could contribute positively to sustainability locally and/or globally provided their required energy sources are also sustainable. We developed a metrics for carbon chemicals by using ethanol equivalents (**EE**),[4] based on first generation corn-based bioethanol technology commercially practiced in the USA in 2008.[41] We reported the definitions of the *sustainability value of resource replacement* (**SV**$_{rep}$) and the *sustainability value of the fate of wastes* (**SV**$_{waste}$) for basic carbon chemicals produced from bioethanol in the USA,[42] which were then merged into the *sustainability indicator* (**SUS**$_{ind}$) to assess overall sustainability.

$$SUS_{ind} = \frac{SV_{rep} \cdot SV_{waste}}{SV_{rep} + SV_{waste}}$$

In general, sustainable production of ethylene, propylene, toluene, p-xylene, styrene, and ethylene oxide from bioethanol still cannot be achieved because of the limited amount of available corn (or maize kernel)-based bioethanol. Similarly, first generation bioethanol was not an option for replacing fossil fuel resources to produce primary energy in the United States, Canada, the European Union, China, and the Russian Federation between 2008–2014. The hope that cellulosic ethanol technology could solve the problem is now also questionable, as a cellulosic ethanol production will be idled soon,[43] but co-production of high-value products could shift the economics for cellulosic ethanol more favorable.[44]

Since the use of **EE** based sustainability metrics is limited to energy and carbon chemicals, we are now developing the *atom equivalent* (**AE**) to represent any elements of the periodic table as constituents of chemicals and materials in resources, products, wastes, etc. The **AE** was defined as the equivalent mass of one of the constituent elements of a chemical or a material based on molar equivalency.[45] The required energy of any given technology must be an intrinsic part of the calculation of SV_{rep} and SV_{waste}, which could be readily calculated by using **EE**s.

V. Green and Sustainable Chemistry

Although *green chemistry* and *sustainable chemistry* are fundamentally different from each other, there are some overlaps and each has pieces of the other. This could be demonstrated by a **yin** and **yang** (陰陽), one of the key principles of Chinese philosophy (Figure 6). Their simultaneous application could lead to their integration to *green* **and** *sustainable chemistry*.

For example, the 6th principle of green chemistry states that "a raw material of feedstock should be renewable rather than depleting wherever technically and economically practicable." Indeed, this is one of the overarching goals of the sustainable chemical industry by switching the production of carbon chemicals from fossil resources to biomass. However, without owning the appropriate size of land and securing the necessary amount of water to produce the required biomass, industrial scale utilization should note be called sustainable. Thus, **sustainable and green chemistry** should have its carefully drafted definition to set appropriated deliverables and achieve accountability and credibility.

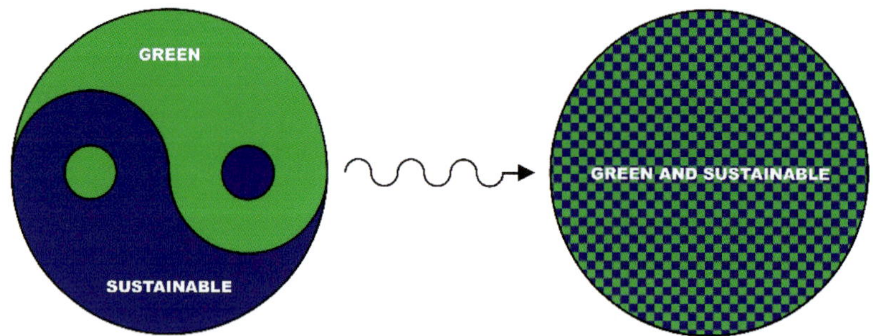

Figure 6. Integration of *green* chemistry with *sustainable* chemistry to *green and sustainable* chemistry.

It should be noted, that some of the challenges we are facing now and in the future, might require the use of molecules, materials, reactions, or processes that are neither green nor sustainable, but suitable to support our life to the maximum length, and maintain our society into the future. Of course, we must use them with extreme care and proper protection of the environment and health.

VI. References

1. Anastas, P. T.; Warner, J. C. *Green Chemistry: Theory and Practice*, Oxford University Press, Oxford, **1998**.
2. World Commission on Environment and Development, *Our Common Future*, Oxford University Press, Oxford, **1987**.
3. Horváth, I. T.; Cséfalvay, E. Sustainability of green synthetic processes and procedures. In Ballini, R. (ed.) *Green Chemistry Series No. 61.*, *Green Synthetic Processes and Procedures*, Royal Society of Chemistry, Cambridge, **2019**.
4. Cséfalvay, E.; Akien, G.; Qi, L.; Horváth, I. T. *Catal. Today* **2015**, *239*, 50–55.
5. Horváth, I. T. *Chem. Rev.* **2018**, *118*, 369–371.
6. Collins, J.; Porras, J. I. *BHAGs: Big Hairy Audacious Goals. In Built To Last: Successful Habits of Visionary Companies*. HarperBusiness, New York, **1994**.
7. Bendix, A. Bill Gates is backing a technology that concentrates sunlight to generate temperatures of 1,000 degrees. It looks like a mosaic of mirrors. *Business Insider*, November 20, **2019** (*https://www.businessinsider.com/solar-power-heliogen-bill-gates-2019-11*).
8. Composition of the crust (*https://www.sandatlas.org/composition-of-the-earths-crust/*, accessed on May 19, **2019**).
9. Brown, T. L.; LeMay, H. E.; Bursten, B. E. *Chemistry: The Central Science*, 8th ed., Prentice Hall, Upper Saddle River, New Jersey, **1999**.
10. Smith, J. M.; Szathmary, E. *The Origins of Life: From the Birth of Life to the Origin of Language*, Oxford University Press, Oxford, **2009**.
11. Cockell, C.; Corfield, R.; Edwards, N.; Dise, N.; Harris, N. *An Introduction to the Earth-Life System*, Cambridge University Press, Cambridge, **2008**.
12. Poldervaart, A. *Crust of the Earth: A Symposium*, GSA Volume 62, Geological Society of America, Boulder, Colorado, **1955**.
13. Lower, S. General Chemistry Virtual Textbook → kinetics/dynamics → collision/activation (*http://www.chem1.com/acad/webtext/dynamics/dynamics-3.html*, accessed on May 14, **2019**).
14. Learner's definition (*http://www.learnersdictionary.com/definition/suitable*, accessed on October 27, **2019**).
15. Punekar, N. S. *Enzymes: Catalysis, Kinetics and Mechanisms*, Springer, **2018**.

16. Peterson, J. B. *12 Rules for Life: An Antidote to Chaos*, Penguin Books Ltd., London, **2018**.

17. Definition of "suitable" by Merriam-Webster Learner's Dictionary, an Encyclopædia Britannica Co.: having the qualities that are right, needed, or appropriate for something (*http://www.learnersdictionary.com/definition/suitable*, accessed on November 22, **2019**).

18. EPA, USA, DDT — A Brief History and Status, (*https://www.epa.gov/ingredients-used-pesticide-products/ddt-brief-history-and-status*, accessed on October 27, **2019**).

19. Stockholm Convention on Persistent Organic Pollutants, UNEP-POPS-PUB-Booklet-16NewPOPs-201706.English-1.pdf (**2017**).

20. Heaton, A. *The Chemical Industry*, Chapman & Hall, London, **1994**.

21. Seyferth, D. *Organometallics* **2003**, *22*, 2346–2357 and 5154–5178, respectively.

22. Japanese Ministry of Economy, Trade and Industry(*https://www.meti.go.jp/policy/chemical_management/ozone/files/pamplet/panel/08e_basic.pdf*, accessed on October 27, **2019**).

23. Woods Hole Oceanographic Institution, Oil in the Ocean: A Complex Mix, Previous Oil Spills (*https://www.whoi.edu/oil/oil-spills*, accessed on October 27, 2019).

24. World Health Organization, Melamin (*https://www.who.int/foodsafety/areas_work/chemical-risks/melamine/en/*, accessed on October 27, **2019**).

25. Muller, J. 05, 09, 2013, CNN (*http://edition.cnn.com/2013/09/05/world/asia/china-river-dead-fish/*, accessed on October 27, **2019**).

26. Anastas, P. T. Origins and Early History of Green Chemistry, *Series on Chemistry, Energy and the Environment, Advanced Green Chemistry, Part 1: Greener Organic Reactions and Processes,* Horvath, I. T.; Malacria, M. (Eds.) World Scientific Publishing, Singapore, **2018**.

27. 42 U.S.C. §13101 et seq., Pollution Prevention Act, **1990**.)

28. Pollution Prevention Act of 1990 (*http://www.epa.gov/p2/pubs/p2policy/act1990.htm*, accessed on October 27, 2019).

29. Pereira, C. *J. Chem. Eng. Sci.* **1999**, *54*, 1959–1973.

30. Rideout, D. C. *J. Am. Chem. Soc.* **1980**, *102*, 7816–7817.

31. Kuntz, E. French Patent 2 314 910 (**1975**).

32. Cornils, B.; Herrmann, W. A.; Eckl, R. W. *J. Mol. Cat. A: Chemical* **1997**, *116*, 27–33.

33. Trost, B. M. *Science* **1991**, *254*, 1471–1477.

34. (a) Sheldon R. A. 13th Seminar on Frontier Technology, Fuji Hakone, Japan, July 23–26, **1991**; (b) 3rd International Conference on Catalytic Chemistry for Global Environment, Hokkaido University, Japan, July 27, **1991**; (c) Interview with R. A. Sheldon in *Chem. Eng. Progr.* **1991**, *87*, 11.

35. Horváth, I. T. *Chem. Rev.* **1995**, *95*, 1.

36. Horváth, I. T.; Anastas, P. T. *Chem. Rev.* **2007**, *107*, 2167–2168; 2169–2173.

37. World Commission on Environment and Development, *Our Common Future*, Oxford University Press, Oxford, **1987**.

38. Alexander, D.; Schwandt, H. The impact of car pollution on infant and child health: Evidence from emissions cheating, *Federal Reserve Bank of Chicago, Working Paper 2019-04*, June 13, **2019** (*https://doi.org/10.21033/wp-2019-04* accessed on November 21, 2019).

39. Division for Sustainable Development United Nations: *http://sustainabledevelopment.un.org/*, accessed on November 27, **2019**).

40. Sustainable Development Solutions Network, Indicators and a Monitoring Framework for the Sustainable Development Goals — Launching a data revolution for the SDGs, Report, 12 June **2015** (*https://sustainabledevelopment.un.org/content/documents/2013150612-FINAL-SDSN-Indicator-Report1.pdf*, accessed on December 1, 2019)

41. Shapouri, H.; Gallagher, P.W.; Nefstead, W.; Schwartz, R.; Noe, S.; Conway, R. *Energy Balance for the Corn-Ethanol Industry*, USDA, 2008, Agricultural Economic Report Number 846, June, **2010**.

42. Horváth, I. T.; Cséfalvay, E.; Mika, L. T.; Debreczeni, M. *ACS Sustainable Chem. Eng.* **2017**, *5*, 2734.

43. Clayton, C. Cellulosic production idled. Citing EPA decisions, POET-DSM converting cellulose plant to research, development. *Progressive Farmer* **2019**, November 19, (*https://www.dtnpf.com/agriculture/web/ag/news/business-inputs/article/2019/11/19/citing-epa-decisions-poet-dsm-plant*, accessed on December 1, 2019).

44. Rosales-Calderon, O.; Arantes, V. *Biotechnol. Biofuels* **2019**, *12*, Article number: 240.

45. Horváth, I.T. Atom equivalents based metrics to bridge green chemistry and sustainability, *7th Asia-Oceania Conference on Green and Sustainable Chemistry (AOC7-GSC)*, Singapore, November 19–21, **2018**.

2 Pd-Catalyzed Sequential Reactions Involving C–H Bond Activation: A Green and Sustainable Tool for Natural and Industrial Product Synthesis

Elena Motti,*[‡] Nicola Della Ca',* Giovanni Maestri* and Max Malacria[†,§]

*Dipartimento di Scienze Chimiche, della Vita e della Sostenibilità Ambientale, Università di Parma, Area delle Scienze 17A, 43124 Parma, Italy
[†]UPMC Sorbonne Université, IPCM (UMR CNRS 8232), 4 place Jussieu, C. 229, 75005 Paris, France
[‡]elena.motti@unipr.it
[§]max.malacria@upmc.fr

Table of Contents

I. Introduction: Green Chemistry and Sustainability

Green chemistry consists in the production of environmentally benign chemicals and in the use of processes which minimise the negative impact that the chemical industry has on human health, safety and the environment.[1]

Green chemistry looks at a process with a "cradle to grave" approach, considering all the steps starting from the raw material acquisition, all the way through the lifecycle of the product, to its final disposal. This includes all inputs in terms of energy and mass, all substances generated during the process such as useful products, co-product and by-product, intermediates or wastes, the use of that product and then the final disposal of all the involved materials.

All these aspects of green chemistry, from raw materials, atom economy, use of catalysts, ligands, auxiliares and solvents, waste and reaction monitoring, are included in the 12 Principles of Green Chemistry, published in 1998, and concern all the branches of chemistry: organic, inorganic, physical, industrial, analytical and computational.[2]

Sustainability, on the other hand, is not a literal synonymous of "green chemistry" but undoubtedly share many similar goals with it. In a sustainable lifecycle assessment, together with the all inputs and outputs of the green chemistry model, it has been to take into account how long it takes for the raw materials to be renewed, looking in this way to truly renewable feedstocks as starting points for the production of commodity chemicals.[3] In order to apply the concept of green chemistry to processes for the production to fine chemicals and pharmaceutical products, Sheldon has introduced in 1992 the E factor, as the ratio between how much waste is produced in a process and the amount of useful products obtained.[4] In terms of E factor, the pharmaceuticals sector is the most wasteful of the whole of the chemicals industry, with E Factors almost 1000 times bigger than the oil refining sector and 2–5 times bigger than fine chemicals industry in general.

This evidence is now directing research and development in the use of new, more efficient catalytic processes for the pharmaceutical and fine chemical sectors.

II. Sequential Reactions

Sequential reactions are able to realize many chemical transformations, regarding different functional groups, in one single sequence, or even in a "one-pot" operation, and because of these features sequential reactions are one of the most powerful tools in modern organic syntheses of complex molecules.

Compared with traditional stepwise synthetic methodologies, that we can call "consecutive reactions," where the conditions change, reagents are added and intermediates isolated and purificated after each step, the sequential approach presents unique step economy features and relevant advantages.

First of all, the use of protocols based on an "one-pot" operation decreases the costs of the processes, simplifying the conduction of the reactions and the purification of the products, which takes place only once at the end of the trasformations. Moreover, usually sequential reactions start from relatively cheap, simple and readily available materials, in order to build very sophisticated targeted molecules.

Clearly, these methodologies are highly atom economic and environmentally benign, as well promoting elegant syntheses in a "green chemistry" fashion.[5]

In 1996, Wender, in a review published in the thematic issue of *Chemical Reviews* entitled "Frontiers in Organic Synthesis,"[6] illustrated the main advances in sequential transformations such as cascade, domino and tandem reactions.

Cascade,[7] domino[8] or tandem[9] reactions have in common that all the starting materials are initially present and the reaction proceeds without modification of reaction conditions; as Faber[10] says, intermediates are not generally isolable.

According to the definition of Fogg and dos Santos,[11] the term "tandem catalysis" describes coupled catalyses in which sequential transformation of the substrate occurs via two (or more) mechanistically distinct processes; in other words, tandem reactions in particular are reactions where almost two different catalytic mechanisms are present at the same time.

Using the definition of Tietze, domino reactions are processes "...*of two or more bond-forming reactions under identical conditions, in which the subsequent transformations take place at the functionalities obtained in the former transformations.*"

In domino/cascade reactions multiple transformations are effected via a single catalytic mechanism, either in intra- or intermolecular transformations. Cascade catalysis is a virtually synonymous term of domino sequence, but is usually reserved for multiple, more than three, domino sequences.[12]

The level of elegance of a domino/cascade/tandem reaction can be directly correlated to the number of bonds formed during the process and the increase of complexity. Although domino reactions can be performed as single- or two-component trasformations, the most intriguing domino reactions are the multicomponent ones, where more than two starting materials react to build a targeted product, incorporating essentially all of the atoms in the adducts.

III. C–H Activation

Through the recent synthetic methodologies, undoubtedly, transition-metal mediated C–H activation processes play an important role and have revolutionized the synthetic field in an unprecedented way.[13]

However, despite the increasing application of such synthetic methodologies in direct functionalization of $C(sp^2)$-H and $C(sp^3)$-H bonds located nearby to suitable directing groups, only recently important examples have reported the activation of remote $C(sp^2)$-H and $C(sp^3)$-H moieties, which had previously been thought inaccessible (Scheme 1).

Unreactive remote C–H moieties have thus become the target to be harnessed by transition-metal catalysis. This goal can be achieved in some cases by the use of suitable cordinating groups, able to activate for example the *meta-* and *para*-C–H of aromatic rings.

A very broad range of directing groups have so far been used in the C–H activation.

If we consider the principles of the green chemistry, the directing groups, as well as the protecting group of the traditional organic chemistry,

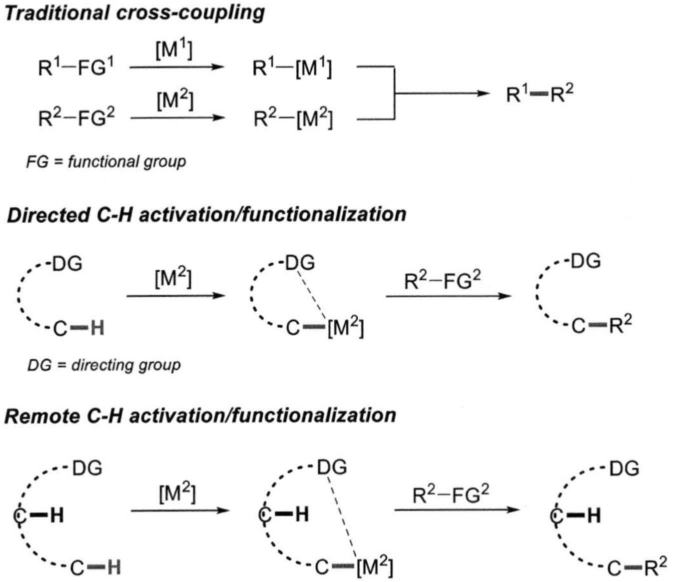

Scheme 1.

that need to be incorporated and then removed from the core unit to be functionalized, imply a considerable synthetic cost, hence lowing the efficiency of the overall functionalization process.

Thus, directing groups which can be reversibly linked to the substrate can be a useful solution; for example, 2-norbornene used as an auxiliary ligand in Pd-catalysis (see Sections IV. D and IV. N–IV. R) has proven to be a versatile directing group for the dual *ortho-* and *ipso*-functionalization of aryl halides.[14]

Keeping into account the principles of the green chemistry, if we focus on an isolated non acidic C–H bond, the "unfunctional group" by definition, the possibility of directly activate the C–H bond could make possible the use of cheap and abundant feedstocks to prepare complex organic molecules, without any previous transformation of the C–H bonds in more reactive groups, such as halides, alcohols and so on.

One of the main problems is that the direct C–H activation is not a selective process if more than one C–H bond is present on the substrate, which is the typically case even if we consider very simple organic molecules.

The activation of C–H bond is not only a topic regarding the fine chemical production, but it involves also very important challenge such as the activation of methane to methanol and so on.

In this chapter, the authors will be focused on green and sustainable sequential reactions involving C–H bond activation, catalyzed by palladium, the most versatile transition metal, for the synthesis of natural products, useful intermediates, pharmaceuticals or other industrial products. In this context, recent advances which we consider particularly interesting in terms of novelty, innovation and perspectives, will be presented. The authors apologize in advance for omissions.

IV. Relevant Selected Examples

A. Flemichapparin C

Gogoi and coworkers developed a palladium-catalyzed cascade reaction of 4-hydroxycoumarins and *in situ* generated arynes for the direct synthesis of coumestans, displaying important biological activities including anticancer, antibacterial, antifungal, antimyotoxic, and phytoalexine effects.[15] A member of this important class of polycyclic lactones, Flemichapparin C, was successfully prepared in 67% yield with excellent regioselectivity

(Scheme 2). This cascade process proceeds via C–H bond activation/C–O and C–C bond formations in a one-pot fashion and tolerates a variety of functional groups including halides.

A plausible reaction mechanism for this reaction is shown in Scheme 2. The sodium salt of 4-hydroxycoumarin **I** and palladium(II)acetate exchange their alkoxide anions to form the Pd-salt of 4-hydroxycoumarin **II**. The Pd species **III** is then formed via the C–H activation at the 3-position carbon of **II**, and further undergoes carbopalladation of the aryne generated *in situ* from the aryne precursor giving species **IV**. Finally, in the presence of a base the intermediate **IV** is converted to the desired product and Pd(0), which is reoxidized to Pd(II) by $Cu(OAc)_2$ to complete the catalytic cycle.

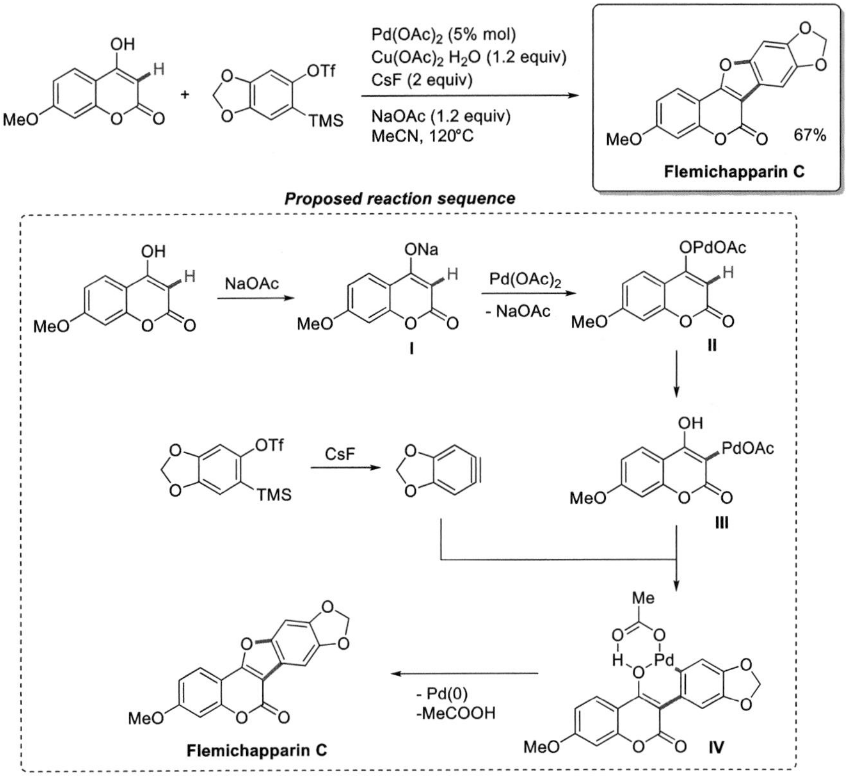

Scheme 2.

In summary, this method has the regiochemical limitations of arynes but enables the use of unfunctionalized starting reagents thus affording complex molecular architectures of biological interest.

B. Trisphaeridine

A facile and practical palladium-catalyzed nucleophilic substitution/C–H activation/aromatization cascade reaction has been employed to build a number of 6-unsubstituted phenanthridines in moderate to good yields (31–85%) from readily prepared N-Ms arylamines and commercially available 2-bromobenzyl bromide derivatives.[16] The potentiality of this methodology was demonstrated by the expeditious synthesis of the natural alkaloid trisphaeridine (Scheme 3).

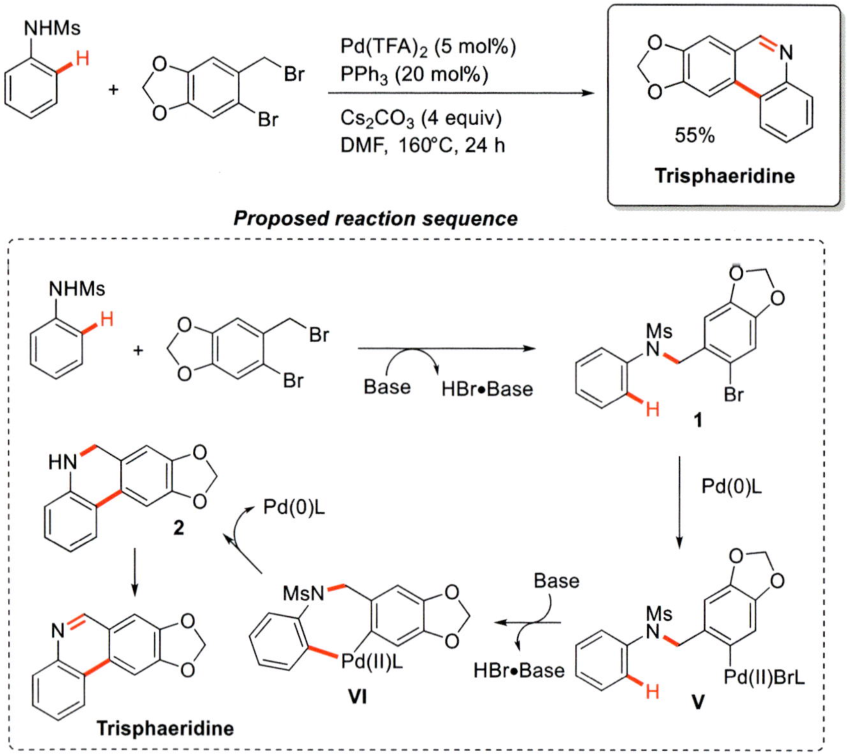

Scheme 3.

Based on experimental findings, a possible catalytic cycle for this cascade reaction can start with the base promoted nucleophilic substitution between N-Ms arylamine and the 2-bromobenzyl bromide derivative producing compound **1**. Then the oxidative addition of **1** to *in situ* generated Pd(0) species occurred, followed by the intramolecular C–H activation to form the intermediate **VI**. Subsequent reductive elimination occurred to afford the annulated product **2**, while Pd(0) species were regenerated to complete the catalytic cycle. Further aromatization of **2** occurred to deliver the natural alkaloid trisphaeridine at high temperature in the presence of a base. Authors do not study the mechanism of this last reaction step although it is known that palladium can readily catalyze oxidative dehydrogenations.

C. Tipifarnib

A practical route to 2-quinolinones via a palladium-catalyzed C–H bond activation/C–C bond formation/cyclization cascade reaction was recently accomplished by Liu and co-workers.[17] The utility of this method was further demonstrated by a formal synthesis of Tipifarnib, which has been found to exhibit potent activity against neoplastic diseases, such as breast cancer. When 4-amino-4′-chlorodiphenylmethane **3** and ethyl 3-chlorocinnamate were used as the starting materials quinolinone **4** was generated in 95% yield (Scheme 4). Treatment of intermediate **4** with *tert*-butyl hydroperoxide in the presence of a catalytic amount of pyridine and iodine afforded quinolinone **5** in 98% yield, which can be subsequently converted to Tipifarnib in three steps according to reported literature.

For this direct ring construction the aryl-Pd complex **VII** is formed via C–H activation at the 2-position carbon atom of the substrate assisted by an *in situ* formed acetyl on N atom. Next, Heck-type coupling of **VII** with acrylate through Pd complex **VIII** provides intermediate **6** and releases the Pd(0) species. This was followed by oxidation of Pd(0) by $Na_2S_2O_8$ to Pd(II) to complete the metal catalytic cycle. Under acidic conditions, the intermediate **6** can readily afford **4**. The last step of this cascade process was successfully demonstrated by a control experiment.

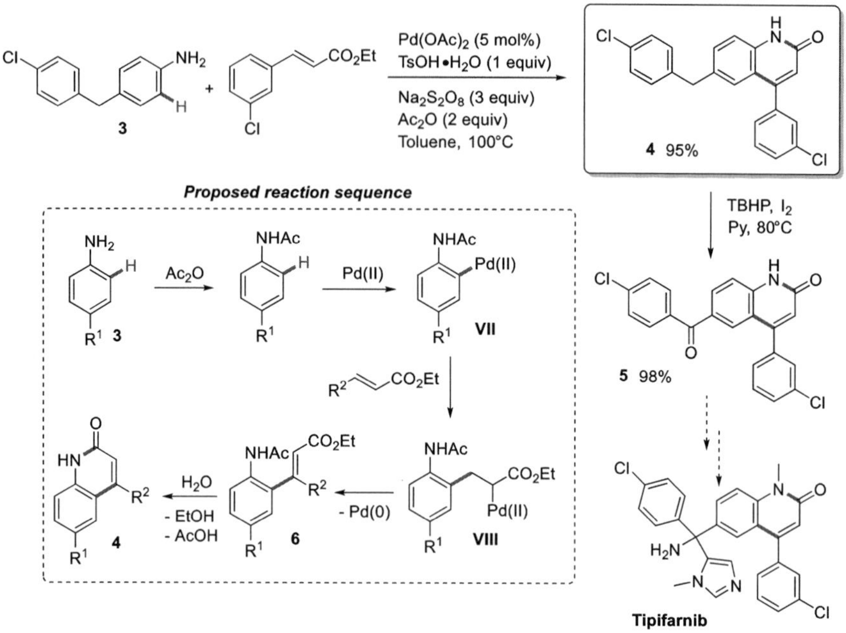

Proposed reaction sequence

Scheme 4.

D. Abilify and Flunixin

The pharmaceuticals Abilify and Flunixin have been recently synthesized by Ritter and co-workers who took advantage of the Catellani reaction.[18] The Catellani reaction is a bisfunctionalization of aryl halides through a palladium/norbornene-catalyzed reaction that proceeds *via* an initial *ortho*-C–H activation by palladacycle formation (see also Sections IV. N–IV. R). In this context, Ritter *et al.* have reported a general method for the formation of a C–B bond at the *ipso* position which can be further differently functionalized with an additional reaction step (Scheme 5). Abilify was obtained in three steps with a 42% total yield. The key step was realized using 5 mol% of $Pd(OAc)_2$, 10.5% of $P(4\text{-}MeOC_6H_4)_3$ and 1 equiv of norbornene, starting from 2-chloroiodobenzene, 1-benzoyloxy-4-Boc-piperazine and B_2Pin_2 in almost equimolar amount.

In a simplified reaction sequence the starting aryliodide gives oxidative addition to Pd(0) and a subsequent electrophilic aromatic substitution, by C–H bond activation, furnishes the palladacycle intermediate. The latter

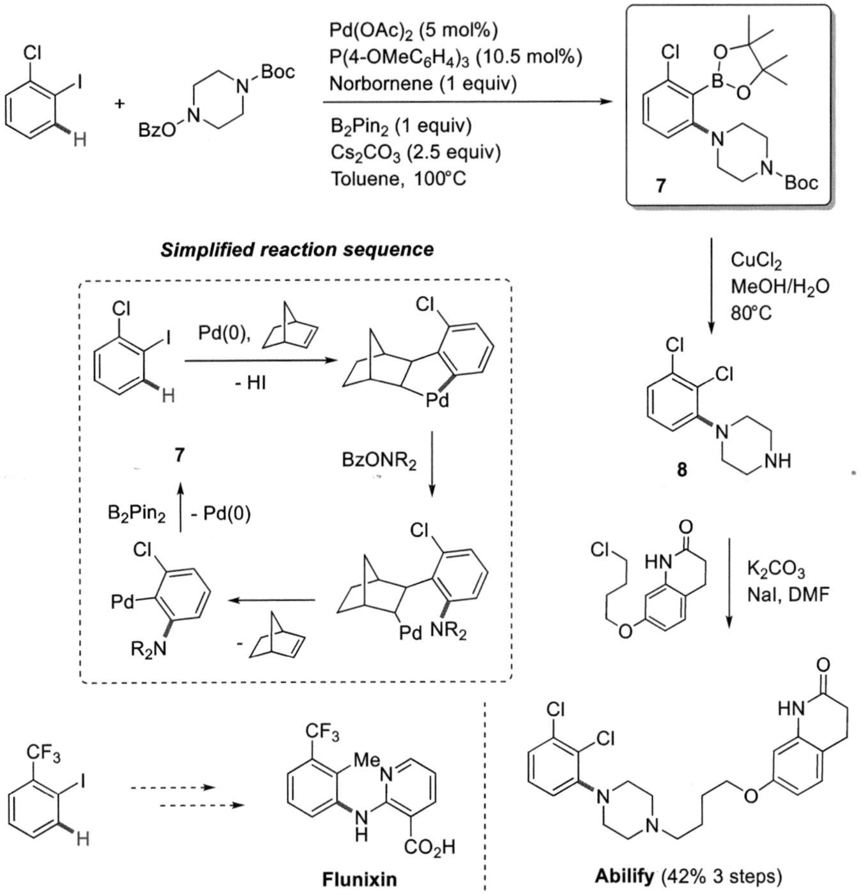

Scheme 5.

reacts with the BzO-amine and after norbornene expulsion the generated arylpalladium(II) complex undergoes a borylation reaction with formation of compound **7**. The intermediate **7** was then easily converted to **8**, which provided Abilify after a simple nucleophilic substitution reaction. Flunixin was similarly prepared starting from 2-CF$_3$-iodobenzene.

E. Xyridin A

Very recently, Della Ca' and co-workers have reported the first general methodology for the direct and regioselective synthesis of isocoumarins through palladium-catalyzed α-arylation of ketones followed by an easy cyclization reaction.[19]

With this methodology, Xyridin A, an important natural product with antibacterial and antifungal activity, was synthesized in two steps with 71% yield starting from the commercial 6-bromobenzo[d][1,3]-dioxole-5-carboxylic acid (Scheme 6).

The addition of iodide anions to the reaction mixture was crucial for the reaction success, being the reaction rate significantly increased when both aryl iodides and bromides where used. Moreover, the experimental tests suggested that iodide anions can act as ligands in this palladium-catalyzed phosphine-free synthesis of isocoumarins. The

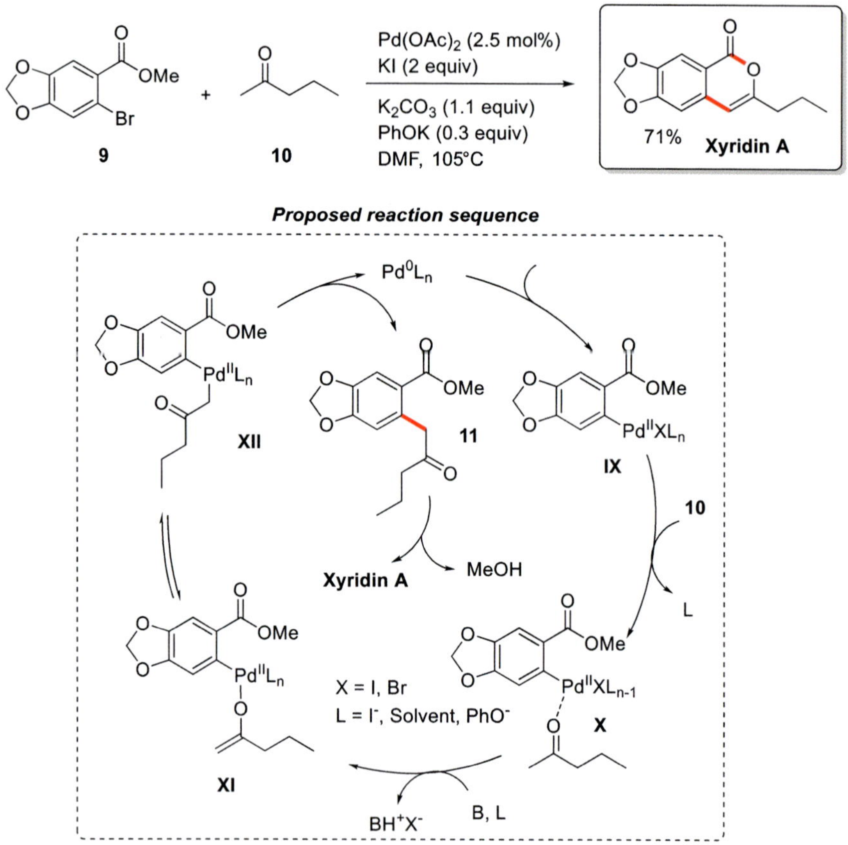

Scheme 6.

formation pathway for Xyridin A is reported in Scheme 6. Oxidative addition of the Pd(0)L$_n$ with 2-bromobenzoate derivative **9** affords the Pd(II) organometallic intermediate **IX**. The pentan-2-one **10** enters into the coordination sphere of palladium (II) and its subsequent deprotonation/enolization by the bases provides the Pd(II) intermediate **XI**, in equilibrium with **XII**. Then, reductive elimination from **XII** gives the α-arylated ketone and regenerates the Pd(0)L$_n$ catalyst. The acyclic intermediate **11** readily provides Xyridin A and methanol as coproduct.

F. Indaziflam

Indane scaffold is frequently found in biologically active compounds. Baudoin and co-workers have recently reported the synthesis of 1-indanols and 1-indanamines by an intramolecular palladium(0)-catalyzed C(sp^3)–H arylation.[20] Following this methodology, *trans* indanamine **13**, intermediate in the synthesis of the herbicide indaziflam, was obtained in 55% yield (Scheme 7).

In the proposed reaction sequence, oxidative addition of the aryl bromide to Pd(0) leads to Pd(II) intermediate **XIII**, which undergoes ligand substitution with potassium pivalate to give complex **XIV**. C(sp^3)–H activation then occurs from **XIV** through a base-induced concerted metalation–deprotonation (CMD) mechanism affording the six-membered palladacycle **XV**. Finally, reductive elimination provides the indane derivative **13** and the catalytically active species. The key step can be successfully carried out on a gram scale. Hydrazine-mediated deprotection of intermediate **13** gives the known racemic 2-aminoindane intermediate **14** in almost quantitative yield. Simple additional steps lead to the target molecule. The present methodology successfully affords the product with complete diastereoselectivity starting from a racemic reagent. Notably, the most simple phosphine ligand (namely triphenylphosphine) turned out to be the best one to efficiently trigger the sequence.

G. Aeruginosins 298A and 98B

The alkaloids Aeruginosins, isolated from dried algae with yields in the range of 0.01–0.05%, exhibit potent inhibitory activity against serine

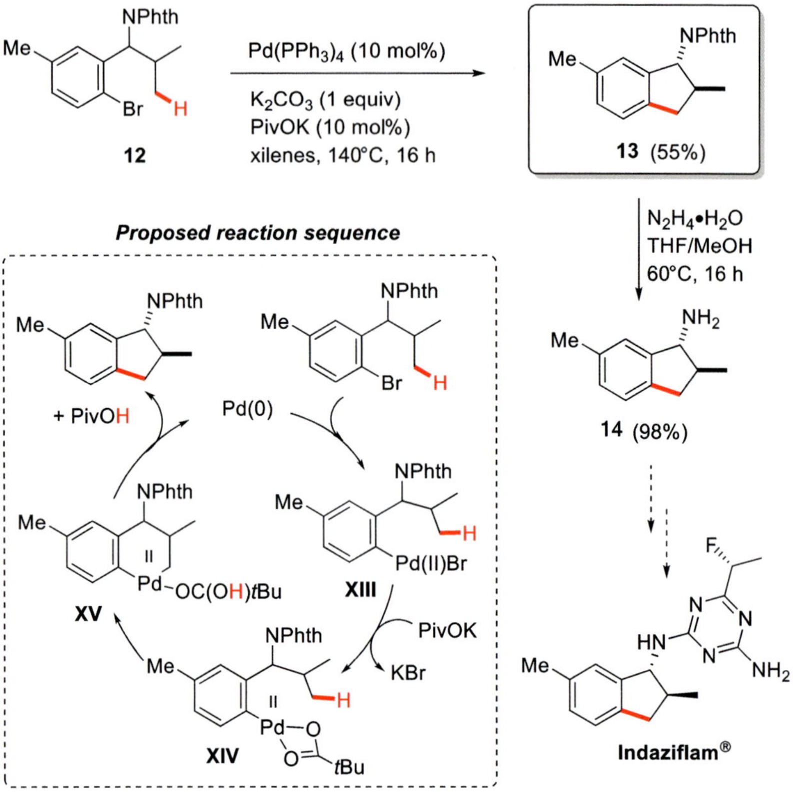

Motti *et al.*

Scheme 7.

proteases. In this context, Baudoin *et al.* have reported an efficient and scalable process for the synthesis of Aeruginosins 298A and 98B.[21] The synthesis was enabled by the appropriate use of two $C(sp^3)$–H activation reactions (Scheme 8). The mechanism is supposed to be the same reported in Scheme 7. Interestingly, $COCF_3$ group protecting the nitrogen atom of the vinylbromide reagent does not cleave under reaction conditions. Hpla and Choi fragments were easily obtained in good yield and further transformations allowed the synthesis of Aeruginosins 98B and 298A, with the latter being performed on unprecedentedly large scale. The same methodology can be probably applied to other family member total synthesis and should enable more advanced pharmacological studies of these potent serine protease inhibitors.

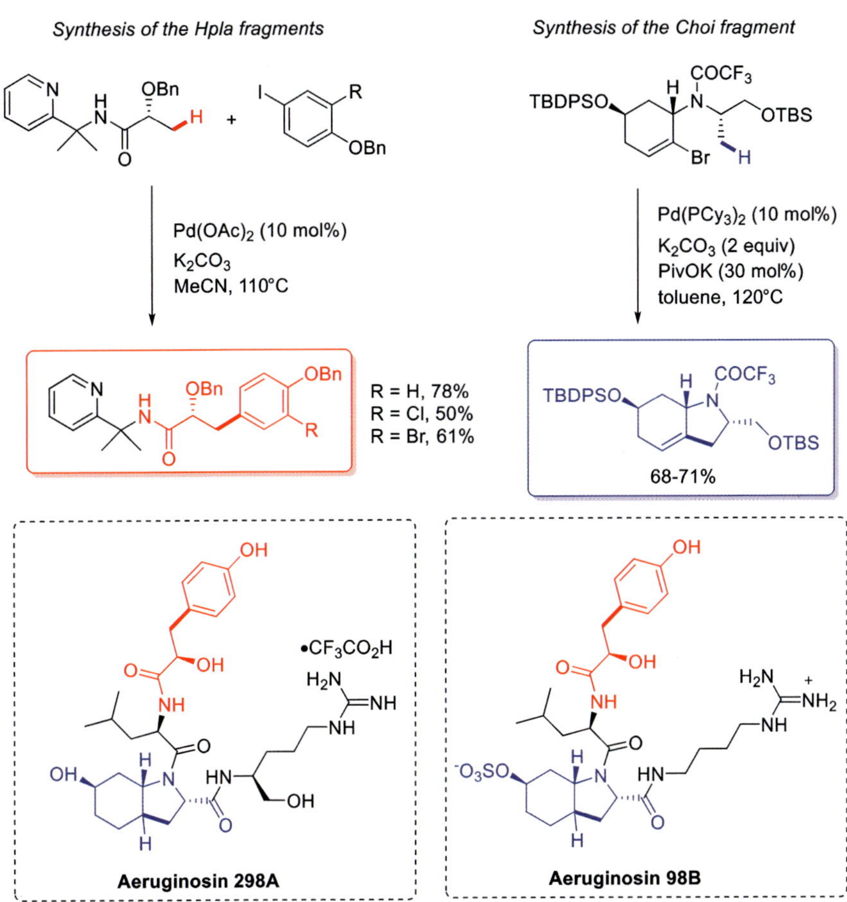

Scheme 8.

H. δ-Coniceine

Baudoin *et al.* have recently reported construction of α-alkylidene-γ-lactams by C(sp^3)–H alkenylation from readily available acyclic bromoalkenes.[22] These lactams are valuable intermediates for accessing various classes of mono- and bicyclic alkaloids containing a pyrrolidine ring. This valuable methodology was successfully applied to the synthesis of δ-coniceine hydrochloride in racemic form with good overall yield. This coupling reaction was successfully scaled up to gram quantities at least. The same protocol can be exploited to access an advanced intermediate for

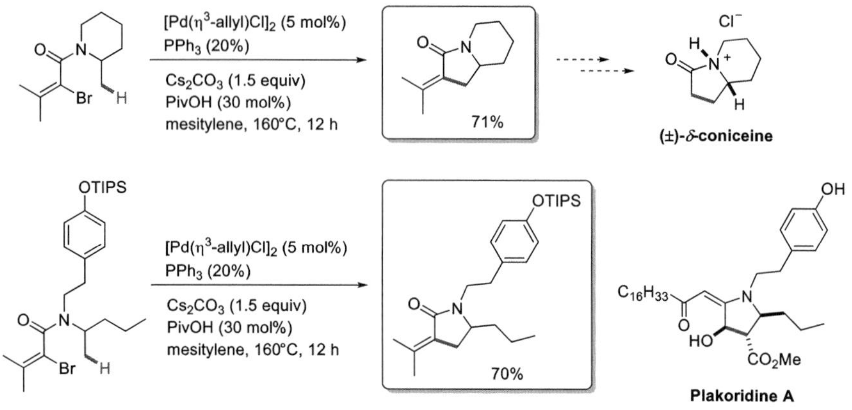

Scheme 9.

the synthesis of the marine natural product Plakoridine A (Scheme 9). In this case the catalyst can selectively activate the C–H of the terminal methyl over that of the corresponding internal methylene.

I. (±)-Goniomitine, (±)-1,2-Dehydroaspidospermidine, (±)-Aspidospermidine, (±)-Vincadifformine, and (±)-Kopsihainanine

The class of monoterpene indole alkaloids, which counts more than 2000 members with broad structural diversity and important bioactivities, has attracted the interest of synthetic chemists for decades. In 2014, Zhu and co-workers have reported an unprecedented strategy to monoterpene alkaloids which have been exploited for the synthesis of (±)-goniomitine, (±)-1,2-dehydroaspidospermidine, (±)-aspidospermidine, (±)-vincadif-formine, and (±)-kopsihainanine.[23] The success of this approach is due to two crucial key steps: (1) a palladium-catalyzed decarboxylative vinyla-tion that provides prompt access to cyclopentene moiety which contains all of the carbons present in the natural products and (2) an integrated oxidation/reduction/cyclization sequence to convert the cyclopentene derivatives to the pentacyclic structures of the natural products.

Goniomitine, presenting an interesting molecular architecture and potent antiproliferative activities, was the first target of this elegant strategy

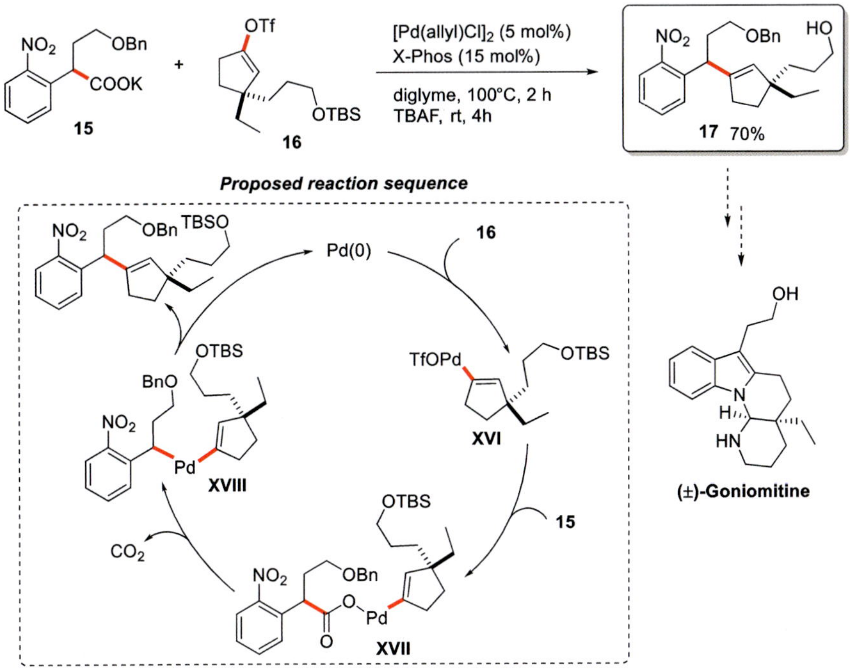

Scheme 10.

(Scheme 10). Using [Pd(allyl)Cl]$_2$ as the palladium source and X-Phos as ligand, intermediate **17** was obtained in 70% yield including the TBS deprotection step.

Based on experimental observations, a possible reaction mechanism was proposed. Oxidative addition of vinyl triflate **16** to the *in situ* generated Pd(0) affords vinyl palladium(II) intermediate **XVI**. A ligand exchange with the potassium salt, gives the carboxylic-acid-ligated intermediate **XVII**. Finally, reductive elimination from **XVIII**, which is generated after CO$_2$ expulsion from **XVII**, furnishes the desired coupling product **17** and the catalytically active palladium species.

J. Podophyllotoxin

Podophyllotoxin is a starting material for the widely used type II topoisomerase targeting drugs etoposide (VP-16) and teniposide (VM-26) used

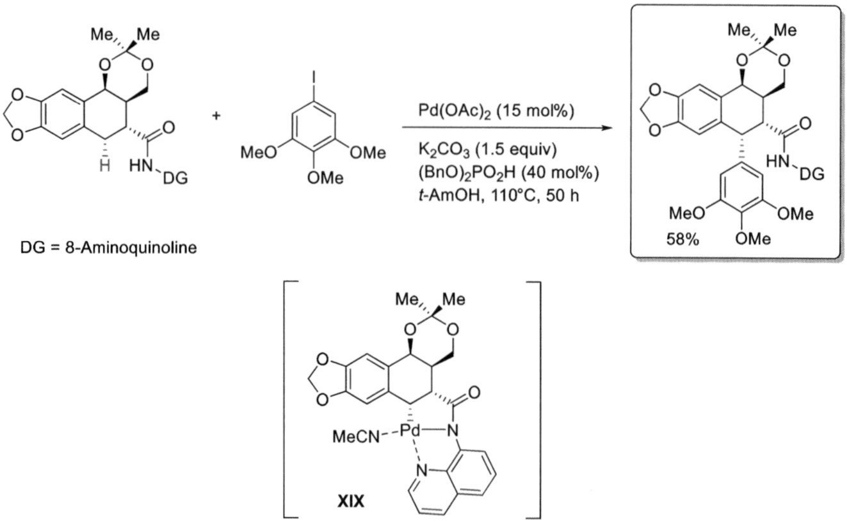

Scheme 11.

for the treatment of lung and testicular cancer, lymphoma, leukemia, and Kaposi's sarcoma. Maimone *et al.* have recently reported a new strategy for the synthesis of this prototypical aryltetralin lignan natural product.[24] The key step is based on a intermolecular Pd-catalyzed $C(sp^3)$–H arylation reaction that is shown in Scheme 11. The use of 8-aminoquinoline as directing group was crucial for the reaction success and palladacycle **XIX**, isolated and unambiguously determined by X-ray diffraction, gave the arylated product under stoichiometric conditions. The result has been used to rationalize the catalytic mechanism which affords exclusively the desired *syn* product.

K. Frutinone A

Frutinone A, isolated from the leaves and root bark of Polygala fruticosa, displays various biological activities, including antibacterial, antioxidant, and potent cytochrome P450 1A2 (CYP1A2) inhibition. C–H activation can also in this case provide a potential alternative for the synthesis of the Frutinone A scaffold. The construction of the chromone-fused coumarin core was achieved[25] through two key transformations involving a C–H

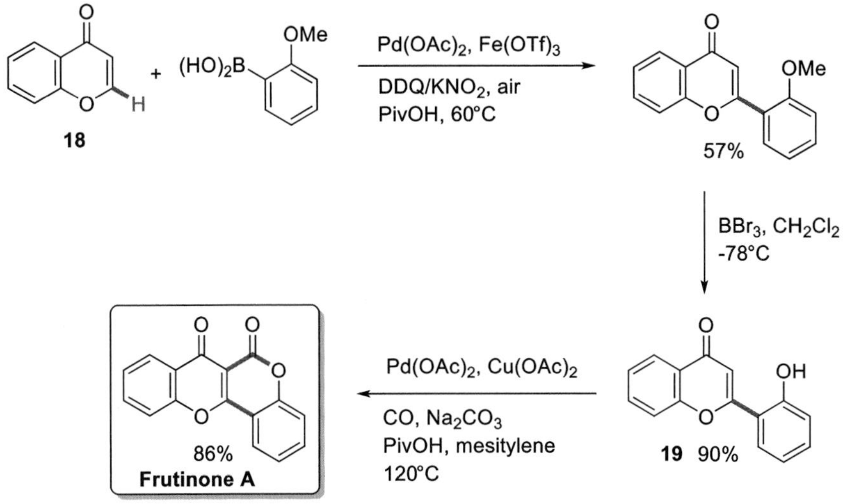

Scheme 12.

activation step: (1) a palladium-catalyzed 1,4-addition of arylboronic acids to chromone **18** and (2) a palladium-catalyzed C–H carbonylation of 2-phenolchromone **19** (Scheme 12). Notably, both protocols required a stoichiometric amount of oxidant in order to restore the active catalytic species.

In the first case an organic oxidant is consumed (DDQ = dicyanodichloroquinone) while the latter requires at least a stoichiometric amount of copper acetate. Possible future developments might thus involve the use of greener and cheaper oxidants such as hydrogen peroxide and molecular oxygen.

L. K-252c

In 2016, Gaunt's group[26] reported the synthesis of K-252c (staurosporinone), a potent inhibitor of protein kinase C, by using the carbonylative C–H functionalization to construct the 5-membered lactam (Scheme 13).

Notably, precursor **20** was obtained by an impressive sequence of three copper-catalyzed C–H functionalizations[26b,26c] and a further

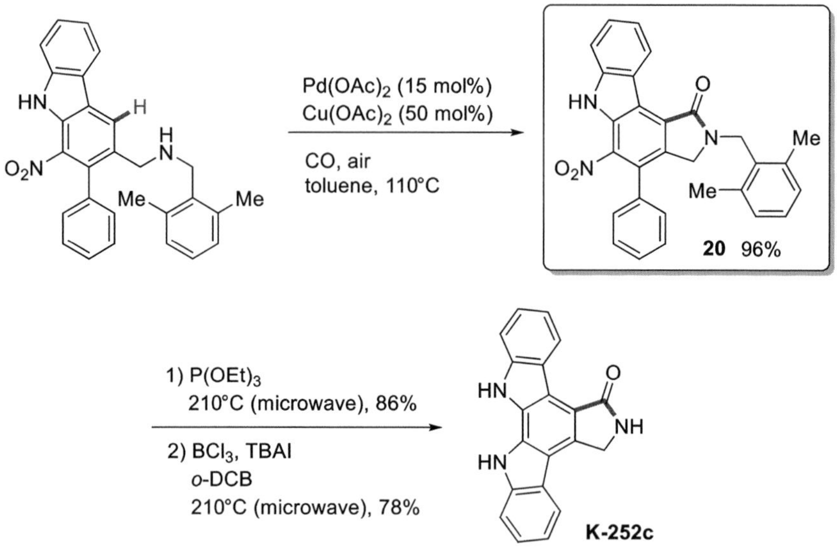

Scheme 13.

C–H activation step was accomplished under microwave conditions to build the target molecule. The installation of the γ-lactam motif was realized through Orito's C–H carbonylation protocol.[26d] The presence of a benzylamine as protecting group displaying *ortho*-substituents, such as 2,6-dimethylbenzylamine, was crucial to block the competitive C–H activation that could occur on this ring. However, this elegant mechanistic feature comes at the cost of an additional deprotection step.

M. Nataxazole, AJI9561 and UK-1

The bis(benzoxazole) natural products are a structurally unique class of *Streptomyces* secondary metabolites exhibiting a plethora of biological activities such as anticancer, antibacterial and cytotoxic activity against various human cancer cell lines. In 2015, Pal and co-workers have reported a concise and efficient syntheses of nataxazole, AJI9561 and UK-1 based on the C–H activation of a benzoxazole ring in the presence of *ortho*-iodoanilides using the palladium(II) acetate/copper(I)

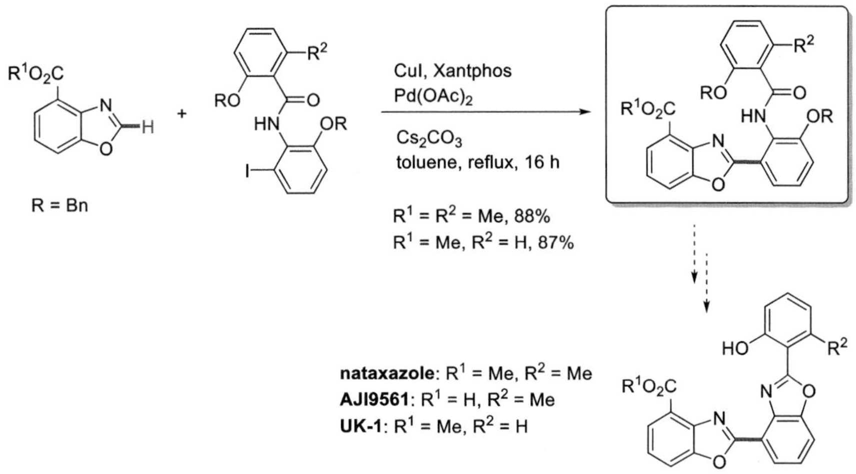

Scheme 14.

iodide/Xantphos/system[27] (Scheme 14). This complex ligand is required to compete with the numerous heteroatoms of both reagents. The presence of a C–H on the aromatic ring (R^2 = H) *ortho* to the amide function is tolerated.

This means that an intramolecular cyclization is not a competitive pathway. The final cyclization to the isoxazole ring was carried out using the dehydrating reagent pyridinium *p*-toluenesulfonate (PPTS) to afford the desired natural product in excellent yield.

N. Carbazomycin A and D

Della Ca' *et al.* have reported a very simple, one-pot and efficient methodology for the direct and regioselective synthesis of carbazoles, based on the combined catalysis of palladium and norbornene starting from *o*-substituted iodoarenes and *N*-sulfonylated or *N*-acetylated bromoanilines.[28]

Using this methodology, carbazomycin A, an important natural product with antibiotic activity, was synthesized in one step with 70% yield (Scheme 15). In contrast to many Catellani couplings this reaction provides best results with substoichiometric amount of norbornene.

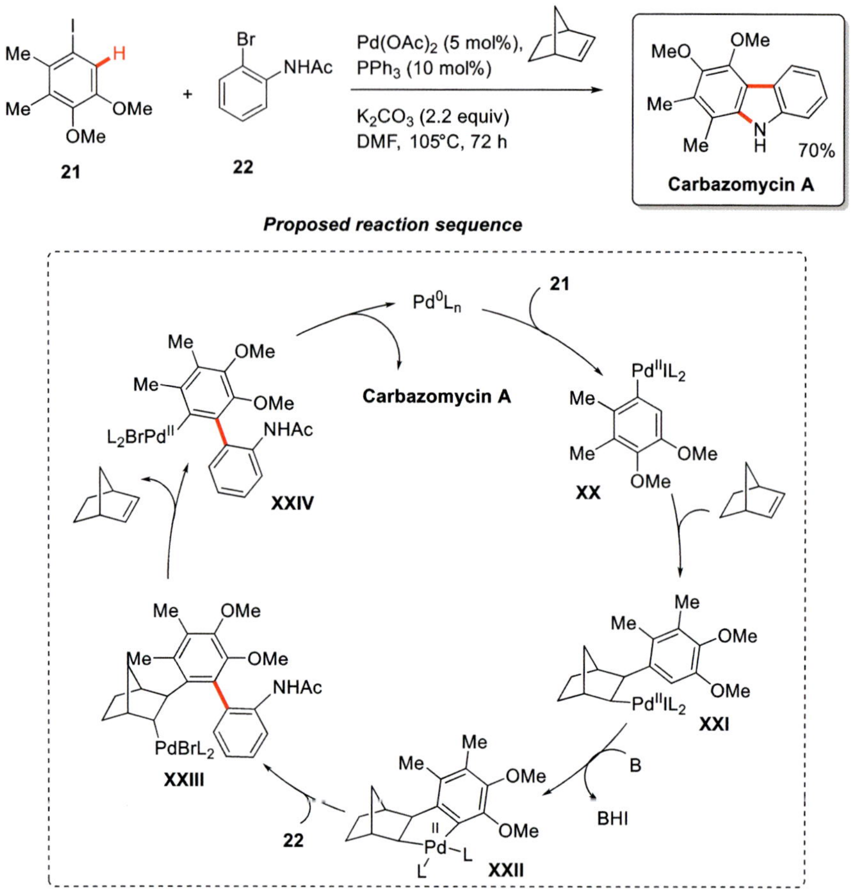

Carbazomycin A   70%

Proposed reaction sequence

Scheme 15.

The palladacycle **XXII** is the key intermediate, formed by (1) oxidative addition of the aryl iodide to Pd(0) to form **XX** followed by (2) stereoselective insertion of norbornene to give the *cis,exo*-arylnorbornylpalladium complex **XXI**, which undergoes (3) ring closure to metallacycle **XXII** by an electrophilic aromatic substitution that involves a C–H activation step. At this point, the bromoderivative reacts with **XXII**, likely through a Pd(IV) species, to afford **XXIII**, which results from selective attack of the aryl group onto the aromatic site of palladacycle **XXII**. Owing to the steric

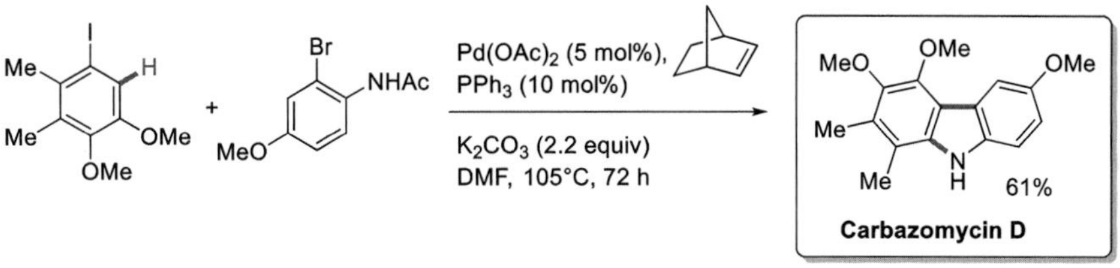

Scheme 16.

effect exerted by the two *ortho* substituents, norbornene is expelled and a biarylpalladium complex **XXIV** is formed, where the N-acetyl group is in an appropriate position for intramolecular amidation reaction leading to carbazomycin A and Pd(0). The aryl bromide wins the competition with the aryl iodide in the reaction with palladacycle **XXII** probably due to a significative chelation effect. Under the same reaction conditions, the antibiotic carbazomycin D was prepared with reasonable yield and good selectivity (Scheme 16).

O. (+)-Linoxepin

Lautens *et al.* developed a remarkable synthetic method to readily access (+)-linoxepin, a lignan which as phytooestrogen exhibits activity against pain, rheumatoid arthritis, and warts and presents interesting potential immunosuppressive activity, tumor growth inhibition, and anti-fungal properties.[29]

Total synthesis of (±)-linoxepin requires several steps, in order to build the condensed cycles of the molecule, and usually do not guarantee enantioselectivity.

The methodology discovered by Lautens allows the preparation of enantiopure (+)-linoxepin in eight steps, where, the key step, is the above mentioned Catellani reaction in its alkylation version, followed by Heck coupling (92% yield). The total synthesis of (+)-linoxepin is protecting group-free and affords the product in a 30% yield (Scheme 17). The Catellani step involves, in this case, two different iodides. The aryl iodide **23** provides with palladium and norbornene the palladacycle **XXV** which

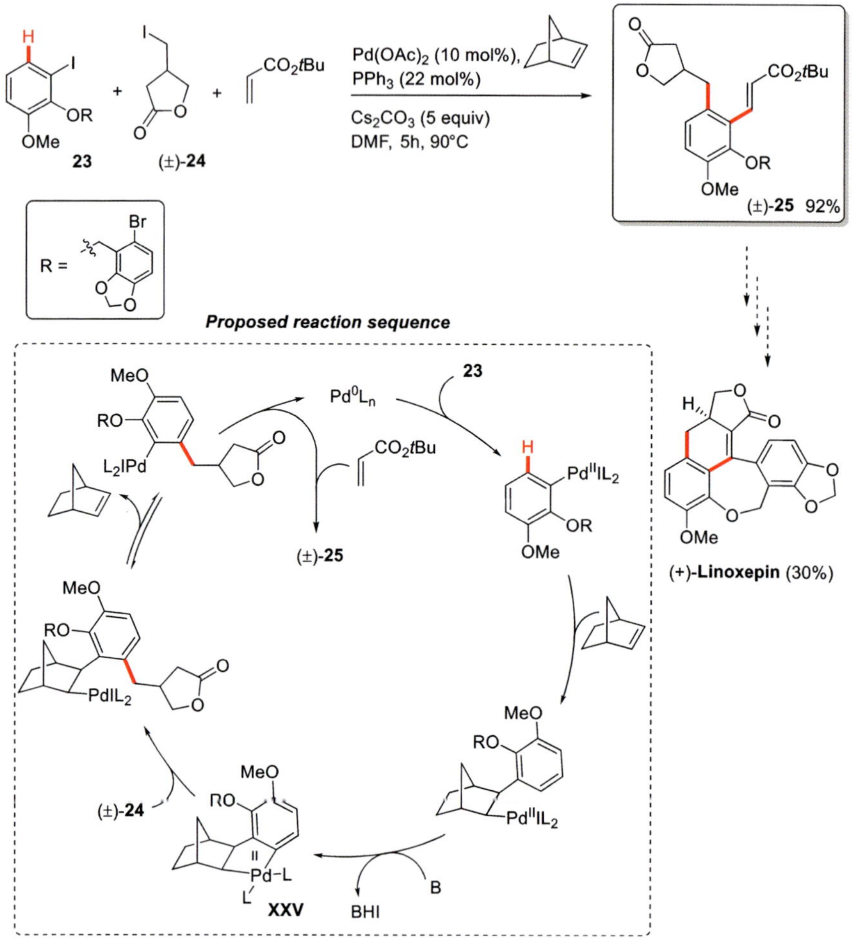

Scheme 17.

reacts selectively with the alkyl iodide **24**, giving rise to a new C–C bond (red bond). After norbornene deinsertion, the Heck coupling with the activated olefin affords the intermediate **25** in its racemic form.

P. (±)-Aspidospermidine and (±)-Goniomitine

The Bach group successfully applied the Catellani methodologies for the alkylation of indole derivatives at the second position.[30] The total synthesis

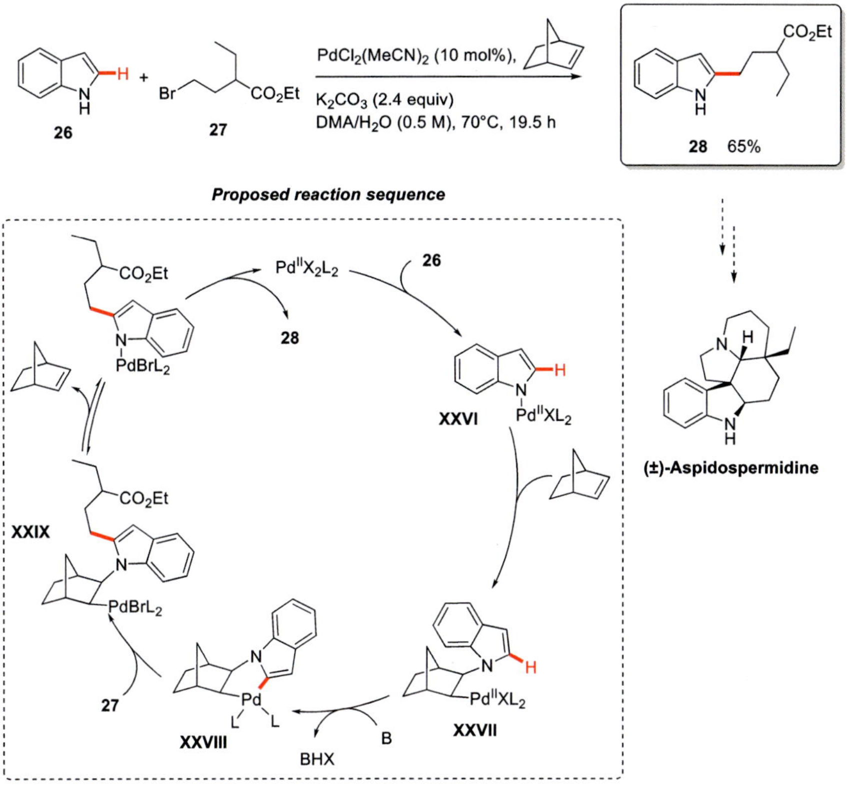

Scheme 18.

of aspidosperma alkaloids, (±)-aspidospermidine and (±)-goniomitine, was in this way accomplished (Schemes 18 and 19). (±)-Aspidospermidine was synthetized in nine steps starting from indole **26** and bromide **27**. The key step corresponds to a Catellani alkylation through the intermediacy of a singular five-membered palladacycle containing the N atom bonded to the norbornyl moiety (Scheme 18). This was the first example of Catellani reaction applied to indole chemistry.

The catalytic cycle begins with a deprotonation of **26** that allows exchange of an anionic ligand on a palladium(II) species. Intermediate **XXVI** inserts norbornene in the Pd–N bond to form **XXVII** which then undergoes C–H activation to metallacycle **XXVIII**. Alkyl bromide **27** oxidatively adds and subsequent reductive elimination affords **XXIX**.

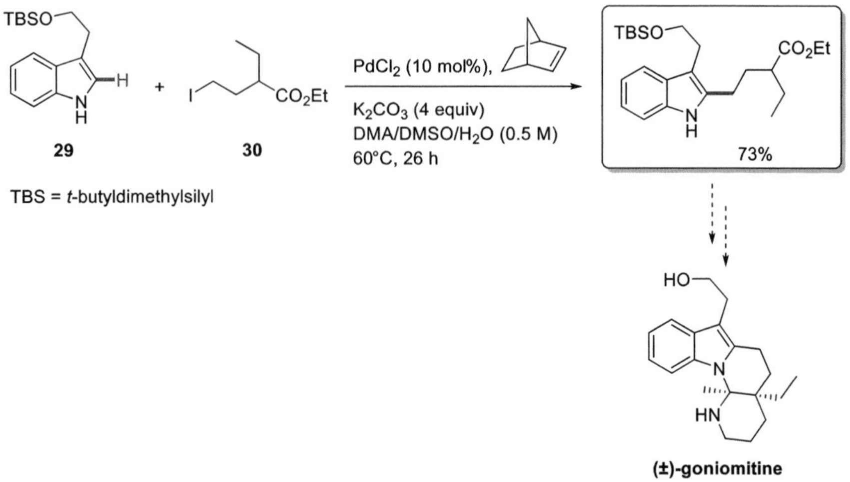

Scheme 19.

Norbornene expulsion and hydrogenolysis steps provide the final 2-functionalized indole derivative.

(±)-Goniomitine was obtained in a similar way, in seven steps, from protected tryptophol **29** and **30** (Scheme 19). In this case an alkyl iodide is used in place of the corresponding bromide.

Q. (±)-Rhazinal

Gu and coworkers published the rapid synthesis of (±)-rhazinal, which is a potent antitumor alkaloid (Scheme 20).[31] Compound **33** was obtained by Pd/norbornene catalysis by sequential *ortho*-arylation and Heck cyclization reaction of iodopyrrole **31** using 1-bromo-2-nitrobenzene **32** as very reactive cross-coupling partner. 2-Iodopyrrole **31** reacts with palladium(0) leading to **XXX** which, after norbornene insertion and C–H activation, delivers palladacycle **XXXI**. Usual palladium(IV) step and norbornene extrusion provide **XXXII**. The trisubstituted olefin arm of **31** can then stereoselectively insert in the Pd-Csp^2 bond to form the spiro carbon and the final β-hydrogen elimination on the terminal methyl gives compound **33**.

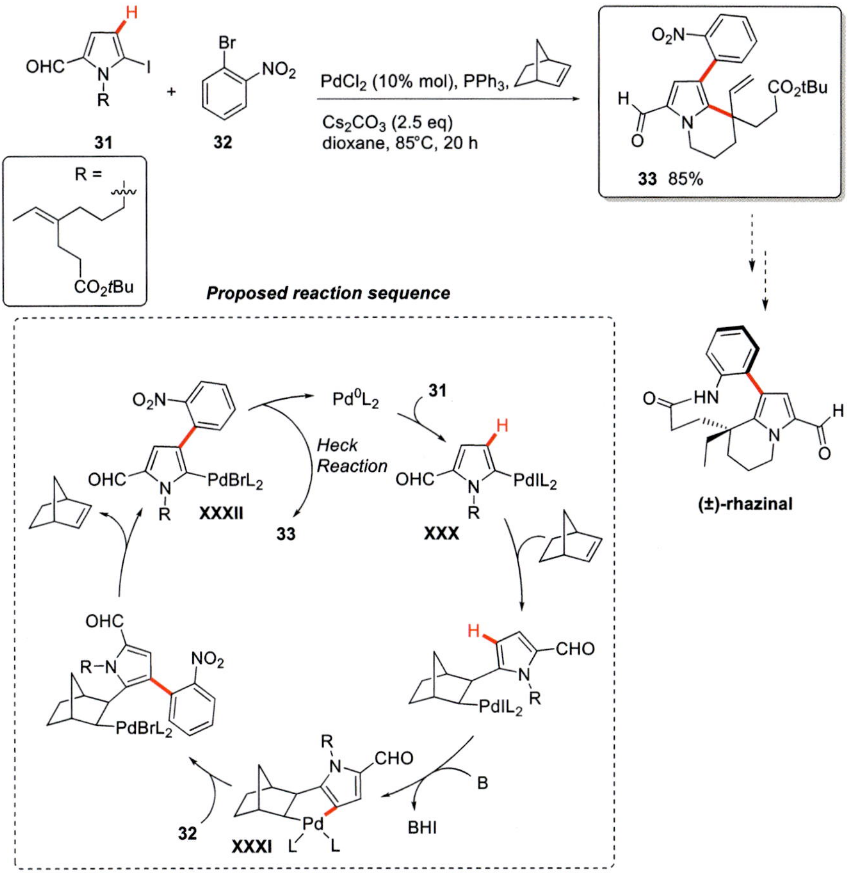

Scheme 20.

R. Ketoprofen

Dong reported the synthesis of ketoprofen, a non-steroidal anti-inflammatory agent (Scheme 21).[32] The key step leading to arylpropionic ester **36** involves the *ortho*-acylation of iodoarene **34** with reactant **35** followed by hydrogen transfer at the *ipso* position. In the mechanism iodide **34** gives **XXXIII** by oxidative addition on Pd(0). Upon formation of palladacycle **XXXIV** by sequential stereoselective norbornene insertion and C–H activation, oxidative addition of anhydride **35** occurs. The isopropyl carbonate

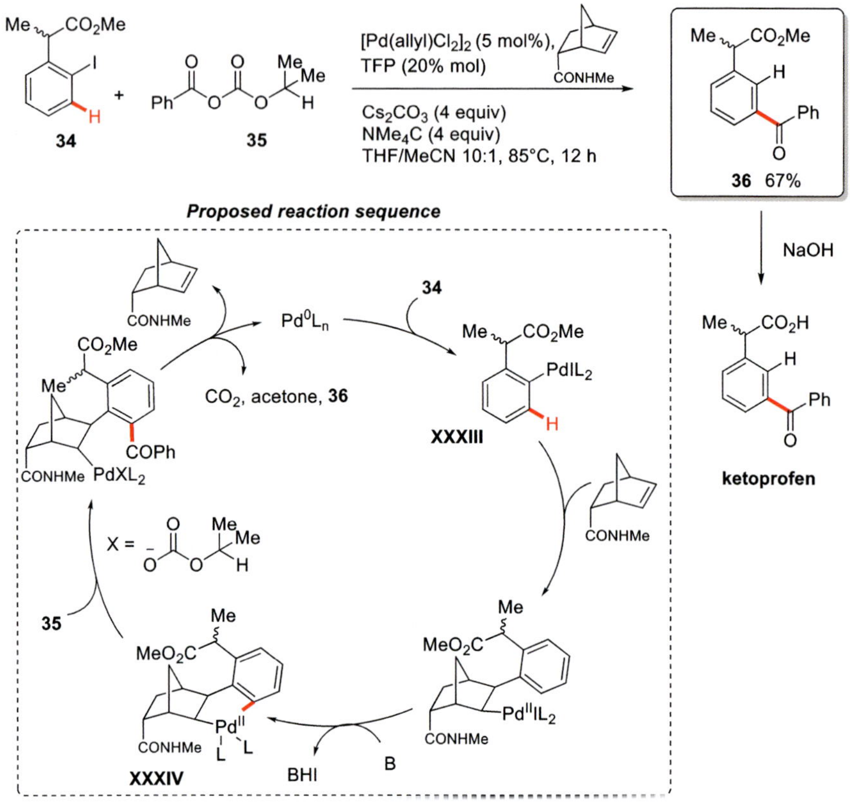

Scheme 21.

anhydrides serve both to deliver an acyl equivalent and to reduce the final organometallic intermediate. Interestingly, the best result is obtained using the norbornene derivative bearing a methyl amide group. This is the first example of an acylation step in the Catellani sequence.

V. Conclusions

Looking at the recent literature, we have picked up and highlighted some significant examples of the synthesis of natural and designed compounds with well-known biological activity, potentially interesting for industrial scale up, based on one or more C–H activation steps.

The research in transition-metal mediated C–H bond activations has recently given rise to a large collection of valuable synthetic procedures, the major part devoted to the preparation of complex, potentially biological active class of molecules. Further advances in both mechanistic studies and development of novel synthetic transformations are expected.

However, it is important to keep in mind that the power of C–H activation should be considered as complementary to the "non-C–H activation" strategies, and it is by their combination that it is possible to realize successful efficient, rational and environmentally friendly synthesis of very complex molecules.

VI. References

1. Clark, J. H.; Macqurrie, D. J. ed. *Handbook of Green Chemistry and Technology*, Blackwell Science, Abingdon, **2002**.
2. Anastas, P.; Warner, J. *Green Chemistry: Theory and Practice*, Oxford University Press, New York, **1998**.
3. Clark J. H.; Deswarte, F. E. I. *Introduction to Chemicals from Biomass*, John Wiley & Sons, Ltd, Chichester, **2008**.
4. (a) Sheldon, R. A. *Chem. Ind.* **1992**, *23*, 903–906. (b) Sheldon, R. A. *Green Chem.* **2007**, *9*, 1273–1283.
5. (a) Horvath, I. T.; Anastas, P. T. *Chem. Rev.* **2007**, *107*, 2169–2173. (b) Tucker, J. L. *Org. Process Res. Dev.* **2006**, *10*, 315–319.
6. Wender, P. A. *Chem. Rev.* **1996**, *96*, 1–2.
7. Selected reviews and papers on cascade reactions: (a) Malacria, M. *Chem. Rev.* **1996**, *96*, 289–306. (b) Nicolaou, K. C.; Edmonds, D. J.; Bulger, P. G. *Angew. Chem., Int. Ed.* **2006**, *45*, 7134–7186. (c) Crone, B.; Kirsch, S. F. *Chem.–Eur. J.* **2008**, *14*, 3514–3522. (d) Padwa, A. *J. Org. Chem.* **2009**, *74*, 6421–6441. (e) Vilotijevic, I.; Jamison, T. F. *Angew. Chem., Int. Ed.* **2009**, *48*, 5250–5281. (f) Williams, F. J.; Jarvo, E. R. *Angew. Chem., Int. Ed.* **2011**, *50*, 4459–4462. (g) Duttwyler, S.; Lu, C.; Rheingold, A. L.; Bergman, R. G.; Ellman, J. A. *J. Am. Chem. Soc.* **2012**, *134*, 4064–4067. (h) Byers, P. M.; Alabugin, I. V. *J. Am. Chem. Soc.* **2012**, *134*, 9609–9614. (i) Hojo, D.; Tanaka, K. *Org. Lett.* **2012**, *14*, 1492–1495. (j) Lin, H.; Tan, Y.; Sun, X.-W.; Lin, G.-Q. *Org. Lett.* **2012**, *14*, 3818–3821. (k) Lu, L.-Q.; Chen, J.-R.; Xiao, W.-J. *Acc. Chem. Res.* **2012**, *45*, 1278–1293. (l) Dhakshinamoorthy, A.; Garcia, H. *ChemSusChem* **2014**, *7*, 2392–410. (m) Muschiol, J.; Peters, C.; Oberleitner, N.; Mihovilovic, M. D.; Bornscheuer, U. T.; Rudroff, F. *Chem. Comm.* **2015**, *51*, 5798–811. (n) Blouin, S.; Blond, G.; Donnard, M.; Gulea, M.; Suffert, J. *Synthesis* **2017**, *49*, 1767–1784.

8. Selected reviews and papers on domino reactions: (a) Tietze, L. F.; Brasche, G.; Gericke, K. *Domino Reactions in Organic Synthesis*, Wiley-VCH: Weinheim, Germany, **2006**. (b) Tietze, L. F. *Chem. Rev.* **1996**, *96*, 115–136. (c) Tietze, L.; Kinzel, T.; Brazel, C. C. *Acc. Chem. Res.* **2009**, *42*, 367–378. (d) de Meijere, A.; Zezschwitz, P. V.; Brase, S. *Acc. Chem. Res.* **2005**, *38*, 413–422. (e) Pellissier, H. *Tetrahedron* **2006**, *62*, 1619–1665. (f) Pellissier, H. *Tetrahedron* **2006**, *62*, 2143–2173. (g) Enders, D.; Grondal, C.; Huttl, M. R. M. *Angew. Chem, Int. Ed.* **2007**, *46*, 1570–1581. (h) Grossmann, A.; Enders, D. *Angew. Chem. Int. Ed.* **2012**, *51*, 314–325. (i) Piou, T.; Neuville, L.; Zhu J. *Org. Lett.* **2012**, *14*, 3760–3763. (j) Newman, S. G.; Howell, J. K.; Nicolaus, N.; Lautens, M. *J. Am. Chem. Soc.* **2011**, *133*, 14916–14119. (k) Cai, S.; Wang, F.; Xi, C. *J. Org. Chem.* **2012**, *77*, 2331–6. (l) Tietze, L. F. *Domino Reactions: Concepts for Efficient Organic Synthesis* **2014**, Wiley-VCH Verlag GmbH & Co. KGaA. (m) Sebren, L. J.; Devery, J. J.; Stephenson, C. R. J. *ACS Catalysis* **2014**, *4*, 703–716. (n) Pellissier, H. *Adv. Synth. Cat.* **2016**, *358*, 2194–2259. (o) Ardkhean, R.; Caputo, D. F. J.; Morrow, S. M.; Shi, H.; Xiong, Y.; Anderson, E. A. *Chem. Soc. Rev.* **2016**, *45*, 1557–1569. (p) Bhar, S.; Ramana, M. M. V. *Curr. Drug Discov. Technol.* **2016**, *13*, 170–187. (q) Zhou, S.; Tong, R. *Org. Lett.* **2017**, *19*, 1594–1597.

9. Selected reviews and papers on tandem reactions: (a) Denmark, S. E.; Thorarensen, A. *Chem. Rev.* **1996**, *96*, 137–166. (b) Wasilke, J.-C.; Obrey, S. J.; Baker, R. T.; Bazan, G. C. *Chem. Rev.* **2005**, *105*, 1001–1020. (c) Taylor, R. J. K.; Reid, M.; Foot, J.; Raw, S. A. *Acc. Chem. Res.* **2005**, *38*, 851–869. (d) Hussain, M. M.; Walsh, P. J. *Acc. Chem. Res.* **2008**, *41*, 883–893. (e) Alba, A.-N.; Companyo, X.; Viciano, M.; Rios, R. *Curr. Org. Chem.* **2009**, *13*, 1432–1474. (f) Haibach, M. C.; Kundu, S.; Brookhart, M.; Goldman, A. S. *Acc. Chem. Res.* **2012**, *45*, 947–958. (g) Gupta, S.; Kumar, B.; Kundu, B. *J. Org. Chem.* **2011**, *76*, 10154–10162. (h) Kang, D.; Park, S.; Ryu, T.; Lee, P. H. *Org. Lett.* **2012**, *14*, 3912–3915. (i) Kieffer, M. E.; Repka, L. M.; Reisman, S. E. J. *Am. Chem. Soc.* **2012**, *134*, 5131–5137. (j) Wang, Y.; Zhao, H. *Catalysts 2016*, *6*, 194/1–194/21.

10. Mayer, S. F.; Kroutil, W.; Faber, K. *Chem. Soc. Rev.* **2001**, *30*, 332–339.

11. Fogg, D. E.; dos Santos, E. N. *Coord. Chem. Rev.* **2004**, *248*, 2365–2379.

12. Tietze, L. F.; Rackelmann N. *Pure Appl. Chem.* **2004**, *76*, 1967–1983.

13. On C–H activation: (a) Yu J.-Q. Catalytic Transformations via C-H Activation **2016**, Vols. 1 and 2, *Science of Synthesis*, Thieme. (b) Dixneuf, P. H.; Doucet H. C–H Bond Activation and Catalytic Functionalization I and II, *Topics in Organometallic Chemistry*, Vols. 55 and 56, **2016**, Springer. (c) Li, J. J. *C–H Bond Activation in Organic Synthesis*, **2015**, CRC Press. (d) Ribas, X. *C–H and C–X Bonds Functionalization: Transition Metal Mediation*, **2013**, RSC, Publishing, London. (e) Pérez, P. J. Alkane C–H Activation by Single-Site Metal Catalysis, *Catalysis by Metal Complexes*, **2012**, *38*, Springer, New York. (f) Chen, X; Engle, K. M.; Wang, D.-H.; Yu, J.-Q. *Angew. Chem. Int. Ed.* **2009**, *48*, 5094–5115. (g) Chen, D. Y.-K.;

Youn, S. W. *Chem. Eur. J.* **2012**, *18*, 9452–9474. (h) Jiang, Y.-Y; Man, X.; Bi, S. *Science China: Chem.* **2016**, *59*, 1448–1466. (i) Mehta, V. P.; Garcia-Lopez, J. A. *ChemCatChem* **2017**, *9*, 1149–1156. (k) Topczewski, J. T.; Cabrera, P. J.; Saper, N. I.; Sanford, M. S. *Nature* **2016**, *531*, 220–224. (l) Chen, G.; Gong, W.; Zhuang, Z.; Andra, M. S.; Chen, Y.-Q.; Hong, V.; Yang, Y.-F.; Liu, V.; Houk, K. N.; Yu, J.-Q. *Science* **2016**, *353*, 1023–1027. (m) Liu, Y.; Ge, H.; *Nat. Chem.* **2017**, *9*, 26–32. (n) He, J.; Wasa, M.; Chan, Q.; Shao, K. S. L.; Yu, J.-Q. *Chem. Rev.* **2017**, *117*, 8754–8786. (o) Davis, H. J.; Phipps, R. J. *Chem. Sci.* **2017**, *8*, 864–877. (p) Roudesly, F.; Oble, J.; Poli, G. *J. Mol. Cat. A: Chem.* **2017**, *426*, 275–296. (q). Kapdi, A. R.; Prajapati D. *RSC Adv.* **2014**, *4*, 41245–41259. (r) Hui, C.; Xu, J. *Tetrahedron Lett.* **2016**, *57*, 2692–2696. (s) Sasmala, S.; Sena, I.; Hallb, R. G.; Pal, S. *Synthesis* **2015**, *47*, 3711–3716. (t) Ronson, T. O.; Taylor, R. J. K.; Fairlamb, I. J. S. *Tetrahedron* **2015**, *71*, 989–1009. (u). Sivanandan, S. T.; Shaji, A.; Ibnusaud, I.; Johansson Seechurn, C. C. C.; Colacot, T. J. *Eur. J. Org. Chem.* **2015**, 38–49. (v) Sui, X.; Zhu, R.; Gu, Z. *Synlett.* **2013**, *24*, 2023–2031. (w) Bai, Y.; Davis, D. C.; Dai, M. *J. Org. Chem.* **2017**, *82*, 2319–2328. (x) Malacria, M.; Maestri, G. *J. Org. Chem.* **2013**, *78*, 1323–1328.

14. (a) Della Ca', N.; Fontana, M.; Motti, E.; Catellani, M. *Acc. Chem. Res.* **2016**, *49*, 1389–1400. (b) Chiusoli, G. P.; Catellani, M.; Costa, M.; Motti, E.; Della Ca', N.; Maestri, G. *Coord. Chem. Rev.* **2010**, *254*, 456–469. (c) Della Ca', N.; Maestri, G.; Malacria, M.; Derat, E.; Catellani, M. *Angew. Chem. Int. Ed.* **2011**, *50*, 12257–12261, S12257/1–S12257/158. (d) Della Ca', N.; Fontana, M.; Xu, D.; Cremaschi, M.; Lucentini, R.; Zhou, Z.-M.; Catellani, M.; Motti, E. *Tetrahedron* **2015**, *71*, 6389–6401. (e) Maestri, G.; Caneque, T.; Della Ca', N.; Derat, E.; Catellani, M.; Chiusoli, G. P.; Malacria, M. *Org. Lett.* **2016**, *18*, 6108–6111. (f) Motti, E.; Della Ca', N.; Xu, D.; Piersimoni, A.; Bedogni, E.; Zhou, Z-M.; Catellani, M. *Org. Lett.* **2012**, *14*, 5792–5795. (g) Maestri, G.; Motti, E.; Della Ca', N.; Malacria, M.; Derat, E.; Catellani, M. *J. Amer. Chem. Soc.* **2011**, *133*, 8574–8585. (h) Motti, E.; Della Ca', N.; Xu, D.; Armani, S.; Aresta, B. M.; Catellani, M. *Tetrahedron* **2013**, *69*, 4421–4428.

15. Neog, K.; Borah, A.; Gogoi, P. *J. Org. Chem.* **2016**, *81*, 11971–11977.

16. Han, W.; Zhou, X.; Yang, S.; Xiang, G.; Cui, B.; Chen, Y. *J. Org. Chem.* **2015**, *80*, 11580–11587.

17. Wu, J.; Xiang, S.; Zeng, J.; Leow, M.; Liu, X.-W. *Org. Lett.* **2015**, *17*, 222–225.

18. Shi, H.; Babinski, D. J.; Ritter, T. *J. Am. Chem. Soc.* **2015**, *137*, 3775–3778.

19. Casnati, A.; Maggi, R.; Maestri, G.; Della Ca', N.; Motti, E. *J. Org. Chem.* **2017**, *82*, 8296–8303.

20. Janody, S.; Jazzar, R.; Comte, A.; Holstein, P. M.; Vors, J.-P.; Ford, M. J.; Baudoin, O. *Chem. Eur. J.* **2014**, *20*, 11084–11090.

21. (a) Dailler, D.; Danoun, G.; Baudoin, O. A *Angew. Chem., Int. Ed.* **2015**, *54*, 4919–4922. (b) Dailler, D.; Danoun, G.; Ourri, B.; Baudoin, O. *Chem. Eur. J.* **2015**, *21*, 9370–9379.

22. Holstein, P. M.; Dailler, D.; Vantourout, J.; Shaya, J.; Millet, A.; Baudoin, O. *Angew. Chem., Int. Ed.* **2016**, *55*, 2805–2809.

23. Wagnières, O.; Xu, Z.; Wang, Q.; Zhu, J. *J. Am. Chem. Soc.* **2014**, *136*, 15102–15108.

24. Ting, C. P.; Maimone, T. J. *Angew. Chem. Int. Ed.* **2014**, *53*, 3115–3119.

25. (a) Kim, D.; Ham, K.; Hong, S. *Org. Biomol. Chem.* **2012**, *10*, 7305–7312. (b) Shin, Y.; Yoo, C.; Moon, Y.; Lee, Y.; Hong, S. *Chem. Asian J.* **2015**, *10*, 878–881.

26. (a) Fox, J. C.; Gilligan, R. E.; Pitts, A. K.; Bennett, H. R.; Gaunt, M. J. *Chem. Sci.* **2016**, *7*, 2706–2710. (b) Ciana, C.-L.; Phipps, R.; Brandt, J.; Meyer, F.-M.; Gaunt, M. J. *Angew. Chem., Int. Ed.* **2011**, *50*, 458–462. (c) Phipps, R.; Gaunt, M. J. *Science* **2009**, *323*, 1593–1597. (d) Orito, K.; Horibata, A.; Nakamura, T.; Ushito, H.; Nagasaki, H.; Yuguchi, M.; Yamashita, S.; Tokuda, M. *J. Am. Chem. Soc.* **2004**, *126*, 14342–14343.

27. (a) Sasmal, S.; Sen, I.; Hall, R. G.; Pal, S. *Synthesis* **2015**, *47*, 3711–3716. (b) for the C–H activation protocol see: Sasmal, S.; Sen, I.; Hall, R. G.; Pal, S. *Tetrahedron Lett.* **2015**, *56*, 1374.

28. (a) Della Ca',N.; Sassi, G.; Catellani, M. *Adv. Synth. Cat.* **2008**, *350*, 2179 – 2182. (b) Catellani, M.; Motti, E.; Della Ca', N. *Top. Catal.* **2010**, *53*, 991–996.

29. (a) Weinstabl, H.; Suhartono, M.; Qureshi, Z.; Lautens, M. *Angew. Chem. Int. Ed.* **2013**, *52*, 5305–5308. (b) Qureshi, Z.; Weinstabl, H.; Suhartono, M.; Hongqiang Liu, H.; Thesmar, P.; Lautens, M. *Eur. J. Org. Chem.* **2014**, 4053–4069.

30. (a) Jiao, L.; Bach, T. *J. Am. Chem. Soc.* **2011**, *133*, 12990–12993. (b) Jiao, L.; Herdtweck, E.; Bach, T. *J. Am. Chem. Soc.* **2012**, *134*, 14563–14572. (c) Jiao, L.; Bach, T. *Angew. Chem. Int. Ed.* **2013**, *52*, 6080–6083.

31. (a) Sui, X.; Zhu, R.; Li, G.; Ma, X.; Gu, *J. Am. Chem. Soc.* **2013**, *135*, 9318–9321. (b) Sui, X.; Zhu, R.; Gu, Z. *Synlett.* **2013**, *24*, 2023–2031.

32. Dong, Z.; Wang, J.; Ren, Z.; Dong, G. *Angew. Chem. Int. Ed.* **2015**, *54*, 12664–12668.

3 Photoredox Catalysis, an Opportunity for Sustainable Radical Chemistry

Christophe Lévêque, Etienne Levernier, Vincent Corcé, Louis Fensterbank,* Max Malacria and Cyril Ollivier[†]

Institut Parisien de Chimie Moléculaire (UMR 8232), Sorbonne Université-CNRS, Paris, France
*louis.fensterbank@sorbonne-universite.fr
[†]cyril.ollivier@sorbonne-universite.fr

Table of Contents

List of Abbreviations

Ac-HMP	4,6-bis[4-(9,9-dimethyl-9,10-dihydroacridine)-phenyl]pyrimidine
BDE	bond dissociation energy
BET	back electron transfer
Boc	tert-butoxycarbonyle
Cbz	carboxybenzyle
4CzIPN	2,4,5,6-tetra(9H-carbazol-9-yl)isophthalonitrile
DIPEA	*N,N*-diisopropyléthylamine
DNA	deoxyriboNucleic Acid
EPR	electron paramagnetic resonance
HOMO	highest occupied molecular orbital
ISC	inter-system crossing
LC	ligand-centered transition
LUMO	lowest unoccupied molecular orbital
LMCT	ligand-to-metal charge transfer
MC	metal-centered transition
MLCT	metal-to-ligand charge transfer
NMR	nuclear magnetic resonance

OLED	organic light-emitting diode
PC	photocatalyst
RISC	reverse intersystem crossover
SCE	saturated calomel electrode
SET	single electron transfer
TADF	thermally activated delayed fluorescence
TEMPO	2,2,6,6-tetramethylpiperidinyloxy
THF	tetrahydrofuran
UV	ultraviolet

I. Introduction

Radicals are chemical intermediates which display one or more atoms with an unpaired electron. The discovery of such species has been reported first by Moses Gomberg in 1900 with evidences of the formation of the triphenylmethyl radical. During the next 50 years, a growing numbers of proofs brought in part by Paneth,[1] Hey[2] and Kharasch[3] and the discovery of EPR[4] allowed to support the existence of radical species. Like the ionic species, radicals exist in nature and play key roles in biological processes. Several enzymes showed radical behaviors pathways such as the ribonucleotide reductase that allows the transformation of ribonucleosides to deoxyribonucleosides, essential building blocks of DNA. An equivalent of this biological reaction has been developed in radical synthesis with the Barton-McCombie reaction (Figure 1).[5] This analogous natural reaction is one of all the methodologies developed by the radical chemistry community.

Although radicals have remained mysterious and considered as out of control by a part of chemists, the contribution of Kochi, Stork, Curran, Giese, Hart and others[6] demystified their use in synthesis and showed as potential alternatives of usual anionic reactions. Indeed, contrary to organometallic reagents, radical precursors can provide highly reactive neutral radicals in smooth conditions, under air atmosphere and non-distilled solvents. In addition, radical reactions are highly chemoselective and can be performed without protected functions. Indeed, several conditions for the generation of radicals are available. Radical initiators (Et_3B, peroxides, azo compounds…) and mediators (stannanes, silanes, thiols…) can provide radicals for chain reactions. Single electron transfers from

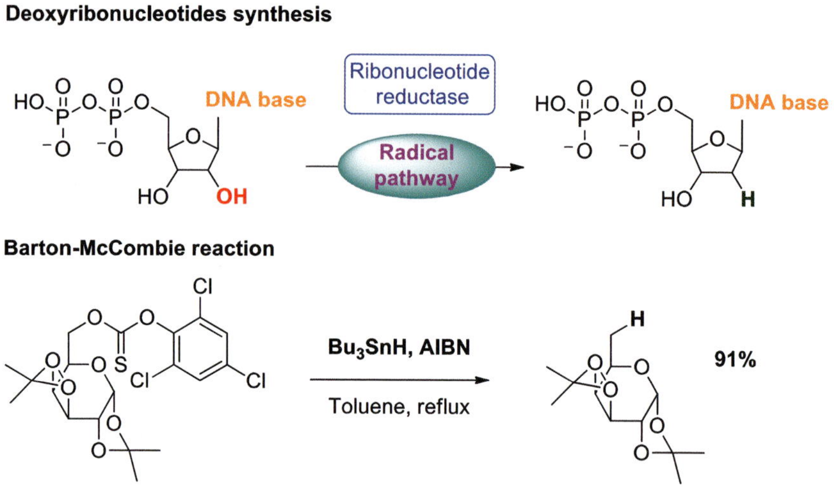

                                    Lévêque *et al.*

Deoxyribonucleotides synthesis

Barton-McCombie reaction

Figure 1.   Bio-catalyzed dehydroxylation of alcohols.

stoichiometric metallic oxidants (manganese, cerium) or reductants (samarium, titanium, zinc, nickel) are as many alternatives for the formation of radicals.

Radical chemistry offers many advantages (*vide supra*) and promising features compared to anionic reactions. However, it suffers of several drawbacks which limit their use in synthesis. Because the reactions are performed in diluted conditions, the scale up is quite complicated. Some of the processes also require an excess of radical acceptor. But, the most problematic aspects are the use of explosive initiators (peroxides, azo compounds) or toxic mediators like the tin (IV) derivatives which are also difficult to remove from the product. In this context, solutions have been proposed to escape from "the tyranny of tin" and to offer the opportunity of using sustainable methods.

In order to progressively substitute the use of tin reagents, methodologies involving only catalytic amounts of these reagents or tin-supported surfaces have been reported.[7] Processes employing stoichiometric mediators such as silanes, phosphines or thiols are also potential alternatives. Organoboranes, as substrates or radical initiators for chain-reactions, showed promising results as well.[6b,8] Less toxic metals responsible of one-electron transfers were also considered as a possible solution. However, methodologies involving metal complexes based on iron, copper, manganese, titanium, samarium… require excess amounts. Unfortunately, in terms of sustainability and eco-compatibility, it is important to develop alternatives even more efficient. At the end of the 70's, some pioneering works[9] mentioned the generation of radicals thanks to a single electron transfer (SET) mediated by a photoactivated metal complex.

II. The Photoredox Catalysis as an Alternative

The radical chemistry allowed chemists to develop a wide range of processes for the carbon-carbon bond formation. Many efforts to avoid the use of toxic tin reagents and to find others alternatives have been done so far, but many reactions are still performed under these conditions. Since the end of the 2000s, photoredox catalysis has emerged as a powerful and versatile eco-compatible approach for the generation of radicals.

A. Nature as a Source of Inspiration

Evolution has allowed living organisms to develop sustainable and highly sophisticated processes. Among them, the photosynthesis of plants was one of the earliest to interest scientists. The chlorophyll present in plant cells absorbs the sunlight in the visible range and initiates the transformation of CO_2 and water to saccharides and oxygen. This natural synthetic process highlights the fundamental and efficient conversion of energy from sunlight into chemical energy.

Taking advantage of this process, increasing efforts from the radical chemistry community have been realized to develop new methodologies using the visible-light as a promotor of redox reactions for the generation of radicals. Visible-light photoredox catalysis has emerged as a powerful methodology for radical formation in terms of selectivity and sustainability. Since the pioneering work on this field, more and more groups have incremented the photoredox catalysis as a part of their research and the number of publications on this topic demonstrates its growing popularity. Inspired by the photoredox process of natural photosynthesis, chemists were interested on the development of photocatalysts absorbing light in the visible region and with a large range of potential values to perform efficient redox transformations. The first man-made photocatalyst mainly reported and still frequently used is the photoactive complex $Ru(bpy)_3Cl_2$. First reported as a photoredox catalyst for organic synthetic purposes by Kellog,[9] Pac[10] and Deronzier,[11] this complex has been essentially used in inorganic chemistry for devices applications[12] or transformation of small molecules[13] (CO_2, H_2O). This is only from 2008 and the advances of MacMillan,[14] Yoon,[15] and Stephenson,[16] that organic photoredox catalysis has definitely taken off.

B. Artificial Redox Photocatalysts

The photosynthesis led to the development of catalytic processes involving chromophores as light harvesters to promote electron transfers for photo-electrochemical cells, photocatalytic water splitting systems or photobiore-actors. In this context, chemists tuned photocatalysts which could act as chlorophyll. The complex $Ru(bpy)_3Cl_2$ first synthesized by Burstall[17] in 1936 has shown photophysical and redox properties to perform single electron processes under visible-light activation. Thus, many efforts have been done to widen the range of transition metal-based photocatalysts. In order to get closer to the principles of green chemistry, organic dyes have also attracted interest. In terms of efficiency, a valuable photocatalyst has to absorb the visible-light, get fitting redox properties and a long lifetime at the excited state to enable electron transfers. Several polypyridyl complexes of metal from the fourth to sixth periods showed these properties. Most of them are ruthenium,[12] rhenium[18] or iridium[19] based complexes (Scheme 1) but some gold,[20] copper,[21] chromium[22] or iron[23] complexes, are also reported. Besides the organometallic complexes, organic dyes are also good candidates for artificial photosynthesis processes (Scheme 1).

1. *Photophysical Properties*

The mainly used metal-based photocatalysts are the ruthenium and iridium complexes due to their photophysical properties and the range of redox potential accessible. Considering the photophysical properties of such complexes, the UV/visible absorption spectrum displays four kinds of absorption bands. The LC (Ligand-Centered) transition, the MC

[Ru(bpy)₃](PF₆) (PF₆)₂

$E_{1/2}(Ru(III)/\mathbf{Ru(II)}^{*}) = -0.81$ V vs. SCE
$E_{1/2}(\mathbf{Ru(II)}^{*}/Ru(I)) = +0.77$ V vs. SCE
$E_{1/2}(Ru(III)/\mathbf{Ru(II)}) = +1.29$ V vs. SCE
$E_{1/2}(\mathbf{Ru(II)}/Ru(I)) = -1.37$ V vs. SCE
$\lambda_{max} = 452$ nm, $\tau = 1100$ ns

[Ru(boz)₃](PF₆) (PF₆)₂

$E_{1/2}(Ru(III)/\mathbf{Ru(II)}^{*}) = -0.26$ V vs. SCE
$E_{1/2}(\mathbf{Ru(II)}^{*}/Ru(I)) = +1.45$ V vs. SCE
$E_{1/2}(Ru(III)/\mathbf{Ru(II)}) = +1.86$ V vs. SCE
$E_{1/2}(\mathbf{Ru(II)}/Ru(I)) = -0.80$ V vs. SCE
$\lambda_{max} = 443$ nm, $\tau = 740$ ns

[Ru(phen)₃](PF₆) (PF₆)₂

$E_{1/2}(Ru(III)/\mathbf{Ru(II)}^{*}) = -0.87$ V vs. SCE
$E_{1/2}(\mathbf{Ru(II)}^{*}/Ru(I)) = +0.82$ V vs. SCE
$E_{1/2}(Ru(III)/\mathbf{Ru(II)}) = +1.26$ V vs. SCE
$E_{1/2}(\mathbf{Ru(II)}/Ru(I)) = -1.36$ V vs. SCE
$\lambda_{max} = 422$ nm, $\tau = 500$ ns

[Ir(dF(CF₃)ppy)₂(bpy)](PF₆) (PF₆)

$E_{1/2}(Ir(IV)/\mathbf{Ir(III)}^{*}) = -0.81$ V vs. SCE
$E_{1/2}(\mathbf{Ir(III)}^{*}/Ir(II)) = +1.32$ V vs. SCE
$E_{1/2}(Ir(IV)/\mathbf{Ir(III)}) = +1.69$ V vs. SCE
$E_{1/2}(\mathbf{Ir(III)}/Ir(II)) = -1.37$ V vs. SCE
$\lambda_{max} = 379$ nm, $\tau = 2280$ ns

fac-Ir(ppy)₃

$E_{1/2}(Ir(IV)/\mathbf{Ir(III)}^{*}) = -1.73$ V vs. SCE
$E_{1/2}(\mathbf{Ir(III)}^{*}/Ir(II)) = +0.31$ V vs. SCE
$E_{1/2}(Ir(IV)/\mathbf{Ir(III)}) = +0.77$ V vs. SCE
$E_{1/2}(\mathbf{Ir(III)}/Ir(II)) = -2.19$ V vs. SCE
$\lambda_{max} = 375$ nm, $\tau = 1900$ ns

[Cu(dap)₂](PF₆) (PF₆)

$E_{1/2}(Cu(II)/\mathbf{Cu(I)}^{*}) = -1.43$ V vs. SCE
$E_{1/2}(\mathbf{Cu(I)}^{*}/Cu(0)) = $ n.d.
$E_{1/2}(Cu(II)/\mathbf{Cu(I)}) = +0.62$ V vs. SCE
$E_{1/2}(\mathbf{Cu(I)}/Cu(0)) = $ n.d.
$\lambda_{max} = $ n.d. $\tau = 270$ ns

Cz = Carabzoyl

4CzIPN

$E_{1/2}(PC^{+}/\mathbf{PC}^{*}) = $ n.d.
$E_{1/2}(\mathbf{PC}^{*}/PC^{-}) = +1.35$ V vs. SCE
$E_{1/2}(PC^{+}/\mathbf{PC}) = $ n.d.
$E_{1/2}(\mathbf{PC}^{+}/PC^{-}) = -1.21$ V vs. SCE
$\lambda_{max} = 435$ nm, $\tau = 5100$ ns

Eosin Y

$E_{1/2}(PC^{+}/\mathbf{PC}^{*}) = -1.60$ V vs. SCE
$E_{1/2}(\mathbf{PC}^{*}/PC^{-}) = +1.18$ V vs. SCE
$E_{1/2}(PC^{+}/\mathbf{PC}) = +0.72$ V vs. SCE
$E_{1/2}(\mathbf{PC}/PC^{-}) = -1.14$ V vs. SCE
$\lambda_{max} = 539$ nm, $\tau = 2.1$ ns

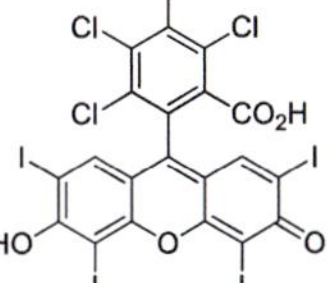

Fukuzumi Catalyst

$E_{1/2}(PC^{+}/\mathbf{PC}^{*}) = $ -n.d.
$E_{1/2}(\mathbf{PC}^{*}/PC^{-}) = +2.06$ V vs. SCE
$E_{1/2}(PC^{+}/\mathbf{PC}) = $ n.d.
$E_{1/2}(\mathbf{PC}/PC^{-}) = -0.57$ V vs. SCE
$\lambda_{max} = 539$ nm, $\tau = 2.1$ ns

Fluorescein

$E_{1/2}(PC^{+}/\mathbf{PC}^{*}) = -1.61$ V vs. SCE
$E_{1/2}(\mathbf{PC}^{*}/PC^{-}) = +1.21$ V vs. SCE
$E_{1/2}(PC^{+}/\mathbf{PC}) = +0.83$ V vs. SCE
$E_{1/2}(\mathbf{PC}/PC^{-}) = -1.14$ V vs. SCE
$\lambda_{max} = 528$ nm, $\tau = 4.2$ ns

Rose Bengal

$E_{1/2}(PC^{+}/\mathbf{PC}^{*}) = -0.68$ V vs. SCE
$E_{1/2}(\mathbf{PC}^{*}/PC^{-}) = +0.99$ V vs. SCE
$E_{1/2}(PC^{+}/\mathbf{PC}) = +1.09$ V vs. SCE
$E_{1/2}(\mathbf{PC}/PC^{-}) = -0.78$ V vs. SCE
$\lambda_{max} = $ n.d., $\tau = $ n.d.

Scheme 1.   Examples of photocatalysts.

Lévêque *et al.*

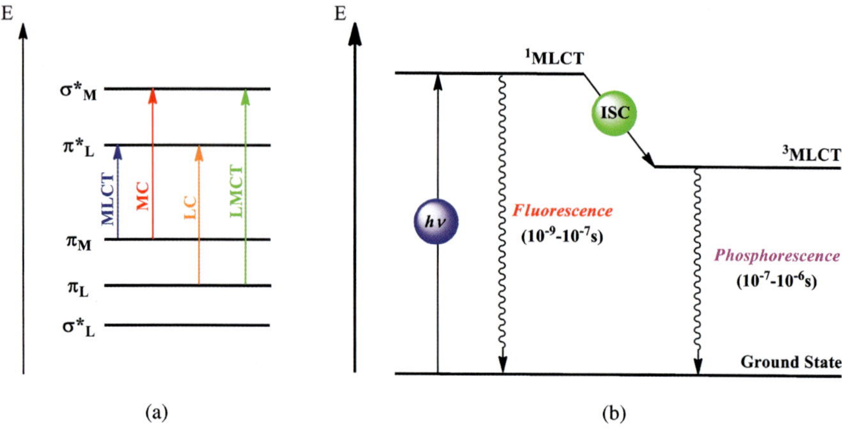

Scheme 2. (a) Simplified molecular orbital diagram for ruthenium and iridium photocatalysts. (b) Simplified energy diagram of absorbing and emitting processes.

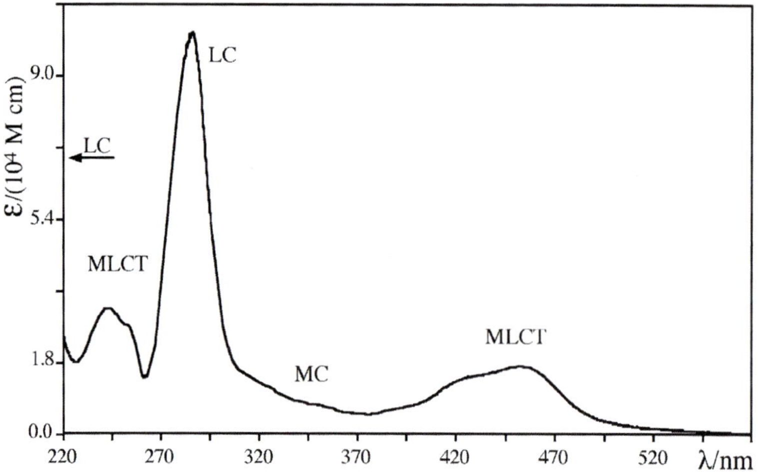

Figure 2. Electronic absorption spectrum of $[Ru(bpy)_3](PF_6)_2$ in EtOH (from Ref. 12a).

(Metal-Centered) transition or d–d transition, the MLCT (Metal-to-Ligand Charge Transfer) transition and the LMCT (Ligand-to-Metal Charge Transfer) transition (Scheme 2a). For example, an electronic absorption spectrum of $[Ru(bpy)_3](PF_6)_2$ in ethanol is mentioned in Figure 2 and showed four bands (LC at 285 nm, MC at 322 and 344 nm, and two MLCT at 240 and 450 nm)

This MLCT transition is actually the most important one since the absorption is in the visible-light region spectra and it allows to reach a ^{3}MLCT after an Inter-System Crossing (ISC) thanks to a strong spin-orbit coupling, usually observed with the heavy metal atom (Scheme 2b). The consequence is a longer luminescence lifetime $(10^{-7}-10^{-6}\text{s})$[12a,19b] compared to 3d metal complexes and so the possibility of electrons transfers.

Constantly trying to evolve toward a greener chemistry, many photoredox processes involve organic dyes as photocatalyst such as fluorescein, Eosin Y or rose bengal.[24] As organic molecules, the strongest light absorption results in the promotion of an electron of a π orbital to a π^* orbital. Because of the lack of heavy atoms and so the inefficient intersystem crossover, the light emission is mainly fluorescence. The consequence for these photocatalysts is a very short lifetime of the S_1 excited state (2 to 20 ns for most of the organic photocatalysts). The envisaged electron transfers must be at least as fast as the deexcitation of the chromophore. Recent work on the development of OLEDs has shown new perspective in terms of organic photocatalysts design.[25] Thanks to a short gap between singlet and triplet state and efficient spin conversion processes, a series of tetracarbazolyl dicyanobenzene with excited state lifetimes of few micro-secondes have shown optimal features for photoredox catalyzed processes.[26]

In this part, we will focus on the description of metal-based photocatalysts and especially on the Ru(bpy)$_3^{2+}$ which is the most studied. However, the mentioned statements are applicable to homo- and heteroleptic iridium photocatalysts. As previously detailed, ruthenium and iridium photocatalysts reach a triplet state (^{3}MLCT) after visible-light absorption. About the Ru^{II}(bpy)$_3^{2+}$, it corresponds to the promotion of an electron from the t$_2$g orbitals of the metal to a π^* orbital of the ligand. The excited complex can be formally written as "[Ru^{III}(bpy)$_2$(bpy)$^-$]$^{2+}$", equivalent to [Ru^{II}(bpy)$_3^{2+}$]* where respectively the metal becomes oxidant and a ligand bpy reductant (Scheme 3).

The photoexcited complex [**M**] can then oxidize or reduce a substrate by a SET giving the opportunity to generate radicals. In order to have an idea on the tolerant substrates for the SET, the redox potentials of photocatalysts at the excited-state have to be determined. From electrochemical and fluorescence data, a qualitative estimation of the excited-state redox potentials

                                   Lévêque *et al.*

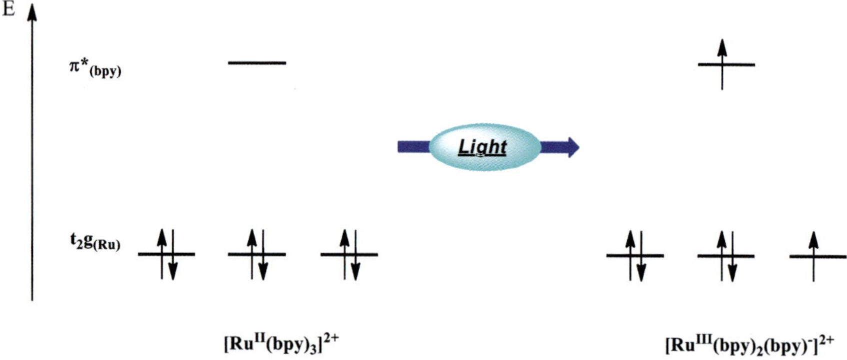

Scheme 3.   Simplified orbital diagram of $Ru(bpy)_3^{2+}$ during light absorption.

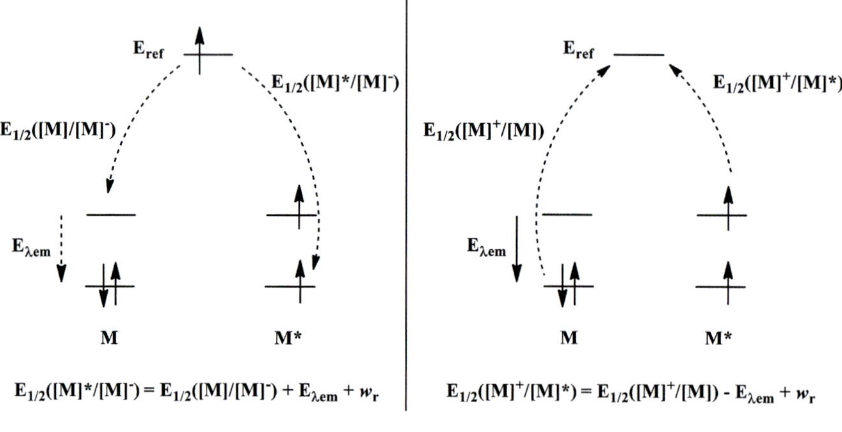

$$E_{1/2}([M]^*/[M]^-) = E_{1/2}([M]/[M]^-) + E_{\lambda em} + w_r \qquad E_{1/2}([M]^+/[M]^*) = E_{1/2}([M]^+/[M]) - E_{\lambda em} + w_r$$

$$E_{1/2}([Ru(bpy)_3]^{2+*}/[Ru(bpy)_3]^+) = +0{,}78V \text{ vs SCE } (+0{,}83V)$$

$$E_{1/2}([Ru(bpy)_3]^{3+}/[Ru(bpy)_3]^{2+*}) = -0{,}87V \text{ vs SCE } (-0{,}79V)$$

Scheme 4.   Calculation of the redox potential of photocatalysts at the exited-state and comparison with the measured values.

can be given. The minimum difference of energy between the ground state of the catalyst and its excited state corresponds to the wavelength of the emission maximum $E_{\lambda em}$. Therefore, the required energy to reduce the photocatalyst in the excited state $E([M]^*/[M^-])$ equals the sum of the required energy to reduce the photocatalyst in the ground state $E([M]/[M^-])$ and $E_{\lambda em}$ (Scheme 4, left). Also, the energy necessary to oxidize the photocatalyst at the excited-state $E([M]^+/[M^*])$ equals the difference between the energy necessary to oxidize the photocatalyst in the ground state $E([M]^+/[M])$ and $E_{\lambda em}$ (Scheme 4, right). However, an electrostatic work term w_r, describing the charge generation and separation within the electron-transfer complex, must be taken in account but is difficult to estimate. Thus the calculated potentials are slightly over- or underrated, but experimental values can be measured[27] by phase-modulated voltammetry, showing a good match with the calculated ones. Fine tuning of the redox potential can be realized changing the ligand (bipyridine, bipyrazine, phenanthroline, etc.) or modifying the metal. In this context, heteroleptic iridium complexes are particularly interesting due to their two ligands type (cyclometalating ligand and bidentate ligand) which can be both modified.

Therefore, the photocatalyst excited state may pick up or give out an electron to a substrate if the redox potentials match. In terms of orbital, if the HOMO of a donor **D** is between the t_2g and π^* orbitals of the Ru(bpy)$_3^{+2}$, a SET can happen giving the oxidized substrate **D**$^{\bullet+}$ and Ru(bpy)$_3^+$ (Scheme 5). In the case of an acceptor **A**, if the LUMO is in the

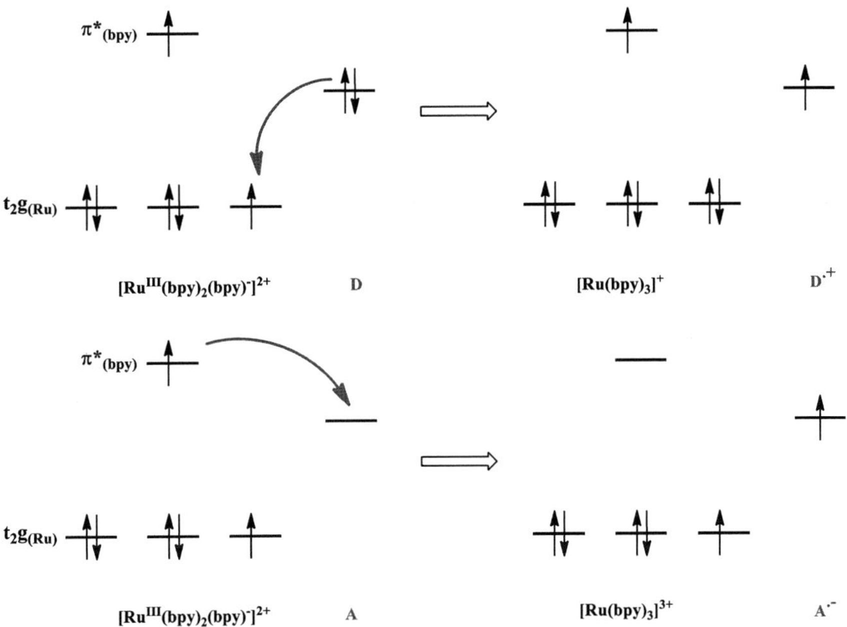

Scheme 5.   SET between excited Ru(bpy)$_3^{2+}$ and donor or acceptor.

same energy range, the electron transfer would form the radical anion $\mathbf{A}^{\bullet-}$ and $Ru(bpy)_3^{3+}$.

2. *Principle of Photoredox Catalysis*

As we have seen above, photoexcited chromophores ($\mathbf{PC}^*$) can transfer one electron to substrates providing radicals which can be involved in radical processes. However, to render the overall process catalytic, the photocatalyst ($\mathbf{PC}^{+/-}$) should undergo another SET with an intermediate or an additive. When the substrate is reduced by the photocatalyst, the process would be named **Photoreduction** and **Photooxidation** if the substrate is oxidized (Scheme 6).

More precisely, after light absorption the excited $\mathbf{PC}^*$ reacts with an electron donor $\mathbf{D}$ so called a "reductive quencher" leading to the reduced photocatalyst $\mathbf{PC}^-$. The altered photocatalyst transfers its excess of electron to an acceptor $\mathbf{A}$, regenerating the starting photocatalyst to end the "reductive quenching cycle." The "oxidative quenching cycle" is the reaction of $\mathbf{PC}^*$ with an acceptor $\mathbf{A}$, which acts as an oxidative quencher. The oxidized photocatalyst $\mathbf{PC}^+$ then reacts with a donor $\mathbf{D}$ and regenerates $\mathbf{PC}$ (Scheme 7).

The radicals engaged in synthetic applications can be generated by photoreduction or photooxidation. Concerning the photoreduction, a substrate (or acceptor in this case) would be directly reduced by the photoexcited catalyst if the redox potentials fit well. But sometimes, the potential

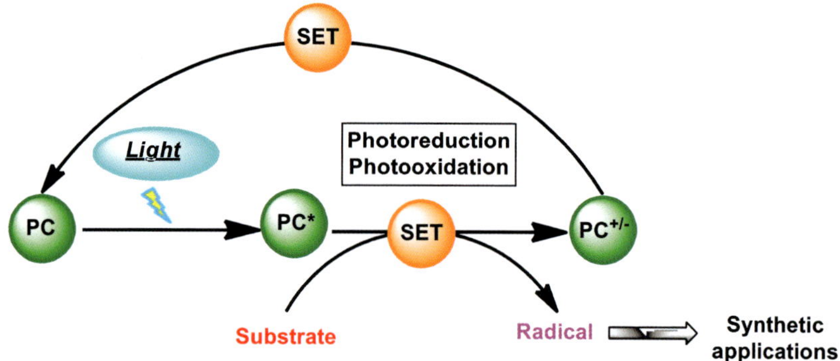

Scheme 6. General catalytic process of photoredox catalysis.

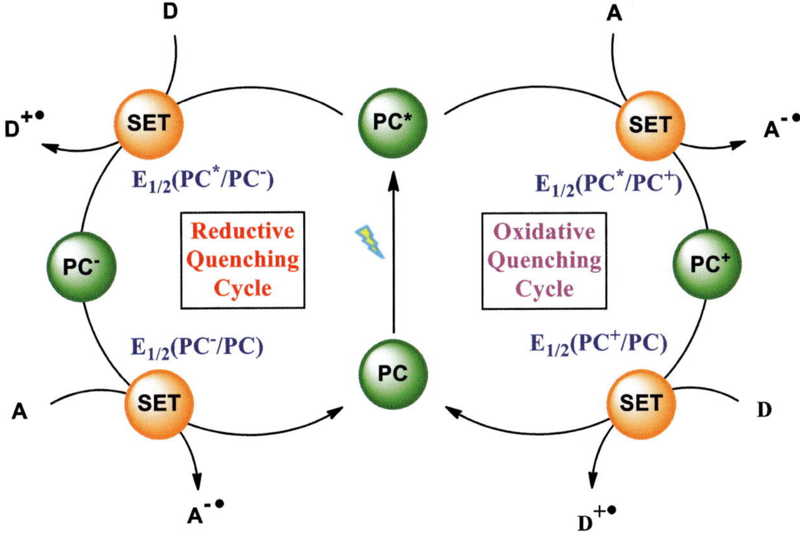

Scheme 7.   Reductive and oxidative quenching cycles.

of the light excited photocatalyst is not enough high to reduce the substrate. A sacrificial electron donor (amines, Hantzsch ester…) would react first with **PC*** to generate **PC$^-$** and provide a stronger reductant capable to reduce the substrate. In photooxidation, the photoexcited catalyst may directly oxidize a radical precursor (or donor in this case). As mentioned previously, if the **PC*** is not enough oxidizing, addition of a sacrificial electron acceptor (oxygen, methylviologen, persulfate agent…) would provide the much stronger oxidant **PC$^+$**. The oxidation of the substrate then should give the radical.

After oxidation or reduction of the substrate by the photocatalyst, the formed radical-ion pairs may undergo back electron transfer (BET) giving back both reactants. After reductive quenching, the spin multiplicity of the system should remain unchanged. If the electron transfer happens when the photocatalyst is in a singlet excited state, the radical-ion pair is also in a singlet state. In this case, the BET results in an overall singlet state without restraint. However, regarding a triplet state, the spin multiplicity must change during the BET which is less favored. Therefore, free radicals are more likely to be formed in the reaction once the photocatalyst is in a triplet excited state (Scheme 8). Metal complexes reach easily the triplet

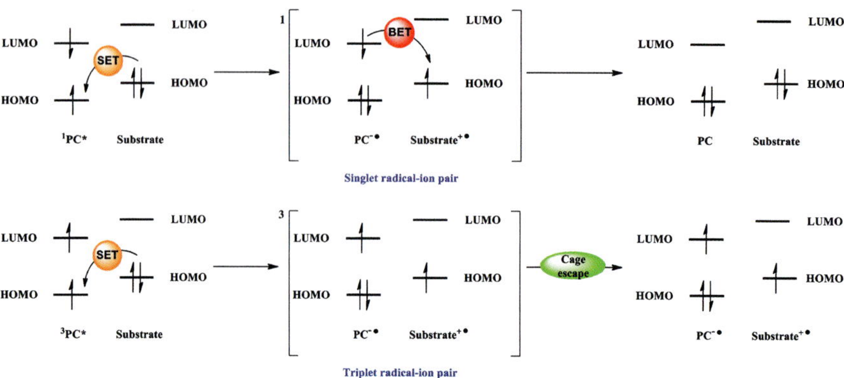

Scheme 8. Cage escape and BET in singlet and triplet ion pairs for photooxidation of a substrate.

excited state thanks to the heavy metal atom which increases the spin-orbit coupling and so the inter-system crossover (ISC). For organic molecules, the most populated excited state is most of the times the singlet state which limits their efficiency in photoredox catalysis. Despite this property, some organic dyes have already attested to be used as photo-catalysts in redox processes.[28] At the moment, no explanation can rationalize these observations inconsistent with the theory.

Since the triplet excited state is more likely to provide radical processes, organic dyes reaching efficiently this state would be more promising for photoredox catalysis. Organic Thermally Activated Delayed Fluorescence (TADF) materials have begun to be exploited in this field. As mentioned in the name category of these molecules, they are chromophores with a very long luminescence lifetime compared to other common organic molecules. Usually, organic dyes or complexes reaching a triplet state after ISC lose their excess of energy by phosphorescence or vibrational relaxation. In the case of TADF materials, a reverse intersystem crossover is possible (RISC) thanks to the thermal activation by the surrounding environment (Scheme 9a). This pathway is possible only if the energy difference between S_1 and T_1 (ΔE_{ST}) is enough low and particularly in the range of the thermal energy. Thanks to a quantum mechanical analysis, Adachi[2b] found that the reduction in the overlap between the HOMO and the LUMO results in a small ΔE_{ST}. In this context, molecules displaying a donor(red)-acceptor(green) scaffold are unavoidable

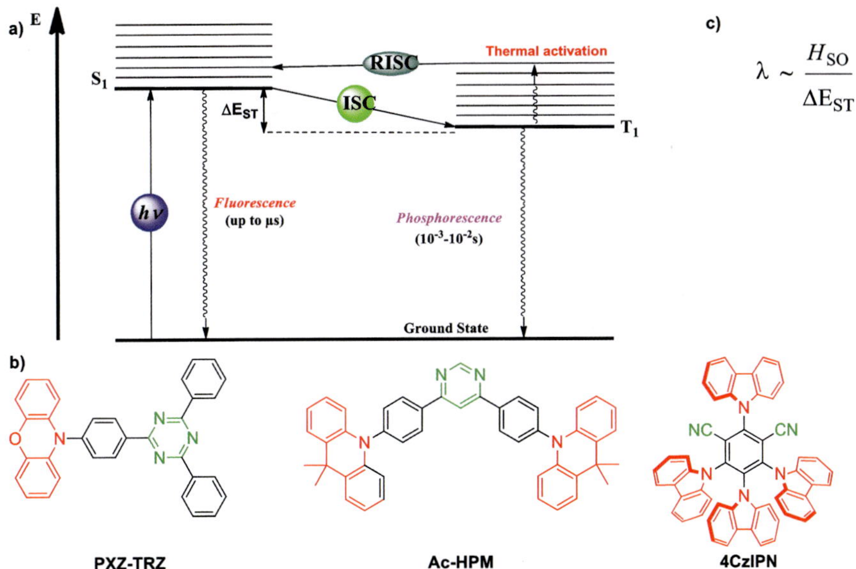

$$\lambda \sim \frac{H_{SO}}{\Delta E_{ST}}$$

Scheme 9. (a) Energy diagram for TADF materials. (b) Examples of organic donor(red)-acceptor(green) TADF materials. (c) First-order spin-orbit coupling parameter.

candidates such as phenoxazine–triphenyltriazine (PXZ-TRZ),[29] 4,6-bis[4-(9,9-dimethyl-9,10-dihydroacridine)-phenyl]pyrimidine (Ac-HMP)[30] or 4CzIPN[2e] (Scheme 9b).

For TADF materials, the kinetics values of ISC, RISC and fluorescence processes are crucial to obtain the enhancement of luminescence lifetime. The fluorescence process is usually in the range of nanoseconds. In order to avoid this direct desexcitation pathway, the ISC kinetics must be in the range of the fluorescence. Also, the luminescence lifetime would be increased if the RISC is slower than the ISC. Actually, for this kind of molecules, the ISC process is rather efficient. The spin conversion efficiency is correlated to the first-order mixing coefficient between singlet and triplet states (λ). This parameter is inversely proportional to ΔE_{ST} and proportional to H_{SO} which is the spin-orbit interaction (Scheme 9c).[31] Because the ΔE_{ST} is low, the overall spin-orbit coupling is important. In consequence, TADF materials can exist in their triplet excited state and perform electron transfers. Particularly, the 4CzIPN proved to be an efficient photocatalyst in photoredox/nickel dual catalysis.[3,32]

3. *Formation of Carbon Centered Radicals*

In terms of synthetic interests, the photoredox catalysis found opportunities to generate various types of radicals by photoreduction or photooxidation of different chemical functions. Radicals centered on carbon, nitrogen,[33] sulphur[34] or phosphorus[35] can be obtained by photoredox catalysis and engaged in the formation of carbon–carbon bonds or heteroatom carbon bonds. The wide diversity of carbon-centered radicals generated (aryl, alkenyl, alkyl) makes this method of great interest. Thus, many works on the generation by photocatalysis of such intermediates have been realized, starting with the formation of aryl radicals.

a. *Formation of aryl radicals by photoreduction*

The pioneering example developed by Deronzier was the reduction of arenediazonium salts with Ru(bpy)$_3$Cl$_2$ as photocatalyst. The excited ruthenium complex enables the formation of an aryl radical by photoreduction of the diazonium salt to perform a Pschorr type reaction (Scheme 10a).[36] More

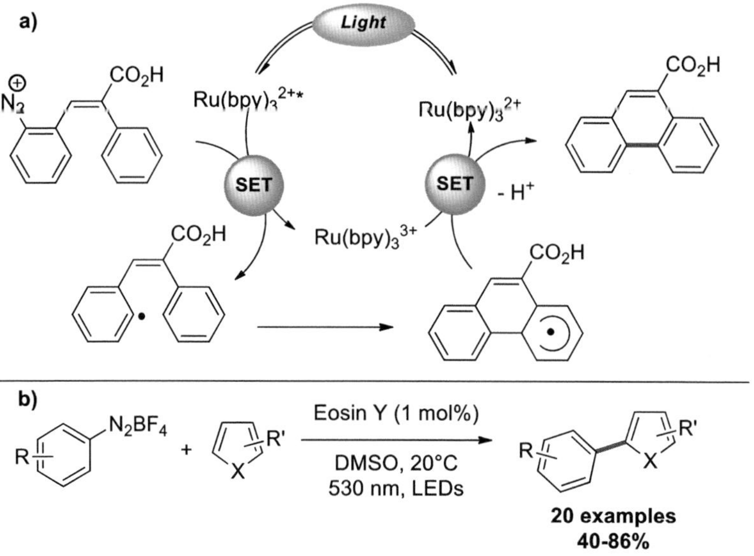

Scheme 10. (a) Photocatalyzed Pschorr reaction. (b) Photocatalyzed Meerwein arylation.

recently, König extended this catalytic process to intermolecular Meerwein type arylation reactions with Eosin Y as photocatalyst (Scheme 10b).[37] Aryl radicals were further obtained by photoreduction of sulfonium[38] or iodonium[39] salts and engaged in intermolecular radical allylation reactions. Although aryl radicals are obtained efficiently, alkyl radicals are more of interest for the formation of molecular scaffolds.

b. *Photoreductive processes for the generation of alkyl radicals*

i. Reduction of alkyl halides

Halides are substrates of choice for reduction processes to generate alkyl radicals. Fukuzumi[40] was the first to report the photoreduction of alkyl halides. Recently, inspired by this work, MacMillan[14] managed to merge photoredox catalysis and organocatalysis, and to perform enantioselective α-alkylation of aldehydes with bromoalkanes as radical precursors. In the presence of a chiral imidazolidinone, an enamine is formed. Reduction of the bromoalkane by $[Ru(bpy)_3^+]$ leads to an alkyl radical which adds to the enamine double bond. The resulting α-amino radical is then oxidized by $[Ru(bpy)_3^{2+}]^*$, generating an iminium intermediate which after hydrolysis liberates the α-alkylated aldehyde (Scheme 11). This methodology was also extended to the reduction of trifluoroiodomethane[41] and benzyl bromide[42] with iridium photocatalysts.

In this field, Stephenson reported the direct reduction of several chloro, bromo and iodo alkanes involving $[Ru(bpy)_3^{2+}]$ as photocatalyst and DIPEA as sacrificial electron donor and formic acid.[43] The presence of formic acid favored the formation of an excellent donor of hydrogen, the ammonium salt $(H^+NEtiPr_2.HCO_2^-)$ In similar conditions (Et_3N instead of DIPEA), radicals obtained from the reduction of bromomalonate could also be engaged in radical cyclizations with alkenes, alkynes,[44] indoles and pyrroles[45] moieties (Scheme 12a). More interestingly, polyene substrates gave polycyclic compounds after cascade cyclizations.

Although the processes look efficient, the main limitation is the nature of the halide. Indeed, reductive potentials of non-activated alkyl iodides or bromides are lower than $[Ru^{2+}]^*$ or $[Ru^+]$ and measured

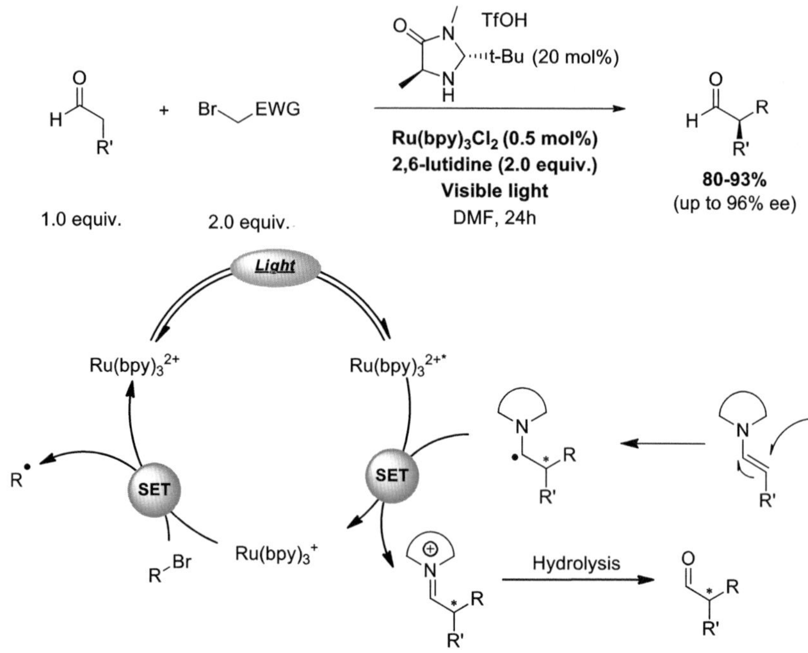

Scheme 11. Enantioselective α-alkylation of aldehydes.

between -1.61 V and -2.5 V *versus* SCE.[46] Stephenson showed that they are reduced with the highly reductant Ir(ppy)$_3$ (-1.73 V *versus* SCE at the excited state) and the generated radical can be engaged in 5-*exo*-trig and 5-*exo*-dig cyclization processes (Scheme 12b).[47] After light absorption, the iridium photocatalyst reduces the alkyl iodide to give the alkyl radical which directly cyclizes. The tributylamine regenerates the photocatalyst by SET, and the cyclized radical abstracts a hydrogen from the ammonium radical cation.

ii. Photocatalyzed Barton-McCombie deoxygenation

A famous reaction generating alkyl radicals is the Barton-McCombie deoxygenation reaction. Secondary or tertiary alcohols are converted to thiocarbonate, thiocarbamate or xanthate derivatives which are afterwards reduced by a tin-hydride reagent (nBu$_3$SnH). During the process, the C–O bond of the modified alcohol undergoes fragmentation, to give an alkyl radical. After a hydrogen abstraction step from the tin-hydride reagent, the

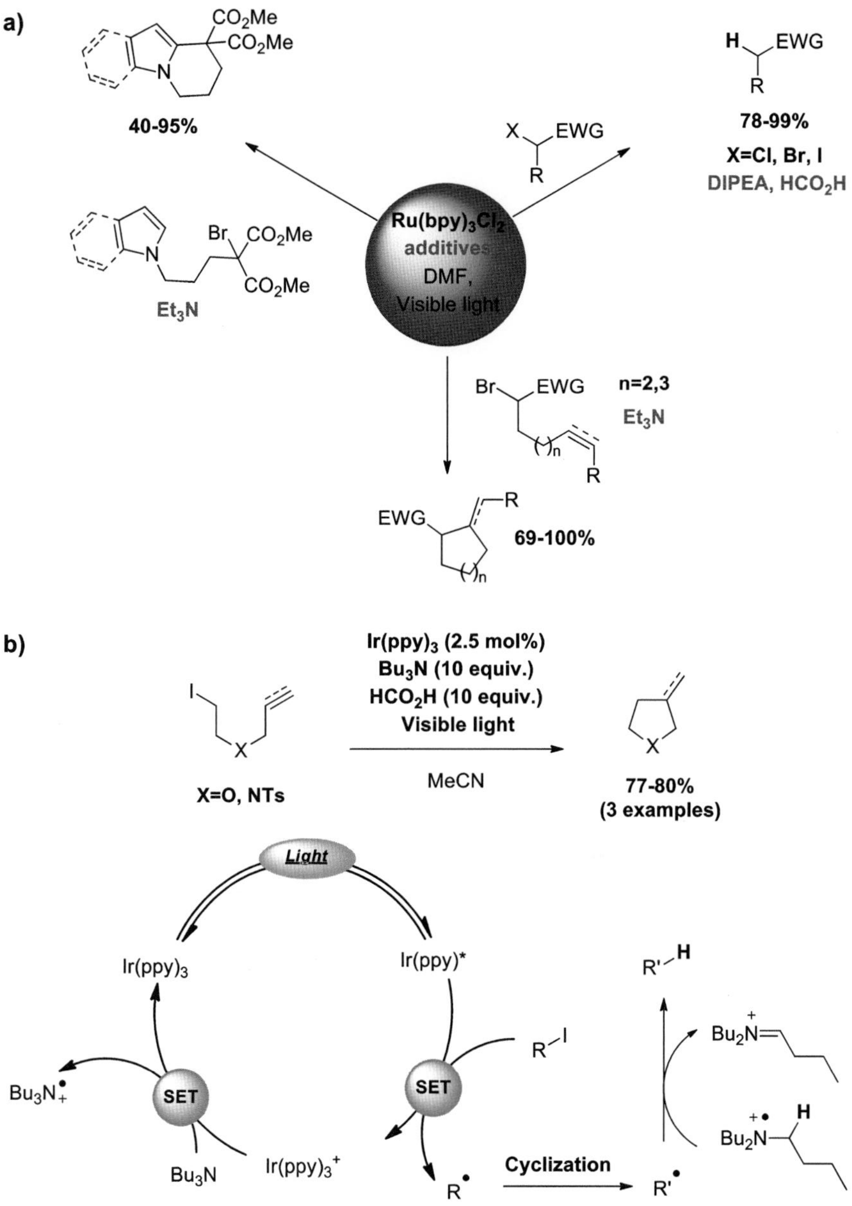

a)

b)

Scheme 12.   Stephenson's photoreduction of halides.

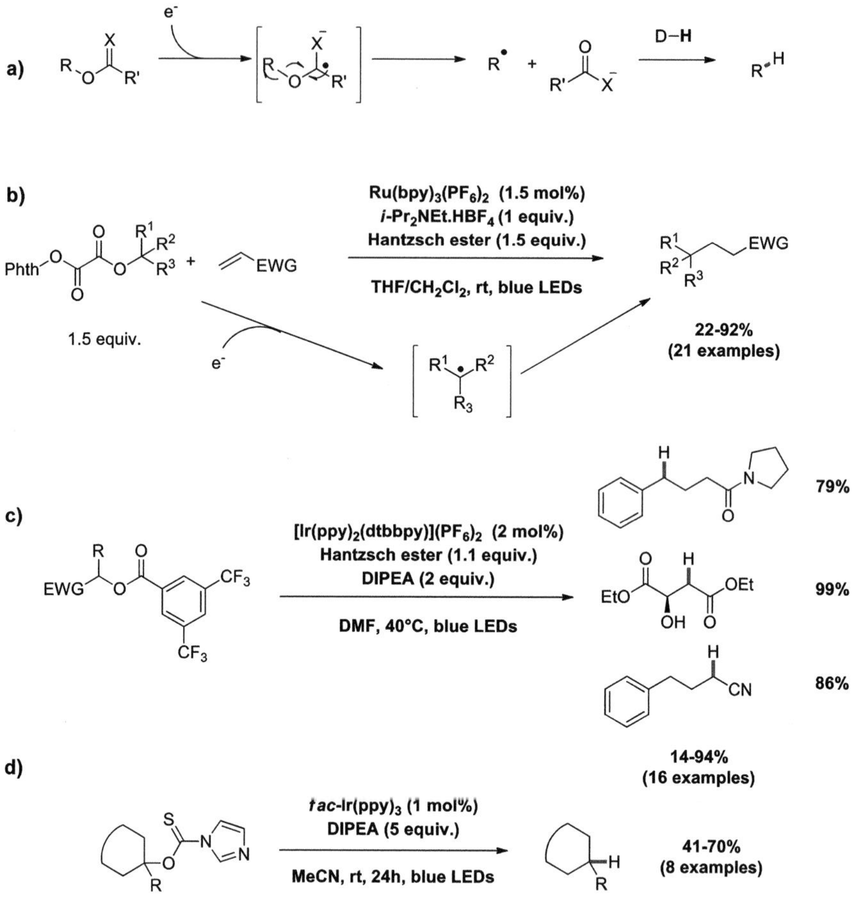

a)

b)

Ru(bpy)$_3$(PF$_6$)$_2$ (1.5 mol%)
i-Pr$_2$NEt.HBF$_4$ (1 equiv.)
Hantzsch ester (1.5 equiv.)

THF/CH$_2$Cl$_2$, rt, blue LEDs

22-92%
(21 examples)

1.5 equiv.

c)

[Ir(ppy)$_2$(dtbbpy)](PF$_6$)$_2$ (2 mol%)
Hantzsch ester (1.1 equiv.)
DIPEA (2 equiv.)

DMF, 40°C, blue LEDs

79%

99%

86%

14-94%
(16 examples)

d)

fac-Ir(ppy)$_3$ (1 mol%)
DIPEA (5 equiv.)

MeCN, rt, 24h, blue LEDs

41-70%
(8 examples)

Scheme 13.   Photoreductive Barton-McCombie type deoxygenation reactions.

deoxygenated product and the reactive tin mediator (*n*Bu$_3$Sn•) are obtained.[48] In order to avoid the use of tin reagents, some groups have tried a photocatalytic version. Similarly to the classical conditions of the Barton-McCombie deoxygenation, alcohols have to be activated as an ester, an oxalate or a thiocarbamate (Scheme 13a).

In 2013, inspired by the *N*-(acyloxy)phthalimides photocatalyzed decarboxylation of Okada,[49] Overman[50] reported the efficient deoxygenation of aliphatic tertiary alcohols by photoreduction of the corresponding *N*-(acyloxy)phthalimide oxalates derivatives in the presence of photocatalyst Ru(bpy)$_3$(PF$_6$)$_2$, Hantzsch ester and *i*-Pr$_2$NEt.HBF$_4$ as a sacrificial electron donor and a H-atom donor respectively. The generated tertiary alkyl radicals could then be then engaged in Giese-type reactions (Scheme 13b). Although no example involving secondary or primary radicals were mentioned, Reiser *et al.* reported the photoreduction of 3,5-bis(trifluoromethyl)benzoate esters.[51] Starting with the reduction of the photoexcited [Ir(ppy)$_2$(dtbbpy)](PF$_6$) by DIPEA, benzyl, α-carbonyl and α-cyano esters were reduced by the [Ir(II)] complex to give the deoxygenated products in good yields (Scheme 13c). At the same time in our group, a procedure involving the reduction of *O*-thiocarbamates was developed.[52] The high reductive potential of such derivatives (−1.56 V to −1.73 V *versus* SCE) required the use of the highly reducing photoexcited *fac*-Ir(ppy)$_3$ photocatalyst (E$_{1/2}$([Ir]$^+$/[Ir]*)= −1.73 V *versus* SCE) and DIPEA as both an electron donor and a H-atom donor. Under these conditions, tertiary and secondary alcohols could be deoxygenated in moderate yields (Scheme 13d).

iii. Reduction of α-ketoepoxides and α-ketoaziridines

Our group reported in 2011 the photoreduction of α-ketoepoxides and α-ketoaziridines[53] to generate an α-carbonyl radical intermediate that can be trapped by allylsulfones. This transformation requires the use of [Ir(ppy)$_2$(dtbbpy)](PF$_6$) that absorbs in the visible range and the Hantzsch ester methylated in position 4 in order to avoid the direct reduction by hydrogen abstraction and favor the allylation process (Scheme 14).

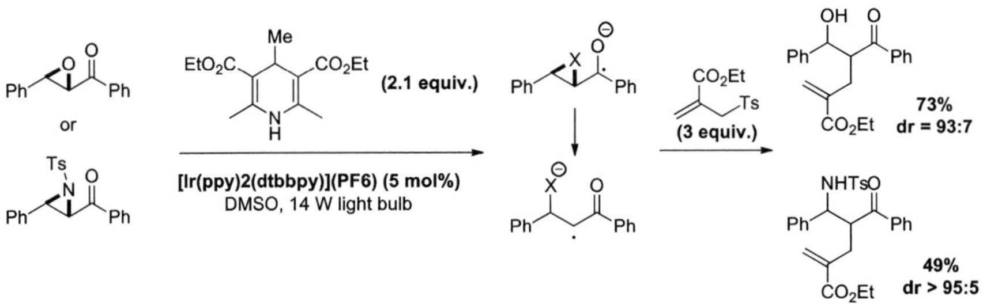

Scheme 14.   Photoreduction of α-ketoepoxides and α-ketoaziridines.

c. *Photooxidative processes for the formation of alkyl radicals*

i. Processes involving oxidation of amines

As previously observed, electron rich tertiary amines can be easily oxidized under photocatalytic conditions for the generation of alkyl radicals. An *N*-centered radical cation is generated which can evolve according two pathways. The first one previously detailed is an H-atom abstraction giving an iminium ion.[54] The second one consists of loss of a proton and generation of an α-amino radical[55] which can also be further oxidized to give the same iminium ion (Scheme 15).[56] Therefore, amines can react as electrophiles in the iminium form but also as nucleophiles in the α-amino radical form.

In 2012, Zheng and co-workers showed that an *N*-centered radical cation can evolve to the formation of carbon–carbon bonds. After oxidation of *N*-cyclopropylanilines, the cyclopropane ring opening led to an intermediate bearing an iminium moiety and a primary alkyl radical. The radical then adds to an alkene or an alkyne substituted by an aryl or an ester moiety.[57] The resulting stabilized radical cyclizes to the iminium to give the *N*-cyclopentyl (or *N*-cyclopentenyl) aniline (Scheme 16).

The same year, Reiser *et al.* reported the trapping of cyclic α-amino radicals by α,β-unsaturated carbonyl compounds.[58] Oxidation of the amine by the excited photocatalyst **PC***⁺ leads to the formation of an α-stabilized carbon-centered radical after the loss of a proton. The addition of the radical to an α,β-unsaturated acceptor gives the radical which

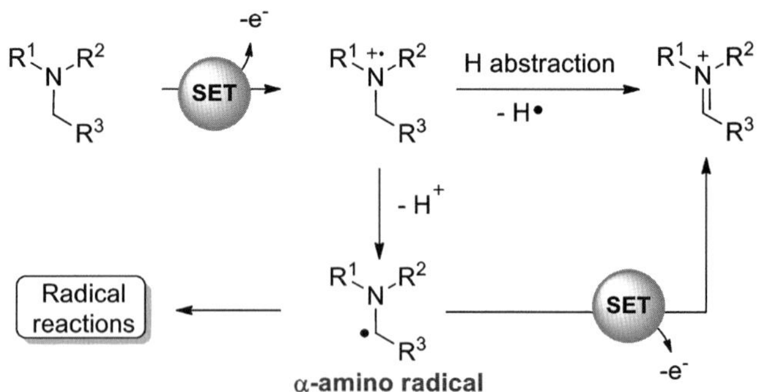

Scheme 15.   Evolution of *N*-centered radical cations.

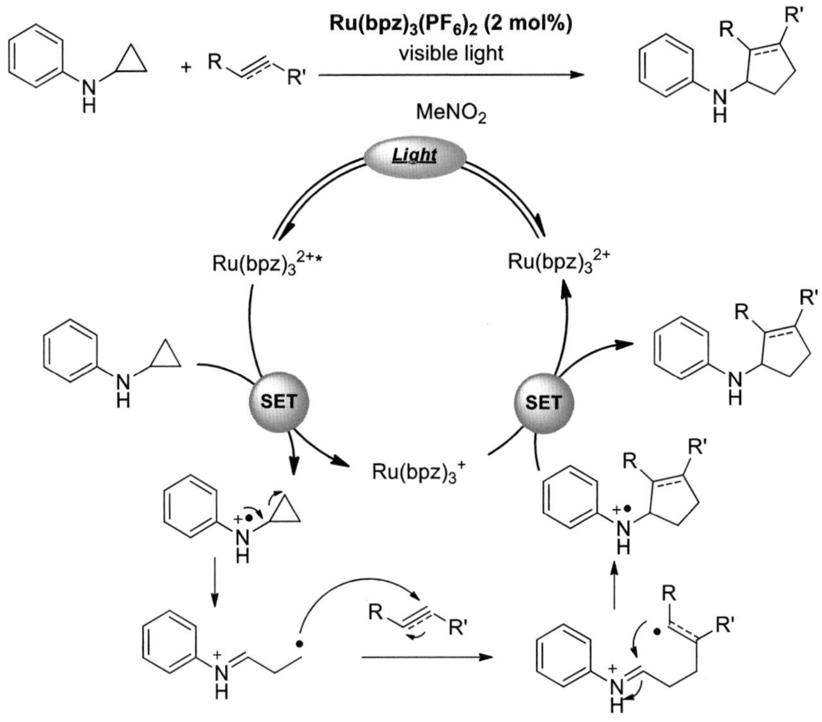

Scheme 16.   Photocatalyzed intermolecular [3+2] cycloaddition of *N*-cylopropylanilines.

is directly reduced by **PC⁻**. The resulting enolate is then protonated to give the final product (Scheme 17). At the same time, the group of Nishibayashi[59] realized a similar process involving *N*-methylaniline or α-silylamine. In this last case, the aminyl radical evolves towards an alkyl radical by releasing a trimethylsilylium cation instead of a proton.

In 2011, MacMillan mentioned a reaction between photogenerated α-amino radicals and benzonitrile.[60] The radicals were obtained by oxidation of *N*-alkyl anilines and added at the *ipso* position of the cyano arene releasing cyanide and the α-amino arene (Scheme 18). The reaction performed with the Ir(ppy)$_3$ photocatalyst was suitable for (non)-cyclic anilines and various cyano (hetero)arenes. Photoexcitation of the iridium photocatalyst leads to the reduction of the cyano arene. The [Ir^{IV}] intermediate then oxidizes the aniline, resulting in the formation of the α-amino

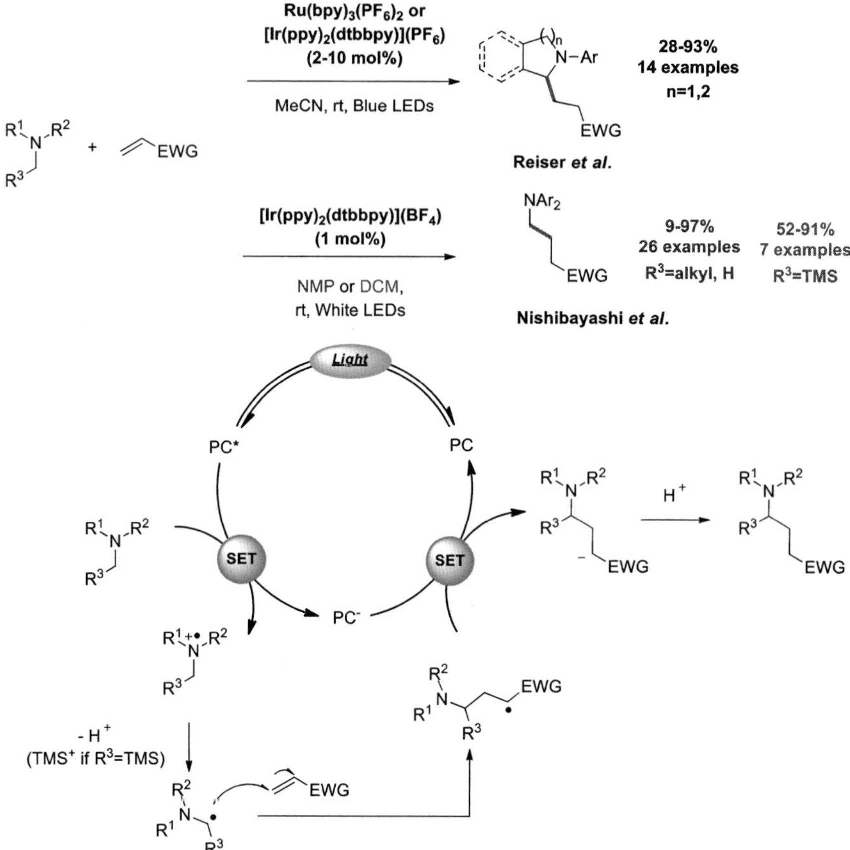

Scheme 17. Photocatalyzed addition of α-amino radicals to Michael acceptors.

Scheme 18. Photoredox amine α-arylation reaction.

radical which couples with the reduced arene. The newly formed anion provides the final formation of the product and cyanide.

ii. Oxidation of α-aminocarboxylates

Anionic species are substrates of choice for oxidation reactions. MacMillan showed the ability of α-aminocarboxylates to undergo photooxidation, generating an α-amino radical by the loss of CO_2. As he showed before, the radicals were able to be coupled with dicyanobenzene in similar conditions.[61] Even if the amine must be protected (Boc or Cbz) to avoid the oxidation of the nitrogen atom, secondary and tertiary amines could be engaged in the process because, these radicals' precursors are quite easy to oxidize ($E_{1/2}$(R-CO$_2^-$/R-CO$_2$.) = +0.95 – +1.16 V *versus* SCE).[38,62] They were also engaged in vinylation reactions.[63] Oxidation of the carboxylate by photoexcited [Ir(dF(CF$_3$) ppy)$_2$(dtbbpy)](PF$_6$) catalyst gives the carbon-centered radical which is directly trapped by a vinylsulfone. After the C-SO$_2$Ph bond fragmentation, the sulfonyl radical ($E_{1/2}$(PhSO$_2$•/PhSO$_2$Na) = +0.50 V *versus* SCE)[64] participates to the regeneration of the photocatalyst *via* a SET with the Ir^{II} intermediate (Scheme 19).

iii. Oxidation of *organoborates*

In the same line, alkyl trifluoroborates have proved to be suitable substrates to generate radicals upon oxidation.[65] Akita and Koike reported their work on the photooxidation of organoborates with the [Ir(dF(CF$_3$) ppy)$_2$(bpy)](PF$_6$) catalyst.[66] In order to prove the photooxidation feasibility, they performed spin-trapping experiments with TEMPO and found that allyl, benzyl and tertiary alkyl trifluoroborates could form the corresponding alkyl-TEMPO adducts. However, secondary and primary alkyl trifluoroborates were not converted to the expected product. Only (triol)borates of 2-(hydroxymethyl)-2-methylpropane-1,3-diol which showed lower oxidation potentials and could provide TEMPO adducts in moderate yields (Scheme 20). Synthetic applications were also extended to Giese type reactions.

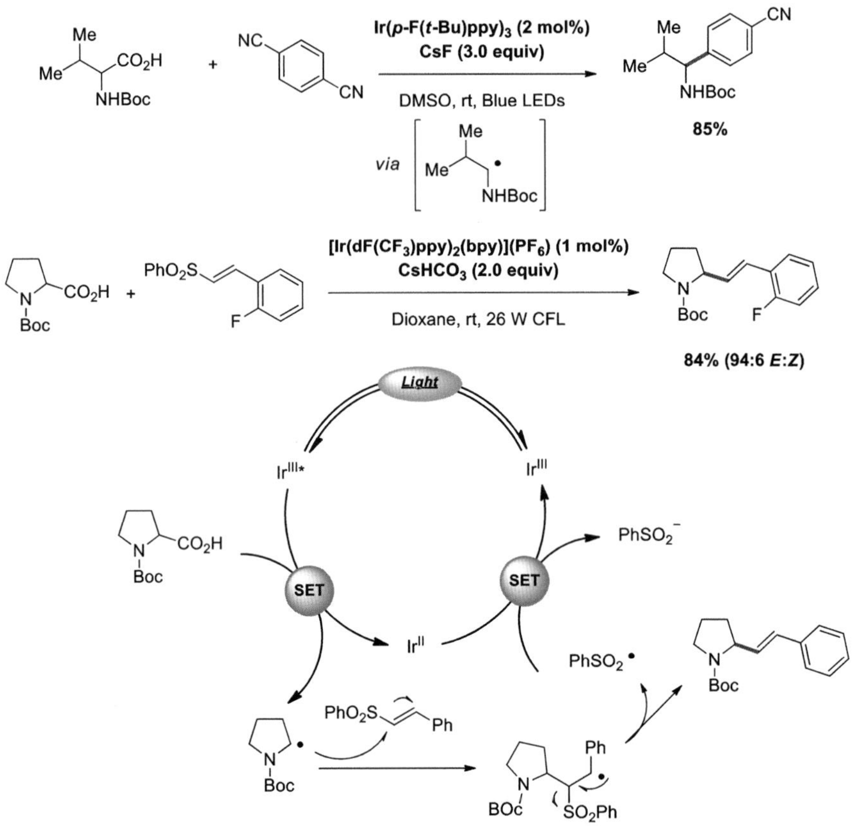

Scheme 19. Photooxidation of α-aminocarboxylates.

Few years after, Molander and co-workers extended the synthetic applications of organotrifluoroborates. They photooxidized α-pyrrolidinyltrifluoroborate ($E_{1/2}^{ox}$ = +0.78 V *versus* SCE) with Eosin Y ($E_{1/2}^{red}$ = +0.83 V *versus* SCE)[67] at 50°C. The generated α-amino radical was engaged in alkenylation reactions with vinyl sulfones and allylation reactions with allyl sulfones. Radical adducts were isolated in moderate to good yields (Scheme 21). Cyanation of primary and secondary alkyl trifluoroborates could be also performed by action of the highly oxidizing Fukuzumi's catalyst ($E_{1/2}^{red}$ = +2.06 V *versus* SCE)[68] and tosylcyanide as radical acceptor. Secondary cyclic trifluoroborates gave the best yields while an unactivated primary aliphatic trifluoroborate provided the expected product in a low yield.

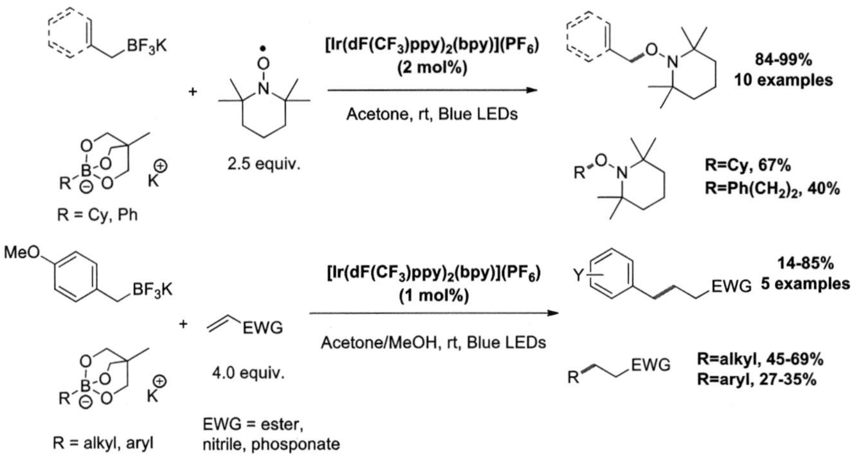

Scheme 20.   *Photooxidation* of organoborates.

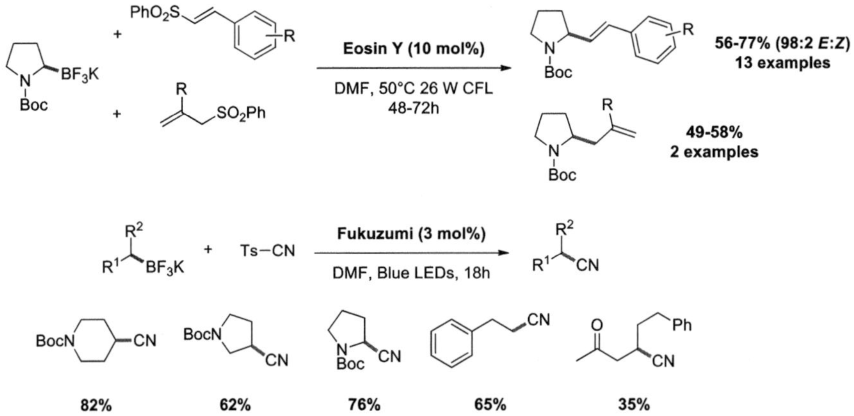

Scheme 21.   Organic *photooxidation* of alkyl trifluoroborates.

III. Merging Photoredox and Organometallic Catalysis for Cross-Coupling Reactions

A. Context

Photoredox catalysis has shown its efficiency to accomplish radical reactions. Methods merging organocatalysis and photoredox catalysis have been also developed.[14,18,19,69] Recently, photoredox catalysis has proved to be compatible with transition metal catalysis. In this type of catalysis, the photocatalyst is necessary to generate a radical and the organometallic catalyst performs the cross-coupling steps between an electrophile and the generated radical. Therefore, metal complexes able to trap radicals are catalysts of choice for such transformations.

B. Radical Trapping by Transition Metals

Transition metal catalyzed reactions have become essential tools for the elaboration of new molecular building blocks. For instance, palladium catalyzed cross-coupling reactions and C–H functionalization reactions are processes well-established and efficient. The mechanisms relying on two electron transfer steps are inherent to the nature of fifth period transition metal. However, concerning 3d transition metals, although they are known to perform oxidative addition in carbon–halogen or carbon–hydrogen bonds, they can also promote single electron transfer[70] and generate radicals. These ones can be further trapped by the metal and involved in organometallic processes.

1. *Cobalt*

Some cobalt complexes showed to be able to reduce alkyl halides to alkyl radicals which could undergo cyclizations. The resulting radical intermediates lead the formation of an alkyl-[Co(III)] complex. Such species can be trapped by iodine, diphenyl disulfide, isonitrile[71] or can undergo cross-coupling reactions with Grignard reagents. Reduction of 5-hexenyl bromide or cyclopropylmethyl bromide lead to the formation of ring-closing or ring-opening products respectively,[72] suggesting the formation of alkyl radicals during the transformations (Scheme 22).

2. *Nickel*

With the development of the Kumada-Corriu[73] reaction in the 70s, nickel catalysis has demonstrated impressive reactivities.[74] Like cobalt, a variety of nickel catalysts have revealed to be remarkably active in the formation of carbon-centered radicals. In the context of $C(sp^3)$-$C(sp^3)$ cross-coupling reactions between electrophiles and Grignard or organozinc reagents,

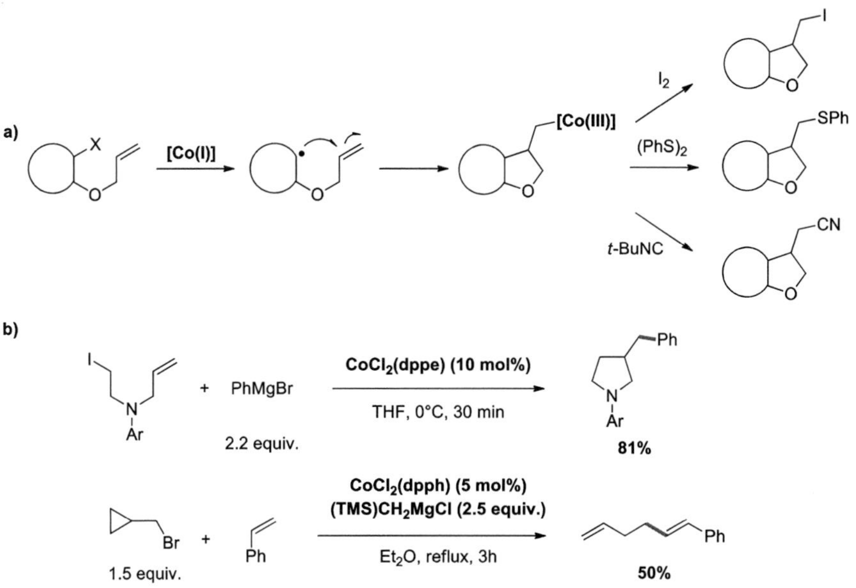

Scheme 22.   Cobalt-mediated radical reactions.

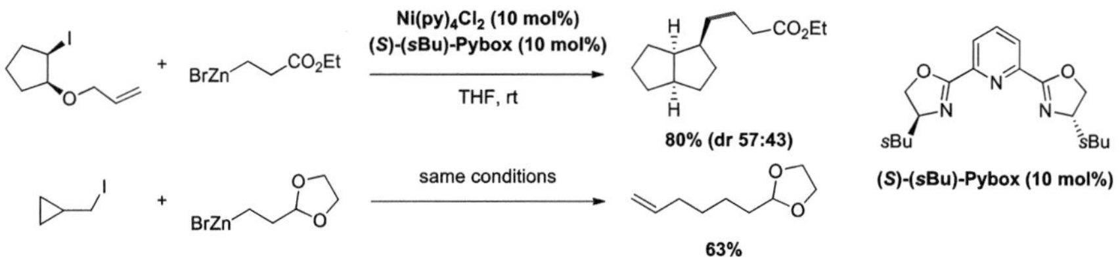

Scheme 23.   Nickel-catalyzed Negishi cross-coupling reaction.

several reports mentioned mechanisms involving radicals. Cardenas[75] accomplished the cross-coupling reactions between iodoalkanes and alkyl zinc halides. He observed that in the presence of a nickel catalyst with a Pybox ligand, a 5-hexenyl iodide type substrate underwent 5-*exo*-trig cyclization before the coupling step and cyclopropylmethyl iodide gave the ring-opening products (Scheme 23). From calculations, Cardenas supposed the formation of the radical by a reduction of the electrophile with an intermediate Ni(I) complex.

Indeed, redox potentials of (terpyridyl)Ni^{I}(alkyl) complexes have been measured between -1.32 V and -1.44 V (*versus* Ag/Ag+ in THF),[76]

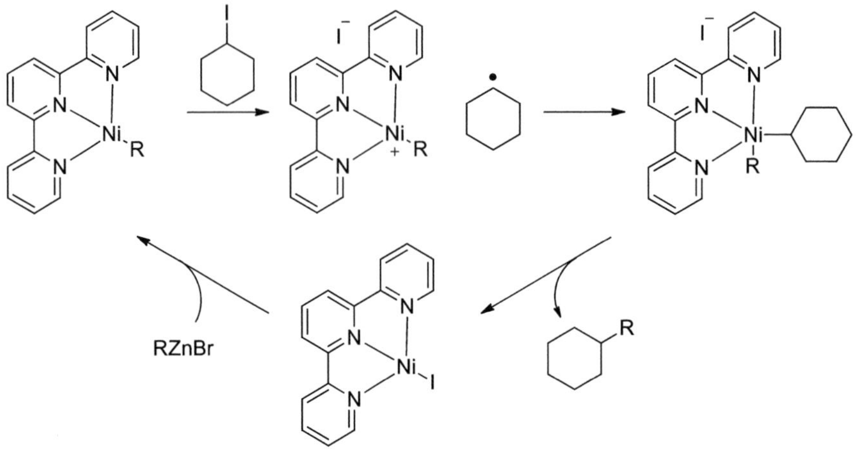

Scheme 24. Mechanism for Nickel-catalyzed alkyl-alkyl Negishi cross-coupling reactions.

allowing the reduction of alkyl halide to a carbon centered radical.[77] Then the radical adds on a Ni(II) complex which after reductive elimination liberates the cross-coupling product (Scheme 24). Furthermore, methodologies using others nucleophiles such as boronic acids, Grignard reagents or organostannanes,[78] involved the formation of alkyl radicals in nickel catalysis.

3. *Copper*

Copper complexes also revealed efficient interaction with carbon centered radicals. Lloris *et al.* reported the arylation of β-diketones at the ε position.[79] In the presence of copper powder, Cu(acac)$_2$ type complexes and aryl diazonium salts could give the expected cross-coupling adducts in low to moderate yields. They proposed the reduction of the diazonium salts by copper(0) giving the aryl radical which adds on Cu(acac)$_2$. The resulting Cu(III) intermediate provides the arylated product and Cu(acac) after reductive elimination (Scheme 25). As a side product, they observed the formation of diphenyldiazene. In their opinion, the product results from the radical coupling between a diazenyl radical and an aryl radical.

In 2001, Xiao demonstrated that (hetero)aryl iodides substrates were trifluoromethylated in the presence of diphenyl(trifluoromethyl)-

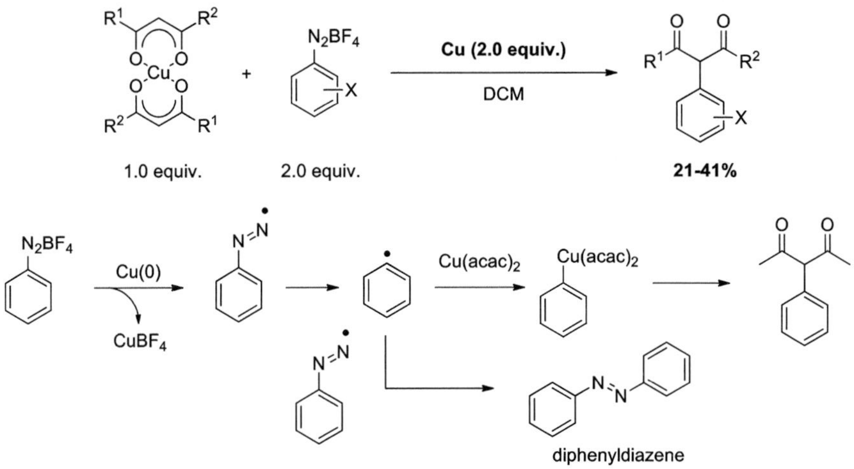

Scheme 25.   Arylation of β-diketones.

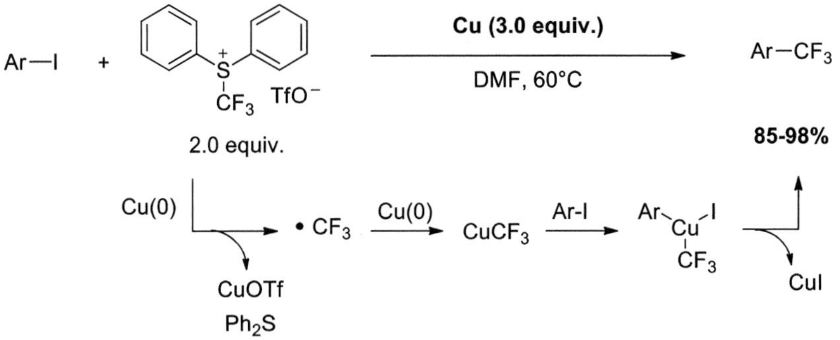

Scheme 26.   Trifluoromethylation of iodo (hetero)arenes.

sulfonium triflate and copper(0).[80] NMR and mass spectrometry studies showed the *in situ* formation of $CuCF_3$. From a SET between Cu(0) and the sulfonium, the trifluoromethyl radical is generated and trapped by another atom of Cu(0) to generate the organocopper $CuCF_3$ which undergoes an oxidative addition of aryliodide followed by a reductive elimination (Scheme 26).

In the same line, previous works in our group showed the formation of a well-defined copper complex displaying a $Cu–CF_3$ bond. Treating the $Cu(L_{SQ})_2$ (L_{SQ} = iminosemiquinone) by the Umemoto reagent lead to the formation of a CF_3 radical which could be trapped by TEMPO.[81] The

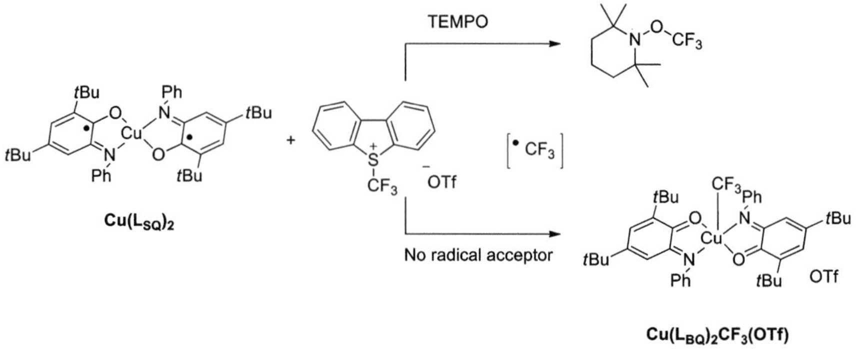

Scheme 27. Formation of Cu(II)-CF$_3$ complex.

redox iminosemiquinone ligand can reduce the Umemoto reagent and generate the radical species. Without any radical acceptor in the reaction mixture, it was observed the formation of the Cu(L$_{BQ}$)$_2$CF$_3$(OTf) complex (L$_{BQ}$ = iminobenzoquinone) by radical addition on the copper. Association of UV-vis data, mass spectrometry values and 2D EPR experiments confirmed the atom sequence around the copper and the formation of the Cu–C bond (Scheme 27).[82] The isolated complex could be further engaged in a Chan–Lam coupling type reaction.[83]

C. Genesis of the Photoredox/Transition Metal Dual Catalysis: Ruthenium and Palladium

The first example of photoredox/transition metal dual catalysis has been realized by the group of Sanford in 2011. Working on the palladium-catalyzed C–H functionalization of arylpyridine with diaryliodonium salts, Sanford showed the reaction to be temperature dependent. Lowering the reaction temperature from 100°C to 25°C decreased the yield dramatically (Scheme 28a). This observation is related to the difficult oxidation step of the dimeric Pd(II) intermediate by the diaryliodonium salt to the Pd(IV) that triggers the reductive elimination.[84]

At the same time, Yu[85] showed that C–H functionalization of the same kind of substrates could be performed with benzoylperoxide (Scheme 28b). At 100°C, the peroxide undergoes fragmentation and CO$_2$ extrusion to provide the phenyl radical which should react with the

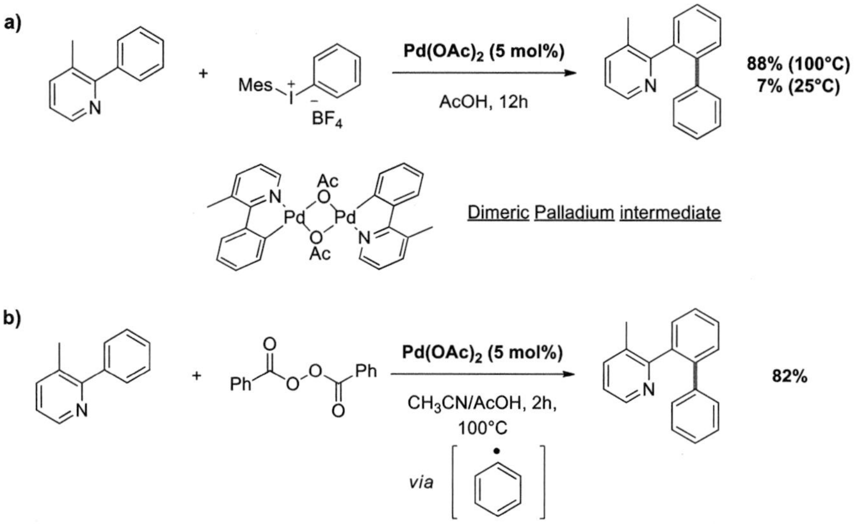

Scheme 28.   Thermally activated C–H arylation of phenylpyridine.

palladium complex intermediate. Based on these results, Sanford[86] envisioned to generate an aryl radical by photoredox catalysis. As precursors of radicals aryldiazonium salts were selected due to their low reduction potential.[13] The diazonium salts were reduced by $Ru(bpy)_3Cl_2$ under visible-light irradiation at room temperature. The generated radical was then involved in palladium catalysis to give the C–H functionalized product. Phenylpyridines and phenylpyrrolidin-2-ones were obtained in moderate to good yields (Scheme 29).

A possible catalytic cycle was then proposed. The photoexcited $[Ru(bpy)_3^{2+}]^*$ complex reduces the diazonium salt to generate the aryl radical and the $[Ru(bpy)_3^{3+}]$ complex. Regarding the palladium catalytic cycle after C–H insertion, the aryl radical adds on the palladium(II) intermediate leading to a Pd(III) species. A SET with $Ru(bpy)_3^{3+}$ regenerates the photocatalyst, and the obtained Pd(IV) undergoes reductive elimination to afford the arylated product and liberates the starting Pd(II) (Scheme 30).

This mild approach for palladium catalyzed-C–H bonds arylation is the first photoredox/transition metal dual catalytic process which demonstrates the feasibility of photoredox mediated cross-coupling reactions and pointed the way towards new synthetic opportunities.

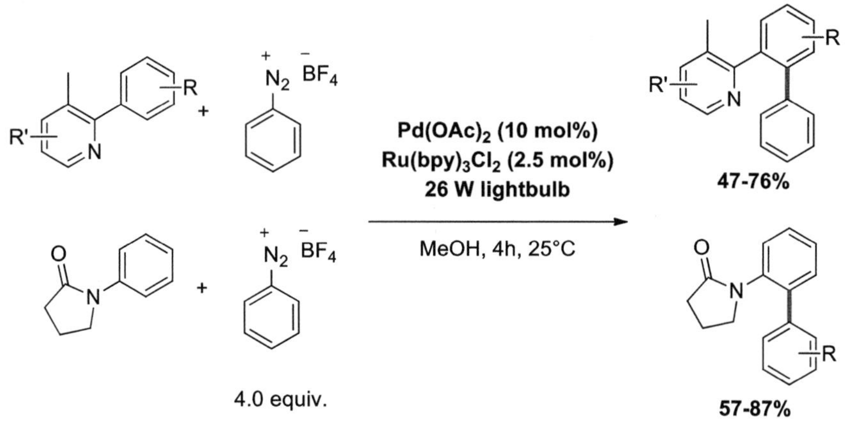

Scheme 29. Photoredox/palladium C–H arylation catalysis.

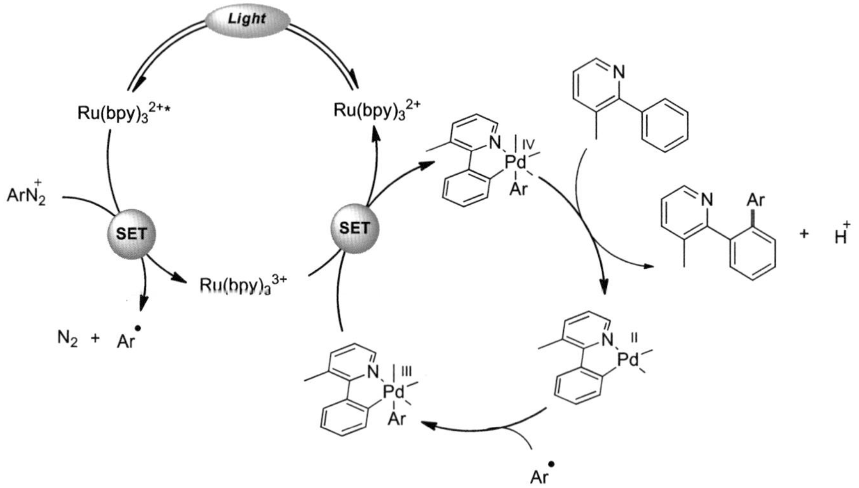

Scheme 30. Proposed mechanism for photoredox/palladium C–H arylation catalysis.

D. Towards Photoredox/Transition Metal Dual Catalysis Processes

Since the seminal work of Sandford, many photoredox/transition metal dual catalysis processes have been developed. One of the most important features of this kind of catalysis is the opportunity offered by the photoredox transformations to modify the oxidation state of organometallic intermediates. This allows

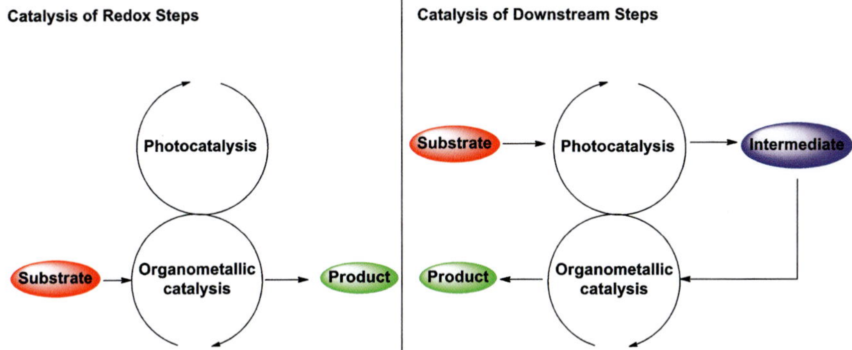

Figure 3.   Modes of photoredox/transition metal catalysis.

tuning the reactivity of the metallic centers for synthetic applications. Two tandem processes can be distinguished:

— Catalytic reactions which do not involve the formation of radicals (except superoxide anion $[O_2]\bullet$-). In this case, the SET takes place only between the photocatalyst and the transition metal.
— Reactions in which a photogenerated radical is trapped by an organometallic complex.

In this case, a second electron transfer between the organometallic catalyst and the photocatalyst occurs to maintain the redox balance. Each process will be therefore named "Catalysis of Redox Steps" and "Catalysis of Downstream Steps" respectively (Figure 3).

1. *Processes without Radical Formation: Catalysis of Redox Steps*

a. *C–H Functionalization*

Among the photoredox/transition metal dual catalytic reactions reported in the literature, C–H functionalization has been particularly studied. Rueping and co-workers developped another approach to oxidative Heck Reaction so called Fujiwara–Moritani reaction under photoredox conditions.[87] Based on an article of van Leeuwen[88] who reported the coupling reaction between anilidines and olefins with $Pd(OAc)_2$ as catalyst and benzoquinone as oxidant, they chose the photooxidizing catalyst [Ir(ppy)$_2$(bpy)](PF$_6$) to perform this catalytic process. Starting from (Z)-(phenylamino)but-2-enoate derivatives in the presence of Pd(OAc)$_2$ and potassium carbonate, they could obtain the formation of indoles in good to excellent yields. Control experiments showed that no reaction occurred under oxygen atmosphere. However, the presence of potassium superoxide proved to be essential. Based on these informations, the authors proposed a mechanism in which Pd(OAc)$_2$ did two consecutive C–H insertions. After reductive elimination, the indole is obtained and a Pd(0) complex re-oxidized simultaneously by superoxide and the photo-excited iridium complex. The Ir(II) photocatalyst transfers its excess of electron to a molecule of oxygen, regenerating the photocatalyst Ir(III) and providing a superoxide anion (Scheme 31).

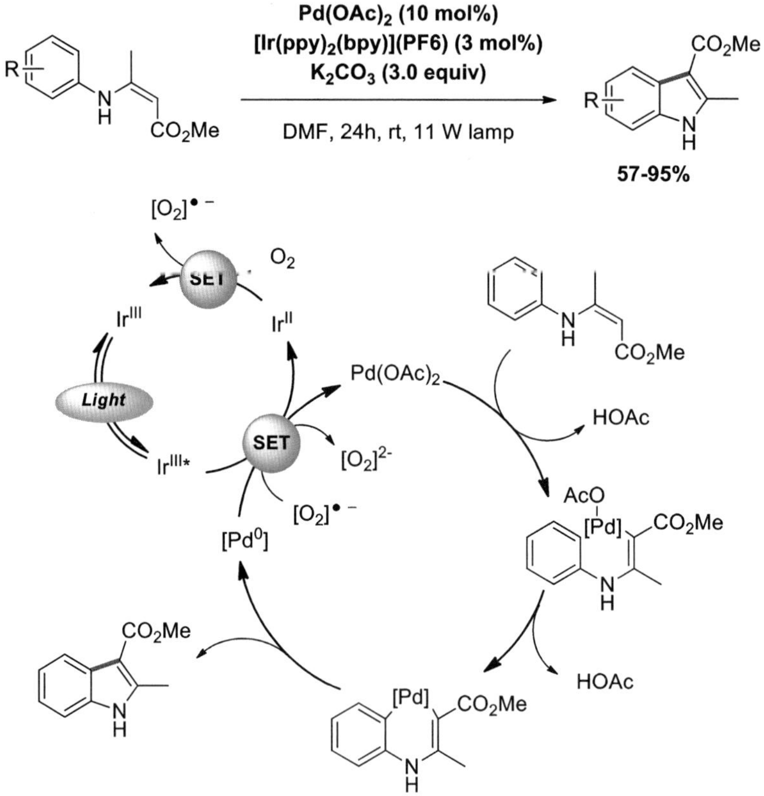

Scheme 31. Photocatalyzed Fujiwara–Moritani reaction.

In the same field, Rueping's group reported the photocatalyzed synthesis of indoles from acetanilides and alkynes. Inspired by the works of Fagnou,[89] they found that the photocatalyst $Ru(bpy)_3^{2+}$ under visible-light irradiation was a suitable photocatalytic oxidant instead of stoichiometric amount of $Cu(OAc)_2$ to regenerate the active rhodium catalyst.[90] Cationization of $[Cp*RhCl_2]_2$ gives the active metalating agent[91] which inserts in the C–H bond in *ortho* position of the acetanilide. After coordination of the alkyne to the rhodium, insertion of the triple bond results in the formation of a six-membered metallacycle intermediate. Then, a reductive elimination gives the *N*-acetyl indole and a Rh(I)complex is released. A similar re-oxidation with superoxide previously detailed with the palladium C–H functionalization may occur to end the catalytic cycle (Scheme 32).

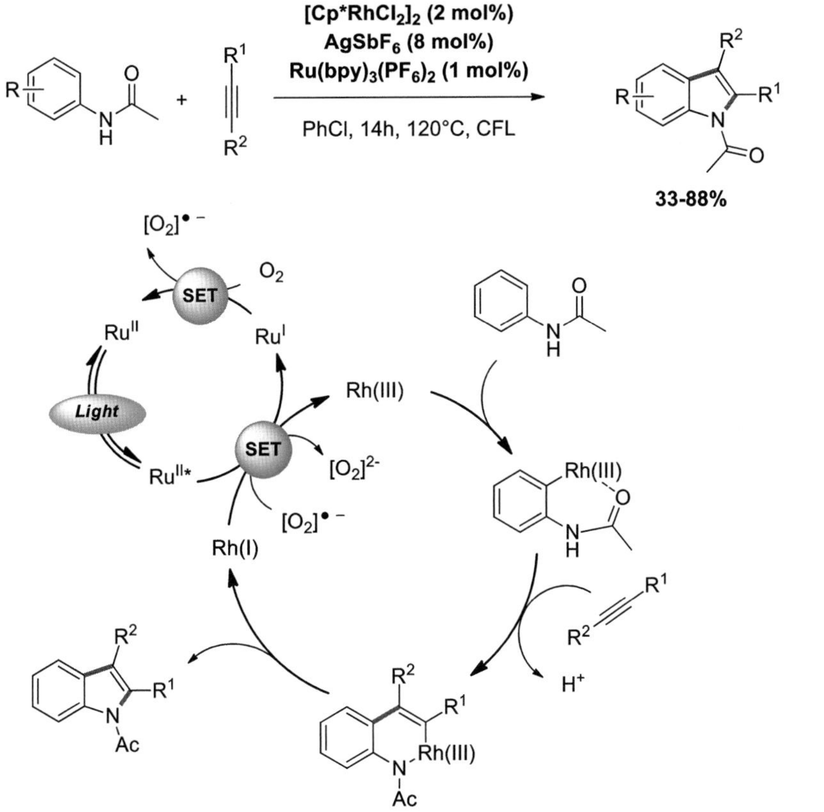

Scheme 32. Combined photoredox/rhodium catalyzed C–H functionalization of acetanilides with alkynes.

b. *Two successive SETs catalysis*

Others photoredox/transition metal dual catalysis processes involving two SETs between both catalysts were developed. Each SET allows the organometallic catalyst to reach the specific oxidation state required for the catalysis. Tunge and co-workers reported a decarboxylative allylation of *para*-amino allyl 2-phenylacetate in the presence of Pd(PPh$_3$)$_4$ as transition metal catalyst and [Ir(ppy)$_2$(bpy)](BF$_4$) as photocatalyst.[92] They could obtain *para*-homoallylaniline products in moderate yields. Alternatively starting the reaction from carboxylic acids and allyl methyl carbonate in stoichiometric amounts provided the products in the same range of yields. During the optimization, the authors noted the formation of a dibenzyl product suggesting the formation of a benzyl radical in the process. In addition, they observed the formation of a 1,5-hexadiene resulting from the presence of an allyl radical. So they suggested a plausible mechanism involving a benzyl-allyl radical coupling. They proposed the oxidative addition of the allyl ester to Pd(0) to form a Pd-π-allyl complex and a carboxylate anion (Scheme 33). The photoactivated iridium catalyst oxidizes the carboxylate which undergoes decarboxylation to furnish the benzyl radical. Thereafter, the Pd-π-allyl cationic complex is reduced by the Ir(II) intermediate to provide the allyl radical and regenerate both catalysts.

Nickel also proved to be a valuable candidate to photoredox/transition metal dual catalysis. MacMillan developed an efficient intermolecular cross-coupling reaction between arylbromides and alcohols involving photoredox/nickel dual catalysis. Studies on the nickel catalysis revealed that the C–O bond formation using this metal is tricky. Reductive elimination of the C–Ni^{II}–O is endothermic[93] and does not favor the C–O bond formation except for nickel metallacycles.[94] However, reductive elimination from a nickel(III) intermediate was suspected to be much easier. In order to reach this oxidation state, the idea of this group was to use a photocatalyst to oxidize the [C–Ni^{II}–O] intermediate. Using NiCl$_2$.dme, a bipyridine type ligand and [Ir(dF(CF$_3$)ppy)$_2$(dtbbpy)]PF$_6$ as photocatalyst, they could obtain the expected aryl ether products in excellent yields (Scheme 34). Regarding the mechanism, they proposed an oxidative addition of the aryl bromide on the Ni(0)(dtbbpy) active

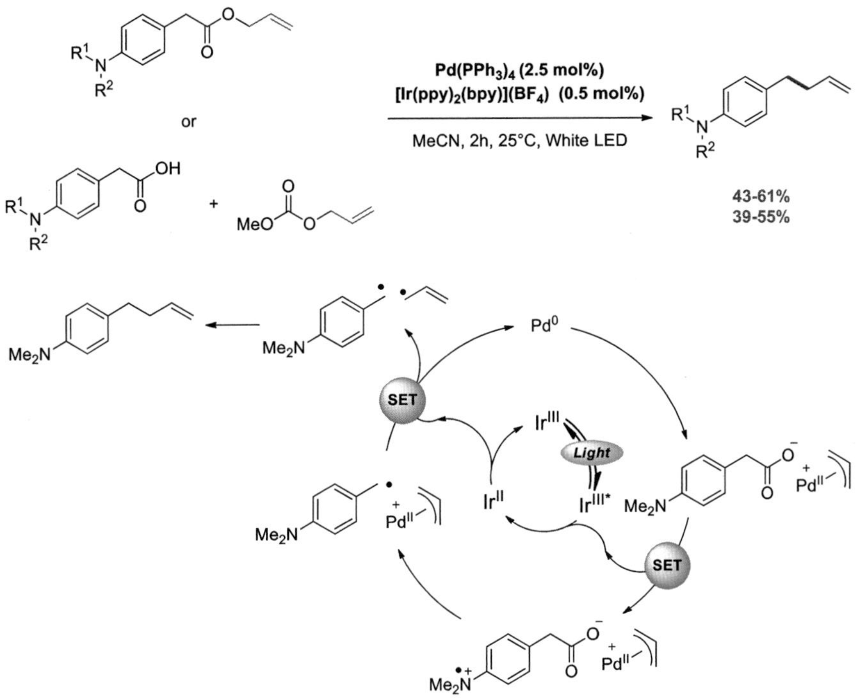

Scheme 33.   Photocatalyzed decarboxylative allylation of phenylacetate derivatives.

catalyst. A ligand exchange with the alcohol would then generate the Ni(II) aryl alkoxide. At this stage, the crucial oxidation to Ni(III) was carried out by the photoexcited iridium complex ($E_{1/2}(Ni^{III}/Ni^{II+})$ = +0.71 V *versus* Ag/AgCl[95] and $E_{1/2}(Ir^{III}*/Ir^{II})$ = +1.21 V *versus* SCE). The Ni(III) aryl alkyloxide complex then underwent reductive elimination to give the aryl ether and a Ni(I) complex reduced by the Ir(II) intermediate to regenerate both catalysts.

2. Processes with Radical Formation: Catalysis of Downstream Steps

Catalysis of downstream steps involves radical formation by photoredox catalysis. This chemical species adds on the organometallic transition metal, generating a reactive metallic intermediate which can undergo

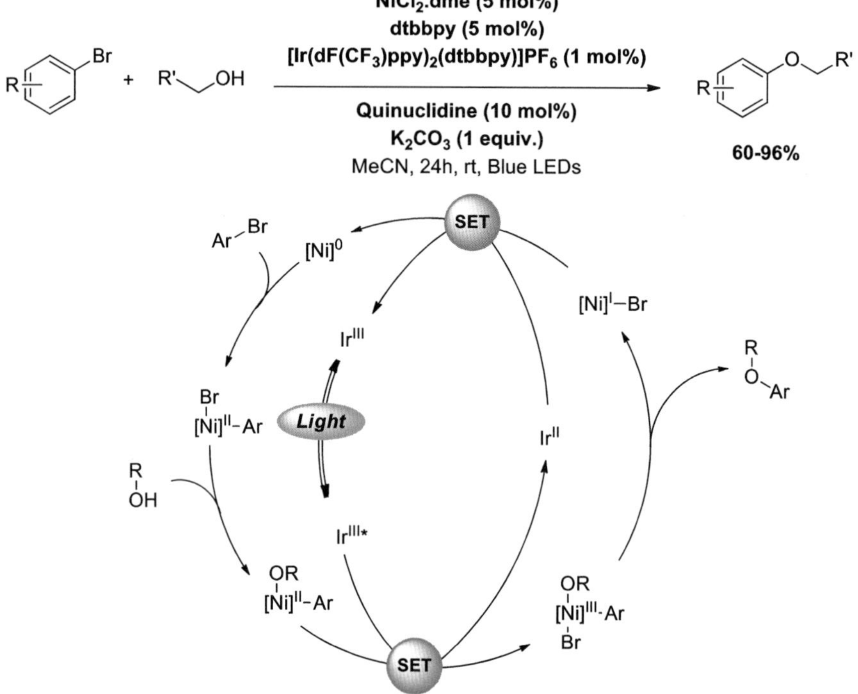

Scheme 34.   Photoassisted nickel catalyzed C–O bond formation.

reductive elimination or participate to electrophilic activation of multiple bonds for instance.

a. *Gold mediated dual catalysis*

Over the last years, gold catalysis has emerged as a powerful tool for the formation of cyclic systems, functionalization of alkenes, alkynes or allenes by nucleophilic addition reactions[96] and cross-coupling reactions. The latter involves a reductive elimination step from a Au(III) intermediate. In 2006, García and co-workers observed that benzyl and aryl radicals react with gold(I) complexes to afford organogold(III) species.[97] This property has been exploited by Glorius to develop the first photoredox/ gold dual catalysis.[98] Aryldiazonium tetrafluoroborate salts were used as aryl radical precursors and Ru(bpy)$_3$(PF$_6$)$_2$ as photocatalyst under visible-light irradiation. The authors showed that the gold catalyst Ph_3AuNTf_2 was the most effective complex in methanol for tandem 5-*exo*-trig cyclization-arylation reactions of 4-penten-1-ol. The scope could be extended to 5-penten-1-ol and 5-penten-1-tosylamide. Various aryldiazoniums bearing substituents such as halogens, esters or methoxy groups led to the product formation of products in moderate to good yields (Scheme 35).

Regarding the mechanism, the starting cationic Au(I) coordinates to the alkene which gives a cyclic alkylgold(I) intermediate. The photoexcited Ru(II)* reduces the aryldiazonium to generate the aryl radical which adds on the alkylgold(I). The resulting Au(II) complex is then oxidized by Ru(III) to afford the key Au(III) intermediate. A reductive elimination step provides the final product and regenerates the initial Au(I) complex (Scheme 36).

On the same line, our group realized a similar tandem process with *o*-alkynylphenols.[99] A neutral gold catalyst was used suggesting another possible pathway for the C–C bond formation. The successive radical addition and oxidation steps would happen directly on the Ph_3PAuCl complex (Scheme 37). Then, the cationic Au(III) intermediate may promote the cyclization step by electrophilic activation of the triple bond and the reductive elimination leading to the C–C bond formation.

This kind of pathway was first proposed by Toste for the arylative ring expansion cascade of alkenyl and allenyl cycloalkanols. Upon treatment with aryldiazonium salts, Ph_3PAuCl and $Ru(bpy)_3(PF_6)_2$ as catalysts, aryl

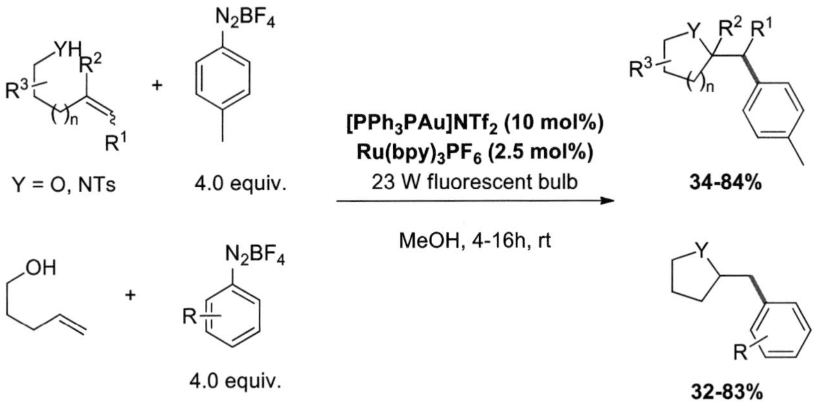

Scheme 35.   Ru/Au catalyzed oxy/aminoarylation of alkenes.

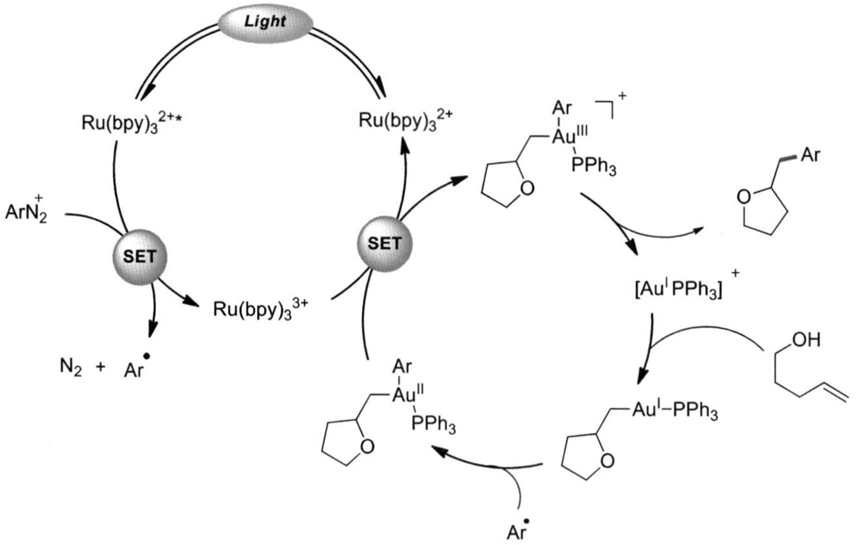

Scheme 36. Proposed mechanism for the Ru/Au dual catalysis.

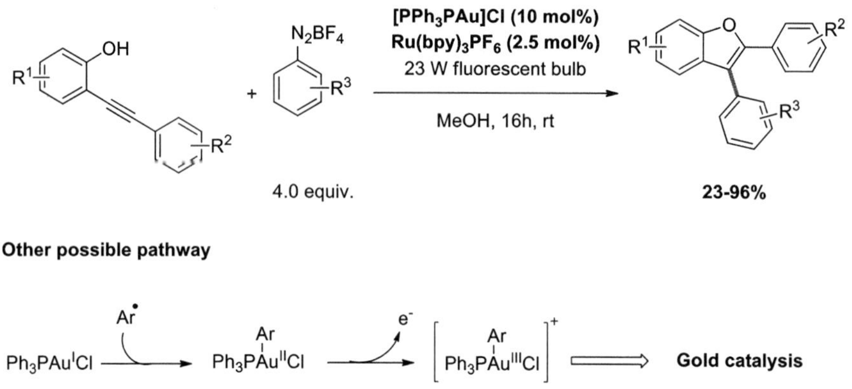

Other possible pathway

Scheme 37. Dual photoredox/gold catalysis arylative cyclization of *o*-alkynylphenols.

substituted cyclic ketones were obtained in low to good yields.[100] Regarding the mechanism (Scheme 38), the photoredox process is similar to the one proposed by Glorius, except that the neutral complex Ph_3PAuCl undergoes an addition of an aryl radical followed by oxidation with Ru(III). The resulting alkylgold(III) intermediate was proposed. Its formation was supported by the observation of $[Ph_3P\text{-}Ph]^+.BF_4^-$ as a side

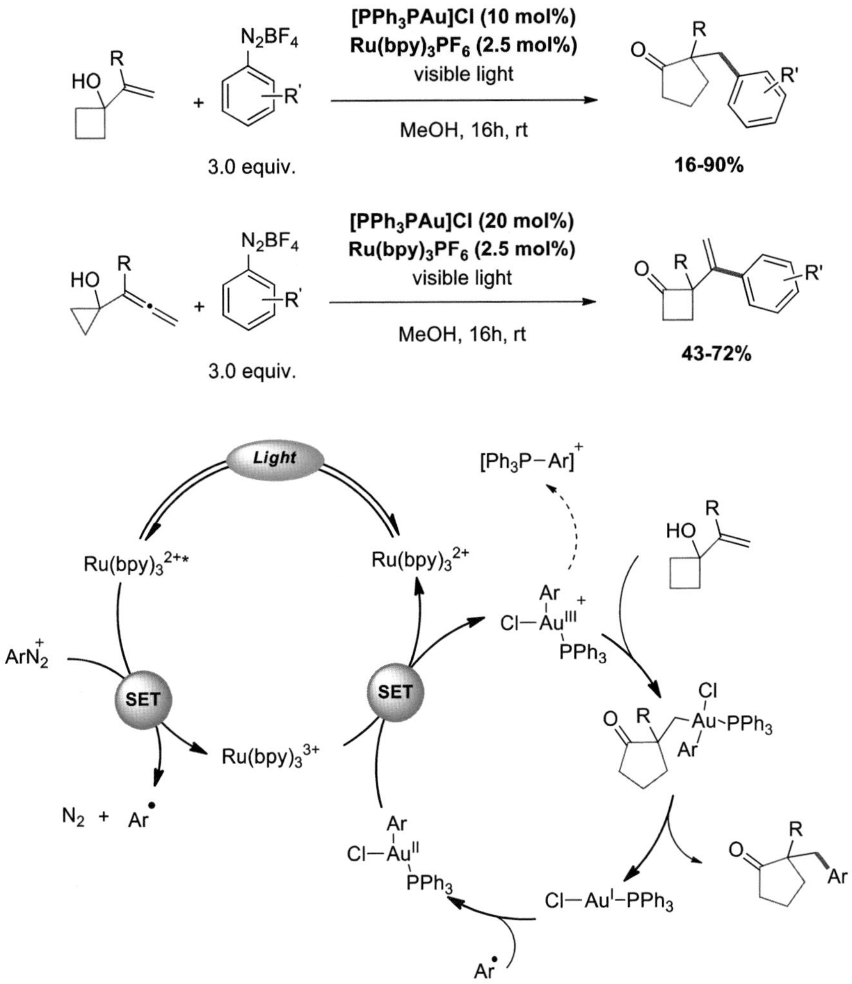

Scheme 38.   Photoredox/gold catalyzed ring expansion-arylation reaction.

product formed by reductive elimination. In the case of a fast coordination of the complex to the alkene or allene, the ring expansion may occur. Then, the reductive elimination affords the corresponding ketone and regenerates the starting Au(I) catalyst.

Gold catalyzed transformation of allenes under photoredox conditions has also been reported by the group of Shin.[101] In the same way, allenoates could be converted to arylated furanone in the presence of a cationic gold

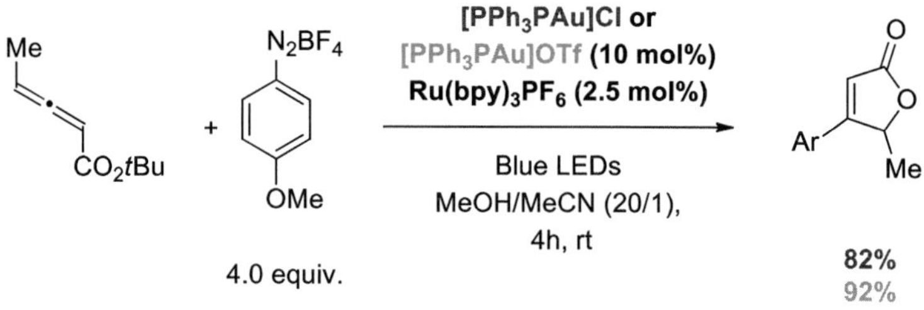

                    Lévêque *et al.*

Scheme 39.   Photoredox/gold catalyzed cross-coupling of allenes with diazonium salts.

complex, aryldiazonium salts and Ru(bpy)$_3$(PF$_6$)$_2$. In the presence of a cationic gold(I) catalyst ([Ph$_3$PAu]OTf), allenoates are converted to a gold(I)-furanone intermediate. Then, the aryl radical addition and the oxidation step would generate the gold(III) complex which after reductive elimination liberates the arylated furanone and the starting gold(I) catalyst (Scheme 39).

The photoredox/gold dual catalysis has demonstrated its efficiency towards cross-coupling reactions with aryldiazonium salts. Moreover, the ability of gold to promote cyclizations or rearrangements and perform cross-coupling reactions depends mainly on the electrophilicity of the catalyst. Two mechanisms have been proposed depending on the nature of the starting gold catalyst (neutral or cationic).

b. *Copper mediated dual catalysis*

As mentioned in Section III.B.3, copper is also an efficient radical trapping agent. After her pioneering works with palladium on dual catalysis, Sanford developed a copper-catalyzed trifluoromethylation of arylboronic acids. This known reaction usually required expensive trifluoromethylating agents or stoichiometric amounts of copper.[102] For these reasons, they choose CF$_3$I which can be reduced in the presence of Ru(bpy)$_3$(PF$_6$)$_2$ as photoredox catalyst to a CF$_3$·radical,[18,103] and a copper catalyst (20 mol% of Cu(OAc)) to obtain the cross-coupling products.[104] Various aryl and heteroaryl (pyridine, furane, quinoline) boronic acids could be trifluoromethylated (Scheme 40). Electron-rich and electron-poor electrophiles were tolerated and converted to the expected products in moderate to excellent yields. In addition, perfluoroalkyliodides were also suitable partners. However, with these materials, the loading of copper must be drastically increased to get satisfactory yields.

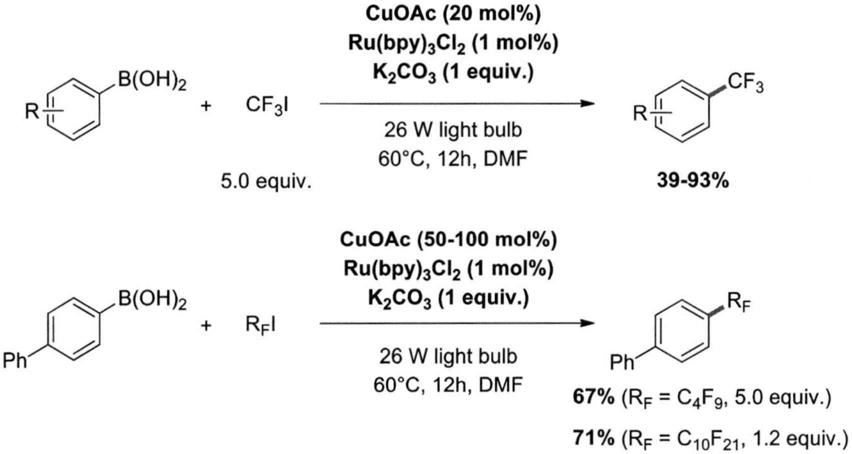

Scheme 40. Photoredox/copper-catalyzed trifluoromethylation and perfluoro-alkylation of boronic acids.

Concerning the mechanism, the authors proposed a SET between the photoexcited Ru(II)* and Cu(OAc) to Ru(I) and Cu(II) complexes. The resulting Ru(I) then reduces CF_3I to produce the CF_3 radical which adds on the intermediate Cu(II). With the new complex Cu(III)–CF_3, the boronic acid may undergo transmetallation giving the transient complex [aryl-Cu(III)–CF_3]. After a reductive elimination, the adduct aryl-CF_3 is released and the starting copper(I) salt is regenerated (Scheme 41).

c. *Nickel mediated dual catalysis*

Dual catalysis using nickel complexes as a catalyst has known some tremendous improvements in the past few years. The discovery of good radical precursors such as carboxylate salts, trifluoroborates and silicates, helped modern chemists to change their way of considering dual catalysis. In this chapter recent literatures about photoredox-nickel catalysis will be discussed.

i. Genesis of the dual catalysis

Seminal papers of dual catalysis merging nickel and a photocatalyst were published in 2014 by Doyle and MacMillan,[105] and Molander.[106] Doyle and MacMillan designed this dual catalysis with two interconnected catalytic cycles: the first one produces the radical and the second one

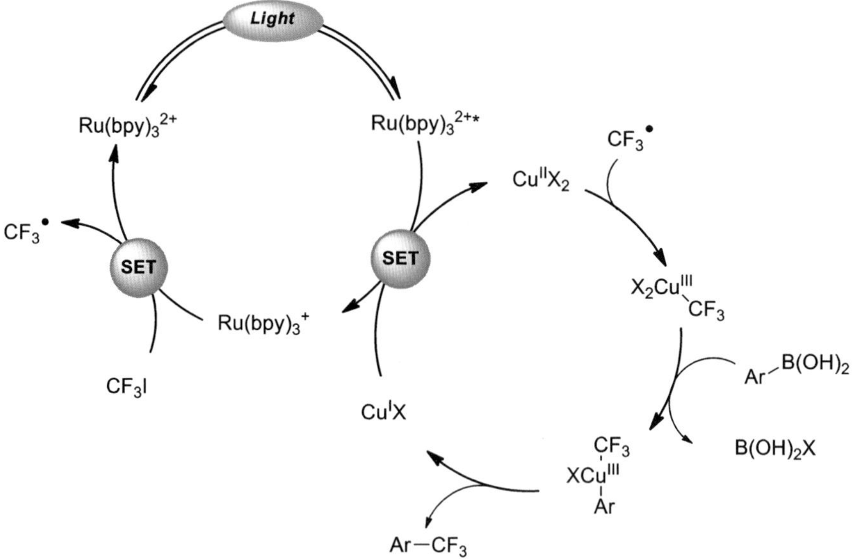

Scheme 41. Possible mechanism for the Cu/Ru-catalyzed trifluoromethylation.

promotes the cross-coupling reaction. For the generation of the radical, they used an elegant decarboxylative pathway. They anticipated that the Boc-protected proline-OH ($E^{1/2}$ = +0.95 V *versus* SCE) can release CO_2 after oxidation by a suitable excited photocatalyst and the produced radical can be added to a Ni(0) catalyst generating a Ni(I) complex. An oxidative addition step with the aryl halide generates a Ni(III) intermediate which can release the product after a reductive elimination step. Both catalysts get back to their ground state by reduction of the Ni(I) to Ni(0) by the reduced photocatalyst (Scheme 42).

The authors showed how broad the scope of this reaction is by coupling Boc-protected proline-OH with several heterocycles bearing multiple functionalities. Indeed, this reaction tolerates electron donating and withdrawing group on the aryl halide (Scheme 43).

Aniline-based substrates can also be used as radical precursors for the formation of α-nitrogen-carbon-based radicals. The commercially available *N,N* dimethylaniline was used as a model and very good yields were obtained for the direct C–H/C–X cross-coupling reactions (Scheme 44).

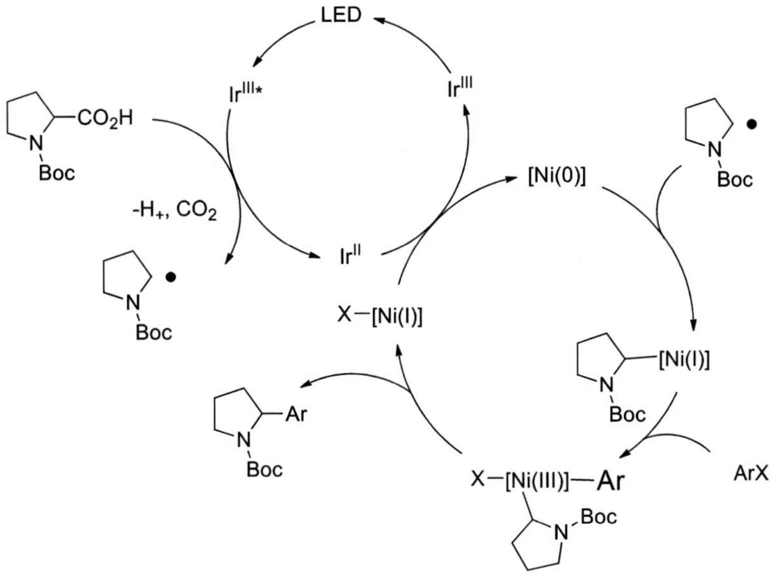

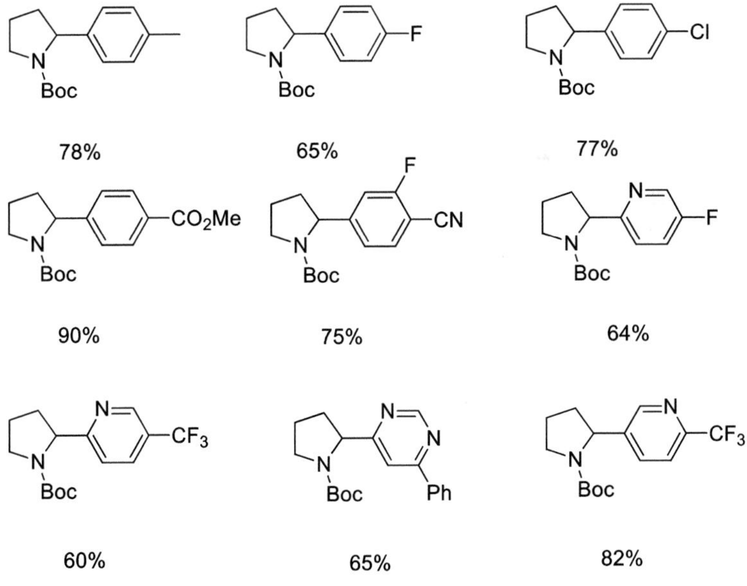

Scheme 42.   Proposed mechanism for the coupling of Boc-protected prolines and heterocycles.

Scheme 43.   Scope of the reaction.

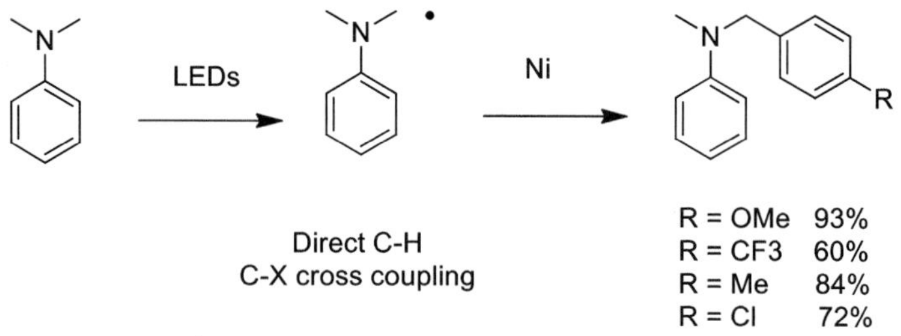

R = OMe 93%
R = CF3 60%
R = Me   84%
R = Cl   72%

Direct C-H
C-X cross coupling

Scheme 44. Aniline-based cross-coupling reaction.

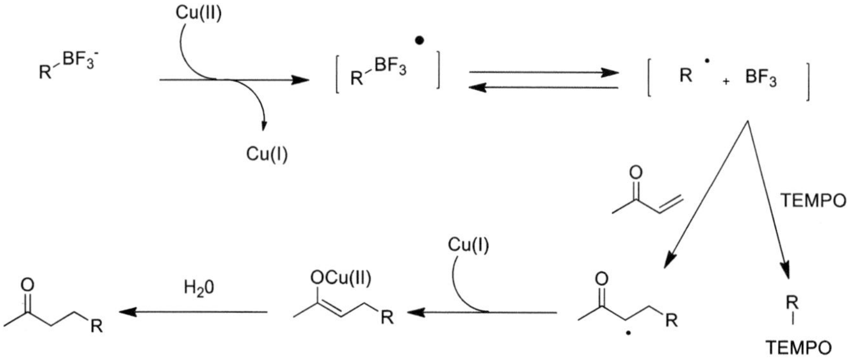

Scheme 45. Copper oxydation of trifluoroborates.

ii. Trifluoroborates as efficient coupling partners

At the same time, Molander described a similar type of reaction using the well-known trifluoborates as radical precursors. Previously, generation of radicals from trifluoroborates was reported in 2010 by Fensterbank's group.[107] This group initially used a stoechiometric copper(II) complex to oxidize these boron species and trap the produced radical with TEMPO or methyl vinyl ketone (Scheme 45).

Two years later, Akita and coworkers described the use of a highly oxidizing iridium photocatalyst to generate radicals from trifluoroborates.[108] They showed that a large variety of boron species could be oxidized and successfully engaged into spin trapping reactions and Giese-type additions (Scheme 46).

After these seminal papers dealing with the radical generation from trifluoroborates, Molander, in 2014, extended their application to dual

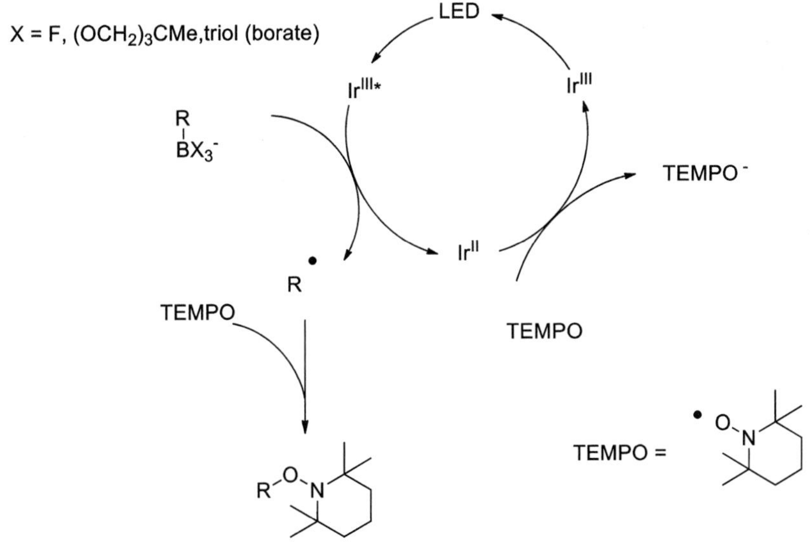

Scheme 46.   Photocatalytic oxydation of trifluoroborates.

catalysis. He showed that, in presence of $Ni(COD)_2$ and $Ir[dFCF_3ppy]_2(bpy)$ PF_6 (written Ir in the this chapter), a cross-coupling reaction could occur between aryl halides and benzyl trifluoroborates (Scheme 47).

Interestingly they also have shown in this paper that the use of a chiral ligand **L1** for the nickel could generate a stereoenriched cross-coupled product starting from the racemic methyl benzyl trifluoroborate (Scheme 48).

Following these stimulating initial results, Molander and co-workers in 2015 reported that activated trifluoroborates, such as benzyl trifluoroborate, possessing a lower oxidation potential afforded the cross-coupling products with aryl bromides in good yields.[109] This methodology is very selective as showed by the combination of the photoredox/Suzuki coupling/Buchwald Hartwig reaction forming poly aromatics species (Scheme 49).

Other activated trifluoroborates could also be used in such catalysis. For example α-hydroxyalkyltrifluoroborates afforded the corresponding benzylic and secondary alcohols in good to excellent yields.[110] For this catalysis, the use of an additive such as 2,6-lutidine is mandatory in order to trap the boron by product.

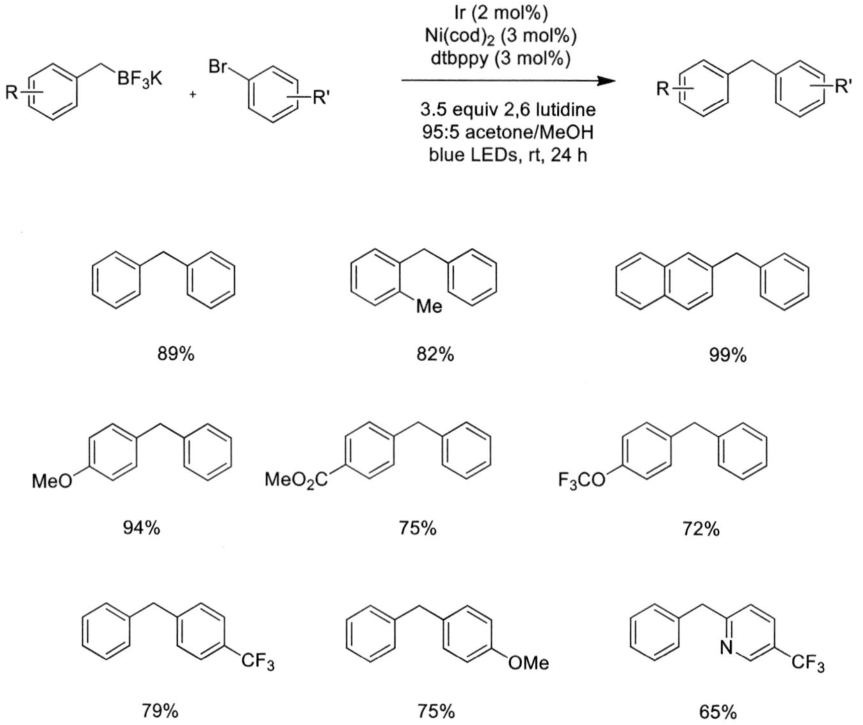

98 Lévêque *et al.*

Scheme 47. Cross-coupling reaction between aryl halides and benzyl tifluoroborates.

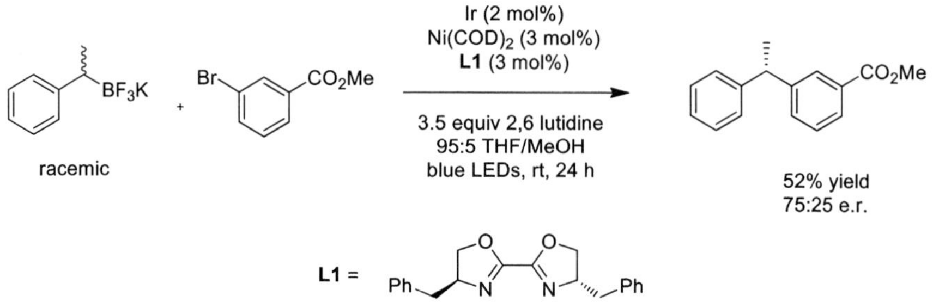

Scheme 48. Stereospecific cross-coupling reaction.

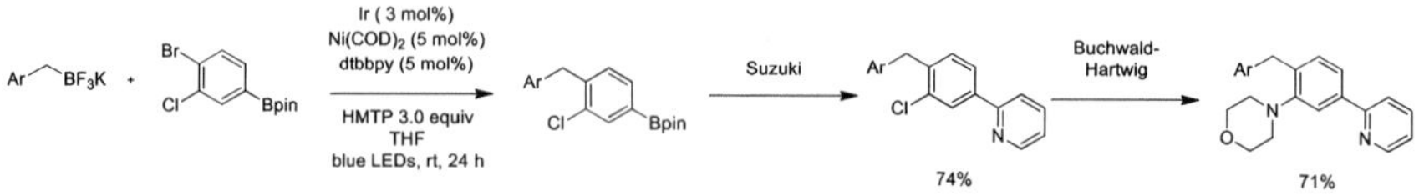

Scheme 49. Combination of photoredox/Suzuki coupling/Buchwald Hartwig reaction.

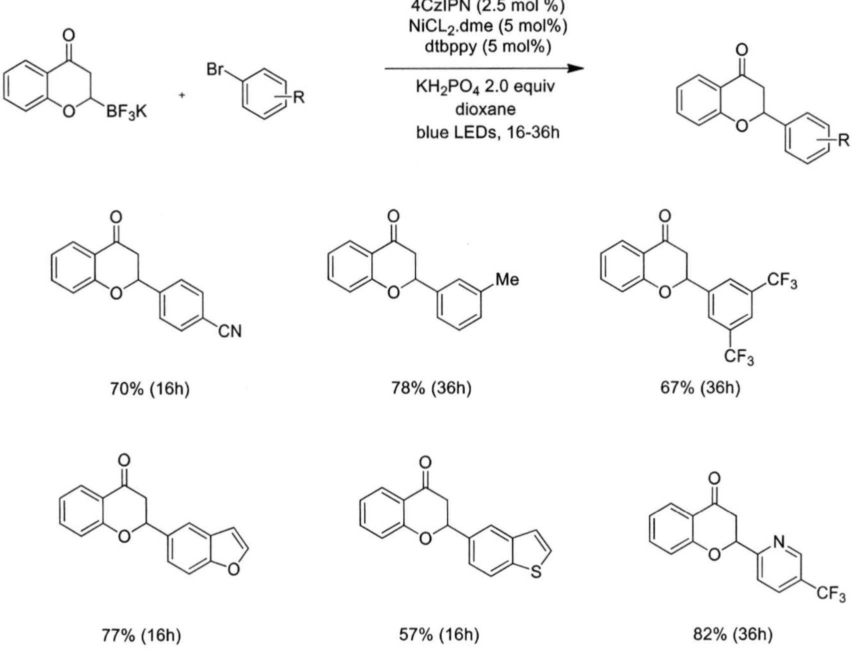

Scheme 50.   Synthesis of new flavanones and flavonoids.

In 2017, Molander and co-workers showed that it was possible to synthesize new trifluoroborates by a β-borylation process and, then they were able to succeed in synthesizing new flavanones and flavonoids.[111] The organic dye 4CzIPN, bearing very interesting properties as a long life time at the excited state was used as a photocatalyst (Scheme 50).[112,113] This photocatalyst was already used in photoredox dual catalysis by Zhang group one year before.[114]

Trifluoroborates can also be used to produce different ketones as demonstrated by Doyle and Rovis. They described an elegant way to desymmetrize cyclic *meso*-anhydrides.[115] Thanks to this reactivity, compounds with an enantioselectivity up to 91% and a diastereomeric ratio up to 99:1 with the BnBF$_3$K can be obtained (Scheme 51). Different chiral ligands can be used in order to improve the stereoselectivity.

In 2016, Molander's group developed a new mode of preparation of alkoxyketones using acyl chlorides and potassium alkoxymethyltrifluoroborates[116] (Scheme 52).

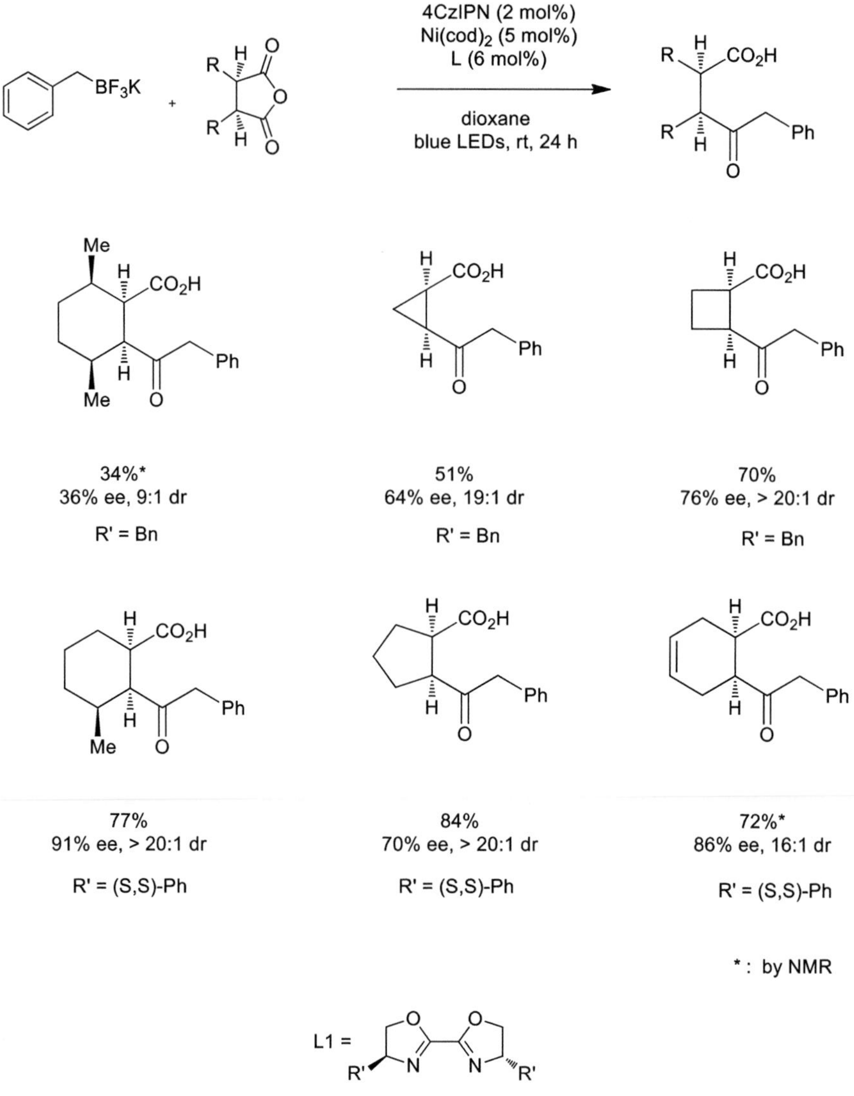

Lévêque et al.

34%*
36% ee, 9:1 dr

R' = Bn

51%
64% ee, 19:1 dr

R' = Bn

70%
76% ee, > 20:1 dr

R' = Bn

77%
91% ee, > 20:1 dr

R' = (S,S)-Ph

84%
70% ee, > 20:1 dr

R' = (S,S)-Ph

72%*
86% ee, 16:1 dr

R' = (S,S)-Ph

* : by NMR

L1 =

Scheme 51. Desymmetrizaytion of cyclic *meso*-anhydrides.

This group also showed that secondary trifluoroborates can be efficiently engaged in this reaction.[117] The same year, they have shown that amides like *N*-acyl succinimide[118] could be an efficient partners in this reaction. Nevertheless, only a few amides can perform this oxidative addition.

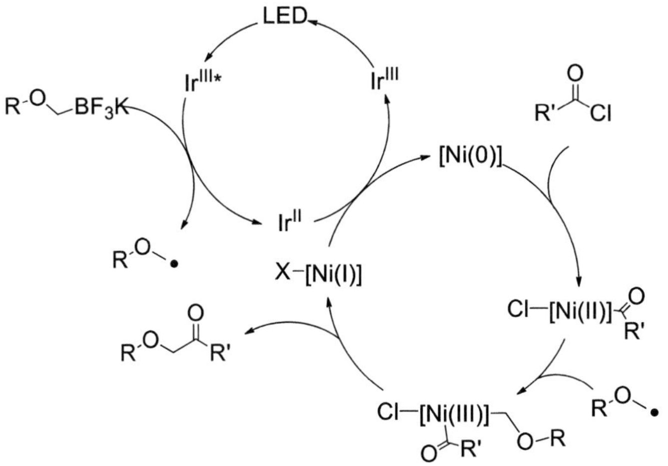

Scheme 52.   Proposed mechanism for the generation of alkoxyketones.

The trifluoromethane substituant is of fundamental interest in drugs due to its modulation of the lipophility but its formation remained sometimes tiedious. In 2016, a mild and efficient synthesis of 1,1-diaryl-2,2,2 trifluoroethanes was reported using dual nickel/photoredox catalysis[119] (Scheme 53).

A large number of trifluoromethane substituted diaryl compounds were obtained thanks to this reaction (Scheme 54). It is worthmentioning that this metalloredox reaction also works efficiently with aromatic heterocycles, interesting result if we consider the large proportion of heterocycles in common drugs.

iii. Carboxylates as efficient and easy accessible radical precursors

The carboxylate is probably one of the simplest radical precursor that can be found in the literature. After the seminal work of Doyle and MacMillan about CO_2 extrusion,[106] MacMillan and co-workers extended this process to vinyl halides.[120]

In 2017, they published a decarboxylative process to perform an enantioselective arylation of α-amino acids.[121] The use of a bis(oxazoline) ligand gave very good enantioselectivity up to 90% ee. This reaction can be successfully applied to the synthesis of PEG_2 receptor antagonist and

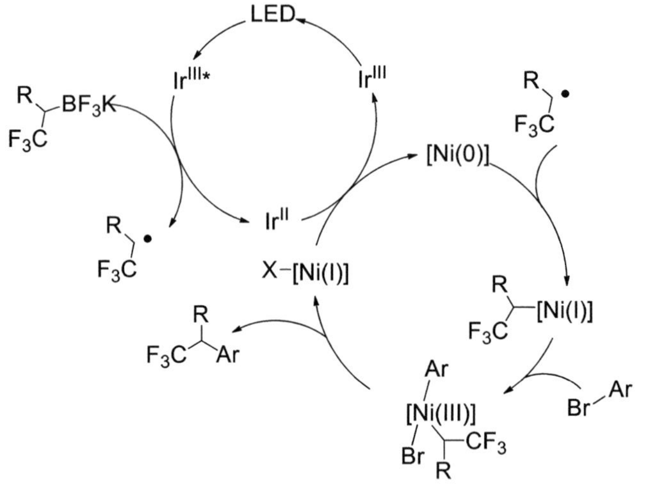

Scheme 53. Proposed mechanism for the synthesis of 1,1-diaryl-2,2,2 trifluoroethanes.

a glucagon receptor antagonist. He also showed two years later that it was possible to realize a macrocyclization *via* this decarboxylative dual catalysis process.[122] This reaction was applied to the synthesis of the bioactive cyclic peptide COR-005 a somatostatin receptor agonist. In 2018, his group also reported the hydroalkylation of alkynes[123] (Scheme 55).

The carboxylates can also be used to achieve the "so desired" sp^3-sp^3 cross-coupling.[124] This reaction can be performed with a large variety of electrophiles with the Boc-protected proline OH which has a low oxidation potential, but also with other carboxylates not as activated as the proline derivatives. This reaction, thanks to his high tolerance to functionnal groups, remains a great breakthrough in radical chemistry (Scheme 56).

iv. Silicates as easy oxidized radical precursors

Trifluoroborates and carboxylates are not the only radical precursors which can be used in dual catalysis. Bis-catecholato silicates, first synthetized by Frye, were used in metallophotoredox catalysis in 2015 by Fensterbank

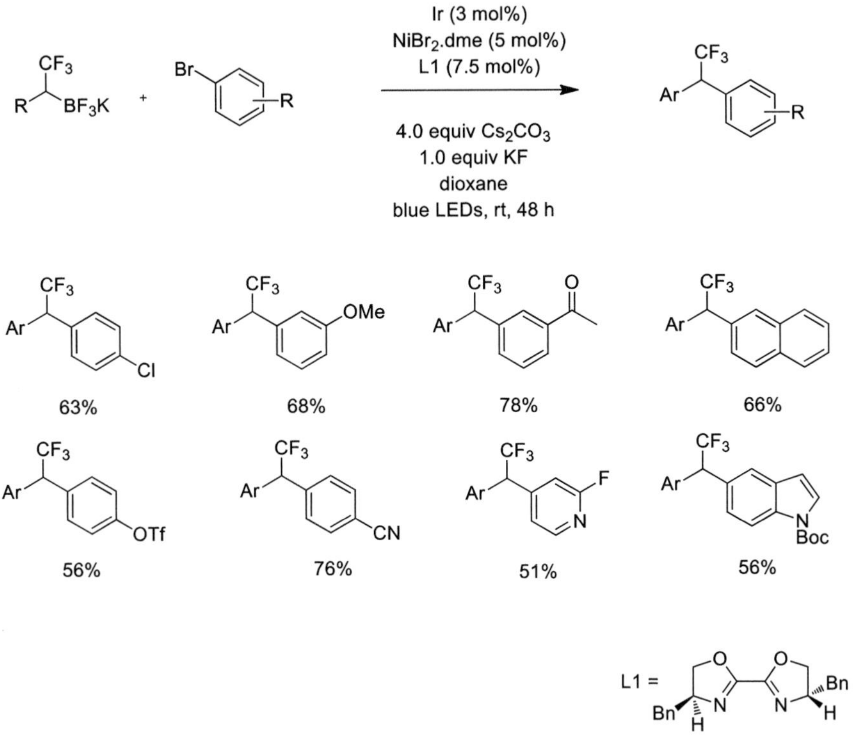
Ir (3 mol%)
NiBr2.dme (5 mol%)
L1 (7.5 mol%)
4.0 equiv Cs2CO3
1.0 equiv KF
dioxane
blue LEDs, rt, 48 h
CF3
R
BF3K
Br
Ar
OMe
Cl
OTf
CN
F
N
Boc
63%
68%
78%
66%
56%
76%
51%
56%
L1 =
Bn
H

Scheme 54. Scope of the reaction forming trifluoromethane diaryl compounds.

et al.[125] This group has shown that silicates are very useful reagents for this type of chemistry thanks to their low oxidation potential (Scheme 57).

Thereafter silicates became a popular partner in dual catalysis. In 2016, Molander published a new coupling reaction using the ammonium salts of the silicates and an aryl halide, triflate, mesylate or tosylate[126] as the electrophilic partner. This reaction is also compatible with an alkenyl halides.[127] In 2017, it has also been showed that the silicates react selectively, as expected, first with an iodide but not with a bromide or a fluoride[128] (Scheme 58).

Molander's group also pointed out some very interesting behaviors of the silicates.[129] They tried to use a thiosilicate in a cross-coupling process but instead of obtaining the desired product they obtained the thioetherified compound (Scheme 59).

　　　　　　　　　　Lévêque *et al.*

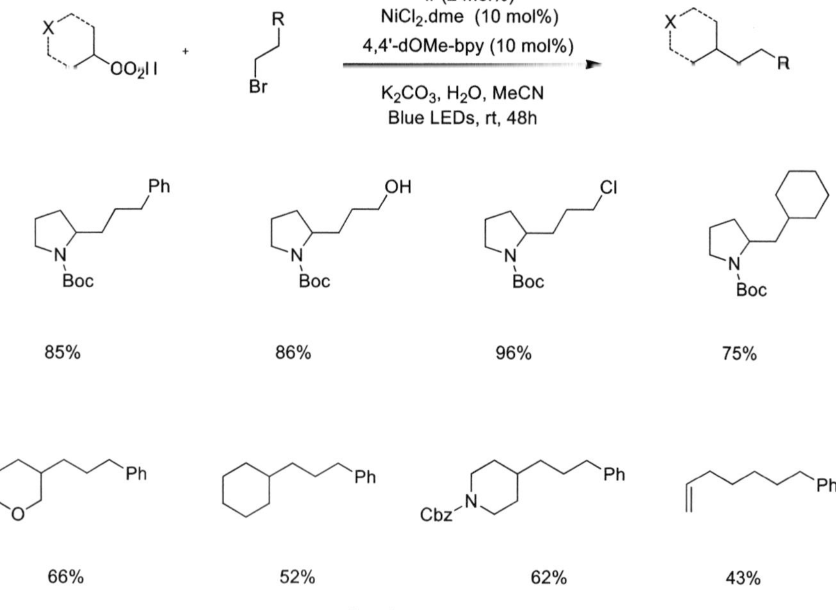

Scheme 55.   Hydroalkylation of alkynes using carboxylates radical precursors.

Scheme 56.   sp^3-sp^3 cross coupling reaction.

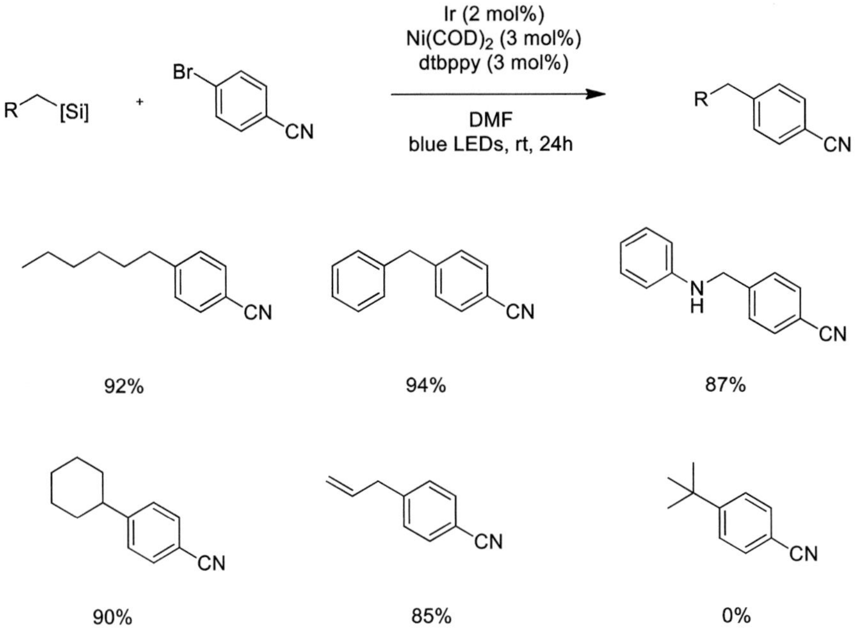

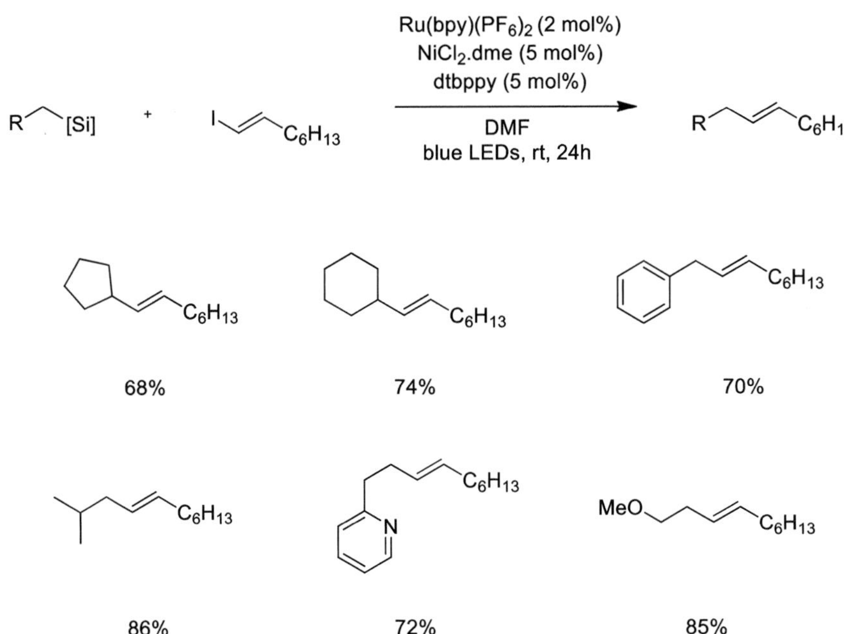

Scheme 57.   Scope of the cross-coupling using silicates and aryl bromides.

Scheme 58.   Cross-coupling reaction between silicates and vinyl iodides.

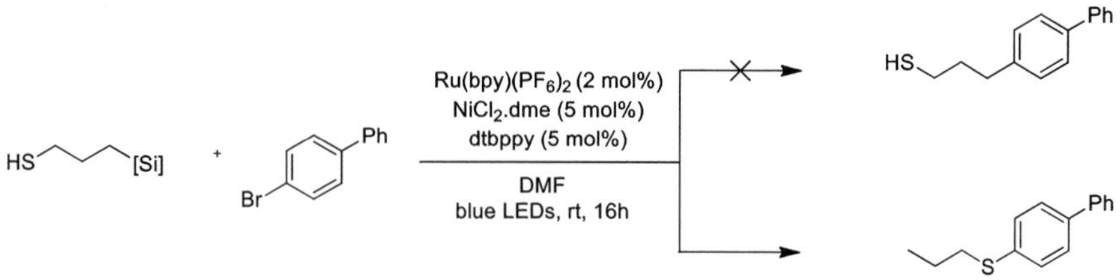

Lévêque *et al.*

Scheme 59. Product of the coupling reaction with a thiosilicate.

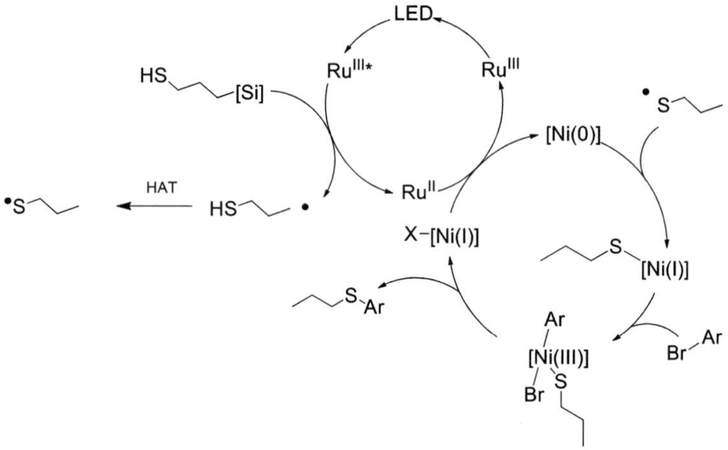

Scheme 60. Proposed mechanism for the thioether formation.

This can be explained by the following catalytic cycle depicted in Scheme 60.

Thanks to this methodology many different thioethers were synthetized (Scheme 61).

Two years later they used this interesting fast hydrogen atom transfer to perform some thioarylations.[130] Nevertheless a sacrificial silicate had to be used which limits the potential of this reaction.

In 2017, Fensterbank and co-workers reported a very interesting reaction about dual catalysis. They were able to perform a sp^3-sp^3 coupling using silicates as radical precursors.[131] The yield remained low but this reaction is definitely of interest for organic chemists because non-activated partners can be used (Scheme 62).

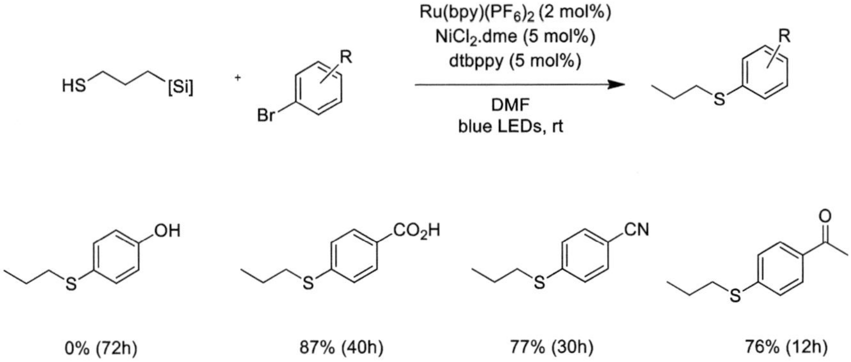
Ru(bpy)(PF6)2 (2 mol%)
NiCl2.dme (5 mol%)
dtbppy (5 mol%)
DMF
blue LEDs, rt
HS
[Si]
Br
R
S
OH
CO2H
CN
O
0% (72h)
87% (40h)
77% (30h)
76% (12h)

Scheme 61. Synthesis of thioethers.

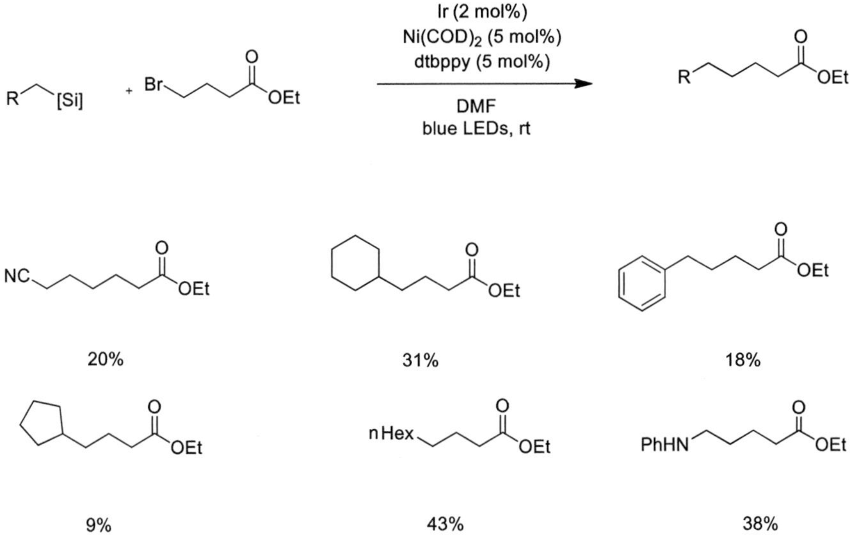
Ir (2 mol%)
Ni(COD)2 (5 mol%)
dtbppy (5 mol%)
DMF
blue LEDs, rt
R
[Si]
Br
O
OEt
NC
20%
31%
18%
9%
nHex
43%
PhHN
38%

Scheme 62. Scope of the sp^3-sp^3 cross-coupling reaction.

v. Other precursors

In 2017, Helal and co-workers reported the cross-coupling of alkyl sulfinate salts and aryl halides[132] (Scheme 63). These reactants are quite interesting because of their relatively low oxidation potential ($E_{1/2}^{ox}$ (sodium-4-tetrahydropyran sulfinate) = 0.64 V *versus* SCE). Thanks to their low oxidation potential, the inexpensive $Ru(bpy)_3Cl_2$, which is not as

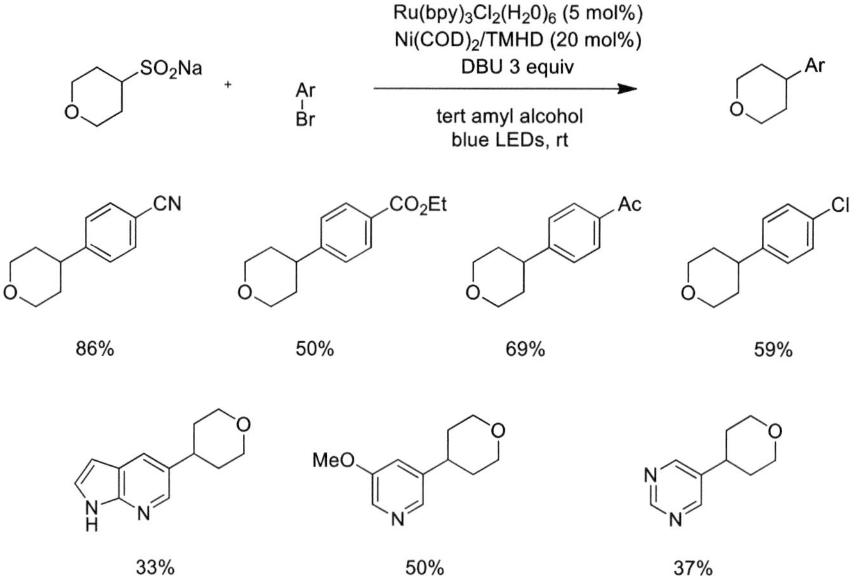

Lévêque *et al.*

Scheme 63. Cross-coupling of aryl halides and alkyl sulfinate salts.

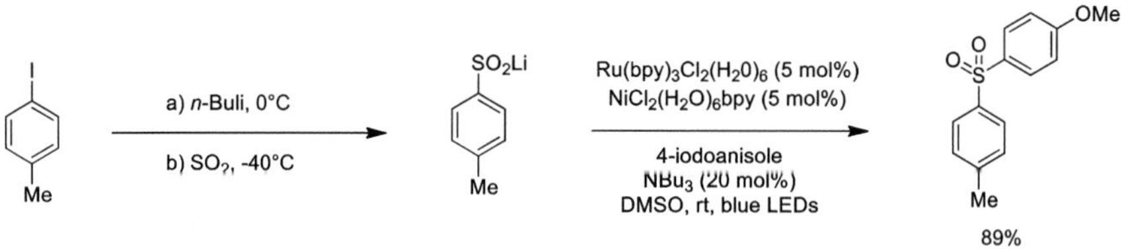

Scheme 64. Utilization of lithium salts for cross-coupling.

oxidant as the iridium photocatalyst, can be used and Helal's group succeeded in synthetizing some analogs of the caseine kinase 1δ inhibitor by using this mild desulfinative metallaphotoredox procedure.

Recently the group of Manolikakes reported a nickel coupling of sodium sulfinates salt and aryl halides but the conditions for the realization of this reaction were harsh[133]: high temperature, the use of an additional reducing agent and an expensive NHC ligand are required. Immediately there after they have successfully succeeded in performing the nickel photoredox catalysis coupling under milder conditions. Moreover, they showed in this paper that lithium salts can also be used (Scheme 64).[134]

This reaction can also be achieved with benzene as a starting material resulting in the formation of the desired sulfone in 28% yield. Rueping's group extended this concept and showed that many types of coupling could be achieved thanks to these sodium sulfinates salts.[135] Vinyl bromides can be used as the electrophilic partner and the product is preferentially the *trans* product (Scheme 65). The vinyl bromide can be bearing a electro donating or withdrawing group on the aromatic without affecting the yield of the reaction.

Other precursors can undergo the CO_2 extrusion developed by Doyle and MacMillan. In 2015, the same group showed that anhydrides can be used for a CO_2 extrusion-recombination reaction (Scheme 66), leading smoothly to a large variety of products in moderate to excellent yields.[136]

vi. Catalytic cycle insight

It is important to note that all of the catalytic cycles reported in this chapter are based on a nickel(0) catalytic cycle even if a nickel (II) catalyst was added at the beginning. In 2016, Johannes and his group have developed a mild, highly chemoselective, and robust photoinduced Ni-catalyzed method for the cross-coupling of aryl, benzyl and alkyl thiols with a wide array of functionalized aryl and heteroaryl iodides. This elegant study is

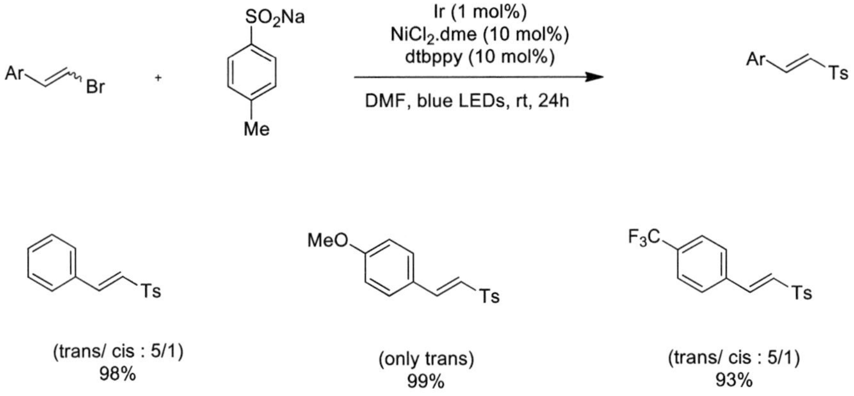

Scheme 65. Cross-coupling of sulfinate salts and vinyl halides.

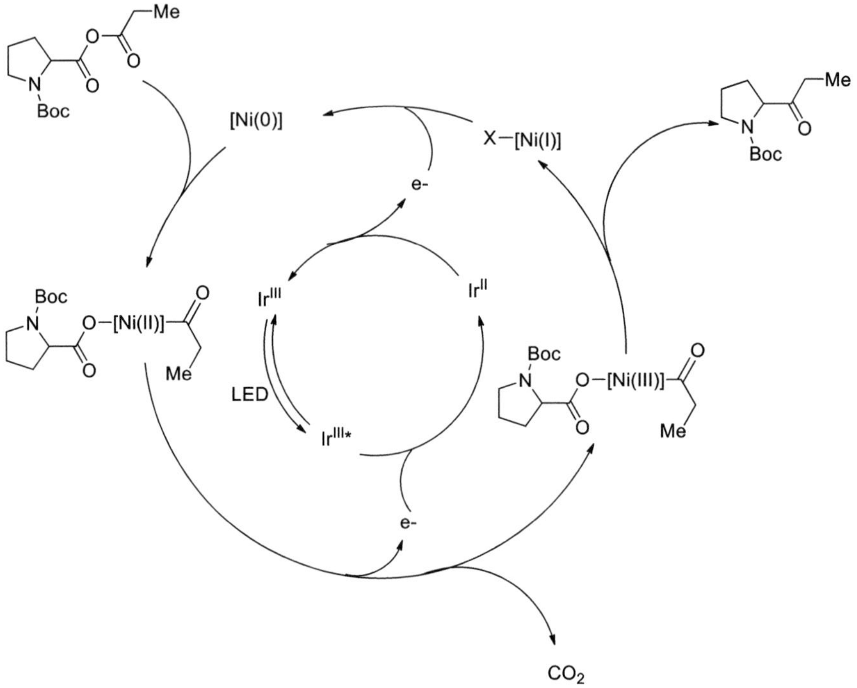

                                   Lévêque *et al.*

Scheme 66.   Proposed mechanism for the cross-coupling reaction using anhydrides.

the first method that allows thiyl radicals or any heteroatom radical to engage in a transition metal catalyzed cross-coupling reactions with iodoarenes. The ability of the catalytic system to operate with high efficiency in the presence of molecular oxygen further enhances the practical utility of this method and moreover gives new insights onto the mechanism. Distinct from nickel catalyzed cross-coupling reactions involving carbon centered radicals, control experiments and spectroscopic studies indicated that this reaction does not involve a Ni(0)-species but is catalyzed by a transient Ni(I)-species.[137] Johannes assumed that the nickel (II) precatalyst was first reduced by the photocatalyst then the radical was added to that newly formed nickel(I) species. This compound was in turn oxidized by the photocatalyst then the aryl halide could perform an oxidative addition and finally a reductive elimination released the desired product (Scheme 67).

Another very interesting result about this catalytic cycle can be found in Doyle's paper.[138] She showed that Ni(III) formed by the oxidative

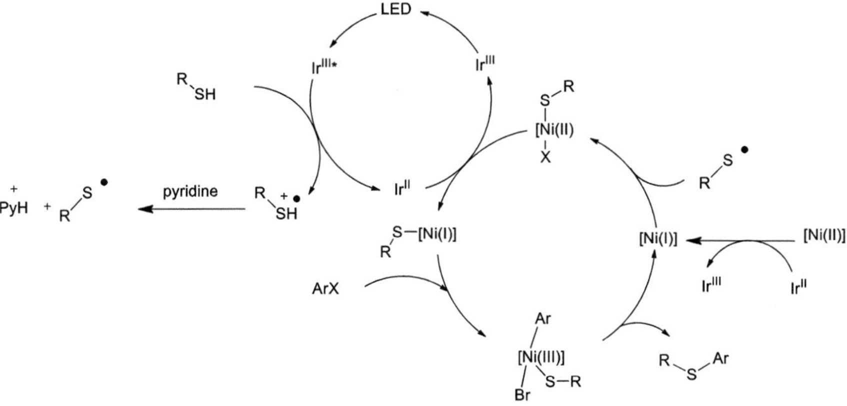

Scheme 67.   Proposed mechanism for dual catalysis by Johannes.

addition of an aryl halide and oxydation by the photocatalsyt, could undergo a photolysis and release a bromine or a chlorine radical. This radical can rapidly abstract a hydrogen from the solvent THF (H–Cl BDE = 102 kcal mol^{-1}, THF BDE = 92 kcal mol^{-1}). The THF radical formed is added to the nickel and the nickel (II) obtained can undergo a reductive elimination to produce the desired product. Thanks to that surprising reactivity of the nickel species, she managed to obtain a large variety of cross-coupling products (Scheme 68).

One year later she extended this concept by using 1,3 dioxolane as coupling reagent.[139] The chlorine radical formed after the photolysis of the nickel complex can abstract an hydrogen on the acetal compound then the cross-coupling can occur. A large variety of compounds, which are hidden aldehydes, have been synthesized showing all the versatility of this reaction (Scheme 69).

vii. Extension of the dual catalytic process

After having shown that a wide array of radical precursors that can be used successfully in this reaction, it is then possible to extend the efficiency of this catalytic process. A third catalytic cycle can be added to the two other ones as MacMillan's group reported it in 2018.[140] Thanks to an hydrogen atom transfer (HAT) they were able to develop a C–C bond formation at the α-carbon of the desired alcohol (Scheme 70).

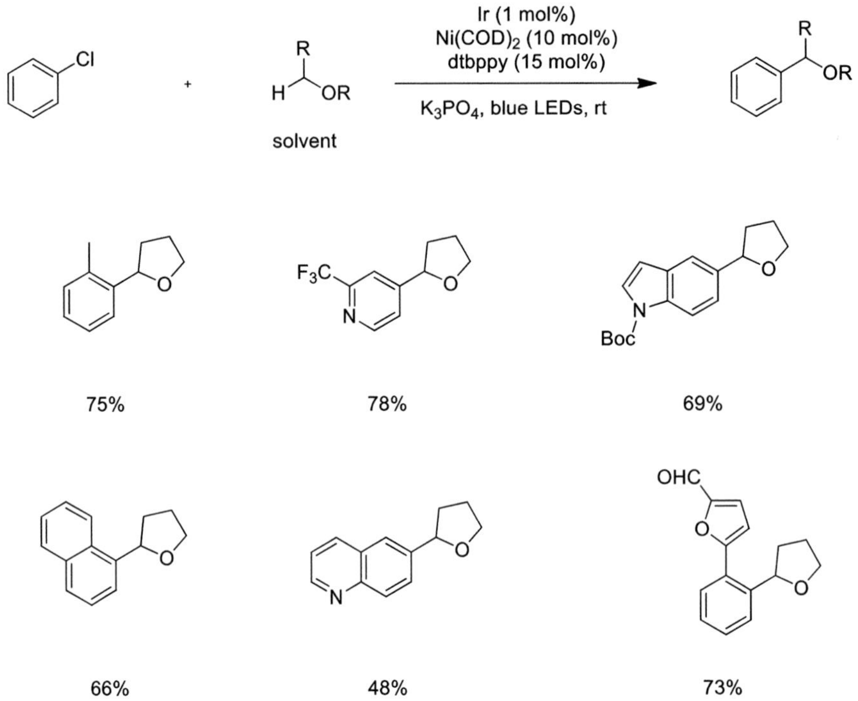

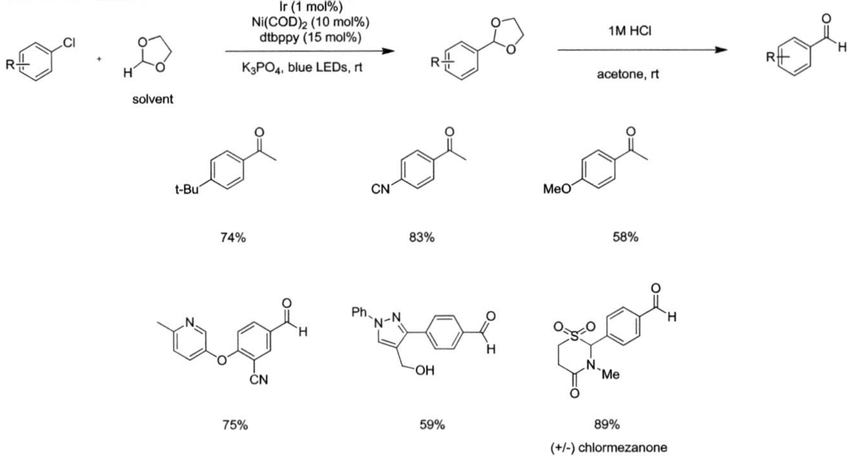

Scheme 68. Cross-coupling of radical based solvent and aryl halides.

Scheme 69. Cross-coupling of radical based 1,3 dioxolane and aryl halides.

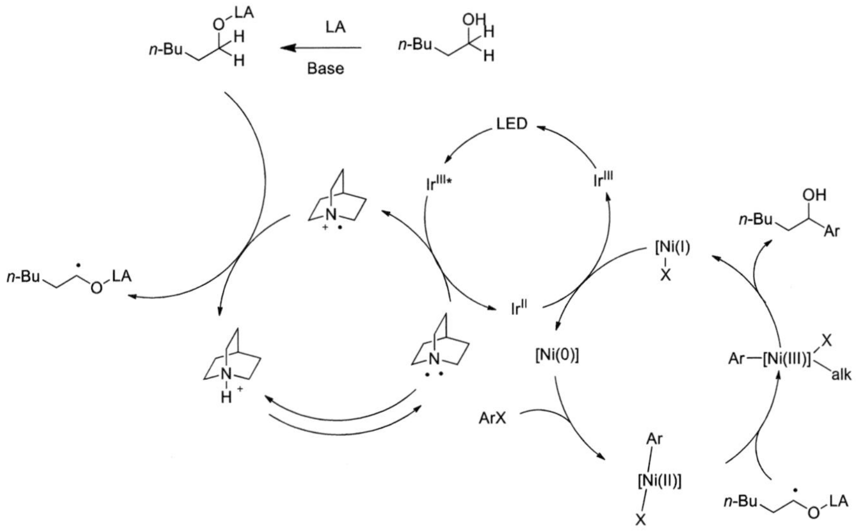

Scheme 70.   Triple catalysis cycle.

This last reaction clearly shows that this concept can be extended even further and one could imagine that a triple, a quadruple or even more catalytic cycles will be reported in the future.

IV. Conclusion

The photoredox catalysis has emerged as a powerful alternative method for the formation of radicals. The diversity of photocatalyst (ruthenium/iridium based complexes or organic dyes) with various redox potentials allows to promote single electron transfers selectively. Many reactions performed in classical conditions, could be realized in a photocatalytic version. The appeal of this field grows year after year and demonstrates its importance in the chemistry community and particularly for the development of a greener radical chemistry.

The combination of photoredox catalysis and organometallic catalysis has shown to be a highly versatile approach for the construction of a wide range of molecular scaffolds. The overall methodology involves either only electron transfers between both catalysts, or with an additional radical trapping step by the organometallic catalyst. Various transition metals including palladium, gold, copper and nickel can be used to perform cyclizations or cross-coupling reactions *via* C–H or C–X bond activation.

V. Acknowledgments

The authors thank CNRS, Sorbonne Université, MSER (ASN PhD grant to C.L.), ANR-17-CE07-0018 Hyper-Silight (PhD grant to E.L.).

VI. References

1. Paneth, F.; Hofeditz, W. *Chem. Ber.* **1929**, *62*, 1335–1247.
2. Hey, D. H.; Walters, W. A. *Chem. Rev.* **1937**, *21*, 169–208.
3. For selected reports see: (a) Brown, Herbert C.; Kharasch, M. S.; Chao, T. H. *J. Am. Chem. Soc.* **1940**, *62*, 3435–3439. (b) Kharasch, M. S.; Kane, S. S.; Brown, H. C. *J. Am. Chem. Soc.* **1941**, *63*, 526–528. (c) Kharasch, M. S.; Jensen, E. V.; Urry, W. H. *Science* **1945**, *102*, 128–128. (d) Kharasch, M. S.; Jensen, E. V.; Urry, W. H. *J. Am. Chem. Soc.* **1946**, *68*, 154–155.
4. Zavoisky, E. *Fizicheskiĭ Zhurnal*, **1945**, *9*, 211–245.
5. Chenneberg, L.; Ollivier, C. *Chimia,* **2016**, *70*, 67–76.
6. For a comprehensive account on all aspect of radical chemistry in synthesis, see: (a) Chatgilialoglu, C.; Studer, A.; in *Encyclopedia of Radicals in Chemistry, Biology and Materials*, Eds. John Wiley & Sons Ldt, Chichester, **2012**. (b) Renaud, P.; Sibi, M. P.; in *Radicals in Organic Synthesis Vols. 1 & 2*, Wiley-VCH, Weinheim, **2001**. (c) Curran, D. P.; Porter, N. A.; Giese, B.; in *Stereochemistry of Radical Reactions*, VCH, Weinheim, **1996**. (d) Gansäuer, A.; in *Radicals in Synthesis I & II, Topics in Current Chemistry*, Springer, Heidelberg, Vols. 263 & 264, **2006**.
7. Studer, A.; Amrein, S. *Synthesis.* **2002**, *7*, 835–849.
8. Ollivier, C.; Renaud, P. *Chem. Rev.* **2001**, *101*, 3415–3434.
9. Hedstrand, D. M.; Kruizinga, W. H.; Kellogg, R. M. *Tetrahedron Lett.* **1978**, *19*, 1255–1258.
10. Pac, C.; Ihama, M.; Yasuda, M.; Miyauchi, Y.; Sakurai, H. *J. Am. Chem. Soc.* **1981**, *103*, 6495–6497.
11. Cano-Yelo, H.; Deronzier, A. *Tetrahedron Lett.* **1984**, *25*, 5517–5520.
12. For reviews on the use of $Ru(bpy)_2^{2+}$ in photoredox processes: (a) Campagna, S.; Puntoriero, F.; Nastasi, F.; Bergamini, G.; Balzani, V. *Top. Curr. Chem.* **2007**, *280*, 117. (b) Kalyanasundaram, K. *Coord. Chem. Rev.* **1982**, *46*, 159–244. (c) Juris, A.; Balzani, V.; Barigelletti, F.; Campagna, S.; Belser, P.; von Zelewsky, A. *Coord. Chem. Rev.* **1988**, *84*, 85–277. (d) Teplý, F. *Collect. Czechoslov. Chem. Commun.* **2011**, *76*, 859–917.

13. (a) Kalyanasundaram, K.; Kiwi, J.; Grätzel, M. *Helv. Chim. Acta* **1978**, *61*, 2720–2730. (b) Kiwi, J.; Graetzel, M. *J. Am. Chem. Soc.* **1979**, *101*, 7214–7217. (c) Hawecker, J.; Lehn J.-M.; Ziessel, R. *J. Chem. Soc. Chem. Commun.* **1985**, 56–58. (d) Kalyanasundaram, K.; Grätzel, M. *Coord. Chem. Rev.* **1998**, *177*, 347–414.

14. Nicewicz, D. A.; MacMillan, D. W. C. *Science* **2008**, *322*, 77–80.

15. Ischay, M. A.; Anzovino, M. E.; Du, J.; Yoon, T. P. *J. Am. Chem. Soc.* **2008**, *130*, 12886–12887.

16. Narayanam, J. M. R.; Tucker, J. W.; Stephenson, C. R. J. *J. Am. Chem. Soc.* **2009**, *131*, 8756–8757.

17. Burstall, F. H. *J. Chem. Soc.* **1936**, 173–175.

18. Campagna, S.; Puntoriero, F.; Nastasi, F.; Bergamini, G.; Balzani, V. *Top. Curr. Chem.* **2007**, *281*, 45.

19. (a) Costa, R. D.; Ortí, E.; Bolink, H. J.; Monti, F.; Accorsi, G.; Armaroli, N. *Angew. Chem. Int. Ed.* **2012**, *51*, 8178–8211. (b) Puntoriero, Campagna, F.; Nastasi, F.; Bergamini, G.; Balzani, V. *Top. Curr. Chem.* **2007**, *281*, 143.

20. McCallum, T.; Rohe, S.; Barriault, L. *Synlett.* **2017**, *13*, 289–305.

21. Campagna, S.; Puntoriero, F.; Nastasi, F.; Bergamini, G.; Balzani, V. *Top. Curr. Chem.* **2007**, *280*, 69.

22. (a) Kane-Maguire, N. A. P.; Kerr, R. C.; Walters, J. R. *Inorganica Chim. Acta.* **1979**, *33*, L163–L165. (b) Pagliero, D.; Argüello, G. A. *J. Photochem. Photobiol. Chem.* **2001**, *138*, 207–211.

23. Zhang, J.; Campolo, D.; Dumur, F.; Piao, X.; Fouassier, J.-P.; Gigmes, D.; Lalevée, J. *J. Polym. Science, Polym. Chem.* **2016**, *54*, 2247–2253.

24. For reviews on the use of organic photocatalysts: (a) Fukuzumi, S.; Ohkubo, K. *Org. Biomol. Chem.* **2014**, *12*, 6059–6071. (b) Romero, N. A.; Nicewicz, D. A. *Chem. Rev.* **2016**, *116*, 10075–10166. (c) Neumann, M.; Fldner, S.; König, B.; Zeitler, K. *Angew. Chem. Int. Ed.* **2011**, 50, 951–954.

25. (a) Baldo, M. A.; Lamabsky, S.; Burrows, P. E.; Thompson, M. E.; Forrest, S. R. *Appl. Phys. Lett.* **1999**, *75*, 4–6. (b) Endo, A.; Sato, K.; Yoshimura, K.; Kai, T.; Kawada, A.; Miyazaki, H.; Adachi, C. *Appl. Phys. Lett.* **2011**, *98*, 083302. (c) Lee, S. Y.; Yasuda, T.; Nomura, H.; Adachi, C. *Appl. Phys. Lett.* **2012**, *101*, 093306. (e) Uoyama, H.; Goushi, K.; Shizu, K.; Nomura, H.; Adachi, C. *Nature* **2012**, *492*, 234–238.

26. Luo, J.; Zhang, J. *ACS Catal.* **2016**, *6*, 873–877.

27. Jones, W. E.; Fox, M. A. *J. Phys. Chem.* **1994**, *98*, 5095–5099.

28. (a) Ravelli, D.; Fagnoni, M. *ChemCatChem.* **2012**, *4*, 169–171 (b) Romero, N. A.; Nicewicz, D. A. *Chem. Rev.* **2016**, *116*, 10075–10166.

29. Tanaka, H.; Shizu, K.; Miyazaki, H.; Adachi, C. *Chem. Commun.* **2012**, *48*, 11392–11394.

30. Komatsu, R.; Sasabe, H.; Seino, Y.; Nakao, K.; Kido, J. *J. Mater. Chem. C.* **2016**, *4*, 2274–2278.

31. Turro, N. J. *Modern Molecular Photochemistry*, 98–100 (Benjamin Cummings, 1978)

32. (a) Lévêque, C.; Chenneberg, L.; Corcé, V.; Ollivier, C.; Fensterbank, L. *Chem Commun.* **2016**, *52*, 9877–9880. (b) Vara, B. A.; Jouffroy, M.; Molander, G. A. *Chem Sci.* **2017**, *8*, 530–535. (c) Stache, E. E.; Rovis, T.; Doyle, A. G. *Angew. Chem. Int. Ed.* **2017**, *56*, 3679–3683.

33. Allen, L. J.; Cabrera, P. J.; Lee, M.; Sanford, M. S. *J. Am. Chem. Soc.* **2014**, *136*, 5607–5610. (b) Qin Q.; Yu, S. *Org. Lett.* **2014**, *16*, 3504–3507. (c) Greulich, W.; Daniliuc, C. G.; Studer, A. *Org. Lett.* **2015**, *17*, 254–257. (d) Hu, X.-Q.; Chen, J.-R.; Wei, Q.; Liu, F.-L.; Deng, Q.-H.; Beauchemin, A. M.; Xiao, W.-J. *Angew. Chem. Int. Ed.* **2014**, *53*, 12163–12167.

34. (a) Tyson, E. L.; Ament, M. S.; Yoon, T. P. *J. Org. Chem.* **2013**, *78*, 2046–2050 (b) Morse, P. D.; Nicewicz, D. A. *Chem. Sci.* **2015**, *6*, 270–274. (c) Sun, J.-G.; Yang, H.; Li, P.; Zhang, B.; *Org. Lett.* **2016**, *18*, 5114–5117.

35. (a) Yoo W.-J.; Shu Kobayashi, *Green Chem.* **2013**, *15*, 1844–1848. (b) Quint, V.; Morlet-Savary, F.; Lohier, J.-F.; Jacques Lalevée, Gaumont A.-C.; Lakhdar, S. *J. Am.Chem. Soc.* **2016**, *138*, 7436–7441. (c) Bu, M.-J.; Guo-ping Lu Chun Cai, *Catal. Sci. Technol.* **2016**, *6*, 413–416. (d) Luo, K.; Chen, Y.-Z.; Yang, W.-C.; Zhu, J.; Wu, L. *Org. Lett.* **2016**, 18, 452–455.

36. (a) Cano-Yelo, H.; Deronzier, A. *J. Chem. Soc. Perkin Trans. 2.* **1984**, 1093–1098. (b) Cano-Yelo, H.; Deronzier, A. *J. Photochem.* **1987**, *37*, 315–321.

37. (a) Schroll, P.; Hari, D. P.; König, B. *ChemistryOpen.* **2012**, *1*, 130–133. (b) Hari, D. P.; Schroll, P.; König, B. *J. Am. Chem. Soc.* **2012**, *134*, 2958–2961.

38. Donck, S.; Baroudi, A.; Fensterbank, L.; Goddard, J.-P.; Ollivier, C. *Adv. Synth. Catal.* **2013**, *355*, 1477–1482.

39. Baralle, A.; Fensterbank, L.; Goddard, J.-P.; Ollivier, C. *Chem. Eur. J.* **2013**, *19*, 10809–10813.

40. Fukuzumi, S.; Mochizuki, S.; Tanaka, T. *J. Phys. Chem.* **1990**, *94*, 722–726.

41. Nagib, D. A.; Scott, M. E.; MacMillan, D. W. C. *J. Am. Chem. Soc.* **2009**, *131*, 10875–10877.

42. Shih, H.-W.; Vander Wal, M. N.; Grange, R. L.; MacMillan, D. W. C. *J. Am. Chem. Soc.* **2010**, *132*, 13600–13603.

43. Narayanam, J. M. R.; Tucker, J. W.; Stephenson, C. R. J. *J. Am. Chem. Soc.* **2009**, *131*, 8756–8757.

44. (a) Tucker, J. W.; Nguyen, J. D.; J. Narayanam, M. R.; Krabbe, S. W.; C. Stephenson, R. J. *Chem. Commun.* **2010**, *46*, 4985–4987. (b) Dange, Nitin S.; Jatoi, A. H.; Robert, F.; Landais, Y. *Org. Lett.* **2017**, *19*, 3652–3655.

45. Tucker, J. W.; Narayanam, J. M. R.; Krabbe, S. W.; Stephenson, C. R. J. *Org. Lett.* **2010**, *12*, 368–371.

46. (a) Hill, H. A. O.; Pratt, J. M.; O'Riordan, M. P.; Williams, F. R.; R. Williams, J. P. *J. Chem. Soc. Inorg. Phys. Theor.* **1971**, 1859–1862. (b) Rondinini, S.; Mussini, P. R.; Muttini, P.; Sello, G. *Electrochim. Acta.* **2001**, *46*, 3245–3258. (c) Poizot, P.; Laffont-Dantras, L.; Simonet, J. *J. Electroanal. Chem.* **2008**, *624*, 52–58.

47. Nguyen, J. D.; D'Amato, E. M.; Narayanam, J. M. R; Stephenson, Corey R. J. *Nature Chem.* **2012**, *4*, 854–859.

48. Barton, D. H. R; McCombie, S. W. *J. Chem. Soc. Perkin 1* **1975**, 1574–1585.

49. (a) Okada, K.; Okamoto, K.; Morita, N.; Okubo, K.; Oda, M. *J. Am. Chem. Soc.* **1991**, *113*, 9401–9402. (b) Okada, K.; Okubo, K.; Morita, N.; Oda, M. *Tetrahedron Lett.* **1992**, *33*, 7377–7380.

50. Lackner, G. L.; Quasdorf, K. W.; Overman, L. E. *J. Am. Chem. Soc.* **2013**, *135*, 15342–15345.

51. Rackl, D.; Kais, V.; Kreitmeier, P.; Reiser, O. *Beilstein J. Org. Chem.* **2014**, *10*, 2157–2165.

52. Chenneberg, L.; Baralle, A.; Daniel, M.; Fensterbank, L.; Goddard, J.-P.; Ollivier, C. *Adv. Synth. Catal.* **2014**, *356*, 2756–2762.

53. Larraufie, M. H.; Pellet, R.; Fensterbank, L.; Goddard, J. P.; Lacôte, E.; Malacria, M.; Ollivier, C. *Angew. Chem. Int. Ed.* **2011**, *50*, 4463–4466.

54. Dinnocenzo, J. P.; Banach, T. E. *J. Am. Chem. Soc.* **1989**, *111*, 8646–8653.

55. Parker, V. D.; Tilset, M. *J. Am. Chem. Soc.* **1991**, *113*, 8778–8781.

56. (a) Chow, Y. L.; Danen, W. C.; Nelsen, S. F.; Rosenblatt, D. H. *Chem. Rev.* **1978**, 78, 243–274. Lewis, F. D.; Ho, T. I.; Simpson, J. T. *J. Am. Chem. Soc.* **1982**, 104, 1924–1929. Lewis, F. D.; *Acc. Chem. Res.* **1986**, 19, 401–405.

57. (a) Maity, S.; Zhu, M.; Shinabery, R. S.; Zheng, N. *Angew. Chem. Int. Ed.* **2012**, *51*, 222–226. (b) Nguyen, T. H.; Morris, S. A.; Zheng, N. *Adv. Synth. Catal.* **2014**, *356*, 2831–2837. (c) Nguyen, T. H.; Maity, S.; Zheng, N. *Beilstein J. Org. Chem.* **2014**, *10*, 975–980.

58. Kohls, P.; Jadhav, D.; Pandey, G.; Reiser, O. *Org. Lett.* **2012**, *14*, 672–675.

59. (a) Miyake, Y.; Nakajima, K.; Nishibayashi, Y. *J. Am. Chem. Soc.* **2012**, *134*, 3338–3341. (b) Miyake, Y.; Ashida, Y.; Nakajima, K.; Nishibayashi, Y. *Chem. Commun.* **2012**, *48*, 6966–6968.

60. McNally, A.; Prier, C. K.; D. MacMillan, W. C. *Science* **2011**, *334*, 1114–1117.

61. Zuo, Z.; D. MacMillan, W. C. *J. Am. Chem. Soc.* **2014**, *136*, 5257–5260.

62. Galicia, M.; F. Gonzalez, J. J. *Electrochem. Soc.* **2002**, *149*, D46.

63. Noble, A.; D. MacMillan, W. C. *J. Am. Chem. Soc.* **2014**, *136*, 11602–11605.

64. Persson, B. *Acta Chem. Scand.* **1977**, *31B*, 88.

65. (a) Nishigaichi, Y.; Orimi, T.; Takuwa, A. *J. Organomet. Chem.* **2009**, *694*, 3837–3839. (b) Sorin, G.; Martinez Mallorquin, R.; Contie, Y.; Baralle, A.; Malacria, M.; Goddard, J.-P.; Fensterbank, L. *Angew. Chem. Int. Ed.* **2010**, *49*, 8721–8723. (c) Molander, G. A.; Colombel, V.; Braz, V. A. *Org. Lett.* **2011**, *13*, 1852–1855.

66. Yasu, Y.; Koike, T.; Akita, M. *Adv. Synth. Catal.* **2012**, *354*, 3414–3420.

67. Prier, C. K.; Rankic, D. A.; D. MacMillan, W. C. *Chem. Rev.* **2013**, *113*, 5322–5363.

68. Benniston, A. C.; Harriman, A.; Li, P.; Rostron, J. P.; van Ramesdonk, H. J.; Groeneveld, M. M.; Zhang, H.; Verhoeven, J. W. *J. Am. Chem. Soc.* **2005**, *127*, 16054–16064.

69. For examples of photoredox-organo dual catalysis: (a) Hamilton, D. S.; Nicewicz, D. A. *J. Am. Chem. Soc.* **2012**, *134*, 18577–18580. (b) Neumann, M.; Zeitler, K. *Org. Lett.* **2012**, *14*, 2658–2661. (c) DiRocco, D. A.; Rovis, T. *J. Am. Chem. Soc.* **2012**, *134*, 8094–8097. (d) Petronijević, F. R.; Nappi, M.; MacMillan, D. W. C. *J. Am. Chem. Soc.* **2013**, *135*, 18323–18326. (e) Zeller, M. A.; Riener, M.; Nicewicz, D. A. *Org. Lett.* **2014**, *16*, 4810–4813.

70. Kochi, J. K. *Acc. Chem. Res.* **1974**, *7*, 351–360.

71. (a) Patel, V. F.; Pattenden, G. *Tetrahedron Lett.* **1987**, *28*, 1451–1454. (b) Patel, V. F.; Pattenden, G. *J. Chem. Soc. Chem. Commun.* **1987**, 871–872. (c) Patel, V. F.; Pattenden, G.; Thompson, D. M. *J. Chem. Soc. Perkin 1* **1990**, 2729–2734. (d) Ali, A.; Harrowven, D. C.; Pattenden, G. *Tetrahedron Lett.* **1992**, *33*, 2851–2854.

72. (a) Wakabayashi, K.; Yorimitsu, H.; Oshima, K. *J. Am. Chem. Soc.* **2001**, *123*, 5374–5375. (b) Ikeda, Y.; Nakamura, T.; Yorimitsu, H.; Oshima, K. *J. Am. Chem. Soc.* **2002**, *124*, 6514–6515.

73. (a) Tamao, K.; Sumitani, K.; Kumada, M. *J. Am. Chem. Soc.* **1972**, *94*, 4374–4376. (b) Corriu, R. J. P; Masse, J. P. *J. Chem. Soc. Chem. Commun.* **1972**, 144a–144a.

74. For more details, see section 4.1.

75. Phapale, V. B.; Buñuel, E.; García-Iglesias, M.; Cárdenas, D. J. *Angew. Chem.* **2007**, *119*, 8946–8951.

76. Jones, G. D.; McFarland, C.; Anderson, T. J.; Vicic, D. A. *Chem. Commun.* **2005**, 4211–4213.

77. Jones, G. D.; Martin, J. L.; McFarland, C.; Allen, O. R.; Hall, R. E.; Haley, A. D.; Brandon, R. J.; Konovalova, T.; Desrochers, P. J.; Pulay, P.; Vicic, D. A. *J. Am. Chem. Soc.* **2006**, *128*, 13175–13183.

78. (a) Zhou, J.; Fu, G. C.; *J. Am. Chem. Soc.* **2004**, *126*, 1340–1341. (b) Guisán-Ceinos, M.; Soler-Yanes, R.; Collado-Sanz, D.; Phapale, V. B.; Buñuel, E.; Cárdenas, D. J. *Chem. Eur. J.* **2013**, *19*, 8405–88410. (c) Powell, D. A.; Maki, T.; Fu, G. C. *J. Am. Chem. Soc.* **2005**, *127*, 510–511.

79. Lorris, M. E.; Abramovitch, R. A.; Marquet, J.; Moreno-Mañas, M. *Tetrahedron.* **1992**, *48*, 6909–6916.

80. Zhang, C.-P.; Wang, Z.-L.; Chen, Q.-Y.; Zhang, C.-T.; Gu Y.-C.; Xiao, J.-C. *Angew. Chem. Int. Ed.* **2011**, *50*, 1896–1900.

81. Jacquet, J.; Salanouve, E.; Orio, M.; Vezin, H.; Blanchard, S.; Derat, E.; Murr, M. D.-E.; Fensterbank, L. *Chem. Commun.* **2014**, *50*, 10394–10397.

82. Jacquet, J.; Blanchard, S.; Derat, E.; Murr, M. D.-E.; Fensterbank, L. *Chem. Sci.* **2016**, *7*, 2030–2036.

83. Jacquet, J.; Chaumont, P.; Gontard, G.; Orio, M.; Vezin, H.; Blanchard, S.; Desage-El Murr M.; Fensterbank, L.; *Angew. Chem. Int. Ed.* **2016**, *55*, 10712–10716.

84. Deprez, N. R.; Sanford, M. S. *J. Am. Chem. Soc.* **2009**, *131*, 11234–11241.

85. Yu, W.-Y.; Sit, W. N.; Zhou, Z.; Chan, A. S.-C. *Org. Lett.* **2009**, *11*, 3174–3177.

86. Kalyani, D.; McMurtrey, K. B.; Neufeldt, S. R.; Sanford, M. S. *J. Am. Chem. Soc.* **2011**, *133*, 18566–18569.

87. Zoller, J.; Fabry, D. C.; Ronge, M. A.; Rueping, M. *Angew. Chem. Int. Ed.* **2014**, *53*, 13264–13268.

88. Boele, M. D. K.; van Strijdonck, G. P. F.; de Vries, A. H. M.; Kamer, P. C. J.; de Vries J. G.; van Leeuwen, P. W. N. M. *J. Am. Chem. Soc.* **2002**, *124*, 1586–1587.

89. (a) Stuart, D. R.; Bertrand-Laperle, M.; Burgess, K. M. N; Fagnou, K. *J. Am. Chem. Soc.* **2008**, *130*, 16474–16475. (b) Stuart, D. R.; Alsabeh, P.; Kuhn, M.; Fagnou, K. *J. Am. Chem. Soc.* **2010**, *132*, 18326–18339.

90. For more details see: Fabry, D. C.; Rueping, M. *Acc. Chem. Res.* **2016**, *49*, 1969–1979; ref 69.

91. Li, L.; Brennessel, W. W.; Jones, W. D. *Organometallics.* **2009**, *28*, 3492–3500.

92. Lang, S. B.; O'Nele K. M.; Tunge, J. A. *J. Am. Chem. Soc.* **2014**, *136*, 13606–13609.

93. Macgregor, S. A.; Neave, G. W.; Smith, C. *Faraday Discuss.* **2003**, *124*, 111–127.

94. (a) Matsunaga, P. T.; Hillhouse, G. L.; Rheingold, A. L. *J. Am. Chem. Soc.* **1993**, *115*, 2075–2077. (b) Matsunaga, P. T.; Mavropoulos, J. C.; Hillhouse, G. L. *Polyhedron.* **1995**, *14*, 175–185. (c) Han, R.; Hillhouse, G. L. *J. Am. Chem. Soc.* **1997**, *119*, 8135–8136.

95. Klein, A.; Kaiser, A.; Wielandt, W.; Belaj, F.; Wendel, E.; Bertagnolli, H.; Záliš, S. *Inorg. Chem.* **2008**, *47*, 11324–11333.

96. For recent reviews on Gold catalysis see: (a) Fensterbank, L.; Malacria, M. *Acc. Chem. Res.* **2014**, *47*, 953–965. (b) Dorel, R.; Echavarren, A. M. *Chem. Rev.* **2015**, *115*, 9028–9072.

97. Aprile, C.; Boronat, M.; Ferrer, B.; Corma, A.; García, H. *J. Am. Chem. Soc.* **2006**, *128*, 8388–8389.

98. Sahoo, B.; Hopkinson, M. N.; Glorius, F. *J. Am. Chem. Soc.* **2013**, *135*, 5505–5508.

99. Xia, Z.; Khaled, O.; Mouriès-Mansuy, V.; Ollivier, C.; Fensterbank, L.; *J. Org. Chem.* **2016**, *81*, 7182–7190.

100. Shu, X.; Zhang, M.; He, Y.; Frei, H.; Toste, F. D. *J. Am. Chem. Soc.* **2014**, *136*, 5844–5847.

101. Patil, D. V.; Yun, H.; Shin, S. *Adv. Synth. Catal.* **2015**, *357*, 2622–2628.

102. (a) Liu, T.; Shen, Q. *Org. Lett.* **2011**, *13*, 2342–2345. (b) Xu, J.; Luo, D.-F.; Xiao, B.; Liu, Z.-J.; Gong, T.-J.; Fu, Y.; Liu, L. *Chem. Commun.* **2011**, *47*, 4300. (c) Zhang, C.-P.; Cai, J.; Zhou, C.-B.; Wang, X.-P.; Zheng, X.; Gu Y.-C.; Xiao, J.-C. *Chem. Commun.* **2011**, *47*, 9516.

103. Pham, P. V.; Nagib, D. A.; MacMillan, D. W. C. *Angew. Chem. Int. Ed.* **2011**, *50*, 6119–6122.

104. Ye, Y.; Sanford, M. S. *J. Am. Chem. Soc.* **2012**, *134*, 9034–9037.

105. Zuo, Z.; Ahneman, D. T.; Chu, L.; Terrett, J. A.; Doyle, A. G.; MacMillan, D. W. C. *Science* **2014**, *345,* 437–440.

106. Tellis, J. C.; Primer, D. N.; Molander, G. A. *Science* **2014**, *345,* 433–436.

107. Sorin, G.; Mallorquin, R. M.; Contie, Y.; Baralle, A.; Malacria, M.; Goddard, J. P.; Fensterbank, L. *Angew. Chem. Int. Ed.* **2010**, *49*, 8721–8723.

108. Yasu, Y.; Koike, T.; Akita, M. *Adv. Synth. Catal.* **2012**, *354*, 3414–3420

109. Yamashita, Y.; Tellis, J. C.; Molander, G. A. *Prot. Natl. Acas. Sci. USA.* **2015**, *112*, 12016–12029.

110. Alam, R.; Molander, G. A. *J. Org. Chem.* **2012**, *82*, 13728–13734.

111. Matsui, J. K.; Molander, G. A. *Org. Lett.* **2017**, *19*, 436–439.

112. Uoyama, H.; Goushi, K.; Shizu, K.; Nomura, H.; Adachi, C. *Nature* **2012**, *492*, 234–238.

113. For a reevaluation of the reduction oxidation potentials of the excited state of 4CzIPN, see: Vaillant, F. L.; Garreau, M.; Nicolai, S.; Gryn'ova, G.; Corminboeuf, C.; Waser, J. *Chem. Sci.* **2018**, *9*, 5883–5889.

114. Luo, J.; Zhang, J. *ACS. Catal.* **2016**, *6*, 873–877.

115. Stache, E. E.; Rovis, T.; Doyle, A. G. *Angew. Chem. Int. Ed.* **2017**, *56*, 3679–3683.

116. Amani, J.; Sodagar, E.; Molander, G. A. *Org. Lett.* **2016**, *18*, 732–735.

117. Amani, J.; Molander, G. *J. Org. Chem.* **2017**, *83*, 1856–1863.

118. Amani, J.; Alam, R.; Badir, S.; Molander, G. A. *Org. Lett.* **2017**, *19*, 2426–2429.

119. Ryu, D.; Primer, D. N.; Tellis, J. C.; Molander, G. A. *Chem. Eur. J.* **2016**, *22*, 120–123.

120. Noble, A.; McCarver, S. J.; MacMillan, D. W. C. *J. Am. Chem. Soc.* **2015**, *137*, 624–627.

121. Zuo, Z.; Cong, H.; Li, W.; Choi, J.; Fu, G. C.; MacMillan, D. W. C. *J. Am. Chem. Soc.* **2016**, *138*, 1832–1835.

122. McCarver, S. J.; Qiao, J. X.; Carpenter, J.; Borzilleri, R. M.; Poss, M. A.; Eastgate, M. D.; Miller, M.; MacMillan, D. W. C. *Angew. Chem. Int. Ed.* **2016**, *55*, 1–6.

123. Till, N. A.; Smith, R. T.; MacMillan, D. W. C. *J. Am. Chem. Soc.* **2018**, *140*, 5701–5705.

124. Johnston, C. P.; Smith, R. T.; Allmendinger, S.; MacMillan, D. W. C. *Nature.* **2016**, *536*, 322–325.

125. Corcé, V.; Chamoreau, L. M.; Derat, E.; Goddard, J. P.; Ollivier, C.; Fensterbank, L. *Angew. Chem. Int. Ed.* **2015**, *54*, 11414–11418.

126. Jouffroy, M.; Primer, D. N.; Molander, G. A. *J. Am. Chem. Soc.* **2016**, *138*, 475–478.

127. Patel, N. R.; Kelly, C. B.; Jouffroy, M.; Molander, G. A. *Org. Lett.* **2016**, *18*, 764–767.

128. Lin, K.; Wiles, R. J.; Kelly, C. B.; Davies, G. H. M.; Molander, G. A. *ACS. Catal.* **2017**, *7*, 5129–5133.

129. Jouffroy, M.; Kelly, C. B.; Molander, G. A. *Org. Lett.* **2016**, *18*, 876–879.

130. Vara, B. A.; Li, X.; Berritt, S.; Walters, C. R.; Petersson, E. J.; Molander, G. A. *Chem. Sci.* **2018**, *9*, 336–344.

131. Lévêque, C.; Corcé, V.; Chenneberg, L.; Ollivier, C.; Fensterbank, L. *Eur. J. Org. Chem.* **2017**, 2118–2121.

132. Knauber, T.; Chandrasekaran, R.; Tucker, J. W.; Chen, J. M.; Reese, M.; Rankic, D. A.; Sach, N.; Helal, C. *Org. Lett.* **2017**, *19*, 6566–6569.

133. Liu, N. W.; Liang, S.; Margraf, N.; Shaaban, S.; Luciano, V.; Drost, M.; Manolikakes, G. *Eur. J. Org. Chem.* **2018**, 1208–1210.

134. Liu, N. W.; Hofman, K.; Herbert, A.; Manolikakes, G. *Org. Lett.* **2018**, *20*, 760–763.

135. Yue, H.; Zhu, C.; Rueping, M. *Angew. Chem. Int. Ed.* **2018**, *57*, 1371–1375.

136. Nawrat, C. C.; Jamison, C. R.; Slutskyy, Y.; D. MacMillan, W. C.; Overman, L. E. *J. Am. Chem. Soc.* **2015**, *137*, 11270–11273.
137. Oderinde, M. S.; Frenette, M.; Robbins, D. W.; Aquila, B.; Johannes, J. W. *J. Am. Chem. Soc.* **2016**, *138*, 1760–1763.
138. Shields, B. J.; Doyle, A. G. *J. Am. Chem. Soc.* **2016**, *138*, 12719–12722.
139. Nielsen, M. K.; Shields, B. J.; Liu, J.; Williams, M. J.; Zacuto, M. J.; Doyle, A. G. *Angew. Chem. Int. Ed.* **2017**, *129*, 7297–7300.
140. Twilton, J.; Christensen, M.; DiRocco, D. A.; Ruck, R. T.; Davies, I. W.; MacMillan, D. W. C *Angew. Chem. Int. Ed.* **2018**, *57*, 1–6.

4 Brønsted Acid as Efficient Catalyst for Synthesis of Biologically Active Natural Products

Guillaume Levitre and Géraldine Masson*

Institut de Chimie des Substances Naturelles, CNRS UPR 2301,
Université Paris-Sud, Université Paris-Saclay, 1, av. de la Terrasse,
91198 Gif-sur-Yvette Cedex, France
*Geraldine.masson@cnrs.fr

Table of Contents

List of Abbreviations

Å	Angström
Ac	acctyl
APTS	*para*-toluenesulfonic acid
Ar	aryl
BINOL	1,1′-binaphtalene-2,2′-diol
Bn	benzyl
Boc	*tert*-butyloxycarbonyl
bpy	bipyridine
Bu	butyl
°C	degreecelsius
d	days
DAR	Diels-Alder reaction
dr	diastereomieric ration
ee	enantiomeric excess
eq	equivalent
Et	ethyl

EWG	electron withdrawing group
Fmoc	fluoromethoxycarbonyl
h	hour
HEH	Hantzsch ester hydride
HFIP	1,1,1,3,3,3-hexafluoroisopropanol
HFIPA	1,1,1,3,3,3-hexafluoroisopropyl acrylate
*i*Bu	*iso*-butyl
β-ICD	*β*-isocupreidine
*i*Pr	*iso*-propyl
MBH	Morita-Baylis-Hilman
Me	methyl
mol%	mole per cent
MOM	methoxymethyl
Ms	Mesyl
MS	molecular sieve
MTBE	methyl *tert*-butyl ether
NBSA	*ortho*-nitrobenzenesulfonic acid
Nu	nucleophile
Ph	phenyl
Pin	pinacol
PMB	*para*-methoxybenzyle
Red-Al	sodium bis(2-methoxyethoxy)aluminium hydride
rt	room temperature
TBS	*tert*-butylsilyl
*t*Bu	*tert*-butyl
Tf	trifluoromethanesulfonyl
TFA	trifluoroacetic acid
TFAA	trifluoroacetic anhydride
THF	tetrahydrofurane
TMG	1,1,3,3-tatramethylguanidine
TMS	trimethylsilyl

I. Introduction

Biological properties of many drugs can be explained by the affinity of these substances for a specific macromolecular biological target such as a protein or a nucleic acid. The binding sites of such biological targets mostly form chiral environments and the configuration of a corresponding chiral ligand constitutes a crucial recognition parameter between the two partners. Consequently, such stereochimical considerations can strongly modulate the biological activity of a drug by eroding the dose-response relationship, cancelling affinity of the ligand for its target or causing antagonist effects.[1] For those reasons, considering achiral compounds as drug candidates lowers chances of working with an optimal structure from the biological activity standpoint. Also, several examples have demonstrated the risk of using a racemic mixture as therapeutic agent. Such approaches can also be a costly and wasteful enterprise if an *a posteriori* resolution process has to discard half of the final target molecule.[2] The development of synthetic operations which directly produce the highest proportion of a desired enantiomer has become a major contribution of organic chemistry to the pharmaceutical field. To reach this goal, enantioselective catalysis proved to be one of the most efficient and atom-economical approach and was extensively exploited to perform asymmetric transformations.[3] Alternatively to organometallic catalysis, the development of organocatalysis has exploded during the last decade and naturally impacted the field of total synthesis.[4] Interestingly, chiral organocatalysts offers various modes of activation have been involved in the preparation of numerous bioactive compounds. Recently, non-covalent organocatalysis for electrophilic activation has known a tremendous interest due to its efficiency in the asymmetric formation of C–C and C–heteroatom bonds.

This approach typically relies on Brønsted acid catalysts that increase the reactivity of electrophilic species by lowering their electronic density.[5] Many privileged hydrogen bond donor scaffolds capable of forming single or double hydrogen bonds with electrophilic substrates have been developed as catalysts over past years.[6] The use of such transformations as key steps in the synthesis of bioactive compounds is highlighted in the present report.[7,8] However, asymmetric syntheses of natural and pharmaceutical products mediated by covalent organic catalysts[3,4] and phosphoric acid[9] have been reviewed recently and are not covered in this chapter. Here, we will attempt to provide an overview of enantioselective total or formal syntheses relying on hydrogen bond organocatalysis. Notably, a focus will be put onto activation by catalyst such as bifunctional thioureas, Cinchona alkaloids, bis-imides, and BINOL derivatives that offer an exceptional entry to the pharmaceutically valuable compounds. In order to demonstrate the versatility of such processes and the diversity of the accessible motifs, this chapter will be organized according to target scaffolds and reaction types.

II. Enantioselective Organocatalytic Michael Additions[10]

The catalytic asymmetric conjugate addition of carbon nucleophiles to electron-deficient olefins is highly attractive transformation to design key steps during the retrosynthetic planning in total synthesis. In the area of bifunctional organocatalysis, a large number of asymmetric conjugate addition reactions have been developed to date. By acting all at once as Brønsted acids and Brønsted bases they have already demonstrated their unique efficiency in a broad range of enantioselective conjugate additions (Figure 1).

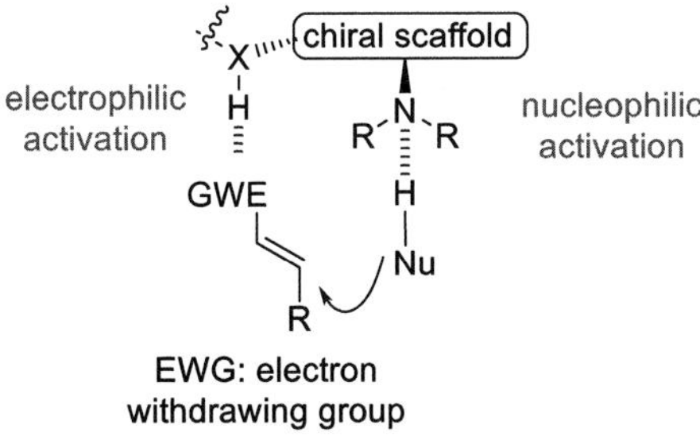

Figure 1. General scheme for the bifunctional activation by chiral bifunctional catalysts in enantioselective Michael additions.

A. Michael Addition to Nitro-Olefins[11]

Nitro-olefins are valuable building blocks in total synthesis as Michael acceptors in C–C bond formation. Additionally, the nitro functionality is easily reduced to amine and can establish hydrogen bonding with appropriate Brønsted acid catalysts with a view to performing enantioselective transformations. In this framework, bifunctional chiral organocatalysts incorporating a thiourea and a tertiary amine moiety, proved to be particularly efficient (Figure 2).[12] Indeed, combined activation of the

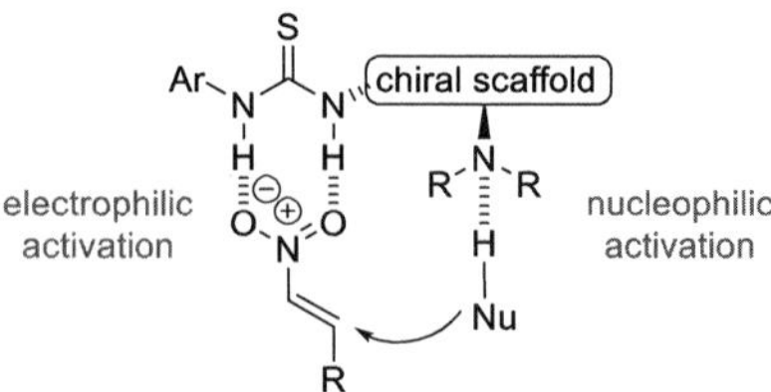

Figure 2. General scheme for the activation of nitro-olefins by chiral bifunctional thiourea-based catalysts.

electrophilic and nucleophilic partners results in high degrees of enantio- and diastereoselectivities.

The enantioselective addition of oxindoles **2** to nitro-olefins **1** is a powerful application of organocatalysis by Brønsted acids and appears as an efficient method for the direct construction of all-carbon quaternary stereogenic centers. Barbas III and co-workers developed a thiourea-catalyzed process permitting the formation of oxindoles **4** in a single step triggered by catalyst **3** (Scheme 1). The two adjacent stereogenic centers, including the quaternary C3 position, were generated in a highly enantio- and diastereoselective fashion. The method offers a privileged access to the hexahydropyrroloindole core encountered in numerous alkaloid derivatives. Hence, this reaction was used as a key step to achieve the formal synthesis of the calabar alkaloid (+)-physostigmine, an Alzheimer's disease therapeutic agent.[13] Based on the same method, catalyst **3** was employed to prepare compound **4a**, isolated in 65% yield and with 96% ee after a single recrystallisation step. Further post-transformations, including the reduction of the nitro substituent to afford amine **6**, ended with the isolation of tricylic (+)-esermethole **7**. As demonstrated by Overman[14] and Brossi,[15] this molecule is an advanced intermediate in the synthesis of of (+)-physostigmine **8**. Finally, the strategy developed by Barbas III constitutes an attractive alternative to the previous syntheses of this target molecule.[16–17]

In 2015, Nagasawa and co-workers[18] reported an enantioselective Michael addition of a phenol **9** with a nitro-olefin **10** in the presence of guanidine/bisthiourea organocatalyst **11** (Scheme 2). An inseparable mixture of diastereomers **12** was isolated at this step. After triflation of the phenol unit, the Michael compound **13** was isolated in good yield with

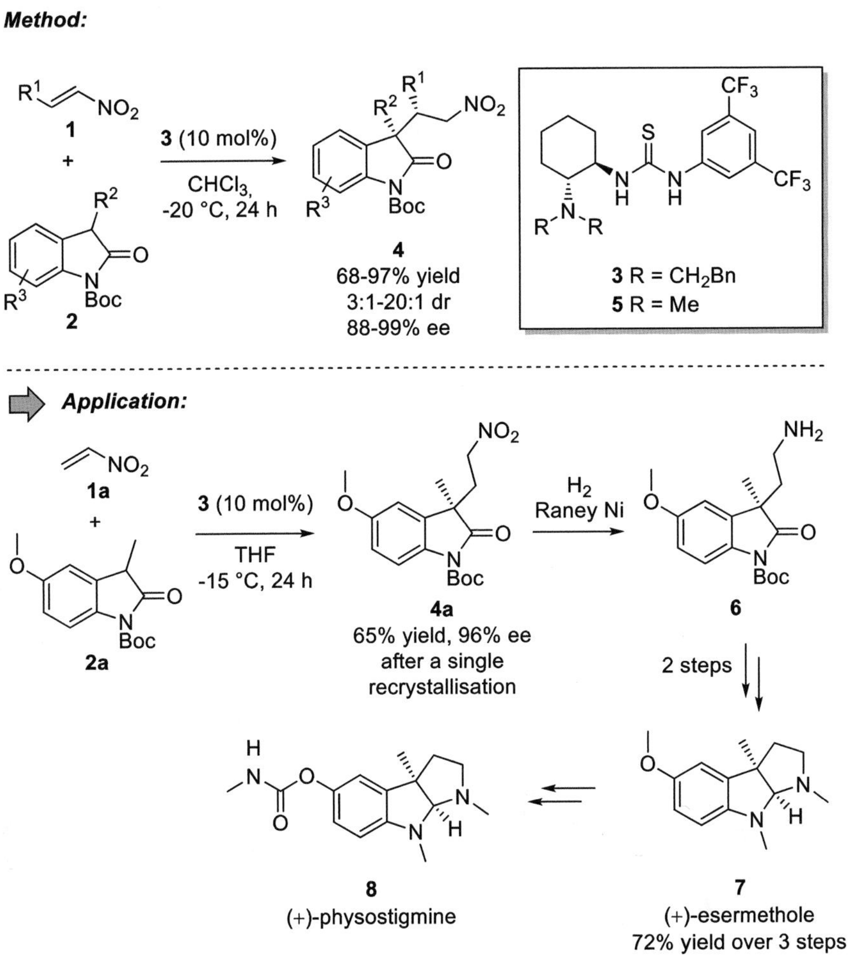

Method:

Application:

Scheme 1.   Synthesis of (+)-physostigmine.

93% ee. The authors proposed a possible transition state to explain the high enantioselectivity, in which the guanidine moiety of catalyst activates the phenol as a nucleophile, and at the same time the thiourea moiety activates the nitro-olefin. The synthetic value of this method was further demonstrated by its involvement in the total synthesis of amaryllidaceae alkaloid (+)-*trans*-dihydrolycoricidine **15**. Several transformations such as olefin oxidative cleavage, intramolecular Henry reaction, CO insertion

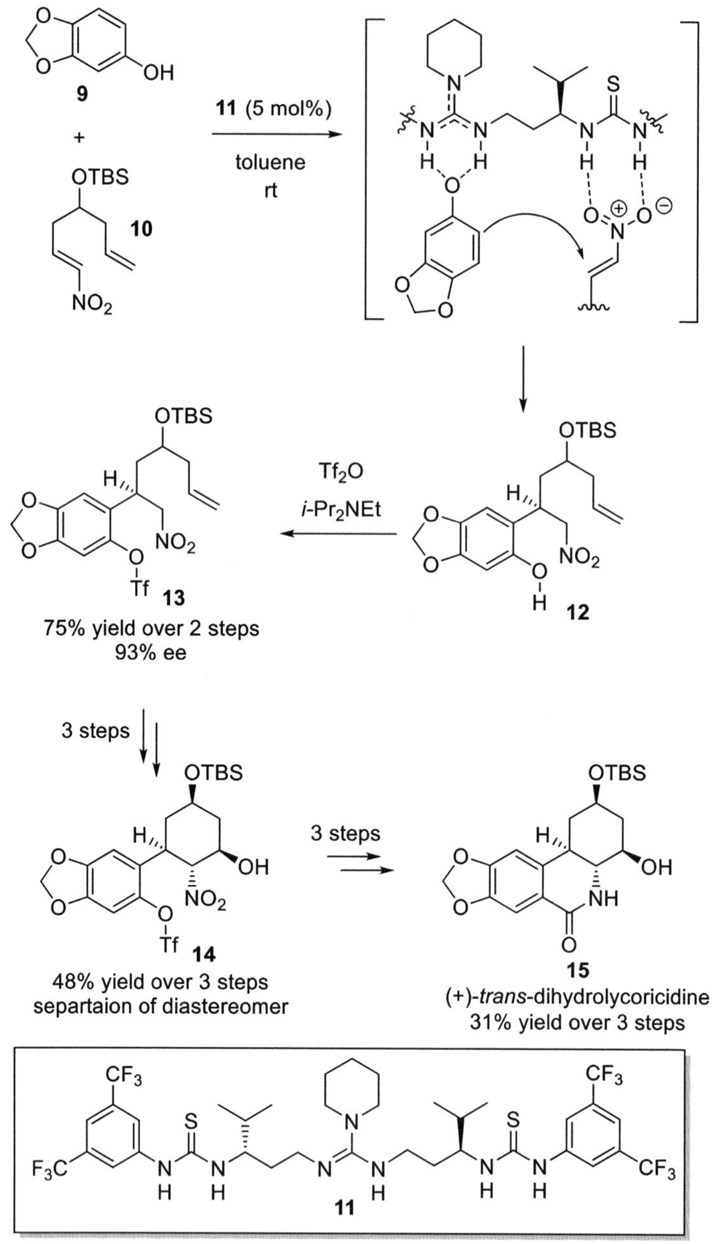

Scheme 2.   Synthesis of (+)-*trans*-dihydrolycoricidine.

reaction and stereoselective hydroxylation were required to reach tricyclic **14**. Targeted product **15** was then reached after three steps.

B. Michael Addition to Acrylates[19]

In 2011, Fan and co-workers reported[20] the total synthesis of three naturals products relying on an enantioselective Michael of α-aryl-α-cyanoketones **16** and 4-iodophenyl acrylate **17**. Moderate level of enantioselectivity was observed using chiral bifunctional thiourea catalyst **18**, however the enantiopure compounds **19** was isolated after recrystallization (Scheme 3). Moreover, the Michael product **21** with the opposite configuration were prepared by using Takemoto catalyst **20**.[21] Interestingly, the reaction affords quaternary stereocenters bearing substituents that were exploited in the synthesis of three natural products like lycoramine **22**, galanthamine **23** and lunarine **24**.

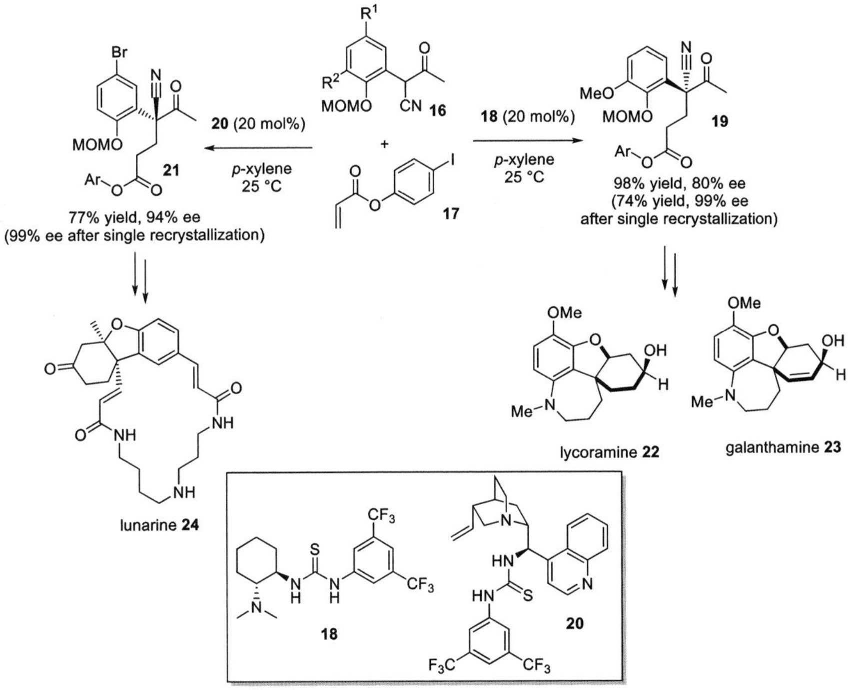

Scheme 3.   Synthesis of lycoramine, galanthamine and lunarine.

C. Addition to Vinyl Phenylselenones

Selenones are important motifs in organic synthesis, because the selenonyl group is a good leaving group, offering an opportunity to undergo further derivatization for the synthesis of natural product. In 2009, Marini and co-workers[22] reported the first enantioselective organocatalytic addition of α-substituted cyanoacetates to vinyl selenones. Two years laters, Liu and coworkers[23] developed an enantioselective Michael addition of 2-oxindoles **25** to α,β-unsaturated selenones **26** catalyzed by a Cinchona alkaloid-based thiourea **20** in room-temperature ionic liquids (Scheme 4). High to excellent yields and enantioselectivities were achieved. The suggested transition state **28** displays a network of hydrogen bonds between the reactions partners initiated by the bifunctional catalyst. This strategy for the preparation of bioactive molecules was highlighted by the synthesis of (–)-physostigmine **29**. Indeed, the target was obtained in three steps from Michael adduct in a 62% overall yield.

Recently, the group of Zhu[24] published a Cinchona-catalyzed method for the preparation of optically active isocyano-bonding quaternary carbon. The reported procedure involves a Michael addition of a

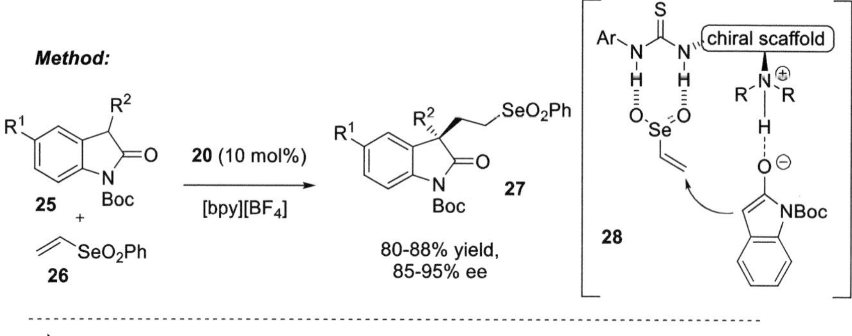

Scheme 4. Synthesis of physostigmine.

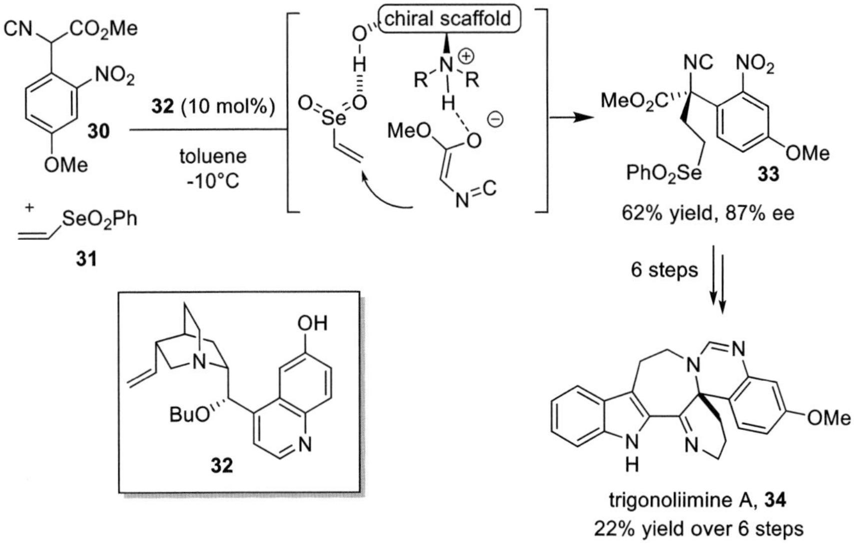

Scheme 5.   Synthesis of trigonoliimine A.

2-isocyanoarylacetate **30** to vinyl phenylselenone **31** (Scheme 5). The catalyst having a free hydroxy group on the C-6′ position was found to be the best in terms of reactivity and enantioselectivity. Once again, the bifunctional aspect of catalyst was invoked to explain the degree of stereocontrol. This elegant strategy was highlighted by the synthesis of trigonoliimine A **34**. Indeed, the target was obtained in six steps from **33** in a 22% overall yield.

D. Double Michael Addition Cascade

Brønsted acids-catalyzed cascades involving nitro-alkenes have been extensively used to quickly increase molecular complexity starting from simple substrates. Notably, double Michael addition sequences[25] have been used to afford, after a unique step, elaborate (poly)cyclic structures displaying multiple stereogenic centers in a one-pot fashion. These transformations rely on nucleophilic partners **35** including a Michael acceptor (Scheme 6). Therefore, the initial Michael addition onto activated nitro-olefin **1** is followed by a second intramolecular conjugate addition. In this context, Takemoto and co-workers reported a chiral thiourea-catalyzed process involving γ,δ-unsaturated β-ketoesters **36** and nitrostyrene **1b** as substrates (Scheme 7).[26]

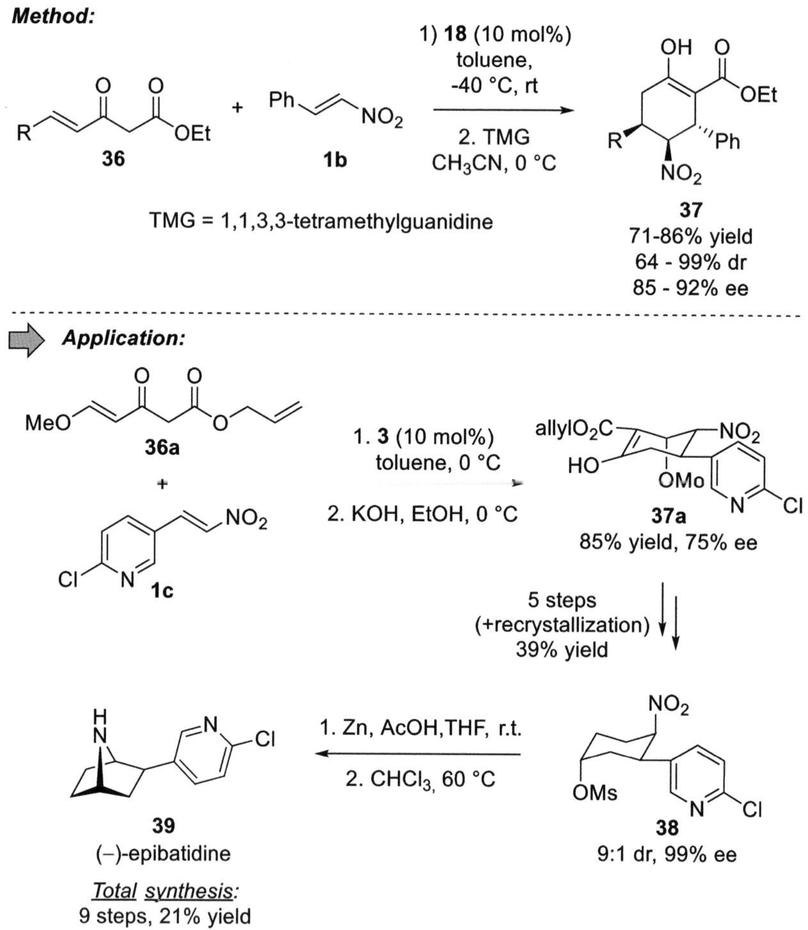

Scheme 6.   General scheme for the Brønsted acid catalyzed double Michael addition of nitro-olefins.

Scheme 7.   Synthesis of (−)-epibatidine.

In this case, the cyclization step requires the subsequent addition of a strong base to proceed. Desired cyclic intermediates **37** were then obtained with good enantio- and diastereoselectivities, since the 3,4-*trans*-4,5-*cis* diastereomers were isolated up to 92% ee. This work constitutes the first description of a nitroalkenes-triggered reaction supporting the formation of three contiguous stereogenic centers. The resulting optically active six-membered rings can serve as chiral scaffolds in the construction of supplementary cyclic structures. Hence, one of these motifs was exploited to achieve the synthesis of (−)-epibatine, a toxic alkaloid isolated from frog skin with a capacity to bind to nicotinic and muscarinic acetylcoline receptors. Slightly modified reaction conditions were involved in the preparation of intermediate **37a**. Indeed, a solution of KOH in EtOH was used to ensure that cyclization occurred without epimerization of the NO_2-substituted stereocenter. Compound **37a** was then isolated as a single diastereomer in 85% yield but with a moderate enantiopurity, while the latter could be increased to 99% by recrystallization of an advanced intermediate of the synthesis. Product **37a** was then converted into molecule **38** in five steps and with 39% overall yield. Interestingly, the presence of the nitro group was ingeniously exploited in the last step as an amine precursor to reach the bridged core of (−)-epibatidine **39**.

As illustrated by the structures of (+)-physostigmine or (−)-epibatidine, alkaloids constitute a highly heterogeneous family of bioactive molecules. Their structural affiliation only relies on the presence of a basic nitrogen atom and, in most cases, a heterocyclic scaffold. However, this loose description remarkably fits with the products recovered after a synthetic sequence including an organocatalyzed double Michael addition of nitro olefins and the reduction of the nitro substituent. Thus, very different alkaloids can be prepared using similar catalysts in similar key transformations by simply varying the nature of the substrates. As a matter of fact, Xu proposed a thiourea-catalyzed double Michael addition between malonate **40** and nitroalkene **1d** to install the C ring of tetracyclic α-lycorane **42** (Scheme 8).[27] In this case, no additional base was required and the desired the six-membered ring **41** was isolated as a single diastereomer (12:1 dr) in 95% yield and 90% ee. As a masked imine, the nitro group had a central

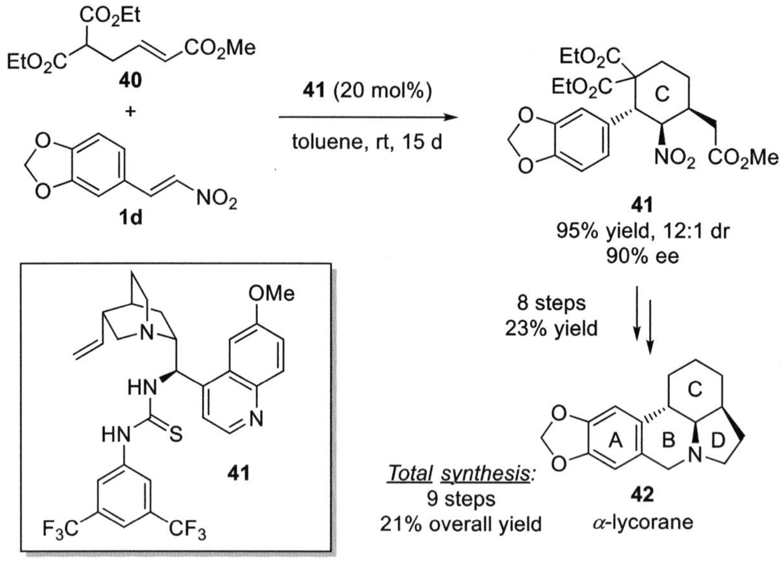

Scheme 8. Synthesis of α-lycorane.

role in the subsequent formation or D and B rings. Impressively, this retrosynthetic analysis allowed the preparation of α-lycorane in nine steps and 21% overall yield.

E. Oxa-Michael Addition

Oxa-Michael additions, in which the nucleophile is an oxygen atom, are a good entry to prepare β-hydroxy ketones, which are found in many natural products. In 2008, the group of Falk described an enantioselective preparation of such derivate starting from α,β-enone and using chiral bifunctional thiourea as catalyst (Scheme 9).[28] In the presence of phenylboronic acid, γ-hydroxy enone **43** formed the corresponding hemi boronate **44** which undergo a hydroxyl transfer through the help of the chiral catalyst **41**. This hydroxyl transfer was believed to proceed *via* the push/pull nature of the catalyst,[29] coordinating the carbonyl of the enone (thiourea moety, pull effect) and exibiting its electrophily. Then, the quinuclidine moiety (tertiary amine, pull effect) coordinates the boron center, increasing the nucleophily of the hydroxyl group. Finally, oxidative cleavage of the boron-oxygen bonds liberated the diol **45a** in both excellent 90% overall yield and 91% ee. Then, the primary alcohol was selectively acetylated to give the natural product (R,12Z,15Z)-2-hydroxy-4-oxohenicosa-12,15-dienyl acetate **46**. In addition, the authors reported the synthesis of (+)-(S)-Streptenol **45b**, using the same catalyst **41**. Electron rich phenyl boronic acid such as 3,4,5-trimethoxyphenyl boronic acid was required to proceed at an acceptable rate and enantioselectivity. Finally, after oxidative cleavage, the natural product **45b** was obtained with a good overall 75% yield and a good 88% ee.

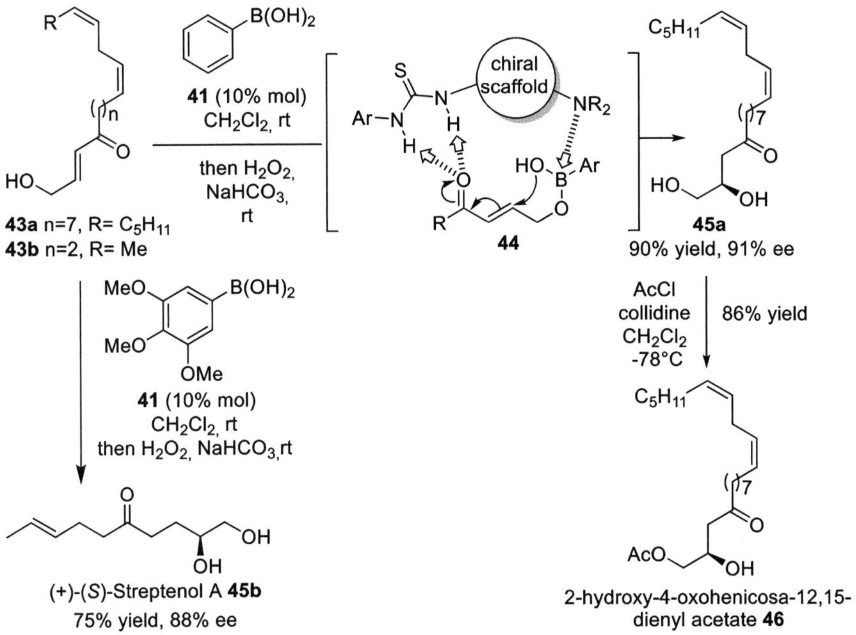

Scheme 9.   Enantioselective synthesis of β- and γ-hydroxy ketone by oxa-Michael addition.

F. Tandem Conjugate Addition–Protonation

In 2007, Deng and co-workers developed an enantioselective entry to 1,3-disubstitued carbon framework bearing a tertiary and quaternary centers, using a cinchona based bifunctional catalyst **49** (Scheme 10).[30] The reaction between 2-chloroacrylonitrile **48** and α-cyanothioester **47** proceeds in two sequential stereo-induction steps. The first step involves asymmetric conjugate addition *via* a well-organized transition state **50** between catalyst and substrates. In the second step, the nucleophilic addition of the

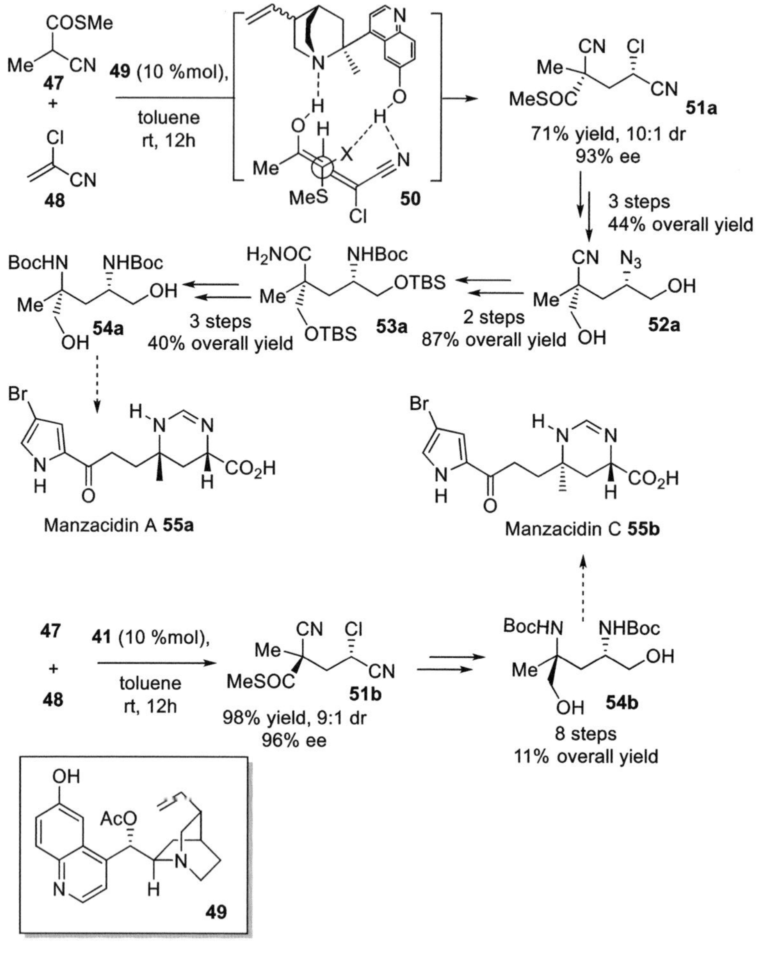

Scheme 10.   Formal synthesis of Manzacidin A and C.

enol on the 2-chloroacrylonitrile followed by protonation give the product **51a** in 71% yield and 93% ee. With this key intermediate in hand, the author set out the formal synthesis of manzacidin A **55a**. The nitrogen was introduced by nucleophilic substitution of chloride to azide. Then, the methanolysis of the less-hindered cyanide group followed by reduction lead to the diol **52a**. Silylation of the diol, hydrolysis of the remaining cyanide, hydrogenation and Boc-protection gave **53a**. Then, the Hoffman rearrangement and deprotection of silyl group led the common intermediate **54a** of synthesis manzacidin A **55a** developed by Ohfune *et al.*

Shortly after the same author reported a method which afforded the complementary sense of diastereoselectivity in this enantioselective tandem asymmetric Michael addition–protonation process by using the cinchona catalyst bearing a thiourea moiety **41** (Scheme 10).[29] They could prepare the Michael adduct **51b**, in 98% yield and 96% ee. According to the synthetic plan described above, the formal synthesis of manzacidin C **55b** was achieved.

G. Michael Addition Aromatization Cascade

C_2-symmetric bis-triflamides were also employed as synthetic tools in the construction of biologically actives molecules. Seminal studies by Jørgensen outlined the ability of bis-sulfonamide **57** to promote enantioselective additions of indoles to nitro-olefins (Scheme 11).[31] Thereafter, the

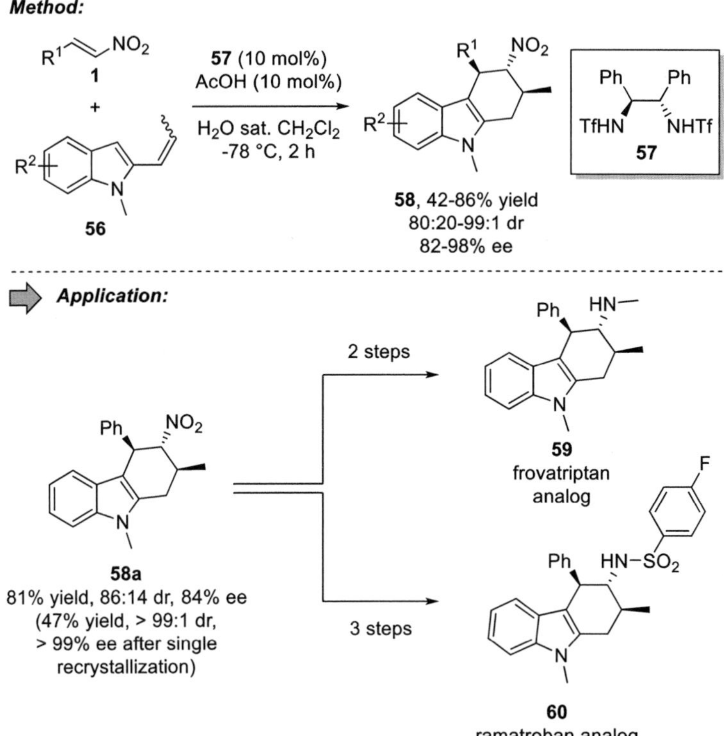

Scheme 11.   Synthesis of frovatriptan and ramatroban analogs.

same catalyst was successfully used, in combination with 10 mol% of AcOH, to perform elaborate double Michael addition-aromatization cascades.[32] This reaction leads to the formation of tetrahydrocarbazoles **58** with *cis, trans* stereotriads. Among the products prepared this way, **58a** was obtained in 81% yield, 86:14 dr and 84% ee. After recrystallization, the same compound was isolated in 47%, > 99:1 dr and >99% ee and was converted in frovatriptan **59** and ramatroban analogs **60** in two and three steps respectively.

III. C–C Bond Formation by Enantioselective Nucleophilic Additions onto Activated Imines[33]

Imines efficiently participate in enantioselective nucleophilic additions to afford nitrogen-substituted stereocenters. Due to their propensity to convert imines into highly electrophilic iminium ions, chiral Brønsted acids represent excellent catalysts for the development of enantioselective reactions (Scheme 12). A broad range of nucleophiles can be used

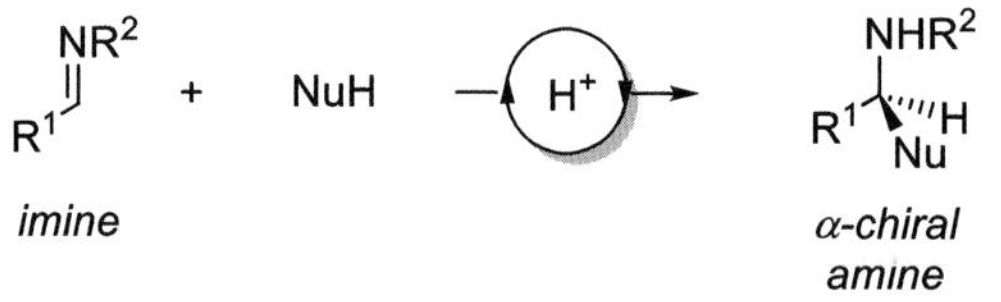

Scheme 12.   General scheme of the Brønsted acid-catalyzed nucleophilic addition onto imines.

in these transformations resulting in the asymmetric formation of new C–C bonds. Hence, Brønsted acid-catalyzed enantioselective additions on imines were extensively considered to choose suitable disconnections when doing the retrosynthetic analysis of α-stereogenic amines.

A. Hantzsch Ester Hydride (HEH) Transfer[34,35]

Direct asymmetric reduction of *N*-alkyl imines by transfer hydrogenation mediated by a Brønsted acid and HEH as hydrogen source has been disclosed by the List group.[36] The *N*-alkyl amines products are important building blocks in the construction of the core of for a plethora of

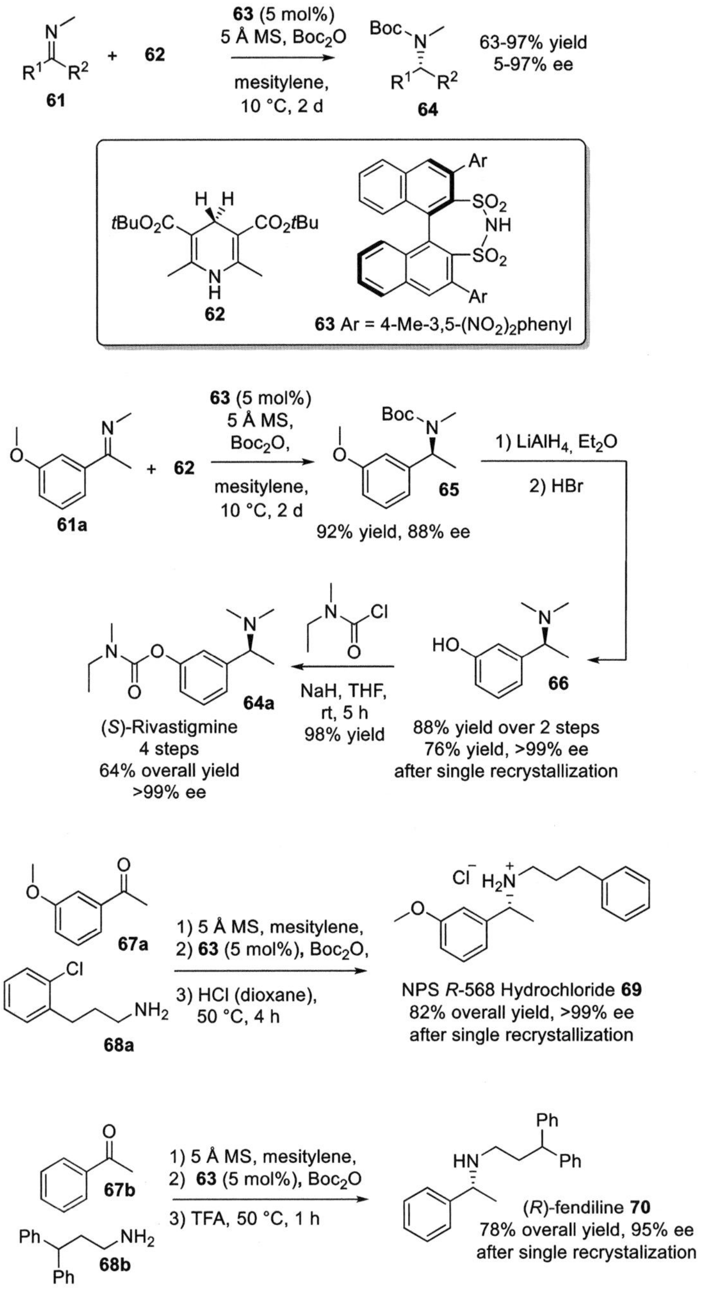

Scheme 13.   Synthesis of (*S*)-rivastigmine, NPS *R*-568. HCl and (*R*)-fendiline.

pharmaceuticals and agrochemicals. The challenge associated with the *N*-alkyl imine substrates consist in the high basicity of the amines products which can inhibit the catalyst with a consequent luck of turnover of the catalytic cycle. Thus the stronger disulfonimide (DSI) **63** has been chosen as catalyst and the reduction proceeds in the presence of Boc_2O, in order to get *in situ* protection of the amines (Scheme 13). The methodology has been successfully applied to the synthesis of three pharmaceuticals. (*S*)-Rivastigmine **64a** is used for the treatment of Alzheimer's disease; the asymmetric reduction of imine **61a** led the *N*-Boc protected amine **65** in 92% yield and 88% ee. Reduction with $LiAlH_4$ delivered the *N,N*-dimethyl amine which was subjected to deprotection of the methoxy group on the aromatic ring to give the corresponding phenol in 76% yield and >99% ee after single recrystallization. Formation of the carbamate with *N*-ethyl-*N*-methylcarbamoyl chloride in the presence of NaH allowed to recover the desired pharmaceutical in only four steps, 64% overall yield and >99% ee. NPS *R*-568 hydrochloride **69** is a calcimimetic compound used for the treatment of hyperparathyroidism, while (*R*)-fendiline **70** is used for the treatment of coronary heart disease. Both are synthesized following a one pot procedure consisting in the *in situ* formation of the imine from the commercially available ketones, followed by enantioselective transfer hydrogenation and Boc protection. NPS *R*-568 hydrochloride **69** and (*R*)-fendiline **70** were achieved after single crystallization in HCl/dioxane (82% overall yield, >99% ee) or TFA (78% overall yield, 95% ee) respectively.

B. Strecker Reaction[37]

The Strecker reaction has attracted considerable attention among synthetic chemists because of the utility of the resulting α-aminonitriles in the synthesis of natural products and biological molecules. Therefore, development of asymmetric, catalytic versions of this Strecker reaction have been the subject of numerous researches. The thiourea catalyst **72** developed by Jacobson is a notable example (Scheme 14). Inspired by his work, Itoh co-workers described enantioselective Strecker reaction for the synthesis of tetrahydroisoquinoline alkaloids. The reaction of cyclic imine **71** in the presence of HCN using

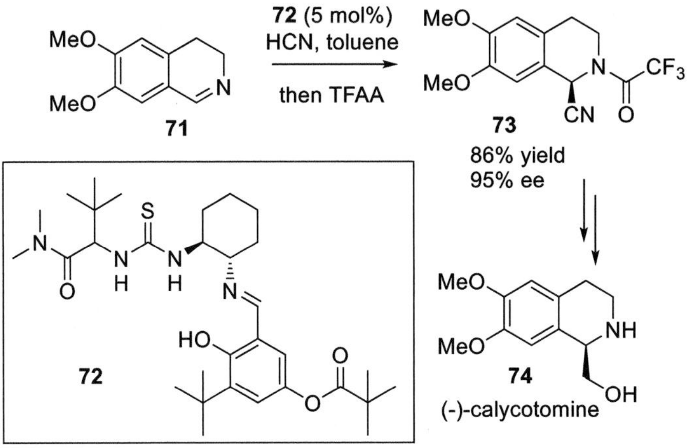

Scheme 14.   Synthesis of (–)-calycotomine.

Jacobsen's thiourea catalyst **72** afforded desired product **73** in 86% yield and with 95% ee. Hydrolysis of the cyano group followed by standard transformations led to (–)-calycotomine **74**.

Fluorine containing group are very attractive in medicinal chemistry, biology and drug discovery due to its unique properties. In the course of the development of new method to prepare enantiopure α-amino acids, Zhou and co-workers developed in 2011,[39] an enantioselective Strecker reaction catalyzed by bifunctionnal thiourea **41** (Scheme 15). Starting to fluorinated protected imine, this methodology delivered fluorinated cyano amine with excellent yield and ee. α-amino acids **77** can be obtained after two steps in 66% overall yield from the cyano precursor **76**. This work not only provided a nice entry to such compounds but also revealed an intriguing reactivity. It has been also

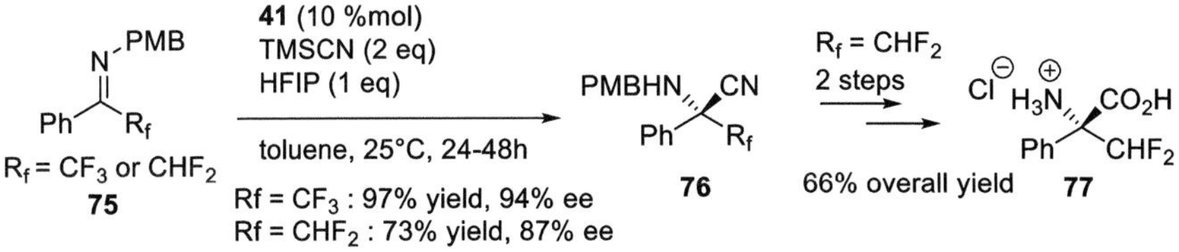

Scheme 15.   Synthesis of fluorinated α-amino acids by enantioselective Streker reaction.

noted that 1 equivalent of HFIP was a good additive to the reaction media, increasing the kinetic of the reaction. An excess of that additive decreased the yield, probably by interfering in the hydrogen bonding network of the catalyst and the imine.

C. Allylation of Imines[38]

Brønsted acid catalyzed allylations of imines result in the stereocontrolled formation of one C–C bond and are particularly interesting for the preparation of non-cyclic chiral amines. A striking example is that of the original BINOL-catalyzed enantioselective allylation of acyl imines set up by Schaus.[39] The method was based on the formation of a chiral allylation agent by combining allyl-diisopropoxyboronate and an appropriate BINOL-derived catalyst **78** (Scheme 16). Resulting

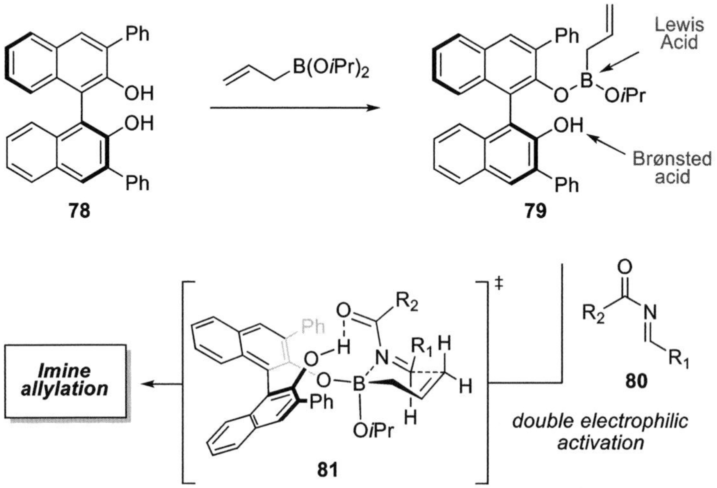

Scheme 16. Proposed transition state for the enantioselective imines allylation catalyzed by BINOL-derived catalysts.

diol-boronate complex **79** is a bifunctional species bearing a Lewis acidic boron center and a Brønsted acidic alcohol, both of them activating acylimine partner **80**. This double interaction provides a highly organized transition state **81** and ensures that the transformation occurs

Method:

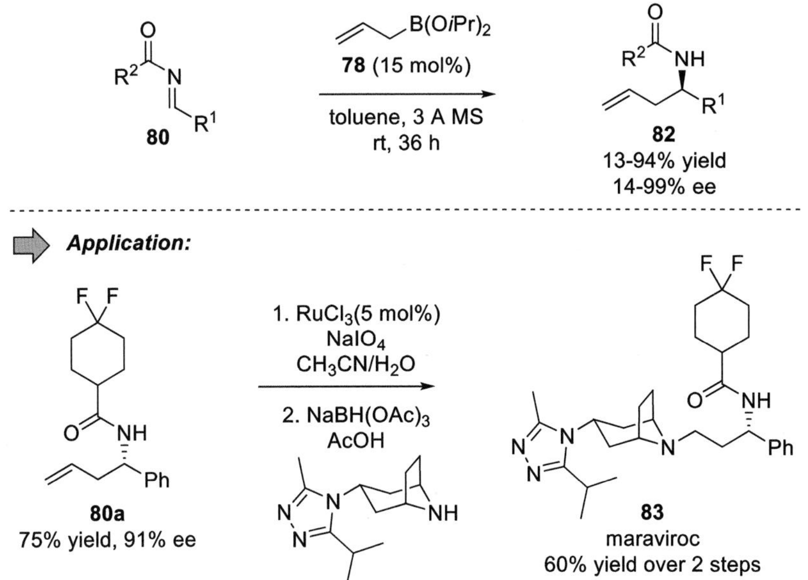

Scheme 17.   Synthesis of maraviroc.

stereoselectively. Homoallylic amines **82** were isolated in poor to excellent yields as well as with variable enantiomeric excesses (Scheme 17). The products form valuable building blocks for the syntheses of numerous natural products. Enantioselective allylation of a well-designed acylimine delivered amine **80a** in 75% yield and 91% ee. From this product, antiviral maraviroc **83** was isolated after only two additional transformations. This enantioselective, protecting group-free synthesis is one-step shorter than the previously reported strategies that introduced chirality by engaging the chiral pool.[40]

In 2013, Hoveyda and coworkers also brought a significant contribution to the field of organocatalyzed allylation of imines and carbonyls derivatives.[41] They reported the use of Brønsted acid catalyst **84** derived from valine able to form an active allylation agent by chelation with pinacolallylborane (Scheme 18). Compared to Schaus' system, boronate complex intermediate **85** displays an additional intramolecular hydrogen bond. The resulting formal positive charge increases Lewis acidity on the

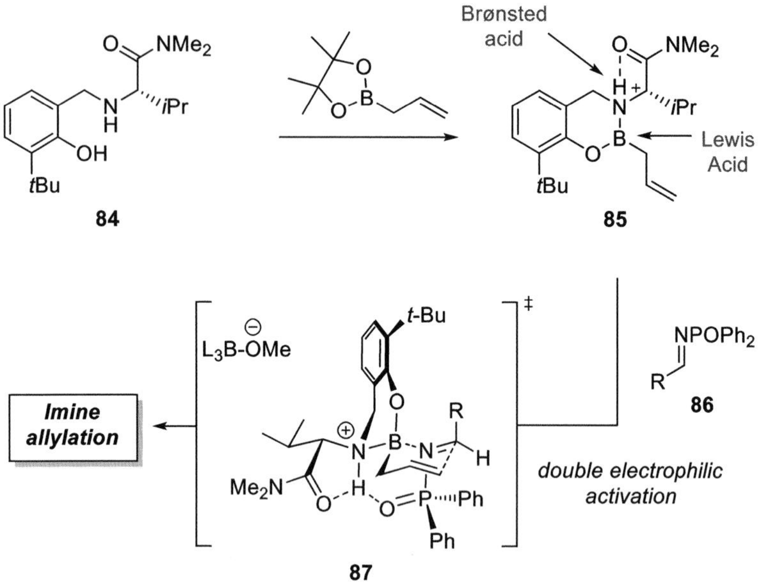

Scheme 18. Proposed transition state for the enantioselective imines allylation catalyzed by a valine-derived catalysts.

boron atom and facilitates coordination of the *N*-phosphinoylimines substrates **86**. These closer interactions between the different partners in transition state **87** result in improved selectivities and shorter reaction times. Indeed, *N*-protected homoallylic amines **88** were delivered in 4–6 h in good yields and with enantiopurities up to 99% ee (Scheme 19). Moreover, the method positively relies on stable organoboron reagents and easily accessible catalysts. The reaction appeared highly versatile and was directly transposed to enantioselective allylation of isatins **89**. Again, the degree of enantiocontrol was excellent and enabled the authors to reach, in few steps, several advanced intermediates in the synthesis of natural products. Optically pure carbinol **90a** was isolated in 94% yield and is a precursor in the synthesis of madindoline A **91** described by Hayashi.[42] For its part, **90b** was already reported as a valuable intermediate to access some representative members **92** of the convolutamydines family.[43]

Method:

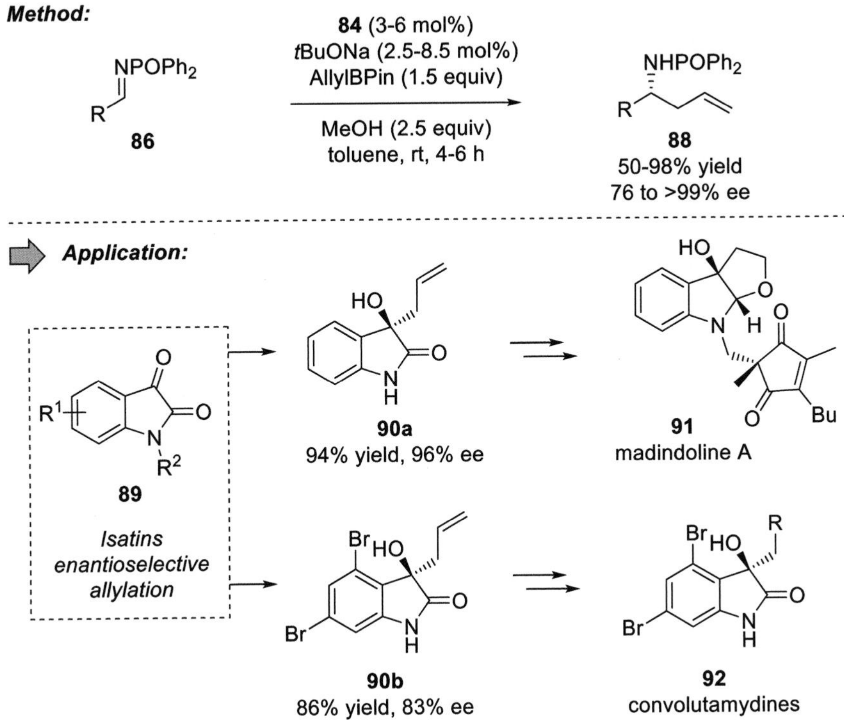

Scheme 19.   Synthesis of madindoline A and convolutamydines.

D. Aza-Sakurai Cyclization

The aza-Sakurai cyclization, which involves the reaction of an iminium ion with an allylsilane, is powerful transformations in the synthesis of indolizidines. In 2016, Jacobsen and coworkers[44] described a thiourea acid-catalyzed aza Sakurai–Hosomi reaction between aminal and allylsilane **93** (Scheme 20). They proposed a transition state model **95** to account for the observed enantioelectivity in which the thiourea catalyst **94** acts as an anion receptor to generate chiral ion pair between *N*-acyliminium and chloride-thiourea as well as an internal Lewis base to activate the allylsilane. Optically enriched indolizidines **96** was isolated in 85% yield and is a precursor in the synthesis of (−)-tashiromine **97**. After oxidative cleavage of the terminal olefin followed by reduction with LiAlH₄ led to natural product in excellent yield.

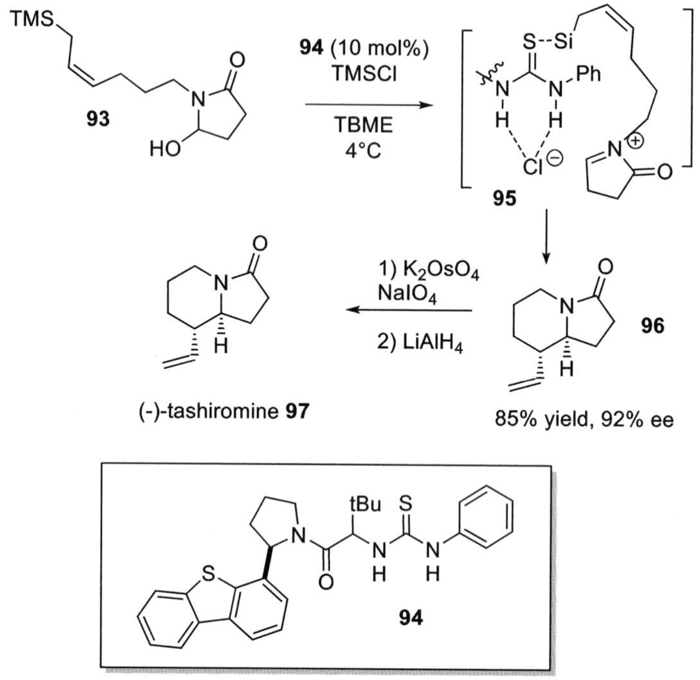

Scheme 20. Synthesis of (–)-tashiromine.

C. Pictet–Spengler Reaction[45]

The Pictet–Spengler cyclization is one of the most potent transformations in the synthesis of alkaloids derivatives, like tetrahydroisoquinolines or tetrahydro-β-carbolines. The reaction proceeds by the formation of an iminium intermediate followed by an intramolecular Friedel-Crafts addition. The overall reaction then results in the creation of two new C–N and C–C bonds. In recent years, the development of enantioselective versions of the Pictet–Spengler reaction was intensively investigated and several Brønsted acid-catalyzed methods were efficiently introduced. In 2004, Jacobsen reported the first asymmetric thiourea catalyzed Pictet-Spengler cyclization between tryptamine derivatives **98** and aldehydes.[46] In this specific case, the imine intermediate was activated as a *N*-acyliminium ion **101** in the presence of acetyl chloride and thiourea catalyst **99** (Scheme 21). The chiral environment required to perform the enantioselective Friedel–Crafts cyclization was provided by the formation of a chiral chloride counteranion. The reaction proceeds with high

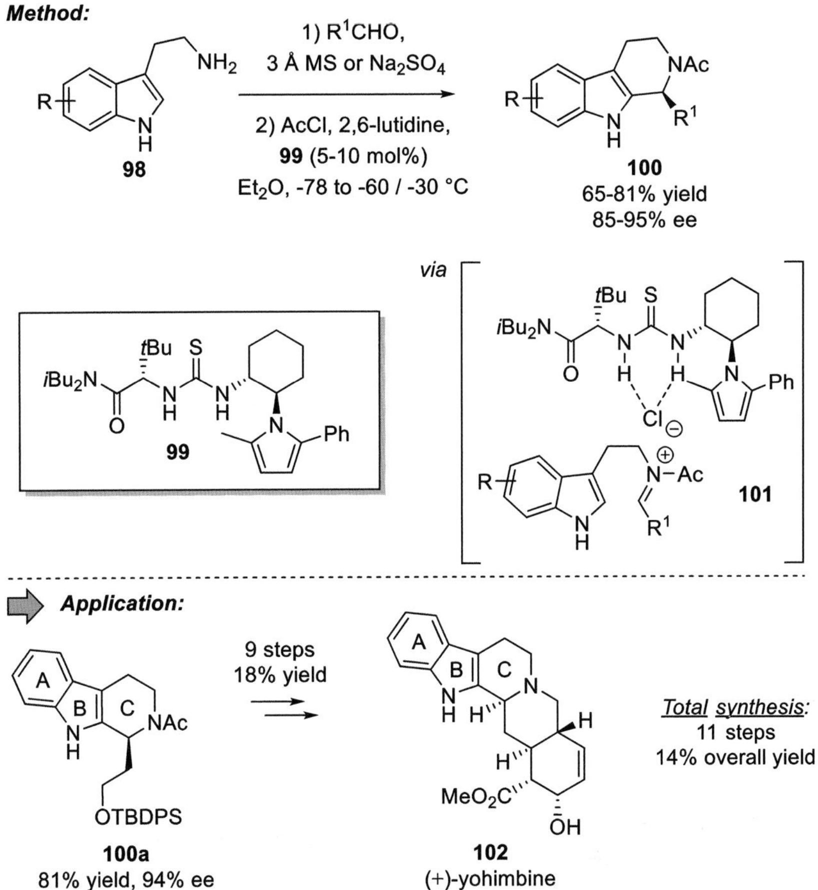

Scheme 21.   Synthesis of (+)-yohimbine.

degrees of enantioselectivity and was rightly used to synthesize (–)-yohimbine **102**.[47] To this end, tryptamine was reacted with 3-((*t*Bu-diphenylsilyl)oxy) propanal in the presence of acylating agent and catalyst **99**. This single acylation/Pictet–Spengler condensation sequence afforded the ABC-core of the target molecule by forming tetrahydro-β-carboline **100a** in 81% yield and with 94% ee. Only nine additional steps were then necessary to complete the synthesis of final product **102**.

Based on the same strategy, the enantioselective Pictet–Spengler cyclization was involved in the asymmetric synthesis of indolizidones and quinolizidones **105** (Scheme 22).[48] In this case, the *N*-acyliminium

Method:

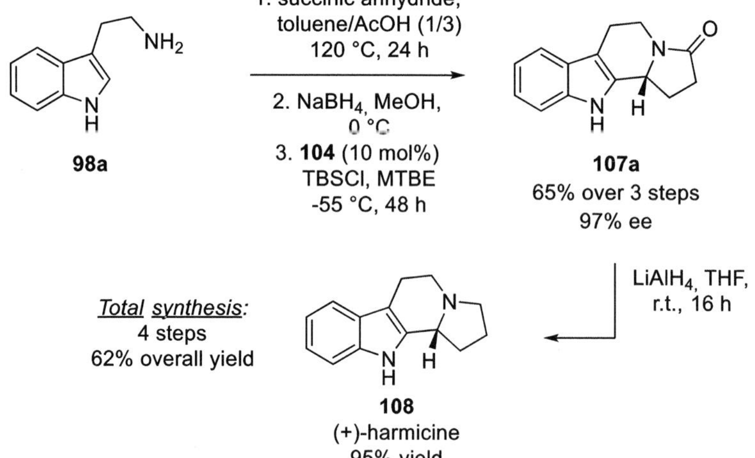

Application:

Scheme 22. Synthesis of (+)-harmicine.

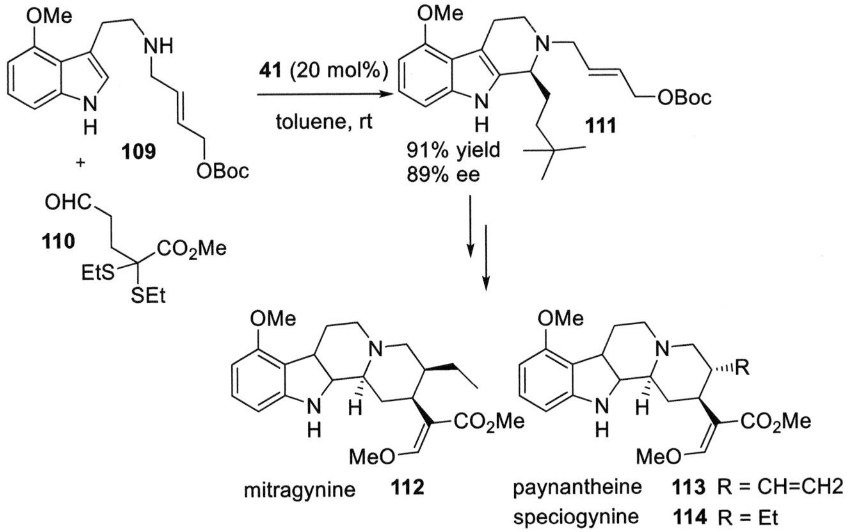

Scheme 23.   Synthesis of mitragynine, paynantheine and speciogynine.

intermediate was generated by dehydration of hydroxylactams **103** in the presence of TBSCl. Chiral ion pair **106** resulting from the association of the chloride anion and thiourea **104** underwent smooth cyclization at low temperature. Building on this method, Jacobsen designed a 4-step synthesis of (+)-harmicine **108** involving intermediate **107a**.

In 2012, van Maarseveen, Hiemstra, and co-workers[49] developed a chiral thiourea-catalyzed Pictet–Spengler reaction between indole **109** and aldehyde **110**. The overall process was efficiently catalyzed by thiourea catalyst **41** derived from a Cinchona alkaloid (10 mol%) and delivered the tetrahydro-β-carboline derivative **111** in 91% yield with high enantioselectivity (89% ee). The synthetic value of this method was further demonstrated with total syntheses of the mitragynine, paynantheine and speciogynine (Scheme 23).

IV. C–C Bond Formation By Enantioselective Nucleophilic Additions onto Carbonyle

Contrary to imines, related few enantioselective additions onto carbonyl groups have been developed. This difference of exemplification can be explained be the difficulty of chiral Brønsted acid catalysts to discriminate

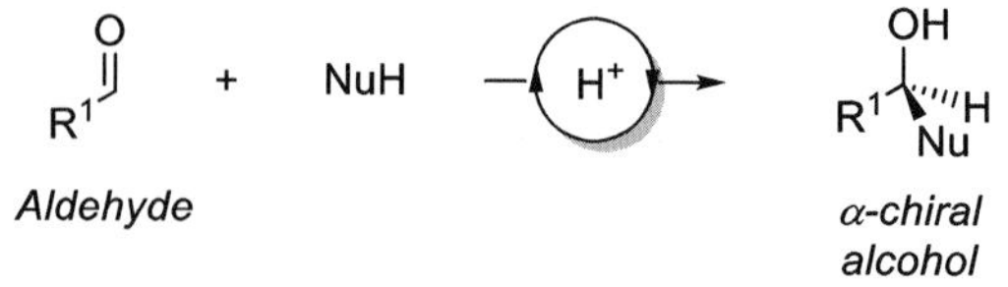

Scheme 24. General scheme of the Brønsted acid-catalyzed nucleophilic addition onto icarbonyles.

the two electronic lone pairs of the oxygen atoms (Scheme 24). However, such transformations remain highly appealing for the asymmetric synthesis of alcohol derivatives and several examples will be reported in the following sections.

A. Henry Reaction[50]

The nitroaldol (Henry) reaction is a valuable tool for the synthesis of β-nitroalcohols and to functionalized intermediates such as nitroalkenes, 1,2-amino alcohols and α-hydroxy carboxylic acids. In 2006, Nagazawa and co-workers used an asymmetric version of the Henry reaction with their guanidinium based bis-thiourea catalyst **117**[51] (Scheme 25) for the total synthesis of (4*S*, 5*R*)-*epi*-Cytoxazone **120**, a type-2 cytokine selective inhibitor.[52] The Henry reaction proceed *via* a controlled *anti* transition state **118**, furnishing syn-nitro alcohol **119** with a good 76% yield and excellent selectivity (95% ee and 90:10 syn/anti ratio). With the new formed nitro alcohol in hand, installation of the cyclic carbamate in a 43% yield over two steps completed the total synthesis, with a 33% overall yield from aldehyde **120**.

In 2011, Wang and coworkers reported the total synthesis of spiro-brassinin **125** (Scheme 26).[53] One of the key step was an enantioselective Henry reaction with simple isatin (**121**) and nitromethane using cupreine as a catalyst affording the desired Henry product in almost quantitative yield with 91% ee. The enantiopure compound **124** was isolated after recrystallization. Once again, the bifunctional aspect of catalyst was invoked to explain the degree of stereocontrol. As depicted in suggested transition state **125**, the ketone undergoes a Brønsted-acid activation while the Brønsted basic nitogen atom of the catalyst activates the nitrometane. Then, the nitro group of **124** was reduced by hydrogenation to give primary amine **126** in 92% yield. From **122**, only three additional steps were required to reach spirobrassinin **127**.

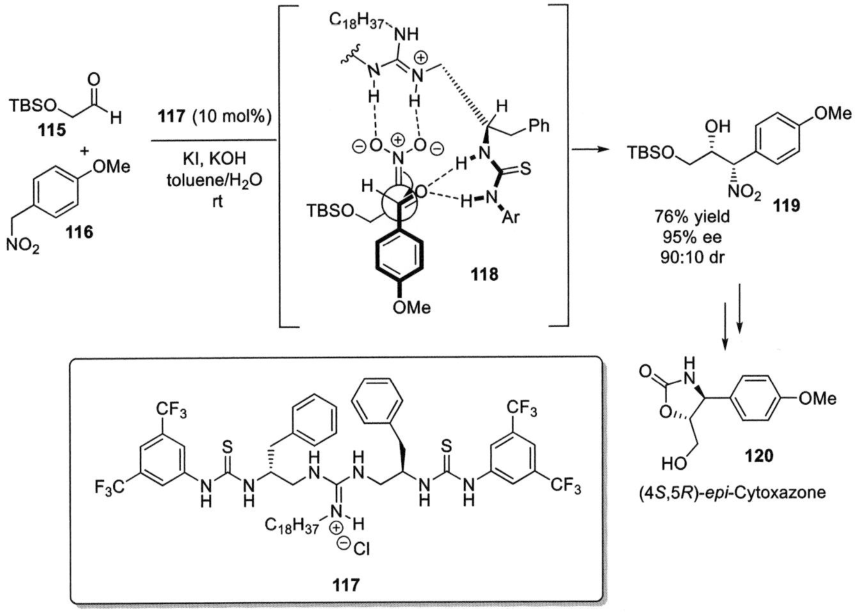

Scheme 25. Enantioselective Henry reaction in the total synthesis of (4*S*, 5*R*)-*epi*-cytoxazone.

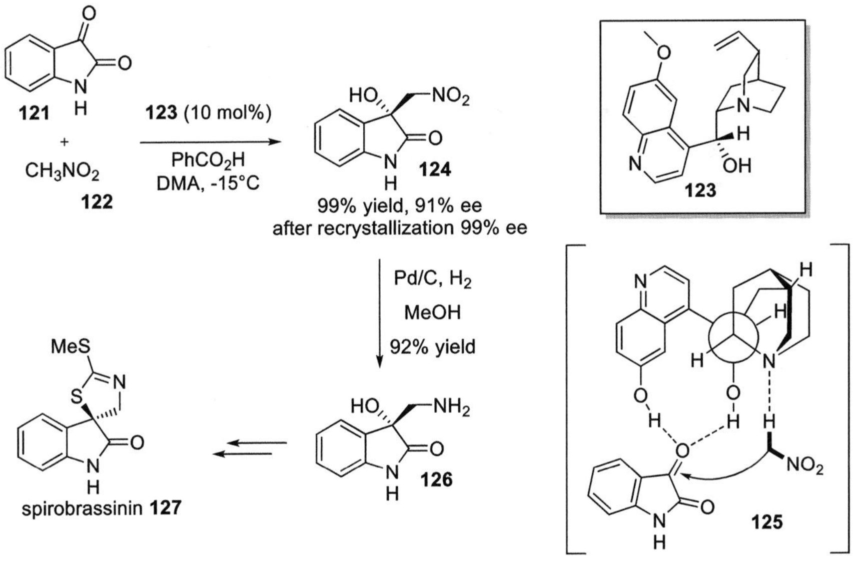

Scheme 26. Enantioselective Henry reaction in the total synthesis of spirobrassinin.

B. Cyanosilylation[54]

Asymmetric addition of TMSCN (TMS = trimethylsilyl) to ketones is the most popular strategy to produce optically active cyanohydrins. Deng and co-workers[55] developed enantioselective cyanosilylation of ketone **128** catalyzed by a modified cinchona alkaloid catalyst **129** providing cyanohydrin (*R*)-**130** in quantitative yield and 92% ee. This intermediate was used for the first enantioselective synthesis of bisorbicillinol **132** (Scheme 27).

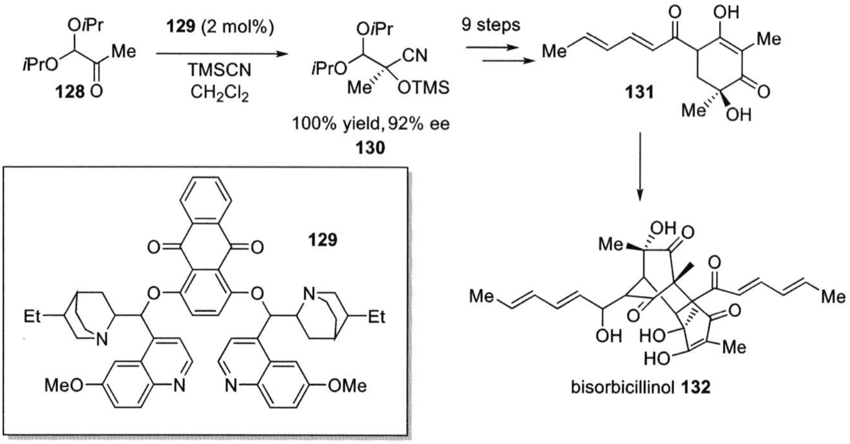

Scheme 27.   Enantioselective synthesis of bisorbicillinol.

C. Morita–Baylis–Hillman Reaction[56]

The Morita-Baylis-Hillman (MBH) reaction has been recognized as one of the most effective method to generate densely functionalized α-methylene-β-hydroxycarbonyl derivatives **135**, which can serve as valuable building blocks for the synthesis of biologically relevant compounds as well as complex natural products. Hatekeyama and coworkers[57] have developed the first highly enantioselective MBR reaction between aldehydes and activated acrylates catalyzed by β-isocupreidine (β-ICD, **134**). The reaction of 1,1,1,3,3,3-hexafluoroisopropyl acrylates (HFIPA) **133** with aldehydes in the presence of a catalytic amount of **134** afforded

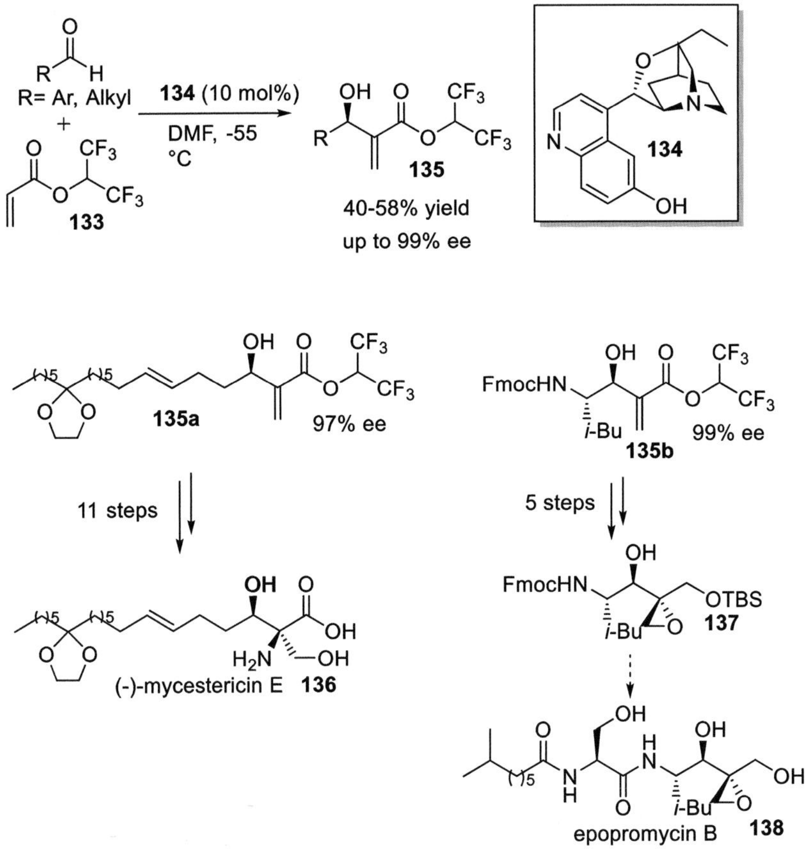

Scheme 28.   Synthesis of mycestericin E (32) and a formal synthesis of epopromucin.

the R allylic alcohols **135** in moderate yields (40–58%), but with excellent enantioselectivities (up to 99% ee, Scheme 28). In 2006, the same authors also reported a highly diastereoselective MBH reaction of chiral N-Boc-α-aminoaldehydes for the synthesis of α-methylene-β-hydroxy-γ-amino acid derivatives **135**.[58] Subsequently, this approach was applied for an enantioselective synthesis of potent immunosuppressant (–)-mycestericin E (**136**) and a formal synthesis of biologically active epopromucin B (**138**, Scheme 28).[59]

V. Diels–Alder Reaction

The Diels-Alder reaction (DAR), is one of the most powerful reactions to build elaborated 6-membered cycle and has been used in many total synthesis of natural products.[60] Many variant has been developed and various asymmetric versions have been developed.

In the course of controlling stereocenters in the DAR, Ricci reported in 2008 an interesting entry to tricyclic indolines with a full control of the stereocenters.[61] Using catalyst **141**, 3-vinylindole **139** and *N*-phenyl maleimide **140** reacted successfully to give the Diels-Alder adduct in an impressive 91% yield and 98% ee. In one step, two carbon–carbon bonds and three stereocenters was created with a total control (Scheme 29). The proposed transition step involves a Brønsted acid activation of the carbonyl of the maleimide (dienophile), lowering

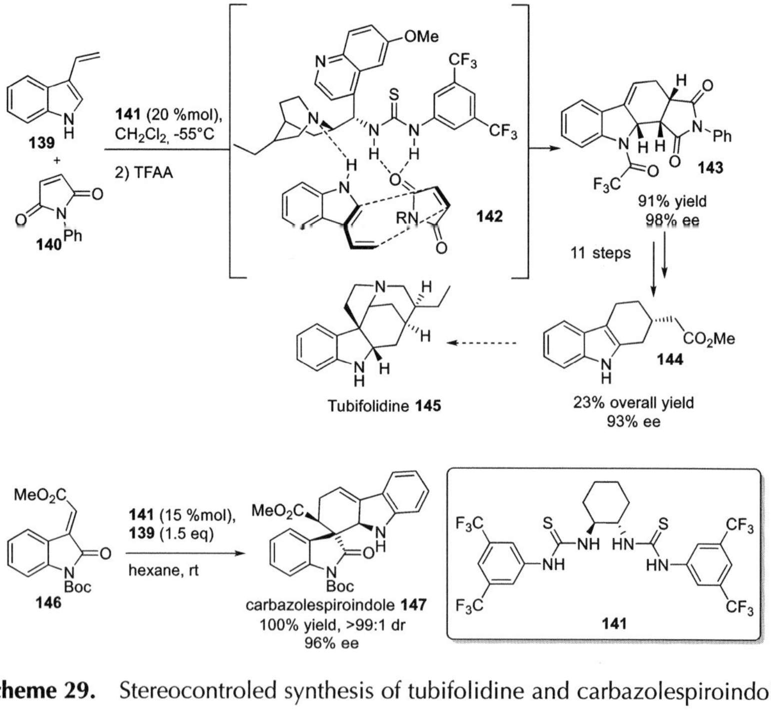

Scheme 29. Stereocontroled synthesis of tubifolidine and carbazolespiroindole.

it LUMO energy. The nitrogen of the quinuclidine ring act as a Brønsted base, activating the 3-vinylindole (diene), increasing it HOMO energy. With the so formed indoline **143** in hand, several chemical manipulation gives the tetrahydrocarbazole **144** in 23% overall yield and with conservation of a good ee (93%). This product is a key intermediate in the synthesis of the natural product Tubifolidine **145** and was used in its total synthesis by Shibasaki *et al.*[62]

In 2013, an efficient entry to carbazolespiroindole was published by Barbas III and co-workers,[63] starting from *N*-Boc protected methyleneindolinone **146** and 3-vinylindole **139**, catalyzed by bis-thiourea **141**. Those substrates possess interesting biological activities and are found in some naturals products,[64] which makes them very attractive for organic chemists.

A. Povarov Reaction[65]

The organocatalyzed strategy to build asymmetric C–C bonds from imines was also efficiently applied to the Povarov reaction. Formally it constitutes an inverse electron-demand Diels–Alder cycloaddition between an electron-poor azadiene and an electron-rich dienophile. The whole process allows the formation of two new C–C bonds and offers a direct and atom-economical route to the versatile tetrahydroquinoline scaffold. With this in mind, Jacobsen and co-workers reported an urea-assisted Povarov reaction between *N*-arylimines **148** and various dienophiles including enamides **150** and enecarbamates **152** (Scheme 30).[66] The optimized reaction conditions were identified as a combination of sulfinamido urea **149** and *o*-nitrobenzenesulfonic acid (NBSA) in a 2:1 ratio. Desired products **151** were formed as a single *exo* diastereomer in good yields and with excellent ee's. When the dienophilic olefin was contained in a cycle, the transformation resulted in complex fused tricyclic moieties **153** with three contiguous stereogenic centers. Interestingly, compound **153a** was reached in one step from **148a** in 76% yield and 95% ee and was described as a key intermediate in Ma's synthesis of martinelline **154**.[67]

Method:

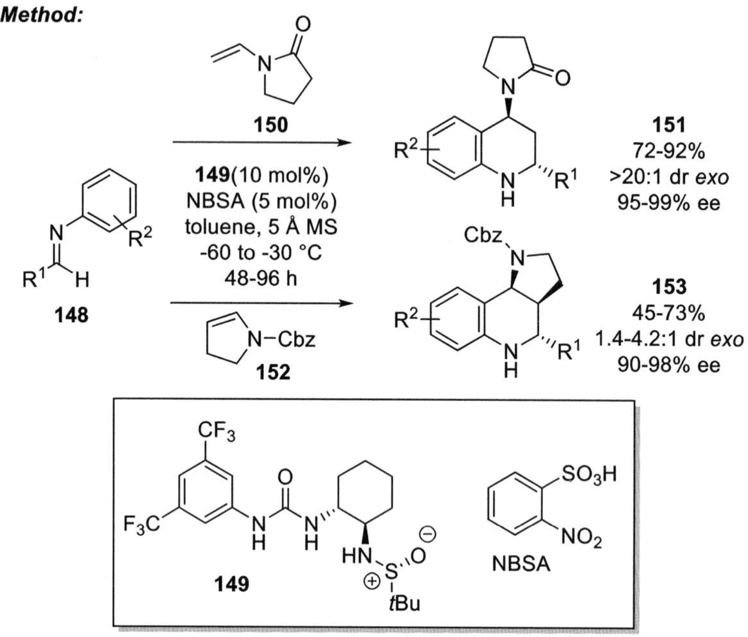

Application:

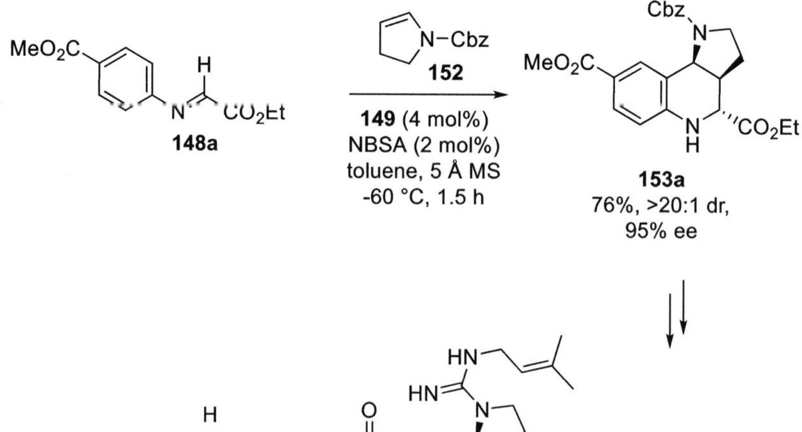

Scheme 30. Synthesis of martinelline derivatives.

VI. Substitution Reaction

Nucleophilic substitution of 3-(1-tosylalkyl)indoles **155**[68] is an efficient method to construct alkaloids. In 2005, Gong investigated this elegant strategy for the purposes of synthesizing optically active spirooxindoles. They explored the use of chiral bifunctional thiourea catalysts with 3-indolyl(aryl)sulfonylmethanes. A combination of thiourea catalyst **41**[69] and K_2CO_3 was observed to be optimal for the addition of oxindoles **156** to the vinylogous imines **157** (Scheme 31). The bifunctional thiourea catalysts acts as Brønsted base, deprotonating the pronucleophile, and a hydrogen bond donor for both the nucleophile and the electrophile, directing the reaction partners to adopt the ternary complex operating in the transition state. The yields and enantioselectivities were generally very good, albeit with moderate diastereoselectivity. Resulting oxindole **157** obtained with 98% ee and 2.5:1 dr, was then converted into trigolute B **158** in seven steps and a 39% yield over the whole synthesis.

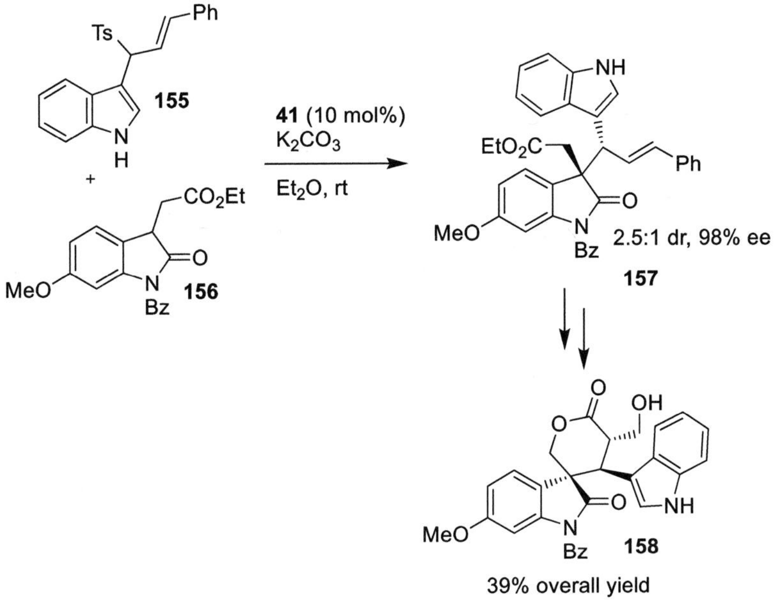

Scheme 31.   Synthesis of trigolute.

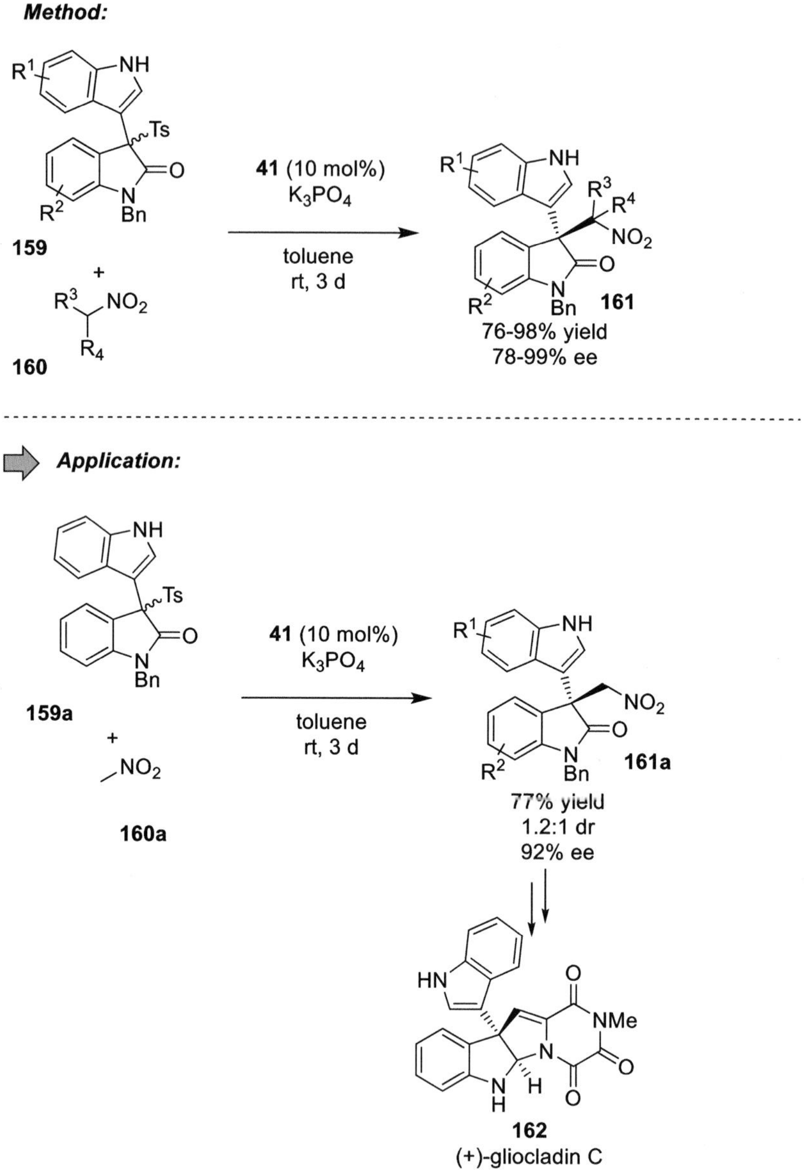

Method:

159

+

160

41 (10 mol%)
K_3PO_4

toluene
rt, 3 d

161
76–98% yield
78–99% ee

Application:

159a

+

160a

41 (10 mol%)
K_3PO_4

toluene
rt, 3 d

161a
77% yield
1.2:1 dr
92% ee

162
(+)-gliocladin C

Scheme 32.   Synthesis of gliocladin C.

Gong group continued their studies towards cyclotryptamine alkaloids.[69] They carried out the addition of nitroalkanes to 3-tosyl-3-indolyloxindoles by the same bifunctional catalyst **41** described above. The addition on nitromethane **160** proceeded in 76–98% yield of **161** and gave enantiomeric excess of 89–99% (Scheme 32). However, poor diastereoselectivity was observed when nitroethane was used (1.2:1). Using the newly developed methodologies, the Gong group was able to synthesize a key intermediate for the synthesis of (+)-gliocladin C **162** in a catalytic enantioselective fashion.

VII. Conclusion

Implementation of chiral organocatalysis in the area of pharmaceutical chemistry to simultaneously introduce one or several of the stereogenic centers present on the target structure has the potential to improve overall selectivity, shorten synthetic routes and increase their convergence. Chiral organocatalysts are often easy to handle compared to other conditions, involving Lewis acids for example, while permitting the use of more readily available starting materials. Examples of use of organocatalytic methods for the construction of diverse natural or synthetic active molecules are already numerous and keep on accumulating. Thereby demonstrating the adequation of these tools to solve problems met when trying to access such chiral, complex and polyfunctional structures. Now that efficiency of chiral organocatalysis at the laboratory scale seems to be established. The next challenge in our opinion is to find ways to increase the adoption rate of this class of processes in development and production contexts. Generalization of this powerful tool would potentially have a great benefit in terms of ease of operation as well as sustainability.

VIII. References

1. Crossley, R. *Chirality and the Biological Activity of Drugs*, CRC Press, **1995**.
2. Caner, H.; Groner, E.; Levy, L.; Agranat, I. *Drug Discov. Today* **2004**, *9*, 105.
3. Gaunt, M. J.; Johansson, C. C. C.; McNally, A.; Vo, N. T. *Drug Discov. Today* **2007**, *12*, 8.
4. For reviews about asymmetric organocatalysis in total synthesis, see: (a) Marqués-López, E.; Herrera, R. P.; Christmann, M. *Nat. Prod. Rep.* **2010**, *27*, 1138; (b) Grondal, C.; Jeanty, M.; Enders, D. *Nat. Chem.* **2010**, *2*, 167; c) Sun, B.-F.

Tetrahedron Lett. **2015**, *56*, 2133; (c) Abbasov, M.; Romoa, D. *Nat. Prod. Rep.* **2014**, *31*, 1318. For a book on this topic: M. Waser, *Asymmetric Organocatalysis in Natural Product Syntheses*, Spinger-Verlag Wien, **2012**.

5. For a review about organocatalytic asymmetric synthesis of chiral nitrogenous heterocycles and natural products, see: Yu, J.; Zhou, Y.; Cheng, D.-F.; Gong, L.-Z. *Pure Appl. Chem.* **2014**, *86*, 1217.

6. (a) Schreiner, P. R. *Chem. Soc. Rev.* **2003**, *32*, 289; (b) Pihko, P. M. *Angew. Chem. Int. Ed.* **2004**, *43*, 2062; (c) Doyle, A. G.; Jacobsen, E. N. *Chem. Rev.* **2007**, *107*, 5713; (d) Phipps, R. J.; Hamilton, G. L.; Toste, F. D. *Nat. Chem.* **2012**, *4*, 603; (e) Brak, K.; Jacobsen, E. N. *Angew. Chem. Int. Ed.* **2013**, *52*, 534; (f) Mahlau, M.; List, B. *Angew. Chem. Int. Ed.* **2013**, *52*, 518; (g) Parmar, D.; Sugiono, E.; Raja, S.; Rueping, M. *Chem. Rev.* **2014**, *114*, 9047; (h) Akiyama, T.; Mori, K. *Chem. Rev.* **2015**, *115*, 9277; (i) Fang, X.; Wang, C.-J. *Chem. Commun.* **2015**, *51*, 1185.

7. Enders, D.; Grondal, C.; Hüttl, M. R. M. *Angew. Chem. Int. Ed.* **2007**, *46*, 1570.

8. For a book on this topic, see: Xu, P.-F.; Wang, W. *Catalytic cascade Reaction*, Wiley-VCH, Weinhein, 2014.

9. Merad, J.; Lalli, C.; Bernadat, G.; Maury, J.; Masson, G. *Chem. Eur. J.* **2018**, *24*, 3925.

10. For general reviews about asymmetric Michael additions, see: (a) Zhang, Y.; Wang, W. *Catal. Sci. Technol.* **2012**, *2*, 42; (b) Pellissier, H. *Adv. Synth. Catal.* **2012**, *54*, 237; (c) Majid, M. H.; Parvin, H.; Hoda, H. *Curr. Org. Chem.* **2014**, *18*, 489.

11. For reviews about asymmetric conjugate additions to nitro-olefins, see: (a) Berner, O. M.; Tedeschi, L.; Enders, D. *Eur. J. Org. Chem.* **2002**, 1877; (b) Almasi, D.; Alonso, A.; Najera, C. *Tetrahedron: Asymmetry*, **2007**, *18*, 299; (c) Sulzer-Mosse, S.; Alexakis, A. *Chem. Commun.* **2007**, *43*, 3123; (d) Roca-López, D.; Sadaba, D.; Delso, I.; Herrera, R. P.; Tejero, T.; Merino, P. *Tetrahedron: Asymmetry* **2010**, *21*, 2561; (e) Somanathan, R.; Chavez, D.; Servin, F. A.; Romero, J. A.; Navarrete, A.; Parra-Hake, M.; Aguirre, G.; Anaya, D. P. C.; González, J. *Curr. Org. Chem.* **2012**, *16*, 2440; (f) Alonso, D. A.; Baeza, A.; Chinchilla, R.; Gómez, C.; Guillena, G.; Pastor, I. M.; Ramón, D. J. *Molecules* **2017**, *22*, 895.

12. (a) Taylor, M. S.; Jacobsen, E. N. *Angew. Chem. Int. Ed.* **2006**, *45*, 1520; (b) Doyle, A. G.; Jacobsen, E. N. *Chem. Rev.* **2007**, *107*, 5713; (c) Zhang, Z.; Schreiner, P. R. *Chem. Soc. Rev.* **2009**, *38*, 1187.

13. Bui, T.; Syed, S.; Barbas III, C. F. *J. Am. Chem. Soc.* **2009**, *131*, 8758.

14. Matsuura, T.; Overman, L. E.; Poon, D. J. *J. Am. Chem. Soc.* **1998**, *120*, 6500.

15. Yu, Q.-S.; Brossi, A. *Heterocycles* **1988**, *27*, 745.

16. Asakawa, K.; Noguchi, N.; Takashima, S.; Nakada, M. *Tetrahedron: Asymmetry* **2008**, *19*, 2304.

17. Trost, B. M.; Zhang, Y. *J. Am. Chem. Soc.* **2006**, *128*, 4590.

18. Kato, M.; Yasui, K.; Yamanaka, M.; Nagasawa, K. *Asian J. Org. Chem.* **2016**, *5*, 380.

19. For a review about asymmetric conjugate additions to acrylates, see: Reyes, E.; Uria, U.; Vicario, J. L.; Carrillo, L. (2016). The Catalytic, Enantioselective Michael Reaction. In *Organic Reactions*.

20. Chen, P.; Bao, X.; Zhang, L.-F.; Ding, M.; Han, X.-J.; Li, J.; Zhang, G.-B.; Tu, Y.-Q.; Fan, C.-A. *Angew. Chem. Int. Ed.* **2011**, *50*, 8161.

21. Okino, T.; Hoashi, Y.; Takemoto, Y. *J. Am. Chem. Soc.* **2003**, *125,* 12672.

22. Marini, F.; Sternativo, S.; Del Verme, F.; Testaferri, L.; Tieccoa, M.; *Adv. Synth. Catal.* **2009**, *351*, 103.

23. Zhang, T.; Cheng, L.; Hameed, S.; Liu, L.; Wang, D.; Chen, Y.-J. *Chem. Commun.* **2011**, *47*, 6644.

24. Buyck, T.; Wang, Q.; Zhu, J. *Angew. Chem. Int. Ed.* **2013**, *52*, 12714.

25. For reviews about double Michael sequences, see: (a) Ihara, M.; Fukumoto, K. *Angew. Chem. Int. Ed.* **1993**, *32*, 1010; (b) Nayak, S.; Panda, P.; Bhakta, S.; Mishra, S. K.; Mohapatra, S. *RSC Adv.* **2016**, *6*, 96154.

26. Hoashi, Y.; Yabuta, T.; Takemoto, Y. *Tetrahedron Lett.* **2004**, *45*, 9185.

27. Wang, Y.; Luo, Y.-C.; Zhang, H.-B.; Xu, P.-F. *Org. Biomol. Chem.* **2012**, *10*, 8211.

28. Li, D. R.; Murugan, A.; Falck, J. R. *J. Am. Chem. Soc.* **2008**, *130*, 46.

29. Okino, T.; Hoashi, Y.; Takemoto, Y. *J. Am. Chem. Soc.* **2003**, *125*, 12672.

30. (a) Wang, Y.; Liu, X.; Deng, L. *J. Am. Chem. Soc.* **2006**, *128*, 3928; (b) Wang, B.; Wu, F.; Wang, Y.; Liu, X.; Deng, L. *J. Am. Chem. Soc.* **2007**, *129*, 768.

31. Zhuang, W.; Hazell, R. G.; Jørgensen, K. A. *Org. Biomol. Chem.* **2005**, *3*, 2566.

32. Wang, X.-F.; Chen, J.-R.; Cao, Y.-J.; Cheng, H.-G.; Xiao, W.-J. *Org. Lett.* **2010**, *12*, 1140.

33. For reviews about asymmetric additions onto imines, see: (a) Doyle, A. G.; Jacobsen, E. N. *Chem. Rev.* **2007**, *107*, 5713; (b) Petrini, M.; Torregiani, E. *Synthesis* **2007**, 159; (c) Yin, B.; Zhang, Y.; Xu, L.-W. *Synthesis* **2010**, 3583; (d) Vesely, J.; Rios, R. *Chem. Soc. Rev.* **2014**, *43*, 611; (e) Kataja, A. O.; Masson, G. *Tetrahedron* **2014**, *70*, 8783. (f) For a book about the synthesis of chiral amines, see: Nugent, T. S. *Chiral Amine Synthesis: Methods, Developments and Applications* Wiley-VCH, Weinhein, **2014**.

34. For reviews on organocatalytic reductions: (a) Rueping, M.; Dufour, J.; Schoepke, F. R. *Green Chem.* **2011**, *13*, 1084; (b) Rueping, M.; Sugiono, E.; Schoepke, F. R. *Synlett* **2010**, 852.

35. For a review on tetrahydroquinolines see: Sridharan, V.; Suryavanshi, P. A.; Menéndez J. C. *Chem. Rev.* **2011**, *111*, 7157.

36. Wakchaure, V.; Kaib, P. S. J.; Leutzsch, M.; List, B. *Angew. Chem. Int. Ed.* **2015**, *54*, 11852.

37. For reviews on catalytic asymmetric Strecker reaction of ketimines, see: (a) Connon, S. J. *Angew. Chem. Int. Ed.* **2008**, *47*, 1176; (b) Spino, C. *Angew. Chem. Int. Ed.* **2004**, *43*, 1764. For selected examples, see: (c) Masumoto, S.; Usuda, H.; Suzuki, M.; Kanai, M.; Shibasaki, M. *J. Am. Chem. Soc.* **2003**, *125*, 5634; (d) Wang, J.; Hu, X.; Jiang, J.; Gou, S.; Huang, X.; Liu, X.; Feng, X. *Angew. Chem. Int. Ed.* **2007**, *46*, 8468; (e) Abell, J. P.; Yamamoto, H. J. *J. Am. Chem. Soc.* **2009**, *131*, 15118.

38. For a review on catalytic asymmetric allylation of carbonyls and imines, see: (a) Yus, M.; González-Gómez, J. C.; Foubelo, F. *Chem. Rev.* **2011**, *111*, 7774; (b) Huo, H.-X.; Duvall, J. R.; Huang, M.-Y.; Hong, R. *Org. Chem. Front.* **2014**, *1*, 303.

39. Lou, S.; Moquist, P. N.; Schaus, S. E. *J. Am. Chem. Soc.* **2007**, *129*, 15398.

40. Price, D. A.; Gayton, S.; Selby, M. D.; Ahman, J.; Haycock-Lewandowski, S. Stammen, B. L.; Warren, A. *Tetrahedron Lett.* **2005**, *46*, 5005.

41. Silverio, D. L.; Torker, S.; Pilyugina, T.; Vieira, E. M.; Snapper, M. L.; Haeffner, F.; Hoveyda, A. H. *Nature* **2013**, *494*, 216.

42. Itoh, T.; Ishikawa, H.; Hayashi, Y. *Org. Lett.* **2009**, *11*, 3854.

43. Cravotto, G.; Giovenzana, G. B.; Palmisano, G.; Penoni, A.; Pilati, T.; Sisti, M.; Stazi, F.; *Tetrahedron: Asymmetry* **2006**, *17*, 3070.

44. Park, Y.; Schindler, C. S.; Jacobsen, E. N. *J. Am. Chem. Soc.* **2016**, *138*, 14848.

45. For a review on the use of the Pictet-Spengler condensation in the synthesis of natural products: Stöckigt, J.; Antonchick, A. P.; Wu, F.; Waldmann, H. *Angew. Chem. Int. Ed.* **2011**, *50*, 8538.

46. Taylor, M. S.; Jacobsen, E. N. *J. Am. Chem. Soc.* **2004**, *126*, 10558.

47. Mergott, D. J.; Zuend, S. J.; Jacobsen, E. N. *Org. Lett.* **2008**, *10*, 745.

48. Raheem, I. T.; Thiara, P. S.; Peterson, E. A.; Jacobsen, E. N. *J. Am. Chem. Soc.* **2007**, *129*, 13404.

49. Kerschgens, I. P.; Claveau, E.; Wanner, M. J.; Ingemann, S.; van Maarseveen, J. H.; Hiemstra, H. *Chem. Commun.* **2012**, *48*, 12243.

50. Luzzio, F. A. *Tetrahedron* **2001**, *57*, 915.

51. (a) Sohtome, Y.; Hashimoto, Y.; Nagasawa, K. *adv. Synth. Catal.* **2005**, *347*, 1643; (b) Sohtomea, Y.; Takemurab, N.; Iguchib, T.; Hashimotoa, Y.; Nagasawa, K. *Synlett* **2006**, *1*, 144.

52. Sohtome, Y.; Hashimoto, Y.; Nagasawa, K. *Eur. J. Org. Chem.* **2006**, 2894.

53. Liu, L.; Zhang, S.; Xue, F.; Lou, G.; Zhang, H.; Ma, S.; Duan, W.; Wang, W. *Chem. Eur. J.* **2011**, *17*, 7791.

54. For reviews on the preparation and application of cyanohydrins, see: (a) Gregory, R. J. H. *Chem. Rev.* **1999**, *99*, 3649; (b) Effenberger, F. *Angew. Chem. Int. Ed. Engl.* **1994**, *33*, 1555; (c) North, M. *Synlett* **1993**, 807; (d) Deng, H.; Isler, M. P.; Snapper, M. L.; Hoveyda, A. H. *Angew. Chem, Int. Ed.* **2002**, *41*, 1009.

55. Hong, R.; Chen, Y.; Deng, L. *Angew. Chem. Int. Ed.* **2005**, *44*, 3478.

56. (a) Masson, G.; Housseman, C.; Zhu, J. *Angew. Chem., Int. Ed.* **2007**, *46*, 4514; (b) Declerck, V.; Martinez, J.; Lamaty, F. *Chem. Rev.* **2009**, *109*, 1; (c) Basavaiah, D.; Rao, K. V.; Reddy, R. J. *Chem. Soc. Rev.* **2007**, *10*, 1581; (d) Basavaiah, D.; Rao, A. J.; Satyanarayana, T. *Chem. Rev.* **2003**, *103*, 811; (e) Ciganek, E. *Org. React.* **1997**, *51*, 201.

57. Iwabuchi, Y.; Nakatani, M.; Yokoyama, N.; Hatakeyama, S. *J. Am. Chem. Soc.* **1999**, *121*, 10219.

58. Nakano, A.; Takahashi, K.; Ishihara, J.; Hatakeyama, S. *Org. Lett.* **2006**, *8*, 5357.

59. Iwabuchi, Y.; Furukawa, M.; Esumi, T.; Hatakeyama, S. *Chem. Commun.* **2001**, *0*, 2030.

60. Nicolaou, K. C.; Snyder, S. A.; Montagnon, T.; Vassilikogiannakis, G. *Angew. Chem. Int. Ed.* **2002**, *41*, 1668.

61. Gioia, C.; Hauville, A.; Bernardi, L.; Fini, F.; Ricci, A. *Angew. Chem. Int. Ed.* **2008**, *47*, 9236.

62. Shimizu, S.; Ohori, K.; Arai, T.; Sasai, H.; Shibasaki, M. *J. Org. Chem.* **1998**, *63*, 7547.

63. Tan, B.; Hernández-Torres, G.; Barbas III, C. F. *J. Am. Chem. Soc.* **2011**, *133*, 12354.

64. (a) Lin, H.; Danishefsky, S. J. *Angew. Chem. Int. Ed.* **2003**, *42*, 36; (b) Galliford, C. V.; Scheidt, K. A. *Angew. Chem. Int. Ed.* **2007**, *46*, 8748; (c) Greshock, T. J.; Grubbs, A. W.; Jiao, P.; Wicklow, D. T.; Gloer, J. B.; Williams, R. M. *Angew. Chem. Int. Ed.* **2008**, *47*, 3573; (d) Miller, K. A.; Tsukamoto, S.; Williams, R. M. *Nat. Chem,* **2009**, *1*, 63.

65. For reviews about Povarov reaction, see: (a) Glushkov, V. A.; Tolstikov, A. G. *Russ. Chem. Rev.* **2008**, *77*, 137; (b) Kouznetsov, V. V. *Tetrahedron* **2009**, *65*, 2721; (c) Bello, D. Ramón, R.; Lavilla, R. *Curr. Org. Chem.* **2010**, *14*, 332; (d) Sridharan, V.; Suryavanshi, P. A.; Menendez, J. C. *Chem. Rev.* **2011**, *111*, 7157; (e) Jiang, X. X.; Wang, R. *Chem. Rev.* **2013**, *113*, 5515; (f) Masson, G.; Lalli, C.; Benohoud, M.; Dagousset, G. *Chem. Soc. Rev.* **2013**, *42*, 902; (g) Jiang, X. X.; Wang, R. *Chem. Rev.* **2013**, *113*, 5515; (h) Fochi, M.; Caruana, L.; Bernardi, L. *Synthesis* **2014**, *46*, 135; (i) Eschenbrenner-Lux, V.; Kumar, K.; Waldmann, H. *Angew. Chem. Int. Ed.* **2014**, *53*, 11146.

66. Xu, H.; Zuend, S. J.; Woll, M. G.; Tao, Y.; Jacobsen, E. N. *Science* **2010**, *327*, 986.

67. Xia, C.; Heng, L.; Ma, D. *Tetrahedron Lett.* **2002**, *43*, 9405.

68. Huang, J.-Z.; Zhang, C.-L.; Zhu, Y.-F.; Li, L.-L.; Chen, D.-F.; Han, Z.-Y.; Gong, L.-Z. *Chem. Eur. J.* **2015**, *21*, 8389.

69. Huang, J.-Z.; Wu, X.; Gong, L. *Adv. Synth. Cat.* **2013**, *355*, 2531.

5   Bio-Sourced Polymers: Recent Advances

Henri Cramail*, Boris Bizet, Océane Lamarzelle,
Pierre-Luc Durand, Geoffrey Hibert and Etienne Grau

Laboratoire de Chimie des Polymères Organiques UMR5629
Université de Bordeaux — CNRS — Bordeaux INP/ENSCBP
16, Avenue Pey-Berland, F-33607 Pessac Cedex, France
*cramail@enscbp.fr

Table of Contents

List of Abbreviations

ADMET	acyclic diene metathesis
AFM	atomic force microscopy
AIBN	azobisisobutyronitrile
AL	Angelica lactone
5-ALA	5-aminolevulinic acid
Aq.	aqueous
ATRP	atom-transfer radical polymerization
BAMF	2,5-bisaminomethylfurfural
BHMF	2,5-bis(hydroxymethyl)furfural
bisCC	bis cyclic carbonates
BPA	bisphenol-A
BTX	benzene, toluene, xylene
CALB	*Candida antartica* lipase B
Cat.	catalyst
CMF	5-chloromethylfurfural
CRP	controlled radical polymerization
CTA	chain transfer agent
DA	Diels-Alder
DBTDL	dibutyl tin dilaurate
DGEBA	diglycidyl ether of bisphenol-A
DMA	dynamical mechanical analysis
DMC	dimethyl carbonate
DMF	dimethylformamide
DMPA	4-(dimethylamino)pyridine
DPA	diphenolic acid
DFF	2,5-diformylfuran
Đ	dispersity
EDA	ethylene diamine
FAME	fatty acid methyl ester
FDA	Food and Drug Administration
FDCA	2,5-furandicarboxylic acid
FF	furfural
GBL	γ-butyrolactone
GVL	γ-valerolactone

HDI	hexamethylene diisocyanate
HDPE	high density polyethylene
HDT	heat distortion temperature
HMF	2,5-hydroxymethylfurfural
HPB	hyperbranched polymer
HBPE	hyperbranched polyester
IA	itaconic acid
IPDA	isophorone diamine
IPDI	isophorone diisocyanate
LA	levulinic acid
LAB	lactic acid bacterias
LCA	life cycle analysis
LDPE	low density polyethylene
MA	methyl acrylate
mcl-PHA	medium chain length poly(hydroxyalkanoate)
MEK	methyl ethyl ketone
MGVL	α-methylene-γ-valerolactone
MDI	4,4′-diisocyanate diphenylmethane
MMA	methyl methacrylate
M_n	number average molar mass
MTT	(3-(4,5-dimethylthiazol-2-yl)-2,5-diphenyltetrazolium bromide)
M_w	weight average molar mass
NIPU	non-isocyanate polyurethane
NMR	nuclear magnetic resonance
P3HB	poly(3-hydroxybutyrate)
P3HHx	poly(3-hydroxyhexanoate)
P3HO	poly(3-hydroxyoctanoate)
P4HB	poly(4-hydroxybutyrate)
PA	polyamide
PAHB	poly(alkylenehydroxybenzoate)
PBF	poly(butylene furanoate)
PDLA	poly(D-lactic acid)
PDLLA	poly(D,L-lactic acid)
PBS	poly(butylene succinate)
PBT	poly(butylene terephtalate)
PCL	poly(ε-caprolactone)

PE	poly(ethylene)
PEA	poly(ester amide)
PEF	poly(ethylene furanoate)
PEG	poly(ethylene glycol)
PEO	poly(ethylene oxide)
PET	poly(ethylene terephthalate)
Ph	phenyl
PHA	poly(hydroxyalkanoate)
PHU	poly(hydroxy)urethane
PHV	poly(β-hydroxybutyrate)
PIC	poly(isosorbide carbonate)
PIT	poly(isosorbide terephthalate)
PLA	poly(lactic acid)
PLimC	poly(limonene carbonate)
PLLA	poly(L-lactic acid)
PMMA	poly(methylmethacrylate)
POM	polyoxometalate
PP	polypropylene
PRic	poly(ricinoleic acid)
PS	polystyrene
PTMO	poly(tetramethylene oxide)
PU	polyurethane
PVA	poly(vinyl alcohol)
PVAc	poly(vinyl acetate)
PVC	poly(vinyl chloride)
RAFT	reversible addition–fragmentation chain-transfer
rDA	retro Diels–Alder
ROcP	ring-opening copolymerization
ROhP	ring-opening homopolymerization
ROMP	ring-opening metathesis polymerization
ROP	ring-opening polymerization
RT	room temperature
SA	succinic acid
scl-PHA	short chain length poly(hydroxyalkanoate)
SPAEK	sulfonated poly(arylene ether ketone)
TBD	triazabicyclodecene

TEMPO	(2,2,6,6-tetramethylpiperidin-1-yl)oxy
TDI	toluene diisocyanate
$T_{dx\%}$	degradation temperature (when x% of the product is degraded)
T_g	glass transition temperature
THF	tetrahydrofuran
T_m	melting temperature
TPU	thermoplastic polyurethane
T_β	β-transition temperature
UPE	unsaturated Polyester
USAXS	ultra small angle X-ray scattering
UV	ultraviolet
wt%	weight percent
5CC	5-membered cyclic carbonate
6CC	6-membered cyclic carbonate

I. General Introduction

Synthetic polymers are of major interest in the chemical industry, with a world production superior to 350Mt in 2016. They find applications in all the branches of industry, from packaging films to the state-of-the-art materials for sports and leisure activities, construction and aerospace industry or medical applications. Half of the amount of molecules produced by the petrochemical industry thus finds itself incorporated in the elaboration of polymer materials. With a production of 160Mt/year, ethylene can be considered as the major example of petroleum-based monomers. Green Chemistry introduced in 1998 by Anastase *et al.* (refer to Chapter 1 of this book), aims at providing solutions to reduce environmental impacts of our society. More precisely, this pursue of a more sustainable chemistry consists in using a source of renewable carbon to both reduce the dependence on fossil resources and thus stabilize greenhouse gas emissions (in particular CO_2) at the end of life. The use of renewable resources is of major interest in the elaboration of bio-sourced polymers. By using them, it is possible to mimick the fossil-based polymers — drop-in structures such as biopolyethylene- or to design new chemical structures, such as poly(lactic acid), PLA. Nowadays, the vegetal-based chemistry (renewable resources) mobilizes less than 0.5% of arable land in the world.

Some reminders of the definitions of the terms biomass, biopolymer, biodegradable polymer, bio-sourced polymer, bioplastic, biorefinery may be necessary:

Biomass: material of biological origin (elaborated by alive bodies) with the exception of the materials of geological or fossil formation.

Biopolymer: polymer developed by alive bodies, extracted from the biomass. Examples: polysaccharides, proteins, bacterial polymers.

Biodegradable polymer: polymer the main degradation mechanism of which can be biotic, by enzymatic way. Under the action of micro-organisms and in the presence of oxygen (aerobic conditions), the organic compound decomposes totally within few months into carbon dioxide, water and mineral salts, with the appearance of a new biomass; in the absence of oxygen (anaerobic conditions), the organic compound decomposes totally within few months into carbon dioxide, methane, mineral salts and creation of a new biomass.

Bio-sourced (or bio-based) polymer: synthetic polymer partially (generally > 20%) or totally obtained from by-products stemming from the biomass. The bio-sourced character of a polymer can be determined in particular from its content in C14, according to the standard ASTM D6866. For the materials of totally fossil origin, the content in C14 is null.

Bioplastic: term of popularization which indicates a 'biobased plastic' (restricted definition) and/or biodegradable (wider definition).

Biorefinery: by analogy with a classic refinery, which works from fossil resources, a bio-refinery handles biomasses to produce, according to the cases, energy, fuels, materials, chemical and polymer products and/ or animal and human feed.

Today, two biopolymers account for a significant part of the world polymer consumption: the 1,4-cis polyisoprene (natural rubber, approximately 12Mt/year) — precursor of elastomers — and the cellulose stemming from wood pulps (regenerated cellulose, ethers and esters of cellulose, for a total about 5Mt/year). As for bio-sourced polymers, the latter reaches approximately 1% of the global production of polymer materials, but their production knows a considerable development and corresponds to the category of polymers the dynamics of which is the most important, both in terms of production and innovation (Figure 1).

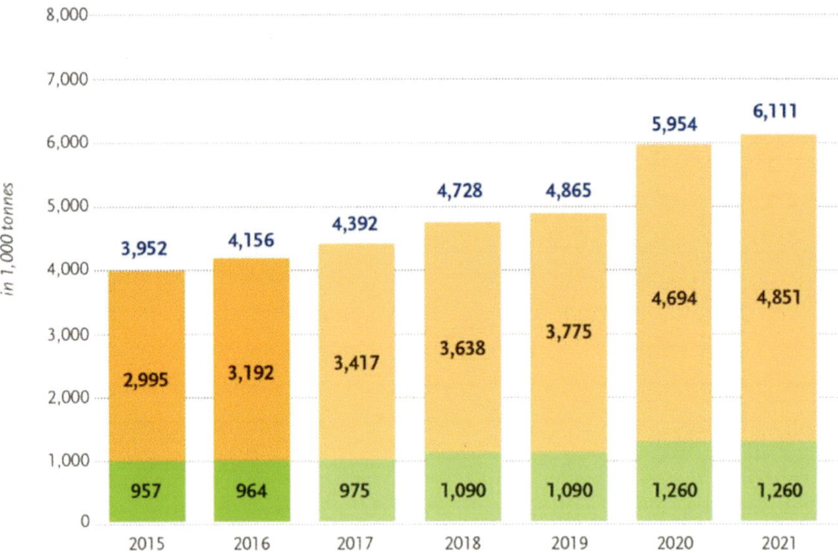

Figure 1. Trend and Capacity of production of "Bioplastics" in Ktons (from 2015 to 2021). Two distinct classes: the non-biodegradable bio-sourced polymers such as bio-polyethylene, BioPE, bio-poly(ethylene terephtalate), BioPET, various bio-based polyamides, PA, etc. (in orange), and the biodegradable bio-based polymers such as the ones based on starch, the polylactide, PLA, etc. (in green). Source: European Bioplastics, Nova-Institute (2016).

The current development of biosourced polymers is first motivated by the search for performances and for new features. Indeed, in a global approach, which has to take into account the price of these new renewable resources, the bio-sourced feature is not enough for the industrial development of these new polymers. Besides, the reduction of impacts, in particular *via* the use of less toxic monomers and catalysts, is a driving force to the development of the renewable resources. The increase of the legislative arsenal — with in particular the regulation REACH — aims indeed at reducing the exposure to dangerous substances of the industry and end-users employees. In the polymers' field, Green Chemistry can be an answer to these challenges, by proposing less dangerous and possibly biosourced monomers (and catalysts). In every case, the danger of a substance does not come from its composition in renewable carbon but from its chemical structure and the qualifier "biosourced" or "natural" does not mean without any danger! Identical substances will bring the same properties and the same dangers. The only advantage, in the short term, is to possess industrial facilities of transformation and existing application sectors. The regulations however consider all the new molecules — may they be biosourced — on an equal basis, and the industrialists have to study their effects on the Man and the Environment before their launch on the market, which is very expensive and limits the development of new products.

Today, the development of bio-sourced polymers is in particular based on the valorization of two main types of resources at the industrial level: oleaginous plants and polysaccharides, in connection with the corresponding agricultural productions.[1] Indeed, numerous development projects and some already marketed products concern the valuation of vegetable oil and glycerin for the elaboration of precursors of polymers such as polyamides, polyurethanes, polyesters and polyethers with world actors as Arkema, Oleon (Avril), Evonik, Elevance, Hunstman in Europe and US Biobased in the USA. Also, the starchy and saccharide-based resources are under industrial development for the elaboration of polyols, polyesters, solvents, etc. with group world-leaders as Roquette, Tereos, Solvay, NatureWork. The United States developed very early (before 2004) a strategy targeted with a strategic vision of the DoE (Department of Energy) for the industrial development of 12 synthons (building blocks), strategic for the American economy (see Figure 2).

Brazil made the industrial choice around the valuation of resources targeted as the sugar cane with, in particular, the production of biosourced polyethylene (Braskem company) by polymerization of ethylene stemming from the dehydration of the ethanol obtained by fermentation. It is in particular this bioethylene which is used to develop bio-poly(ethylene terephtalate), bioPET. Other actors are interested in the valuation of extractable compounds stemming from biomass, such as terpenes for applications as adhesives notably. Finally, numerous research efforts are dedicated today to the valuation of lignin stemming from various plentiful ligno-cellulosic resources — in particular by deconstruction of this one — to propose aromatic synthons of interest.

                    Cramail *et al.*

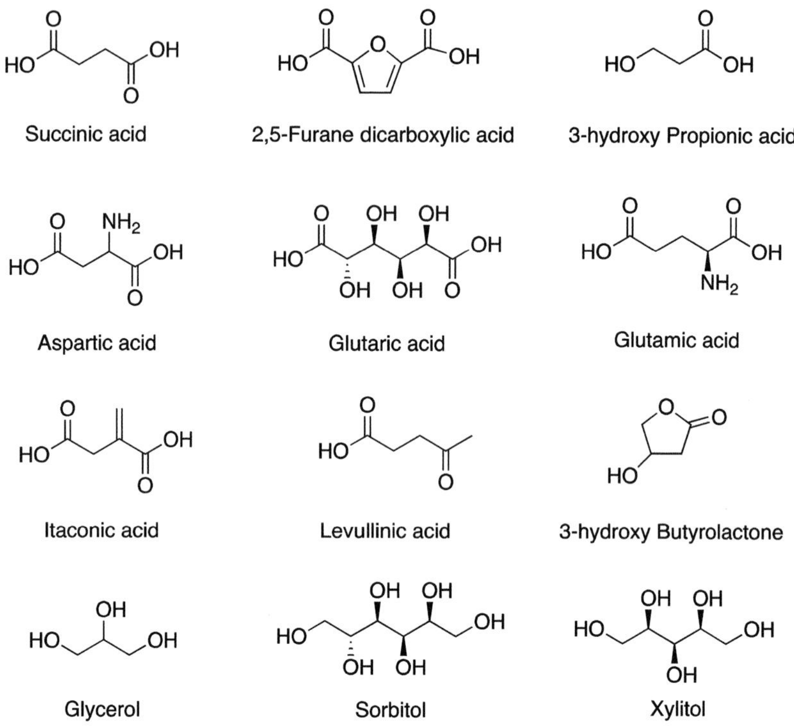

<table>
<tr><td>Succinic acid</td><td>2,5-Furane dicarboxylic acid</td><td>3-hydroxy Propionic acid</td></tr>
<tr><td>Aspartic acid</td><td>Glutaric acid</td><td>Glutamic acid</td></tr>
<tr><td>Itaconic acid</td><td>Levullinic acid</td><td>3-hydroxy Butyrolactone</td></tr>
<tr><td>Glycerol</td><td>Sorbitol</td><td>Xylitol</td></tr>
</table>

Figure 2. List of the 12 strategic "building blocks" targeted by the US Department of Energy (DoE).

II. Polymers from Ligno-Cellulosic Resources

A. Furan-based Polymers

The use of furans in polymer chemistry has attracted a tremendous attention in the last decades. This growing interest can be attributed to several parameters, such as the relatively easy access to sugar sources as well as political incentives depending on each country. Directly accessible from carbohydrate resources, furan derivatives are widely available and can thus be considered as serious candidates for the synthesis of commodity and/or advanced value-added products. However, the apparent easy access to sugar sources must not prevent the fact that care has to be devoted not to enter in competition with the food industry. As such, the selective depolymerization of sugar-originating biopolymers such as cellulose, hemicelluloses, high-fructose corn syrups (among others) has become a potential solution to yield interesting platform molecules.

Due to the extremely high amount of available literature in the field, it seems almost impossible to perform a thorough review of all the results. It is however worth mentioning the work of Gandini and Naceur Belgacem[2] who extensively reviewed a wide scope of monomers and polymers which can originate from furan-derivatives. It is strongly encouraged to refer to this work, should complementary information be needed.

1. *Monomers from Furan*

As described before, furan-based monomers can be accessed through the depolymerization of sugar-originating biopolymers. 5-Hydroxymethylfurfural (HMF) and furfural (FF) are the main platform molecules that can be synthesized from sugar-sources. Acidic dehydration of cellulose yields HMF whereas hemicelluloses yield FF (Figure 3).[3–5] From them, an almost unlimited number of interesting monomers can be accessed, depending on the employed synthetic pathway.

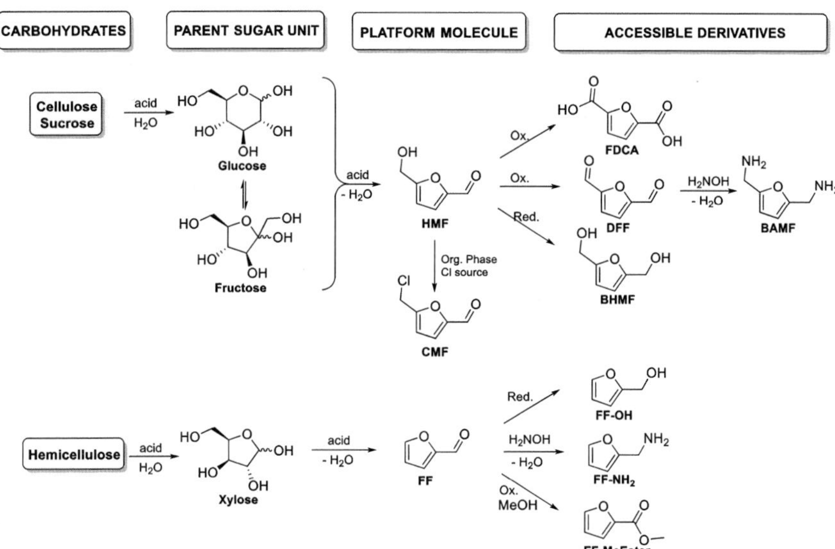

Figure 3. Possible derivatives from the furfural platform.

Describing the different pathways from cellulosic resources to FF and HMF is out of the scope of this chapter, but extensive information can be found in the previous cited reviews. However, an important aspect of the production of HMF, still slowing down its industrial implementation, has to be mentioned. The acidic dehydration process in aqueous phase usually requires harsh to very harsh conditions (such as high temperatures and high-to-very-high acidic concentrations). While reacting, cellulose and hemicelluloses are breaking up to yield glucose, fructose and xylose moieties, which can rearrange in acidic conditions to produce so-called humins (or humic matter), a high-carbon-content brown material, that has — to our knowledge — no existing valorization process (except burning). In the case of HMF, it was found that this reaction was favored within the aqueous phase, which is not desirable. The use of biphasic reaction condition with the addition of an *in situ* extracting immiscible organic solvent during the reaction exhibited superior results in terms of decreased amounts of humins. Recent discoveries[6,7] showed that using a chlorine source (HCl) favored the formation of 5-chloromethylfurfural (CMF) as opposed to HMF. Let alone leaving a very reactive methyl chlorine moiety in place of the hydroxyl group, this molecule possesses the advantage of having a very poor solubility in water so that, when using an aqueous-organic biphasic system, *in situ* extraction could occur right after formation of CMF in the aqueous phase. The molecule is thus protected from the very acidic aqueous phase and less humic matter is formed. Depending on both the sugar source and the utilized process (batch vs flow chemistry), high yields could be reached, leaving a lot of room for improvement towards process economics and further chemical reactivity. However, very few examples of CMF conversion[8–10] are encountered in the literature. This is why only HMF derivatives will be addressed in the following sections.

a. *2,5-Furandicarboxylic acid (FDCA)*

The formation of 2,5-furandicarboxylic acid (FDCA) relies on the oxidation of HMF to yield its diacid derivative.[5] The rate, as well as the selectivity of the reaction, obviously is catalyst-dependant and the number of possible processes is very high. The homogeneous catalytic oxidation of

HMF with O_2 is performed thanks to the Amoco Mid-Century process, in which a mixture of Co/Mn/Br salts is utilized in acetic acid. Depending on both the use of co-catalysts and the process conditions, up to 90% yield of FDCA could be obtained, even if the majority of the obtained yields were reported to be around 60%. However, the acidic conditions are a limitation to this reaction since it favors the degradation of the native HMF (to humins for instance). This is a reason why other pathways, involving noble metals in addition to an external base, were investigated. As far as noble metals are concerned, Pt, Au, Pd are the most studied catalysts. They all gave the best yields when utilized with an external base in water, which is an improvement as it prevents from the degradation of HMF (> 90% yield). Supported, meaning heterogeneous, catalysis was also proven to be suitable for such a process, which is an improvement in terms of recyclability. As a matter of comparison, supported-Au was demonstrated to give better yields as opposed to Pt and Pd-supported catalysts. Another advantage of such a process is its versatility, because a lot of freedom is left to the operators with the choice of the base or the combination of catalysts. However, it has to be noted that FDCA is never obtained pure but under its (Na_2/K_2/Mg-) salt version (depending on the utilization base). This salt formation is the limiting step towards industrialization since it requires complicated purification steps to form and isolate pure FDCA from water. A potential solution is to perform this reaction in methanol, yielding the methyl-ester form of FDCA. Interestingly, this protocole can also be transferred to furfural (FF) in the production of the corresponding methylester (FF-MeEster, see Figure 3). Finally, little amount of literature described the use of supported base metals for the production of FDCA. Fe and Co-based catalysts were mostly investigated. Yields up to 90% could be obtained depending on the process conditions and the used oxidant. Other alternatives, such as the use of ionic liquids have also been reported.[11]

b. *Aldehyde- and aminomethyl-derivatives of HMF and FF*

The selective oxidation of HMF to yield 2,5-diformylfuran (DFF, Figure 3) has also been reported. One of the advantages of making DFF as opposed to FDCA is the higher solubility of DFF in organic solvents. As such, some examples of forming DFF directly from cellulosic resources in a 1 pot-1 step process can be encountered. In a similar fashion with CMF, this proceeds through the aqueous formation of HMF followed by the *in situ* extraction into the organic solvent, in which the oxidation step can be performed.

In order to oxidize HMF to DFF, specific metal catalysts have been investigated. Ru-supported catalysts gave good selectivity towards DFF and yield higher than 95%. Vanadium-based catalysts were also utilized even if they demonstrated lower yields (<85%). Finally, Mn- and Cu- based catalysts, in combination with TEMPO as the oxidizing component, have also been highlighted as potential solutions in the oxidation of HMF to DFF.

Starting from DFF, it is then possible to form the diamine form of HMF: 2,5-bis(aminomethyl)furan (BAMF, Figure 3). A possible process, described in a patent[12] involves a two-step process: a first step in which DFF reacts with hydroxylamine $H_2NOH.HCl$ to form the bis-oxime derivative of HMF, followed by a reduction step *via* Ni catalyzed reaction of H_2 onto the bis-oxime to yield BAMF. Other process involving supported Ni, Cu and Ru catalyst have been described for the production of BAMF from DFF. In a first step, DFF is reacted with ammonia in the gas phase to form the corresponding bis-imine compound. This compound is then hydrogenated with the help of the aforementioned catalysts to yield BAMF with high conversions and selectivities.[13]

c. *2,5-Bis(hydroxymethyl)furan (BHMF) and furfuryl alcohol (FF-OH)*

In a similar fashion to oxidation process, the reduction of HMF and FF can yield a large number of products depending on the reaction conditions. As such, if a selective reduction of the aldehyde moiety can yield BHMF and FF-OH from HMF and FF respectively, it is worth keeping in mind that the reduction of the furan ring and its successive ring-opening are still doable if too harsh reaction conditions are encountered.

Ru-based catalysts exhibited a good selectivity towards BHMF.[14] However, it is to be noted that a lot of other metal catalysts can be suitable for the reaction when adapted reaction conditions are employed. Sn-, Cu-, Pt- Pd- and Au-based catalysts were proven to be efficient.[15–19]

Interestingly, an Ir-based catalyst was highlighted as a promising candidate for the hydrogenation reaction of HMF and FF into BHMF and FF-OH.[20]

Finally, the Cannizzaro reaction was also proven to be a potential route towards the alcohol-derivatives of both HMF and FF. It consists in making a base, such as NaOH, that reacts with the aldehyde moiety of the furfural compounds. A good selectivity in water at 0°C, with a maximum yield of 86%, was reported.[21]

2. *Furan-based Polymers*

a. *Polyesters*

Among the accessible classes of furan-based polymers originating from step-growth polycondensation, polyesters certainly are the most studied. When analyzing the different derivatives that can be obtained from the 5-HMF platform, the presence of diols and diacids justifies why a lot of studies have been performed on the synthesis of various polyesters.

FDCA has particularly attracted a lot of attention as a bio-based alternative to terephtalic acid. This aromatic diacid, classified by the DoE as one of the 12 top value-added chemicals originating from biomass (see Figure 2), has been earmarked for its ability as substitute of terephtalic acid. Since 1951 and the publication of the first patent claiming the preparation of a polyester from FDCA and ethylene glycol,[22] several pathways have been investigated towards the synthesis of polyesters by reacting FDCA (or its dimethyl-ester derivative) with diols — glycols in particular — (Figure 4). While the first pathway consists in a direct

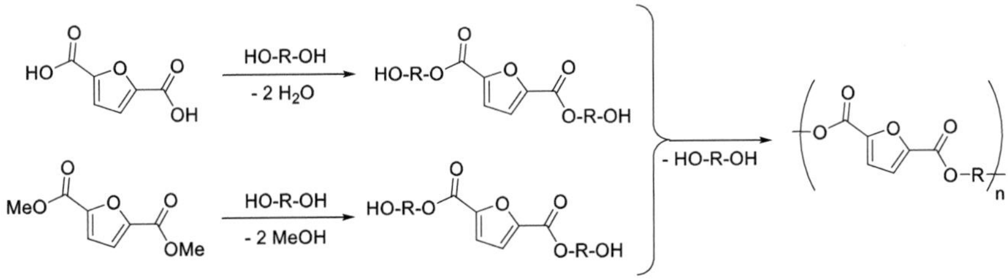

Figure 4. Synthesis of polyesters from 2,5-furan dicarboxylic acid (FDCA).

polycondensation, the second is based on a transesterification process. The choice of the diol has obviously an influence on the resulting properties of the polyesters so-formed, in terms of dimensions and thermomechanical properties.[23–33]

It is worth noting that transesterification reaction is usually performed in the presence of a catalyst following a procedure close to the one of PET synthesis. Lewis catalysts (such as Sb_2O_3, $Ca(OAc)_2$, $Zn(OAc)_2$ or titanium(IV) alkoxides) have proven their efficiency.[5]

Poly(ethylenefuranoate) (PEF) and poly(butylenefuranoate) (PBF) are particular targets of interests. Both are indeed 100% bio-based polyesters that have already been implemented at an industrial scale (Aventium, Coca-Cola).[34,35] Burgess *et al.* demonstrated that PEF exhibited superior properties to its petrochemical analogue (PET) notably in terms of gas permeability.[36–38]

It is interesting to mention that from a historical point of view, industrials showed a huge interest in HMF and FDCA for their ability to be converted into bio-sourced terephthalic acid for the production of bio-sourced PET. The corresponding processes have been extensively described in a review co-authored by five different companies.[39] Two main pathways emerge from their description, both of them relying on a reaction of HMF or FDCA with ethylene through a Diels–Alder reaction (Figure 5).

The first pathway relies on the reduction of HFM to dimethyl furfural and has been the first to be studied. Numerous research teams optimized the process, which led Lin *et al.* to study, in 2013, its techno-economic aspects.[40] The authors concluded that the production cost would be about

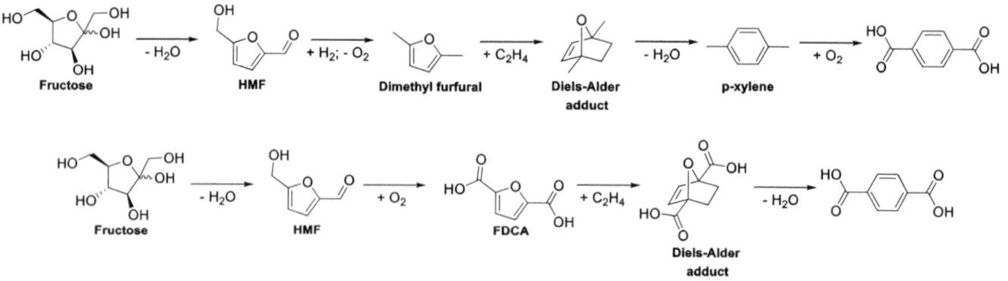

Figure 5. Synthesis of bio-based terephthalic acid from 5-hydroxymethyl furfural and 2,5-furan dicarboxylic acid.

2.5 times the recent price of petroleum-derived *p*-xylene (in 2014), with around 80% of the cost originating from the needed raw materials such as HMF, hydrogen and ethylene. The second pathway relies on the use of FDCA. As a matter of comparison, this process takes advantage of the high oxygen amount contained in FDCA to yield terephthalic acid. This allows avoiding the reduction step observed in the first pathway and thus decreases the number of steps required to form the final product. Both processes have a theoretical weight yield of 61%. To our knowledge, it is today unfortunately impossible to compare the technico-economical aspects of this process as opposed to the first pathway.

Achieving the formation of polyesters can also be possible when starting with the diol-derivative of 5-HMF. 2,5-bis(hydroxymethyl)furan (BHMF) can thus be reacted with diacids to yield the corresponding polyester (Figure 6).

Yoshie *et al.* thus demonstrated in a series of papers the possibility to perform the polycondensation reaction of BHMF with succinic acid ($n = 2$) under basic conditions.[41–44] Very interestingly, self-healing capability was observed with the resulting polyester upon addition of bismaleimide crosslinker; more details will be provided in Section II.A.2.e.

Using lipases through an enzymatic catalytic process, Loos *et al.* also described the esterification process of BHMF with a series of diacids of various chain lengths as described in Figure 6.[45] Relatively low molecular weights ranging from 1600 to 2400 g.mol^{-1} were obtained.

Finally, it is worth mentioning the possibility to form bio-sourced furan-derived polyesters originating from dimer forms of furan-carboxylic acids. This two-step process first requires an oxidation of furfural (FF). As described previously, extensive work has been undertaken in order to find an industrially attractive way of oxidizing HMF to 2,5-FDCA, the most promising results of which being in methanol over gold catalyst.[46,47] FF can be

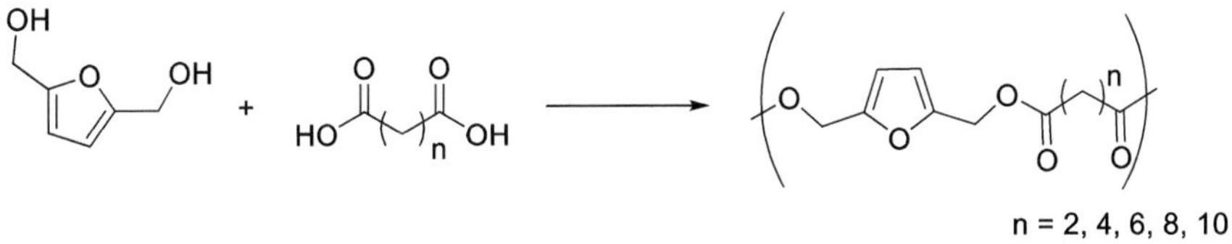

Figure 6.   Synthesis of polyesters from 2,5-bis(hydroxymethyl)furan (BHMF).

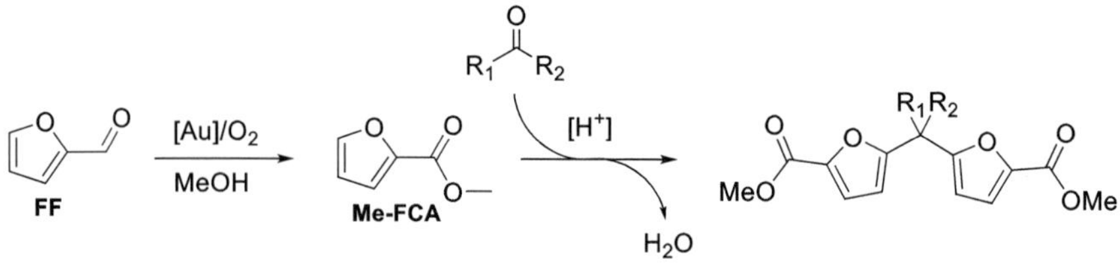

Figure 7. Synthesis of diacid (diester) compounds originating from a dimerization process of furfural.

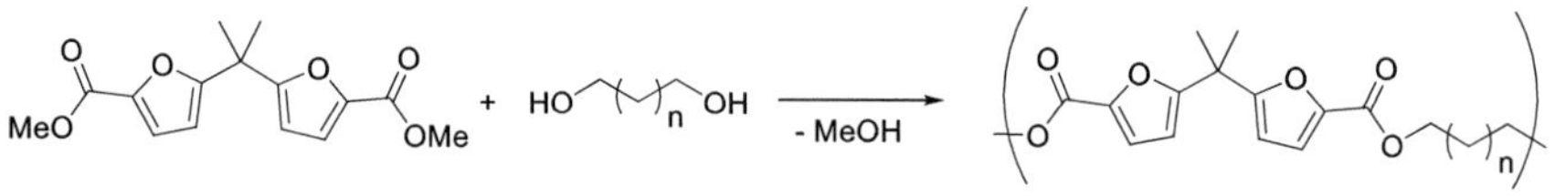

Figure 8. Synthesis of polyesters from furfural-dimer-based diesters.

also oxidized in a very similar fashion.[48–51] The obtained product can thus be dimerized with the help of a carbonyl compound (aldehyde or ketone) through a condensation step in concentrated H_2SO_4 at 0°C (see Figure 7).

Good yields of 80–84% were obtained when formaldehyde (R_1, R_2 = H) and acetaldehyde (R_1 = CH_3, R_2 = H) were used. Stronger reaction conditions, namely a temperature of 60°C, were necessary to achieve the formation of a dimer with acetone (R_1=CH_3, R_2=CH_3) and a moderate yield of 65% was obtained.[52,53] Such dimers are particularly interesting as they can be used as bio-sourced bis-aromatic monomers, providing specific mechanical properties to the final polyesters (Figure 8).

Khrouf *et al.* used the potentially bio-sourced ethylene glycol, 1,3-propanediol and 1,4-butanediol as co-monomers in the transesterification reaction with the furan dimers.[54] Melt polymerization can be employed if desired due to the very high stability of the monomers. It is also interesting to mention that the diacyl chloride form of the dimers can also be advantageously utilized.[55] Finally, such dimers have more recently been employed in the synthesis of copolymers such as block copoly(esters) with PET.[56]

b. *Polyurethanes and polyureas*

Due to a very broad scope of applications, the polyurethane industry nowadays has a great interest in substituting current petroleum-based polymers with bio-sourced products in order to address a growing sustainability issue. Traditionally, polyurethanes originate from the polycondensation reaction between a diol and a diisocyanate. Even if the use of diisocyanates (needing toxic phosgene) can be questionable from a bio-sourced mindset, it is worth mentioning that a profitable process today relies on the use of diisocyanate at an industrial scale. This is why isocyanate-dependant routes will be presented in this section. Given the fact that polyureas can be formed in a very similar fashion as opposed to polyurethanes, their formation will also be discussed thereafter.

When considering the panel of platform molecules accessible through the dehydration process of cellulosic resources, it is easy to notice that 2,5-bis(hydroxymethyl)furan (BHMF) can be used straightly by means of reaction with diisocyanate to form the corresponding polyurethane products (Figure 9).

The Quaker Oats Co. thus patented in 1982 and 1984 the use of BHMF in order to form polyurethane foams. BHMF, methylated-BHMF, BHMF homopolymer or BHFM-propylene oxide copolymer could be reacted with a wide scope of polyisocyanate compounds (Papi® supplied by Dow®) to yield foams exhibiting interesting flame-retardency ability.[57–59]

Finally, and following the similar pathways than those observed for polyesters, FF can be dimerized and transformed to get a diisocyanate, that can further be reacted with polyols to yield polyurethanes (Figure 10).

By converting the furfuryl amine dimer into their diisocyanate counterparts, Cawse *et al.* synthesized polyurethane materials. The resulting polymers exhibited similar properties when compared to the bis(4-isocyanatophenyl)methane (MDI)-based petroleum analogs. The authors thus displayed a bio-sourced alternative for the synthesis of industrially relevant polymeric materials.[60–62] Attempts were also performed by the team of Gandini for the synthesis of 100%-furanic compounds.[63,64] The isocyanate-derivated dimer of FF was thus reacted with various furan-based diols for making final polymers with tunable properties.

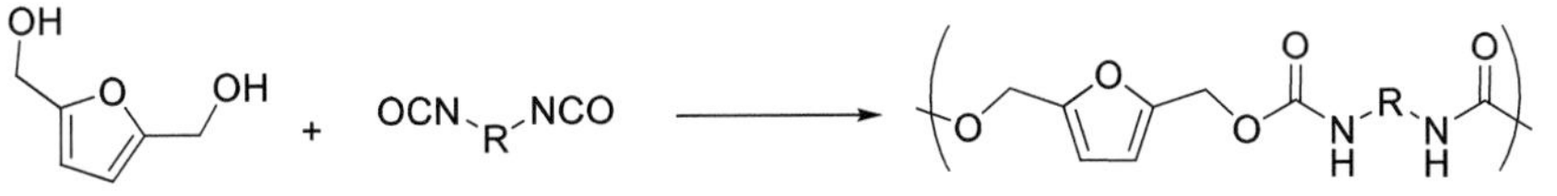

Figure 9.   Synthesis of polyurethanes from 2,5-bis(hydroxymethyl)furan (BHMF).

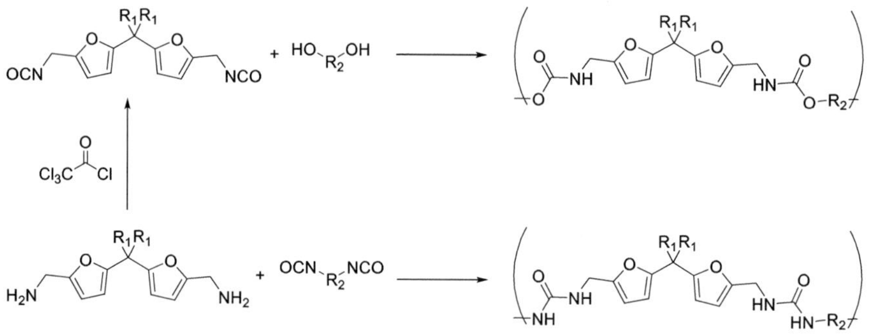

Figure 10. Synthesis of polyurethanes and polyureas from furfurylamine dimers and derivatives.

It is finally worth mentioning that furfurylamine dimer can also be converted into various polymers. Its reaction with diisocyanates yields for instance polyureas, but many other kinds of polymeric materials can be designed thanks to the versatility of the monomer.

c. *Polyamides*

Regarding the synthesis of bio-sourced polyamides from furanic resources (Figure 11), quite scare examples have been highlighted in the scientific literature. Bis(aminomethyl)furan (BAMF) thus emerges as one of the most logical compound to utilize.

Gandini *et al.* successfully copolymerized BAMF in the presence of 2,5-bis(carboxyl chloride)furan.[65] However, polyamides with low molecular weights were obtained. Moreover, the yield was found not to exceed 60%. This was attributed to side reactions occurring onto the CH_2-groups attached to the furan ring. A patent application from Rhodia also claims the preparation of polyamides from BAMF.[66] In the process, a stoichio-

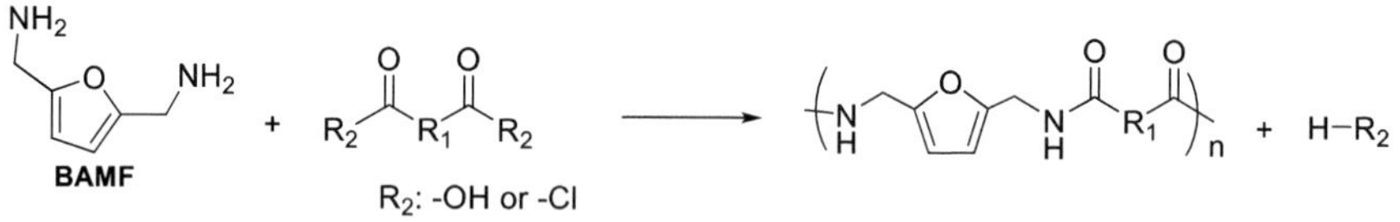

Figure 11. Polyamide synthesis from BAMF.

metric mixture of BAMF is performed in the presence of a diacid to yield the corresponding carboxylic ammonium salt. Polyamides are finally obtained thanks to a melt polymerization process.

d. *Poly(imines/Schiff bases) and polyvinyls*

As for polyamides, quite few examples of furan-based poly(imine) and poly(vinylene)s are encountered in the literature. Most of them describe the use of 2,5-diformylfuran (DFF) as a precursor (Figure 12).[5,67]

As far as the formation of poly(imine)s is concerned, several diamines — namely 1,4-phenylenediamine, 1,6-diaminohexane, 2,5-bis(aminomethyl)furan (BAMF)[68] and hydrazine[69] were used to investigate the structure-properties relationship. In the case of the three first diamines, low molecular weight, soluble polyimines were obtained (M_n were in the range of 1500–2500 g.mol^{-1}). It is however to be noted that interesting thermal stability and optical properties were observed, in particular when 1,4-phenylenediamine was utilized as a co-monomer. Such results were mostly attributed to the conjugation of the resulting polymer.

A more detailed study was conducted with hydrazine (Figure 13) with interesting properties such as a good thermal stability (decomposition started at 270°C without any sign of melting) and a conductivity of 10^{-5} to 10^{-4} S.cm^{-1} after doping. However, the obtained molecular weight was of similar fashion as opposed to the other diamines (M_n was of 2000 g.mol^{-1}).

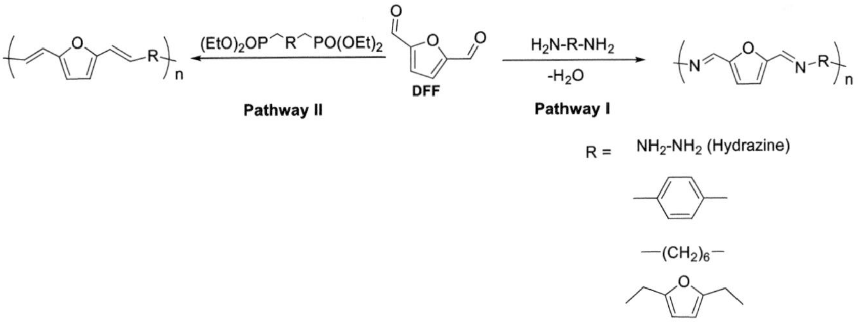

Figure 12. Synthesis of poly(imines) (pathway I) and poly(vinylene)s (pathway II) from 2,5-diformylfuran (DFF).

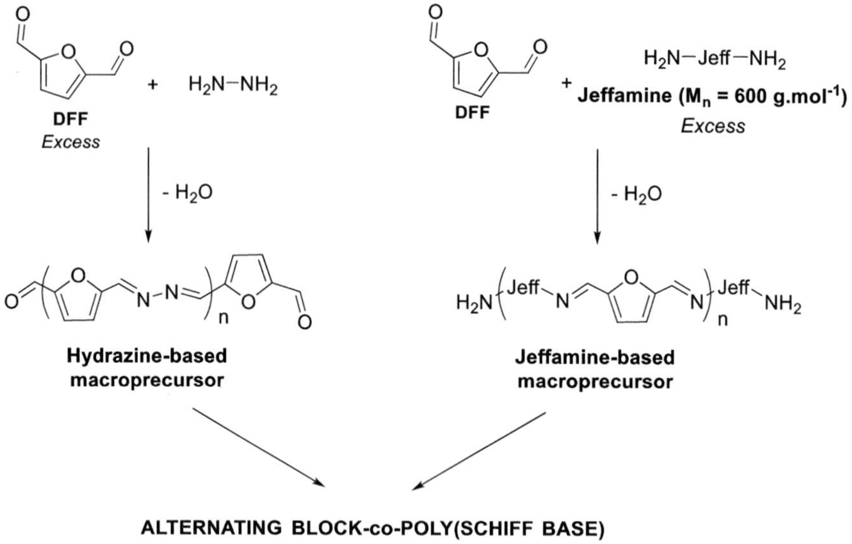

Figure 13. High molecular weight furanic Poly(Schiff bases) by block-copolymerization with soft segments.

Those encouraging, but not satisfactory, — results led to further attempts in order to increase the final molecular weights as well as foster the processability of such polymerization processes.

The introduction of soft segments in the polymer backbone was thus performed (Figure 13).[69] The addition of Jeffamine (M_n = 600 g.mol⁻¹) in a two-step process thus successfully yielded a plastic-like soluble multiblock co-polymer with a molecular weight close to 20,000 g.mol⁻¹.

As far as the formation of poly(vinylene)s is concerned, reductive homocoupling reactions of DFF did unfortunately not yield polymers of interesting chain-length; the formed polymers were too insoluble to undergo further reaction. A solution was proposed by using the Horner–Wadsworth–Emmons Wittig coupling of phosphonate esters with aldehydes (Figure 13, Pathway II). Thanks to a wise selection of bifunctional aromatic phosphonate linkers, it was possible to form more soluble polymers with increased molecular weights along with interesting electronic properties thanks to the conjugation.[70–72]

e. *A reversible polymerization via a Diels–Alder process*

In a context in which the concepts of sustainability, durability and recyclability are emerging, it seems important to mention the growing use of the Diels–Alder reaction (DA) for the formation of thermally reversible macromolecular architectures. Such a system has gained increased attention due to the possibility to form re-crosslinkable, re-mendable, re-shapable and finally recyclable polymers.

The thermoreversible furan/maleimide system in particular perfectly applies to lignocellulosic-based polymers DA processes (Figure 14).

The furan/maleimide relies on the equilibrium between the DA reaction, occurring at a temperature around 60°C and its reverse rDA reaction, occurring at a temperature around 110°C. Both the temperature gap between the two reactions as well as the relatively low temperature range in which both events occur made them popular among the polymer community since it allows synthesizing thermally responsive and recyclable polymers without degradation. This allows for instance for the design of materials capable of self-healing.

Several pathways to incorporate the furan moieties have been presented in the scientific literature: the incorporation of the furan moiety in the inner-backbone of the polymer chain, the modification of an existing monomer/polymer prior to polymerization *via* a DA reaction or the formation of a DA reaction-based monomer prior to polymerization *via* another pathway (Figure 15).

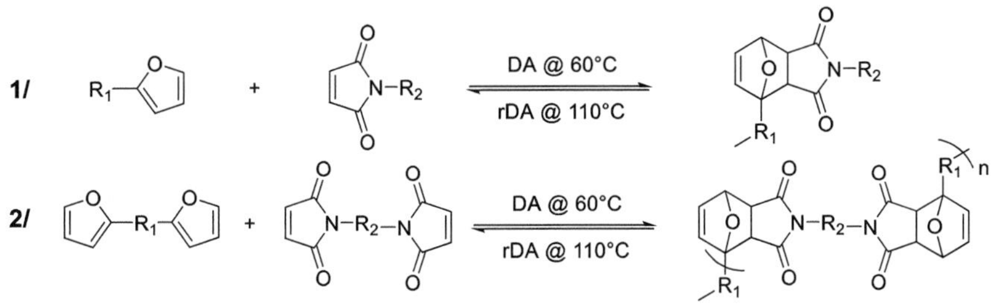

Figure 14. Equilibrium between the Diels–Alder reaction (DA) and retro Diels–Alder reaction (rDA) of 1/ furan- and maleimide-based monomers and of 2/bis-furan- and bis-maleimide-based monomers.

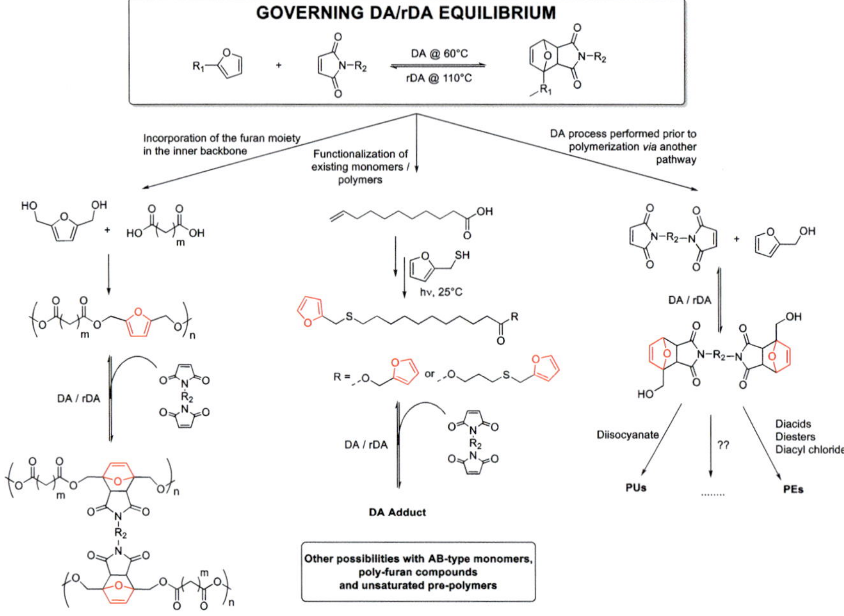

Figure 15. Several chemical pathways using the furan/maleimide DA system upon polymerization.

Yoshie *et al.* studied the effect of incorporating the furan moiety in the inner backbone of polyesters prior to cross-linking *via* a DA process with a bis-maleimide-derived compound.[42,43] The polyesters were synthesized by reacting together BHMF with succinic acid before cross-linking. Several bis-maleimides were tested and it was discovered that mendability was achievable, even with furyl telechelic prepolymers.[73]

The team of Gandini used thiol-ene chemistry on vegetable oil-derived compounds in order to attach 2-furylmethanethiol (obtained from FF) and thus a bio-sourced reactive diene for further DA/rDA process. Several strategies for making AA-type or AB-type monomers, targeting linear polymers, were investigated. The effect of increasing the functionality of the starting materials (A_3-type, AB_2/A_2B-type monomers, see Figure 16), then yielding branched polymers, was also studied to yield *in fine* a broad spectrum of polymers with various molecular weights and thermal properties.[74,75]

Finally, it is worth mentioning the possibility to form a diol by means of reacting bis-maleimide with two equivalents of FF–OH (Figure 15).

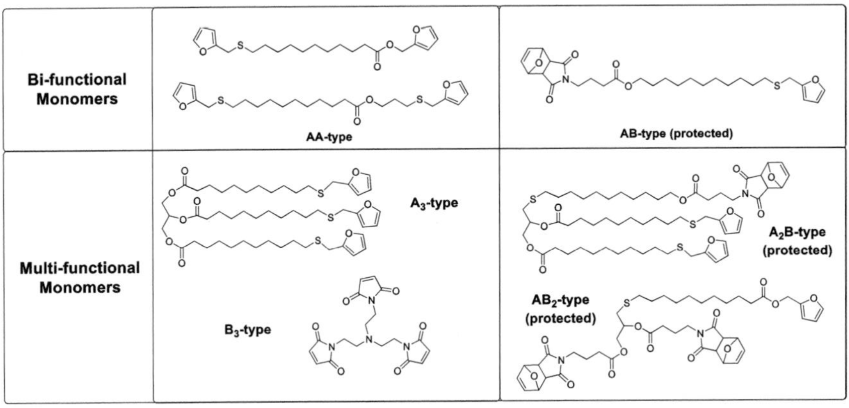

Figure 16. Different furan- and/or maleimide-based monomers with different functionalities as proposed by Gandini *et al.*[74,75]

This monomer can thus be polymerized into a lot of polymers (PUs and polyesters were reported) depending on the applications.[76] Very interestingly, the authors reported the possibility to play with the DA/rDA equilibrium in order to form original and quite controlled polymer hybrids, thanks to a simple heating-then-cooling process. When PUs and polyesters were blended together and heated to allow for the rDA to proceed, the polymer chains were sliced and then recombined each other upon cooling to form PU-Polyester hybrids thanks to the DA reaction. Let alone the easy process at stake (successive temperature ramps), the full potential of such a process lies in the possibility to control the polymer sequence thanks to the kinetics of the rDA/Da processes. Varying the heating and/ or cooling times made possible the control of the extent of depolymerization/repolymerization so that the scope of achievable products (and properties) becomes very large.

B. Levulinic Acid-based Polymers

1. *Accessible Monomers from Levulinic Acid*

Levulinic acid (LA) and levulinic acid-derived monomers are other kinds of very promising reagents. LA is earmarked by the DoE (see Figure 2) as one of the top value-added chemicals originating from biomass.

As described in Figure 17, LA can be obtained through the de-hydration process in acidic medium of HMF and FF. Once formed, it is then possible to convert this platform molecule into different other derivatives. Such derivatives have been extensively described by Isikgor and Becer[77] but we will only focus on specific monomers in the course of this chapter.

It is interesting to note that a large scale production of levulinic acid from lignocellulosic biomass has already been achieved. Maine BioProducts with its "Biofine Process" or Aventium's YXY technology claim to possess a large scale process for the production of levulinic acid. While the first is based on the catalyzed dehydration of lignocellulosic feedstocks into LA in a two-stage process,[78–81] the second company has developed a process to convert carbohydrates into high purity methyl levulinate. It is however to be noted that the cost-competitiveness of those production lines still has to be improved if a complete replacement of petroleum-based monomers is desired. Indeed, according to Becer, the current prices of LA remain between \$5 and \$8/kg.[77] This price has to be decreased below \$1/kg to envision a cost-competitive process in which LA would play a key role in the replacement of fossil-based monomers. This constitutes one of the main challenges for researchers nowadays.

As exemplified in Figure 17, LA can be converted into various monomers, which correspond to diverse final applications. LA can be esterified then converted into levulinic ketal upon reaction with a diol (generally originating from vegetable oils). This process has been developed and led to an industrially viable process at Segetis.[78] The corresponding products can be used for the production of polyurethanes and plasticizers.[82]

Amino-derivatives of LA can also be of high interest, in particular when it comes to their use in photomedicine.[83] However, if the synthesis of 5-aminolevulinic acid (5-ALA) is possible at the laboratory scale by means of bromination agents, it remains a challenge to form 5-ALA at an industrial scale, because a safer and cheaper pathway has to be found to allow for a viable industrial process.[77,84]

2-Butanone (methyl ethyl ketone — MEK) is accessible *via* a catalyzed decarboxylation step of LA. Among others, MEK is known for its use as a common industrial solvent.[85]

4-Hydroxypentanoic acid can be obtained by means of a reduction reaction with hydrogen. This monomer is part of the hydroxyalkanoates

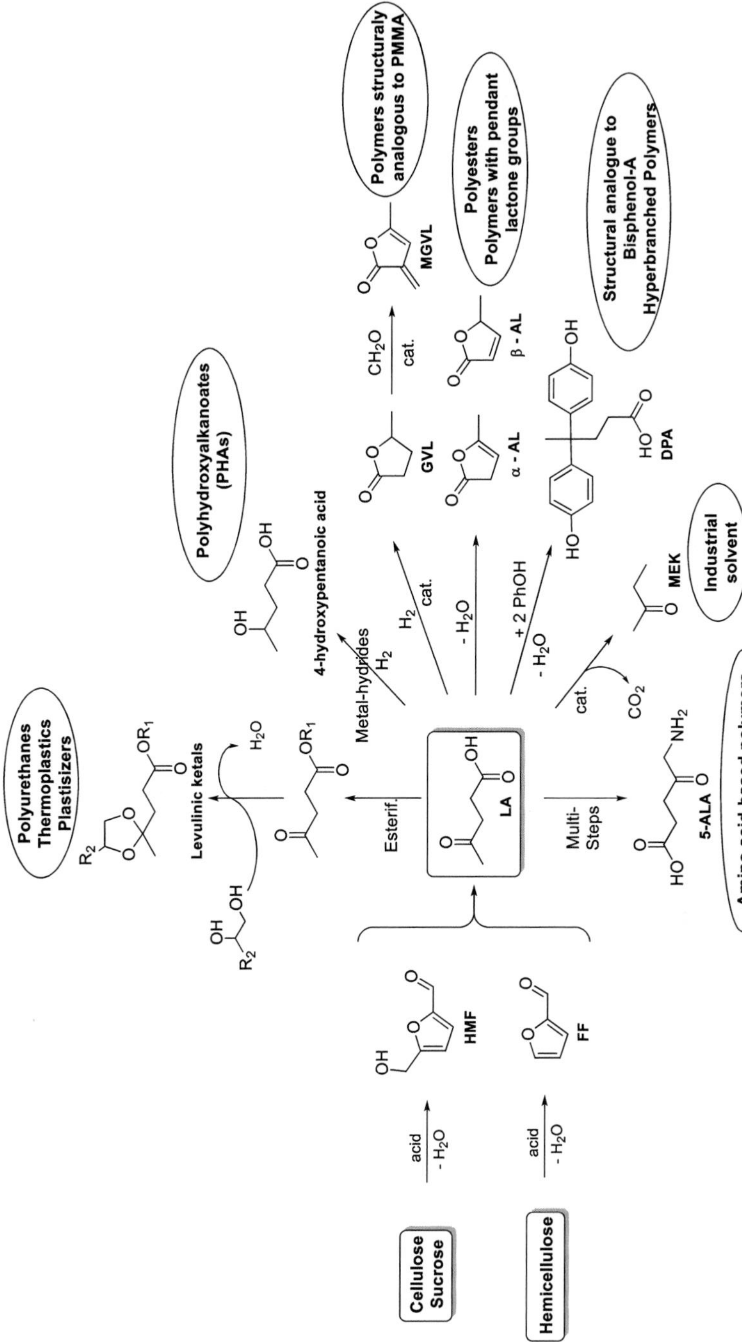

Figure 17. Possible derivatives from the levulinic acid platform and their use thereafter.

family and can yield after polymerization the corresponding polyhydroxyalkanoate (PHAs). Further details about this kind of polymers is given in Section II.D.4.

Three other monomers: α-angelica lactone (AL), α-methylene-γ-valerolactone (MGVL) — also known as γ-methyl-α-methylene-γ-butyrolactone and finally diphenolic acid (DPA), are particular targets from this platform and their polymerization will specifically be addressed in the next paragraphs. The latter are the most studied LA-derived monomers in the literature.

γ-Valerolactone (GVL) can be obtained in high yields through a catalyzed hydrogenation process. Ruthenium-based catalysts were found to be the most suitable for this reaction.[86,87] Unfortunately, the difficulty to find cheap catalysts or to recover the expensive ones prevents the implementation of GVL as a commodity chemical. GVL can thus be converted into MGVL. In 2004, Manzer *et al.* demonstrated that GVL could be reacted with formaldehyde yielding MGVL. This monomer has attracted interest due to its ability to substitute the common fossil-based methacrylate monomers. Two main processes were described: a continuous-gas-phase process using supported catalysts[88] and a two-step cascade process requiring supercritical or near-supercritical CO_2 as a solvent, the second process yield 90% of MGVL.[89]

Angelica lactones (AL) are dehydration products of LA in presence of acidic catalyst. Acetic anhydride or phosphoric acid were found to be suitable for this purpose.[90,91] Finally, diphenolic acid (DPA) can easily be obtained, thanks to the reaction of one mole of LA with two moles of phenol at 100°C under acidic conditions.[84,92] Another pathway, using ionic liquids was also reported in 2013.[93]

2. *Polymers from Levulinic Acid Platform*

Different polymerization processes emerge from the LA platform. The structures of AL and GVL (and their derivatives) provide them versatility in the way of reacting: either by ring-opening polymerization (ROP) or by reaction of alkene moieties contained in or attached to the ring. Finally, DPA is a very particular monomer of interest due to its structure, very close to the chemical structure of bisphenol-A (BPA). This similarity made scientists think it could be a promising candidate for a bio-sourced replacement of BPA.

a. *Ring opening polymerizations*

GVL has been studied in processes involving ring-opening as exemplified in Figure 18.

GVL is a very stable 5-membered ring. This explains why to our knowledge, ring opening homopolymerization (ROhP) of GVL has not been described. This can be compared to its 6-membered ring counterpart, δ-valerolactone, which has been successfully reacted through ROhP. It is often desired to pay careful attention to literature data as confusion might exist between these two monomers.

Ring opening copolymerization (ROcP) of GVL was thus performed in view of forming potentially biodegradable bio-sourced polymers. For instance, the copolymerization of GVL with diglycidyl ether of bisphenol-A (DGEBA) yields semi-bio-sourced thermosets.[94] Copolymers of GVL and β-butyrolactone were also synthesized through a catalytic process.[95]

ROcP is a very interesting solution to force the opening of GVL. This process is however challenging and it prevents as a consequence the amount of available literature in the field. For this reason, researchers investigated other possible pathways to take advantage of this platform molecule. Quite an extensive work has thus been conducted in searching for suitable ways of ring-open GVL prior to polymerization. By opening the ring, the thermo-dynamical stability of the obtained intermediates — such as gamma-hydroxyamines, 1,4-pentanediol, 1,4-pentanenoïc acid and methyl pentenoate — is reduced and allows for further chemical modifications such as polymerization. Without trying to be exhaustive in the amount of presented examples, herein will be presented several occurrences of interest in the literature (Figure 18).

Thanks to an aminolysis process, GVL can ring-open to form γ-hydroxyamides. This process has been earmarked for its versatility to yield monomers that can be employed to produce relatively high molecular weight polymers, especially polyurethanes. Novel monomers can indeed be formed in high yields, in particular when non-sterically hindered amines were utilized in the ring-opening process.[96] When amino-alcohol is used

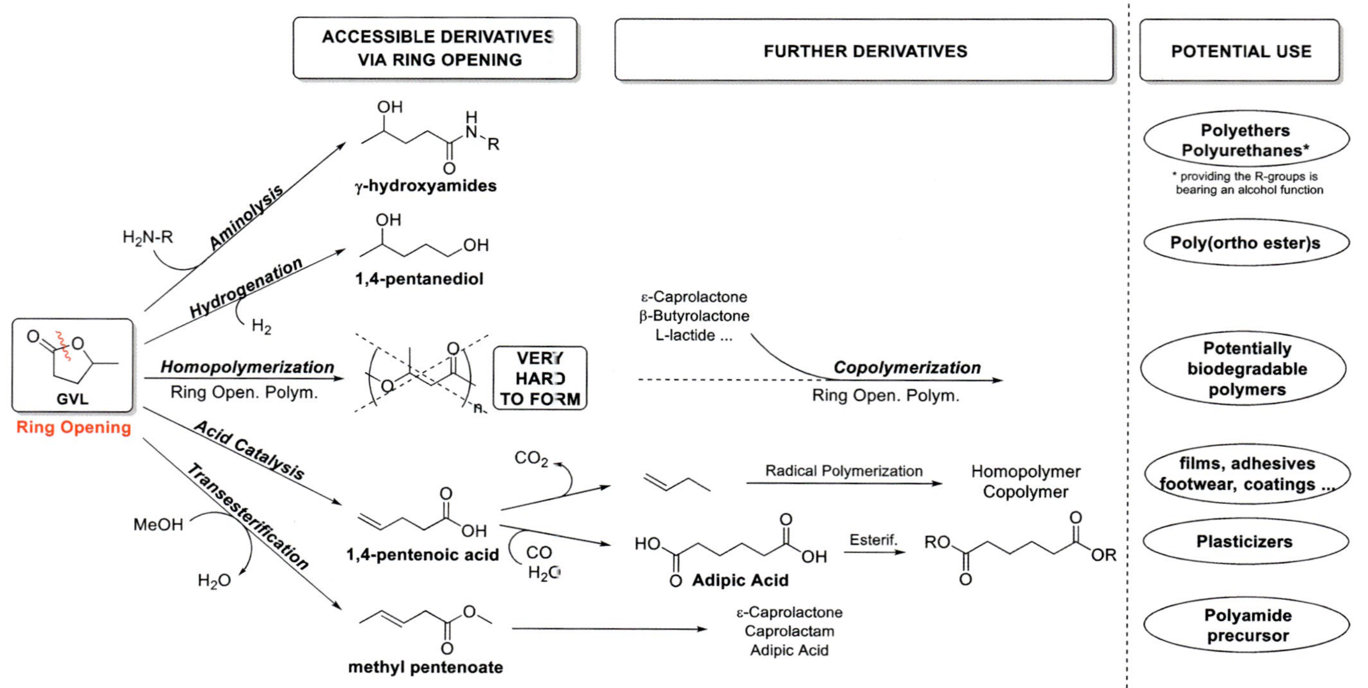

Figure 18. Possible derivatives from the γ-valerolactone platform and their use thereafter.

(for instance 1,2-aminoethanol), polyurethane could be formed by reaction with a diisocyanate.

By hydrogenation process thanks to a copper-based catalyst, GVL can yield 1,4-pentanediol.[97] There is unfortunately scarce information about any polycondensation process of this monomer. The synthesis of poly(ortho ester)s has though been investigated in view of application in the pharmaceutical field.[98]

GVL can also be ring-opened with the help of a transesterification process. Lange *et al.* published in 2007 a transesterification process of GVL with a yield of 95% in which methanol is fed in a catalyzed reactive distillation process.[99] Based on the different boiling points between GVL (207°C) and methylpentanoate (127°C), the latest is progressively isolated through successive distillation processes to yield a product as pure as possible.

Among the accessible monomers from levulinic acid, α-Angelica Lactone (AL) can also be ring-opened. As described for GVL, the small ring strain energy of this 5-membered ring monomer inhibits its ROP. This is the reason why the successful homopolymerization of AL to the corresponding polyester is noteworthy to mention (Figure 19).

Lewis acid needs to be used to induce the ROP of AL. Several examples can be found in the literature, such as sodium isopropylate.[100] Polyester molecular weights were measured around 20 kg.mol^{-1} after reaction at 60–65°C for 40–60 minutes. It is also interesting to mention that 20 to 32% of the C=C double bonds reacted during the ROP process. It is assumed that they got involved in a radical reaction process, which could account for this disappearance. Stannous octoate was employed in toluene at 130°C for 30 hours for the ROP of AL, yielding a corresponding polyester with a M_n of 29 kg.mol^{-1}.[101]

However, the presence of the double bond makes the final polyesters quite unstable, especially when exposed to daylight. The latter can also

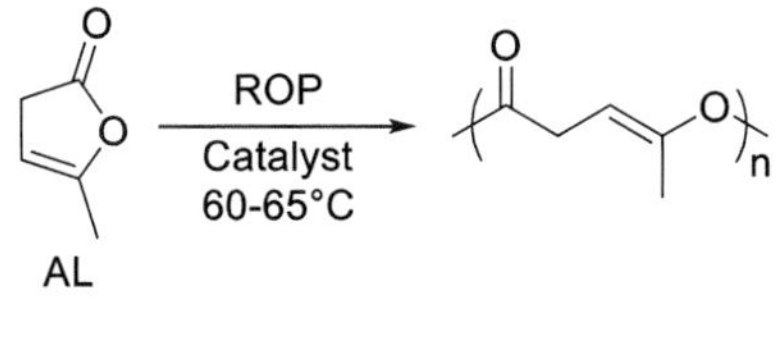

Figure 19. Ring-opening homopolymerization of α-angelica lactone.

decompose in acidic or basic conditions. Thus, post-functionalization of the AL-homopolymer *via* the chemical modification of the C=C bond seems highly desirable if up-scaling of such processes are envisioned.

b. *Polymerization through reactive alkene moieties*

Some LA derivatives contain reactive double bonds that can be reacted to yield original polymers. Among them, α-methylene-γ-valerolactone (MGVL) has attracted a lot of interest as it could be a potential bio-sourced candidate for the replacement of methyl methacrylate (MMA). As described in Figure 20, MGVL can be synthesized through the reaction of GVL with formaldehyde. Due to the difficult handling of formaldehyde in the gas phase, it was found that dissolving GVL in formalin (37% aqueous formaldehyde) in the presence of Barium-based catalysts supported on Silica could be a suitable solution for the production of MGVL.[88]

Early work focused on the polymerization of MGVL in free-radical emulsion process, radical, anionic and group-transfer polymerizations. However, those mostly led to incomplete conversion of MGVL.[102] Chen's group published a series of papers tackling radical polymerization of MGVL using catalysts and provided extensive solutions through the use of frustrated Lewis pairs. In particular, alane-based frustrated Lewis pairs were found to be particularly suitable to promote the radical polymerization of MGVL.[103–110] Fast processes could thus be developed with polymerization in the range of 10 minutes at room temperature.

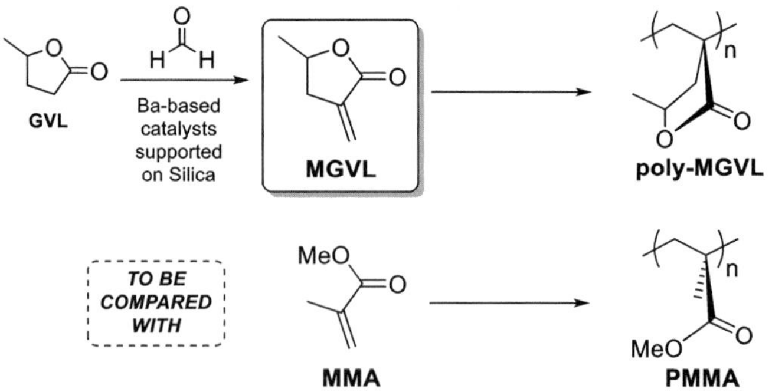

Figure 20. Radical homopolymerization of MGVL as a potential bio-sourced replacement for PMMA.

PolyMGVL molecular weights up to 500 kg.mol^{-1} could be reached when adapting the reaction conditions. As far as thermal properties are concerned, polyMGVL exemplified enhanced properties as opposed to PMMA with a glass transition temperature of 225°C (120°C higher than PMMA) and with similar degradation temperatures. Resistance to heat, solvent and scratch could also be observed.[111] This is attributed to the presence of the ring, providing more rigidity. However, it is to be noted that the obtained polymers were atactic and amorphous. Today, major challenges rely in finding both an economically attractive catalytic parthway to form polyMGVL as well as mastering the control of the tacticity of the final product to fully understand its influence on the resulting mechanical properties.

c. *Diphenolic acid-based polymers*

Main examples of polycarbonates are based on the reaction of ester forms of Diphenolic Acid (DPA). DPA has attracted a lot of attention due to its similar structure with bisphenol A (BPA) making it a potential bio-sourced alternative to the petroleum-based monomer. As described in Figure 21, DPA is produced when reacting LA with two moles of phenol, thus forming a tri-functional compound containing two alcohol and one carboxylic acid moieties. Following the example of BPA, DPA was reacted to form polycarbonates. This however requires the acid group to be protected to avoid branching upon reaction.

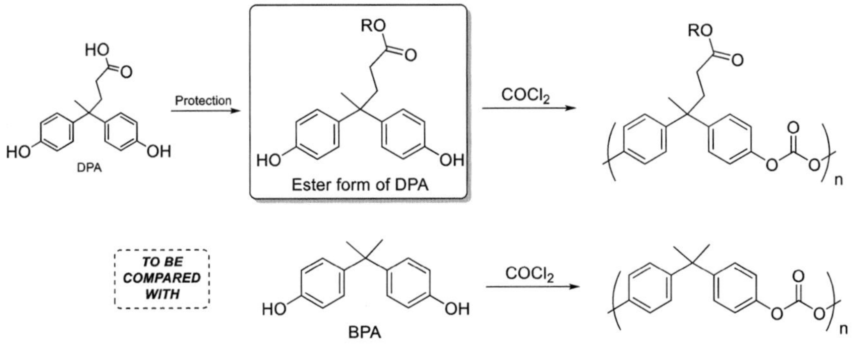

Figure 21.   Formation of DPA-based polycarbonates.

200   Cramail *et al.*

Polycarbonates formed with ester-DPA and phosgene were described in the literature. The final polycarbonate can be formed thanks to several chemical pathways. For instance, interfacial polymerization, using phosgene in gas phase and ester-DPA in a pyridine solution, or the use of phase-transfer catalysts gave interesting results. Polymers containing the ester form of DPA noteworthy exhibited similar molecular weight (around 140 kg/mol) and dispersity as opposed to their BPA-containing counterparts. As far as physical properties are concerned, ester-DPA-containing polymers have lower glass transition temperatures than BPA-containing polycarbonates (108°C *versus* 150°C). Ester-DPA-based products were also soluble in organic solvent. It is finally interesting to mention that if the ester function is hydrolyzed, a dramatic change of the glass transition temperature can be observed as it increases up to 148°C. This behavior was attributed to the formation of hydrogen bonds, inhibiting flexibility of the global system.[112–114]

DPA can also be used to form membranes. Zhou and Kim produced for instance sulfonated poly(arylene ether ketone) (SPAEK) using this monomer in the formulation. The pendant carboxylic acid groups were then used as crosslinkers to form crosslinked SPAEK membranes that exhibited promising properties when compared to the industrial Nafion® 117.[115]

Gross and co-workers have also reported the formation of bio-based thermosets through the synthesis of DPA-derived bis epoxy monomers, and their subsequent curing (Figure 22).[116,117] By means of ester-protection of the acid moiety with short alkyl-chains, the addition of glycerol-devired epichlorohydrin under alkaline conditions formed diglycidyl ethers of diphenolate esters. Tuning of the monomer viscosity could be observed by varying the chain length of the alkyl chain of the ester moieties. Bio-based isophorone diamine was used to cure the so-formed

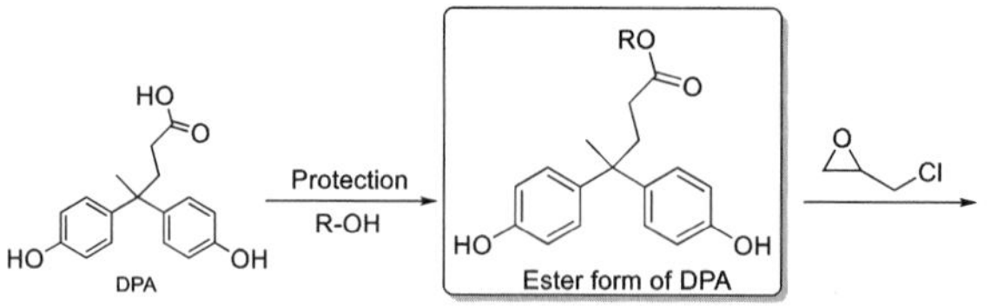

R= CH_3, C_2H_5, C_3H_7, C_4H_9, C_5H_{11}

Figure 22. Synthesis of diphenolate ester-based glycicyl ethers.

bis epoxy monomers. Thermal and mechanical properties exhibited similar valued as opposed to "classical" DGEBA-derived thermosets, thus making DPA-based resins promising alternatives for petroleum-based epoxy-systems.

Hyperbranched polymers (HPBs) nowadays are gaining growing attention due to both their specific properties and their ability to be quite easily transferable at an industrial scale. The synthesis of HPB-polymers requires multifunctional monomers (at least tri-functional). This can allow for one-pot syntheses *via* intermolecular reactions. The obtained polymers, because of their highly branched architecture, usually have interesting physic-chemical properties, lying in between those of linear polymers and dendrimers. DPA contains a carboxylic acid moiety as well as two phenol units. It thus raises as an interesting and promising candidate for the production of HPB polyesters through AB_2-type intermolecular polyesterification (Figure 23).

DPA-based HPB have been synthesized thanks to a melt condensation process, under reduced pressure with the help of catalysts. It is to be noted that a good control of the reaction temperature is needed (temperature ramps are used). Mw values ranging from 47 to 128 kg.mol^{-1} were obtained. No cyclization was observed during the reaction. The obtained polyesters were amorphous, with a Tg around 100°C and a degree of branching close to 49%.[118,119] Properties of the DPA-based HPB

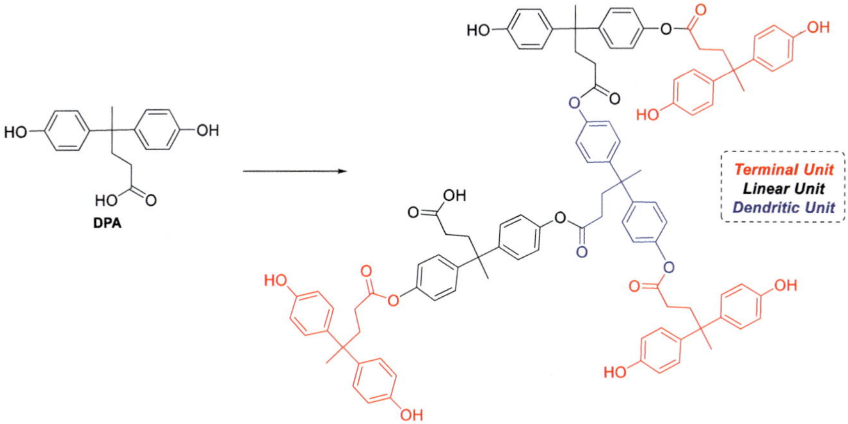

Figure 23. DPA-based hyperbranched polyesters.

polyesters can also be tuned by means of copolymerization. As an example, Serra's team published in 2012 a DPA-polyethylene glycol (PEG)-based triblock HPB copolymer, with a degree of branching of 50%, Mn values of 10 kg.mol^{-1} and self-assembly properties in solution of acetone. The same group used those compounds as chemical modifiers for photo- and thermal-curing of epoxy resins.[120]

C. Isohexide-based Polymers

Multiple mono-saccharide-based monomers can be used in the preparation of polymers having sugar units incorporated into the main chain. These monomers named alditols, $HOCH_2$-$(CH_2OH)_n$-CH_2OH, aldonic acid, $HOCH_2$-$(CH_2OH)_n$-COOH or aldaric acid, HOOC-$(CH_2OH)_n$-COOH (Figure 24) were reported for the synthesis of

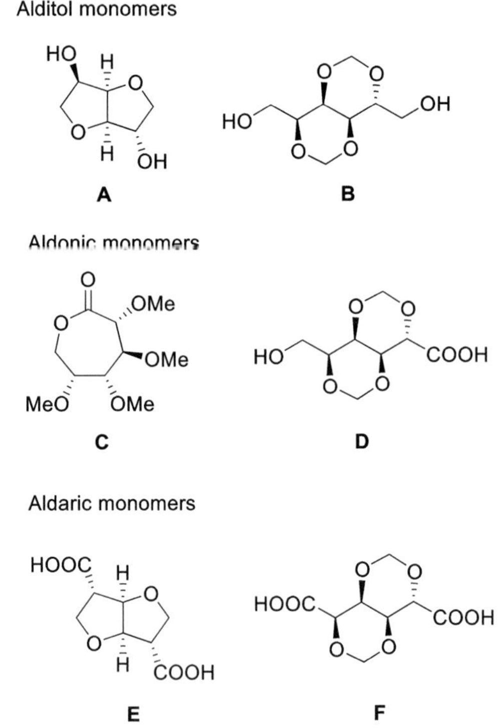

Figure 24. Examples of hexose-based monomers.

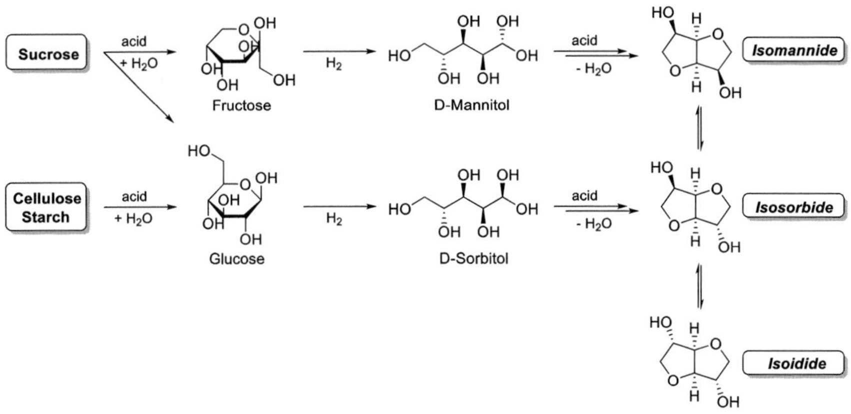

Figure 25. Molecular structures of isosorbide, isomannide and isoidide.

polyesters, polyamides, polycarbonates and polyurethanes as recently reviewed by Galbis *et al.*[121]

Herein we will mainly focus on isosorbide (Figure 24A) and derivatives also called isohexides and polymers thereof. Isosorbide and its stereoisomers, isomannide and isoidide (Figure 25) are thermally stable. The first two are currently commercially available and isoidide can be obtained from isosorbide.[122] By chemical modification of these diols, isohexide-based dicarboxylic acid,[123] diamine[124] and diisocyanate[125] can be produced. The most important feature of these monomers is linked to their rigidity, chirality and non-toxicity. However, the latter exhibits relatively low reactivity due to the nature of secondary alcohol moiety.

Among isohexide derivatives, isosorbide is the most studied monomer. It does not possess any symmetry contrary to isomannide and isoidide and its two alcohols do not possess the same reactivity.[126] Indeed the hydroxyl in C3 position exhibits an endo configuration generating a weak hydrogen interaction with the intracatenar oxygen of the connected ring (isomannide alcohols are in this configuration, too) while the configuration of the hydroxyl in C6 position is exo and thus more accessible (isoidide alcohols have also such a configuration), see Figure 26.

Isosorbide can be obtained by double dehydratation of D-sorbitol or directly from starch or more recently from cellulose (Figure 25).[127] Isomannide is synthesized in similar conditions from another highly available

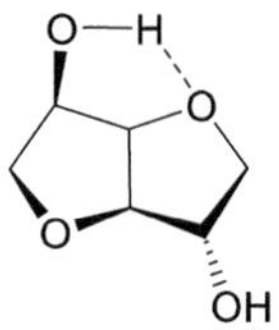

Figure 26.   3D molecular structure of isosorbide.

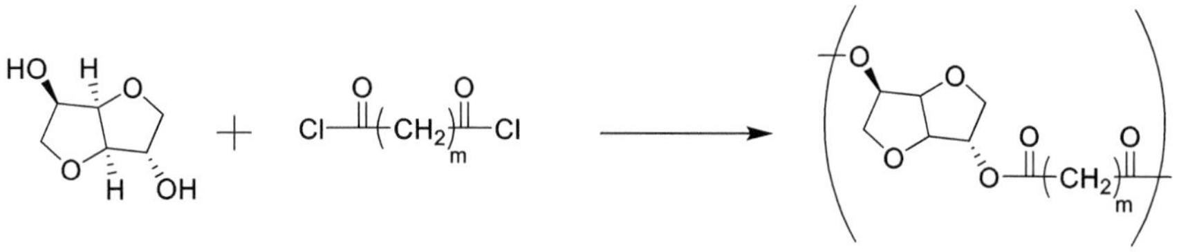

Figure 27.   Synthesis of polyesters based on diacyl chloride and isosorbide.

sugar, i.e. D-mannitol. Isoidide, which possesses the most reactive alcohol functions, can be produced from L-idose but the latter rarely exists in Nature and thus remains nowadays an expensive monomer. Routes need to be developed to isomerize sugar to L-idose in order to decrease its production cost.

The most representative polymers synthesized from these three monomers are polyesters, polycarbonates, polyethers and polyurethanes. They generally exhibit high Tg, good thermomechanical resistance, have specific optical properties and can be biodegradable. Another important industrial application of isohexides is their use as diesters as substitutes to phthalate plasticizers.[128]

Isohexide-based aliphatic polyester can be obtained by molten polymerization using various diesters or their corresponding diacylchlorides (Figure 27). As a general trend, polyesters with a methylene sequence higher than six are semi-crystalline except for isomannide-based polyesters that remain amorphous.[129] Moreover, isoidide-based polyesters exhibit higher melting temperature due to the symmetry of isoidide. As an example, polyesters from sebacic acid possess a Tg of 0°C for all three isohexides and a melting temperature of 60°C and 130°C for isosorbide and isoidide, respectively.[130] Accordingly, the enzymatic degradation of polyesters based on isohexides and sebacic acid decreases in the order isosorbide > isomannide > isoidide.[131]

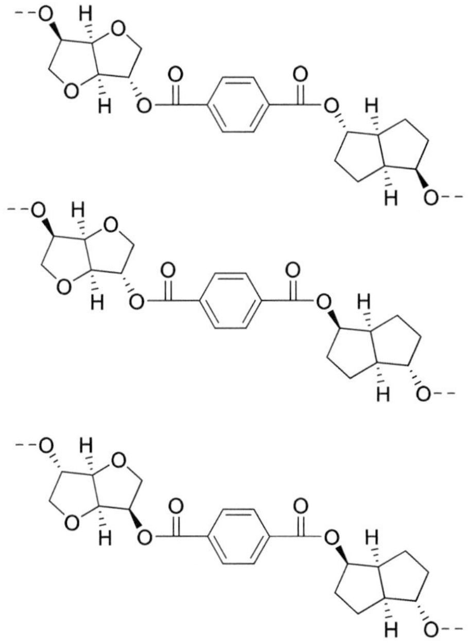

Figure 28. Chemical structures of isosorbide-based terephthalic acid-centered triads, PIT.

Isohexides can also be copolymerized with aromatic diacids. Poly(isosorbide terephthalate) (PIT) (Figure 28), exihibits a Tg of 200°C and its degradation temperature is over 360°C; however limited polyester molecular weights are obtained due to the low reactivity of isosorbide. In the case of isoidide, a Tg of 210°C and a melting temperature of 260°C are obtained while, in the case of isomannide, only very low molecular weights are obtained.[132]

One should note that PIT possesses three different stereo-sequences since the two hydroxyl functions are not equivalent and there is still a lack of report investigating the impact of the microstructure on PIT properties. Indeed, stereoregular PIT could exhibit even higher Tg and this polymer could compete with other high Tg thermoplastics such as poly(ether imide) for instance. Incorporation of isosorbide in a PET backbone (leading to PEIT) can increase the Tg of the polymer up to +120°C.[133] PEIT and PIT are now allowed for food packaging and have

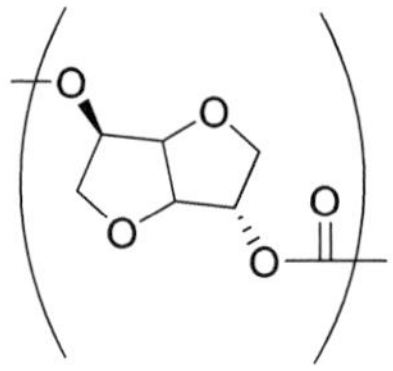

Figure 29.   Poly(isosorbide carbonate), PIC.

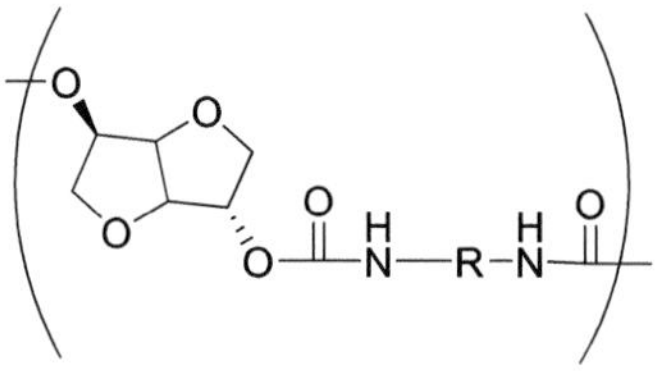

Figure 30.   General structure of polyurethanes based on isosorbide as diol.

the advantage to allow hot-filling contrary to standard PET. FDCA can also be copolymerized with isohexide leading to polyesters with Tg of 194°C, 191°C and 196°C for isosorbide, isomanide and isoidide, respectively.[134]

Another important class of polymers obtained from isohexides is the one of polycarbonates (Figure 29) as substitute to poly(carbonate of bisphenol A). Poly(isosorbide carbonate), PIC, can be obtained by polycondensation of isosorbide with various phosgene analogs leading to a highly transparent polymer with a Tg around 170°C.[135]

Similarly, copolycarbonates can be obtained, from 1/1 isosorbide/bisphenol-A mixture. The polymers so-formed have been reported with a Tg of 135°C and 160°C, for the random and alternate copolymers, respectively.[136] The insertion of isosorbide in a bisphenol-A polycarbonate is now commercially available as Durabio®.

As far as polyurethanes are concerned, isohexides can be used as hard segment diol. Isosorbide has been polymerized with various diisocyanates leading to polymers with Tg of 110°C, 135°C and 190°C for HDI, TDI and MDI, respectively (Figure 30).[122] Isoidide gives similar Tg while isomannide leads to polyurethanes with a 20°C to 50°C lower Tg. Using diamino analog of isosorbide, fully isohexide-based polyurethanes can be synthesized.[137] Low PU molecular weights were obtained (around

3000 g/mol) and Tg were in the range of 120°C to 135°C depending of the isohexide used.

D. Synthons and/or Polymers Originating from Biotechnological Routes

1. *Lactic Acid-based Polymers (PLA)*

Poly(lactic acid), PLA, represents one of the most mature bioplastic as evidenced by its significant production and commercialization.[138] PLA is a linear aliphatic polyester derived from lactic acid (2-hydroxypropionic acid), a chiral organic hydroxyacid that exists in two different forms. Indeed, depending on the stereochemistry of the tertiary carbon of lactic acid molecule, the L-lactic acid and the D-lactic acid isomers are accessible. Lactic acid can be obtained by chemical synthesis or fermentation. Chemical synthesis of lactic acid is mainly based on the hydrolysis of lactonitrile by strong acids (Figure 31). Other routes are however suitable such as base-catalyzed degradation of sugars, oxidation of propylene glycol, reaction of acetaldehyde, carbon monoxide, and water at elevated temperatures and pressures, hydrolysis of chloropropionic acid and nitric acid oxidation of propylene. Main drawbacks of the chemical routes

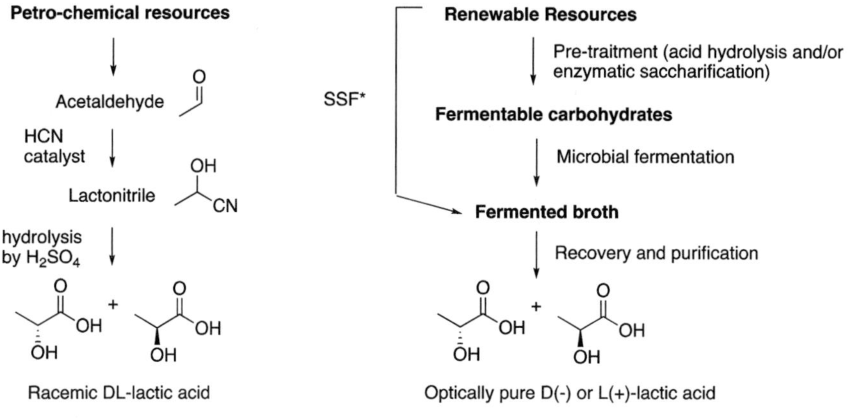

Figure 31. Production of lactic acid from petro-chemical resources or renewable resources (starch).

include the lack of cost effectiveness and the non-stereoselectivity.[139] Indeed lactic acid produced by the petrochemical route exists as a 50/50 optically inactive mixture of L- and D- forms. Due to the above mentioned reasons and environmental concerns, the fermentative pathway to lactic acid has gained widespread interest and is now industrially developed by most of lactic acid producers.[140]

Indeed, the biotechnological production of lactic acid offers various advantages over the chemical pathway like low cost of substrates, low production temperature and low energy consumption. It is also noticeable that, depending on the micro-organism used in the biotransformation, L- or D- isomers are preferentially formed (Figure 31).[141] Lactic acid bacterias (LAB) can be classified into two groups: homofermentative and heterofermentative. While the homofermentative LAB convert glucose almost exclusively into lactic acid, the heterofermentative LAB catabolize glucose into ethanol and CO_2 as well as lactic acid. Only the homofermentative LAB are available for the commercial production of lactic acid.[139] The carbon substrates used for microbial production of lactic acid include various sugars, either pure (e.g glucose and sucrose) or impure (e.g starch). Thus, a large variety of raw materials can be used such as molasses, sugar cane, bagasse and starchy materials from potato, tapioca, wheat, barley, corn, etc.[139,142,143] Starchy materials like sugarcane bagasse, cornstarch, corn cob, wheat bran are the most economical substrates for the production of competitive lactic acid.

By using such starchy substrates, several steps are needed to produce lactic acid. In conventional biotechnological processes, a pretreatment step for gelatinization and liquefaction of the biomass is required, which involves the use of high temperatures in the range 90–130°C. Then, enzymatic saccharification to glucose is carried out followed by subsequent conversion of glucose into lactic acid by fermentation. However, the conversion of starch or cellulose to sugar consumes energy during liquefaction or saccharification and increases the cost of production. In addition, sugar concentration in the hydrolyzate highly affects fermentation as bacterial cells cease to produce lactic acid when the sugar concentration is high.[144] To overcome these limitations, direct fermentation processes were developed. The direct conversion of complex starchy or cellulosic substrates to lactic acid includes three different routes. In the first one, the lactic acid producing fungi can directly convert starch to lactic acid with the help of enzymes. In another pathway, amylolytic lactic acid bacteria allows direct fermentation of the substrate.[144] The last possibility is to simultaneously treat the carbohydrate substrate with degrading enzymes and lactic acid producing bacteria. These solutions then offer the controlled release of sugar in the optimum growth temperature. Glucose inhibition on the enzyme is therefore minimized.

Due to considerable improvements, almost all the produced lactic acid is manufactured through fermentation routes. Thus current global production is between 300,000–400,000 tons per year with the United States being the largest regional market for lactic acid, accounting for a significant share of the worldwide market. Western Europe and Asia-Pacific make up the other major lactic acid producers and consumers on a global scale. The major manufacturers of fermentative lactic acid include Natureworks LLC (USA), Purac (Netherlands), Galactic (Belgium), Cargill (USA) and several Chinese companies.

a. *Synthesis of PLA*

Two main routes are generally referred to convert lactic acid to PLA (Figure 32). Direct polymerization by polycondensation leads to poly(lactic acid) while the indirect route *via* lactide (LA), a cyclic dimer of lactic acid, leads to poly(lactide). Both products are referred to as PLA.[145] Owing to the fact that lactic acid bears both a hydroxyl and a carboxylic acid function, polycondensation is inevitably the most direct route to PLA. However, molecular weights obtained by this pathway are not high enough to consider various applications of the resulting polymer. The low reactivity of the secondary hydroxyl function of lactic acid, combined with the need to efficiently removed water to shift the equilibrium toward the formation of the polymer constitute the main drawbacks of this pathway.

Polycondensation route

It is however possible to obtain higher molecular weights using different reaction conditions or *via* chain extension of pre-polymers. Indeed, a first

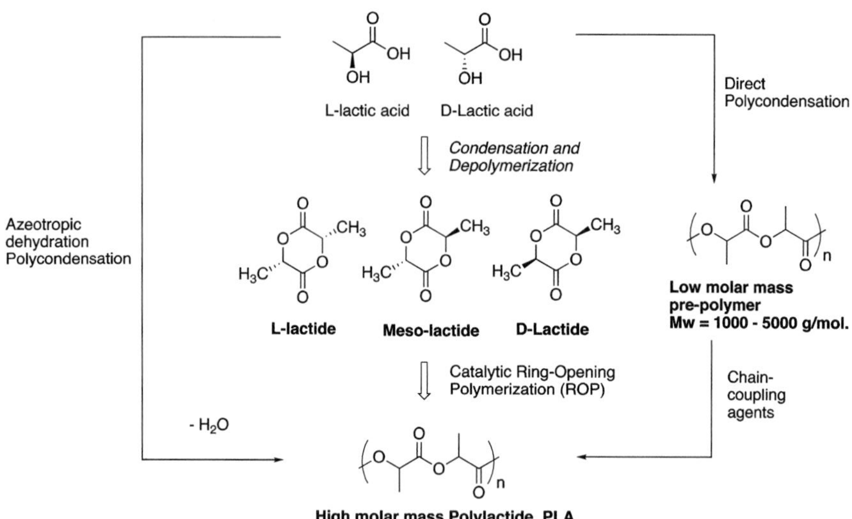

Figure 32.   Routes to high molecular weight PLA.

strategy involves the polymerization of lactic acid in appropriate solvent that allows for azeotropic distillation of the condensate.[146,147] Generally reaction time is in the range 30–40 hours at 130°C making this process unsuitable for low cost PLA production. Moreover, high concentration of catalyst is needed to have access to high molar mass PLA, which facilitates degradation and hydrolysis during processing.

Another approach to synthesize high molecular weight PLA is based on the polycondensation of lactic acid in the presence of difunctional monomers (e.g. diacids or diols) ending up in telechelic prepolymers. These resulting terminal functional groups of the prepolymers can then be reacted by using chain extenders such as diisocyanate,[148] bis(amino-ether).[149] However, increase in both cost and complexity of the multi-step process, as well as possible unreacted chain-extending agents, which alter the properties of the final material, are the main drawbacks limiting the industrial use of this method.

An alternative route consists in the melt/solid polycondensation of L-lactic acid. In this process, a polycondensate with a low molecular weight of around 20 kg.mol^{-1} is first prepared by ordinary melt polycondensation. Crystallization of the obtained poly(L-lactic acid) (PLLA) is then carried out by heat-treatment at around 105°C followed by heating at temperatures above Tg, yet below Tm of PLLA. Thus polymerization is performed in the amorphous phase, where all the reactive end groups reside, allowing the synthesis of high molar mass PLA in the range 100–500 kg.mol^{-1}.[150]

The advantages of solid-state polymerization include low operating temperatures, which control over side reactions as well as thermal, hydrolytic, oxidative degradations along with reduced coloration and degradation of the polymer. Moreover, there is practically no environmental pollution, because no solvent is required.

PLA synthesis by ring-opening polymerization (ROP) of lactide

The most common way to obtain high molar mass PLA is through ring-opening polymerization (ROP) of lactide, Figure 32. The first step consists in the formation of the lactide, a cyclic dimer of lactic acid, by oligomerization of L-lactic acid, D-lactic acid or mixtures permitted by removal of the water condensate followed by subsequent catalytic depolymerization through internal "back-bitting" transesterification.

Depending on the feed ratio of L-lactic acid and D-lactic acid, three stereoforms of lactide can be obtained: L-lactide (L-LA), D-lactide (D-LA) and meso-lactide (D,L-LA). The key point is the separation between each stereoisomer by vacuum distillation in order to control the final PLA structure. In a second step, the purified lactide is converted to high molecular weight poly(lactide) by catalytic ROP.

Polymerization through lactide formation is the current method used by most of PLA producers such as Natureworks, Purac, etc. The ROP of lactones is an attractive method to synthesize aliphatic polyesters because it enables living polymerizations to be conducted and therefore provides a route to control the physical and chemical properties of the polymers. The driving force of this polymerization method is the relief of ring strain. The polymerization mechanism involved in ROP can be anionic, cationic, coordination-insertion or an activated monomer mechanism depending on the initiating system used.[151] A large range of catalytic species have been used to mediate ROP of lactide, including metal-based complexes, enzymes and small molecule organic catalysts.[152–157]

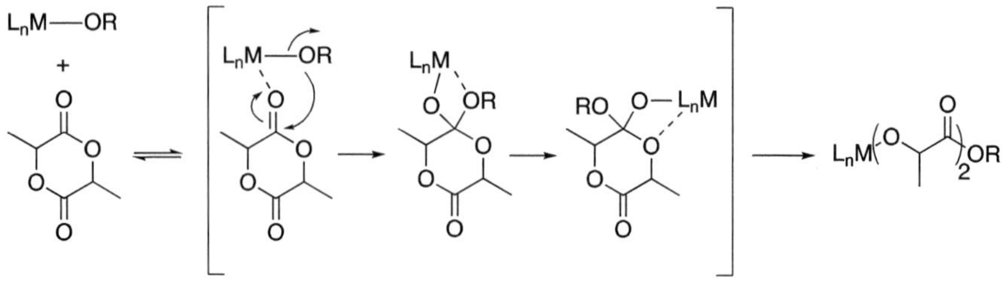

212   Cramail *et al.*

Figure 33.   Coordination/insertion polymerization of lactide by metal-based catalysts.

A large range of simple metal salts and coordination compounds have been reported as catalysts for the ROP of lactide (Figure 33). The catalysts mainly used consist of metal powders, Lewis acids, Lewis bases, organometallic compounds and different salts of metals. Particularly metallic compounds such as alkylmetals and metal halides, oxides and carboxylates are very effective. Among them, tin(II) octanoate presents several advantages that made it a suitable catalyst for industrial production of PLA by ROP.[158] It is soluble in organic solvents and molten lactide, is stable on storage, allows for polymerization up to 180°C and it has been approved by the food and drug administration (FDA).

The applications of polyesters, and more particularly PLA, in packaging and biomedical applications have motivated efforts to develop metal-free organic catalysts in order to suppress the traces of metal in the final polymeric materials. The development of organic catalysts for the ROP of lactide has resulted in very high levels of polymerization control.[152,159] Moreover, organic catalysts allow polymerization of lactide under milder conditions as evidenced by low temperature and reaction times needed. Organic catalysts are postulated to operate *via* one of the three mechanisms: monomer-activated, pseudo-anionic (chain-end activated or general base catalysis) or cooperative dual activation.

The particularity and potential of PLAs reside in the various stereochemical microstructures enabled by the presence of two stereocenters per lactide monomer (Figure 34). The ROP of enantiopure L- or D-lactide results in the isolation of isotactic PLA in which all stereocenters are identical. However, ROP of either rac-lactide (racemic mixture of L- and D-lactide) or *meso*-lacitide (lactide possessing both L- and D-stereocenter) leads to an atactic PLA if the utilized catalyst lacks stereospecificity. Some

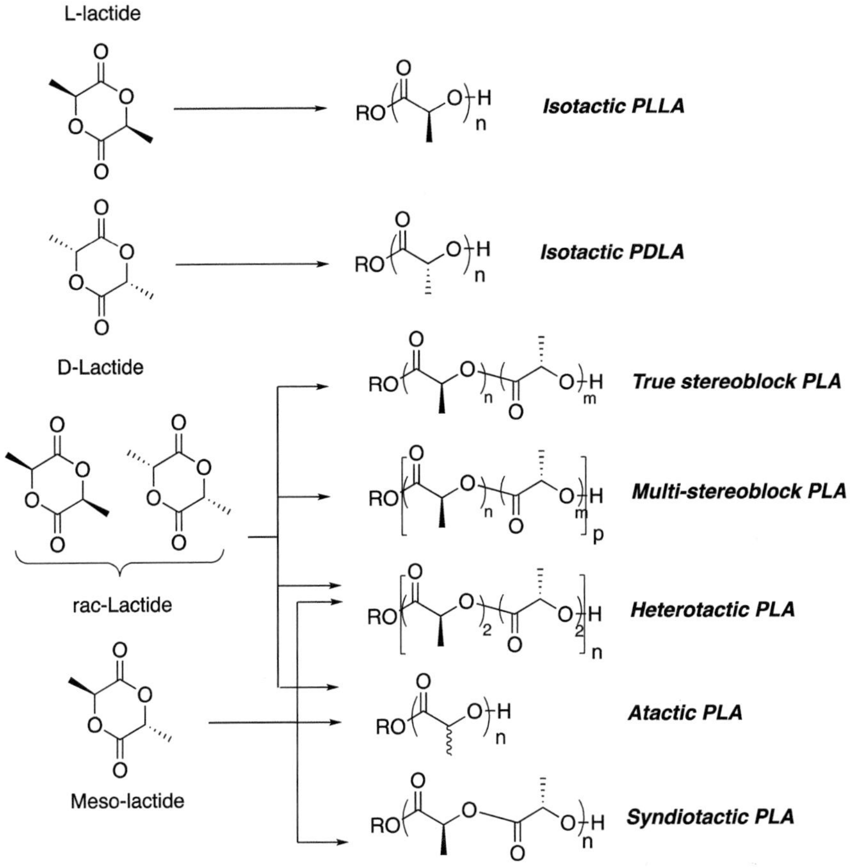

Figure 34.   Stereochemistry of PLA materials.

reviews summarize the different catalytic systems used to synthesize stereoregular PLAs.[154,160] Two types of stereospecific catalysts can be distinguished. First, catalysts that display a *syndio* preference will result in the synthesis of heterotactic PLA -the latter will display doubly alternating stereocenters (i.e. -LLDDLLDD-) from *rac*-lactide by alternating insertion of monomers of opposite stereochemistry or syndiotactic PLA (alternating L- and D- stereocenters i.e. –LDLDLD-) from *meso*-lactide. Second, catalysts that display an *iso* preference (i.e., a preference to ring-open at the same stereocenter as the one of the propagating polymer chain end) will result in the synthesis of heteroatactic PLA from *meso*-lactide but will mediate the synthesis of stereoblock PLAs from *rac*-lactide.

Special emphasis was made on isotactic PLLA due to its interesting mechanical and thermal properties provided by semi-crystallinity. Isotactic PLLA was preferred over isotactic PDLA due to the higher availability of L-LA from biomass in comparison to D-LA.

b. *PLA properties and compostability*

Structure, thermal and physical properties

Due to the three forms of lactide monomers, various natures of PLA can be achieved. High purity L- and D-LA form stereoregular isotactic poly(L-lactide) (PLLA) and poly(D-lactide) (PDLA) respectively. Both are semi-crystalline polymers with a melting point of 207°C and a glass transition temperature in the range 55–60°C.[161,162] However typical melting points are in the range 170–180°C due to small and imperfect crystallites, slight racemization and impurities. The meso-lactide or a racemic mixture of L- and D-LA, on the other hand form atactic poly(D,L-lactide) (PDLLA) which is completely amorphous and shows a Tg in the range 50–60°C.

Another family of polylactides includes stereocomplex-type polylactides (sc-PLA) consisting of both enantiomeric PLLA and PDLA. In 1987, Ikada and co-workers discovered that a mixture of enantiomeric PLLA and PDLA forms stereocomplex (or racemic) crystals with a melting temperature at around 240°C.[163] Stereocomplex-type polylactides were first generated from solution and later from melt mixtures.[164,165] Various factors affect the stereocomplexation such as the molar mass of PLLA and PDLA, the optical purity of the homopolymers and the process of formation.

For amorphous PLA, the glass transition temperature (Tg) is one of the most important parameters since dramatic changes in polymer chain mobility take place around this temperature. Various physical properties are reached depending on the temperature. Below the β-relaxation temperature (240 K), T_β, PLA is completely brittle. Between T_β and T_g, the amorphous PLA undergoes physical aging and can show brittle or ductile fracture. Above Tg, amorphous PLA is rubbery and becomes viscous in the range 110–150°C. Finally, amorphous PLA decomposes between 215°C and 285°C.

For semi-crystalline PLA, physical properties are also highly dependent on the processing conditions. Indeed, the processing conditions will dictate the degree of crystallinity while the stereochemistry will govern the melting point of the polymer. Below Tg, semi-crystalline PLA is a brittle material while, above Tg, it is tough and undergoes a ductile fracture.

Crystallization behavior and crystallographic properties

Isotactic PLA can crystallize in α-, β-, or γ-forms depending on the processing conditions.[166–169] The most common and stable, which can be developed from the melt or solution under normal conditions, is the α-form with a 10_3 helical chain conformation where two chains are interacting in an orthorhombic (or pseudo-orthorhombic) unit cell.

PLLA is a slow-crystallizing material with fastest rates of crystallization observed in the temperature range of 110–130°C, which yields spherulitic crystalline morphology. However, it is difficult to make use of the semi-crystalline character of PLLA because the high speed of mass production (e.g. in injection moulding) implies quenching, leading to amorphous structures. One particularity of quenched amorphous PLA is to crystallize upon reheat. This phenomenon is allowed by the enhanced mobility of PLLA chains observed above Tg. This enhanced mobility allows reorganization and better packing of the chains, which leads to crystallization. However, similarly to the crystallization upon cooling, the cold-crystallization upon heating highly depends on the heating rate. The crystallinity can be either an advantage or a drawback depending on the targeted application. For instance, high crystallinity will not be optimal for packaging solutions due to lack of optical clarity while, in contrast, increased crystallinity will be desirable for injection moulded articles for which good thermal stability is important (related to the increase of heat distortion temperature (HDT) with crystallinity). The lack of crystallinity due to fast cooling in industrial PLLA articles can be overcome by several techniques. Post-process heating treatment higher than the Tg and below the Tm can be carried out leading to improved flexural elasticity, Izod impact strength, and heat resistance.[170]

Physical and mechanical properties

The specific gravity of amorphous PLA has been reported as 1.248 g.cm^{-3} and for crystalline PLLA as 1.290 g.cm^{-3} which is lower than PET (1.34 g.cm^{-3}) but higher than many of other conventional thermoplastics which have a specific gravity in the range 0.8 to 1.1 g.cm^{-3} (e.g. PP, PS, LDPE).[171] The optical properties of PLLA are highly dependent to additives and fabrication effects. Indeed, a low degree of crystallinity will induce transparency while highly crystalline PLLA has poor optical properties, even if transparency can be preserved by reducing the crystallite size.

Similarly to the thermal properties, mechanical properties of PLA can be tuned to a large extent depending on the molecular weight, the stereochemical composition and the thermal history of the polymer.[170] Semicrystalline PLA is generally preferred to the amorphous counterpart when higher mechanical properties are desired. Perego and colleagues demonstrated that the modulus of elasticity is almost 20% higher for PLLA than for PDLLA for samples having molar masses in the range 35,000–55,000 g.mol^{-1}.[170] In addition, the same authors showed that the impact strength was also strongly influenced by the crystallinity. Annealing PLLA increases impact strength due to crosslinking effects on the crystalline domains.[9] Furthermore, high molecular weight PLA is generally needed to have access to suitable mechanical properties.

Heat deflection is also a key parameter that needs to be underlined as one of the main drawbacks of PLLA. Poly(98% L-LA) looses 50% of its elastic modulus and storage modulus at 80°C and 87% of the storage modulus at 100°C. Above the glass transition, non-annealed PLLA has not sufficient film strength to target a wide range of applications. This lack of film strength at temperature above the Tg can be overcome by increasing the crystallinity of the material.

Rheological properties and processing

Melt processing of PLLA materials is essential for the conversion into industrial products. Extrusion is the most important technique for continuously melt processing of PLLA. Extrusion is generally linked to a subsequent processing step such as thermoforming, injection molding, fiber drawing, film blowing or extrusion coating. From these, injection molding is the most widely used converting process for PLLA articles. The different processing technologies for PLLA were reviewed by Rubino and coworkers.[147] The processing of PLLA is a crucial step in the formation of valuable PLLA in order to keep the structural, thermal and mechanical properties of the starting polymer. Indeed, PLLA has a narrow processing window due to its tendency to undergo thermal degradation in the molten state through many mechanisms (hydrolysis, zipper-like depolymerization, oxidative main-chain scission, intermolecular transesterification and intramolecular transesterifications).[172]

Environmental degradation of PLA

PLLA represents a promising polymer due to its compostable behavior (Figure 35). PLLA degradation occurs in two stages. First, high molecular weight PLLA is subject to non-enzymatic chain scission of the ester functions by hydrolysis. Second, low molecular weight PLLA is transformed

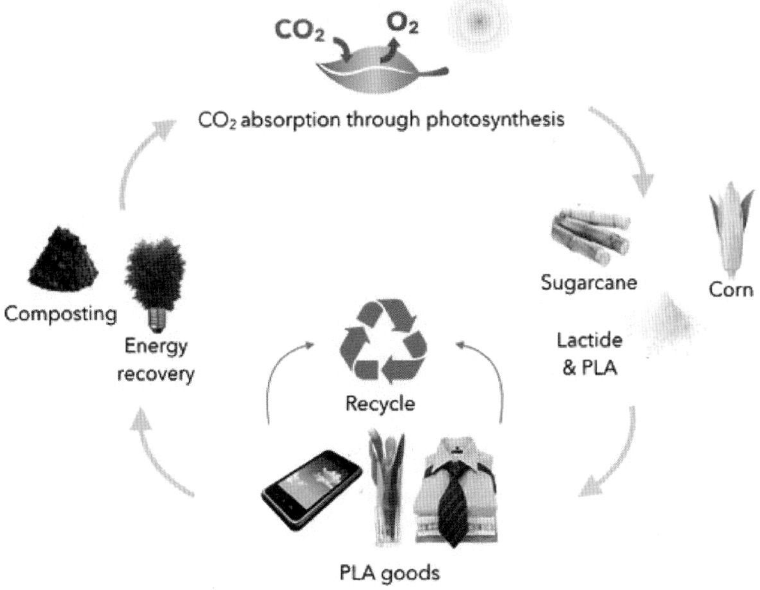

Figure 35.   PLA life cycle.

to lactic acid and shorter oligomers that are naturally metabolized by microorganisms to yield CO_2 and water.

Various factors have an influence on the biodegradation rate. Indeed, exposure conditions can be distinguished from the polymer characteristics. Concerning exposure conditions, moisture, acidity, temperature, aerobic or anaerobic conditions and enzyme specificity are the most influent factors. Other parameters have to be taken into account; i.e. the polymer microstructure that will dictate the chain flexibility and thus the accessibility of the ester functions to water, the crystallinity, the molecular weight, the residual lactic acid concentration and the size and shape of the polymer (surface area). More details about the influence of all these factors on the biodegradation rate can be found elsewhere.[171,173–175]

Compostability and recyclability of PLLA imply an environmental friendly lifecycle, and the industrial are targeting a neutral carbon footprint. Indeed, PLLA presents a cradle-to-cradle lifecycle where products are produced from sustainable and natural resources. In this type of lifecycle, the polymer is used, re-used and recycled as much as possible. At their end-of-life, when the products do not satisfy their functional requirements anymore, these products are transformed back into feedstock for a new product lifecycle (Figure 35).

c. *Summary of PLA development and remaining challenges*

PLLA is a very promising and versatile compostable polymer. One issue that restricts the development of PLLA in a wide range of applications seems to be its price (around 2€/kg). However, this trend could be reversed in the next years due to improvement of biotechnological processes. Nevertheless, there are a number of areas in which PLA-based materials drawbacks still need to be overcome to consider efficient and feasible substitution of existing thermoplastics. For instance, in packaging applications where high barrier protection is essential for the conservation of food, efficient solutions need to be found as barrier properties of PLLA are not in comparison with those of PET or PP. Examples of current investigations deal with the addition of nanoparticles such as nanoclays.[176] Another drawback that needs to be significantly improved is the enhancement of PLLA crystallization rate. Numerous solutions were already found; however, finding a highly efficient solution fitting with the high-speed mass production is still a challenge. The compostable character of PLLA represents an important advantage in environmental applications; nevertheless, aging studies related to the specific conditions of use still need to be conducted in a more precise way in order to anticipate the biodegradation rate. Another area that needs to be studied and improved is the energy and raw materials consumption of PLLA production and its impact on the environment. Related to this, Vink and coworkers published an interesting work presenting the life cycle assessment (LCA) of NatureWorks PLA.[140] In particular, it was mentioned that Cargill Dow's objectives are to decrease the fossil energy use from 54 MJ/kg PLA down to about 7 MJ/kg PLA and to decrease the greenhouse gases emission from +1.8 down to –1.7 kg CO_2 equivalents/kg PLA. Finally, the inherent brittleness of PLLA is a major bottleneck for its use in applications where toughness and impact resistance are critical.

2. *Succinic Acid-based Polymers*

Succinic acid (SA) is an important chemical building block available from biochemical transformation of biorefinery sugars that can also be obtained by chemical transformation from bio-based molecules such as maleic acid and furfural, Figure 36.[177,178] Bio-succinic acid and diesters thereof can be produced through fermentation using a wide range of bacterial strains including *Actinobacillus succigogenes, Anaerobiospirillum succinicipro-ducens, Propprionibacterium succiniciproducens, Mannheimia succinic-iproducens*.[179] Glucose and glycerol are the main carbon sources used but more recent developments using lignocellulosic biorefinery streams (e.g. wood hydroxylate) as carbon source are investigated. The strain efficiency is also CO_2-dependent. Indeed, increasing CO_2 concentration usually results in higher SA yields and reduces by-product concentration (lactate, ethanol).

Using *Anaerobiospirillum succiniciproducens* as the fermentative organism and a three-stage continuous cell recycle bioreactor, optimized processes producing 10.4 $g.L^{-1}.h^{-1}$ and a final concentration of 83 $g.L^{-1}$, equivalent to 1.35 mol succinic acid per mol of sugar, have been reported.[180] Currently, four major commercial producers are active

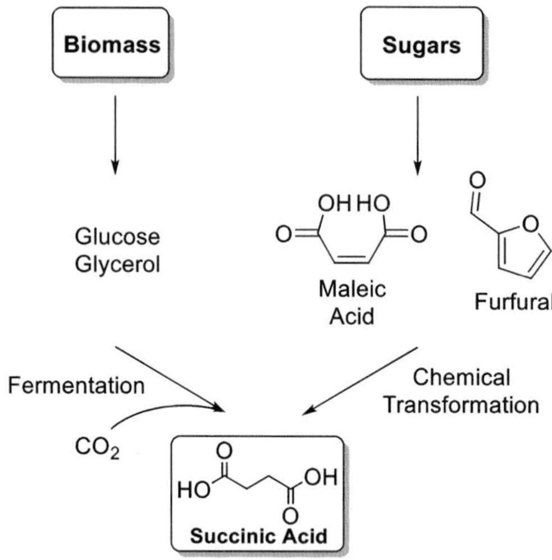

Figure 36.　Conversion of biomass into succinic acid (SA).

(BioAmber, Myriant, Succinity GmbH and Reverdia) using high grades glucose sources with a production cost almost equal to fossil-based SA, i.e. around $0.5/kg.

Succinic acid offers strong potential as a chemical platform for polymer synthesis (see Figure 37).[5] Succinate esters are precursors for known petrochemical products such as 1,4-butanediol, 1,4-diamino-butane, tetrahydrofuran, γ-butyrolactone (GBL) or various pyrrolidinone derivatives (vinyl pyrrolidinone).

From SA platform, various semi-crystalline polyesters (PEs) and polyamides (PA) can be produced. In particular, the copolycondensation of succinic acid with 1,4-butanediol and 1,4-diamino-butane has been reported.[181]

The so-formed aliphatic polyesters either based on succinic acid (PEs-n,4) or 1,4-butane diol (PEs-4,n) exhibit melting points ranging from –24°C to 113°C. PEs-4,4, or poly(butylene succinate) (PBS) and PEs-2,4 are already commercialized due to their polyethylene-like behaviors combined with a good bio-degradability.[5] Moreover, copolyesters have also been investigated using two sets of diacids and or diols to tune their mechanical properties and degradability.[182–184] In particular, copolyesters of ethylene glycol and 1-4-butanediol together with succinic and adipic

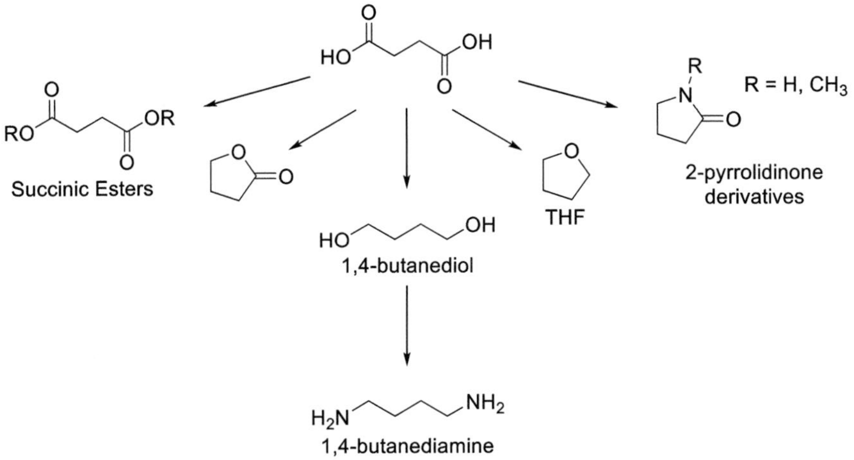

Figure 37. Derivatization of succinic acid into platform molecules and polymers thereof.

acids are commercialized.[185] It should be noted that high molecular weight polyesters are usually obtained in a two-step procedure starting first by the synthesis of a low molecular weight telechelic polyester diol followed by a chain extension using a diisocyanate (usually hexamethylene diisocyanate).[186] In the case of 1,4-butanediol as a diol, poly(butylene terephtalate) (PBT) represents also a polymer of interest with its properties close to PET ($Tg = 60°C$ and $Tm = 223°C$) with a good processability especially for fiber industry.

As far as polyamides are concerned, aliphatic polyamides either based on succinic acid (PA-*n*,4) or on 1,4-butanediamine (PA-4,*n*) exhibit high melting point ($Tm > 200°C$). PA-2,4[187] and PA-4,2[188] exhibit excellent mechanical properties similar to natural silk and could be used as fibers. PA-4,6 and PA-4,10 are actually commercialized as high performance polyamides for automotive industry.[181]

Biodegradable polyesters based on succinic platform have attracted industrial attention as environmentally degradable thermoplastics. However, their thermal and mechanical properties are not optimal for some applications. The properties of such polyesters can be improved by the introduction of amide groups into the main chain.[189] Different biodegradable poly(ester amide)s, PEAs, were commercialized. The synthesis is based on the amide-ester inter-exchange reaction that takes place when a polyester and a polyamide are heated together at a temperature close to 270°C. PEAs based on PA-6 and PE-4,10 or PA-6,10 and PE-4,10 are the most investigated.[190]

Finally, other polymers can be obtained from the succinic platform. Poly(4-hydroxybutyrate), P4HB, and poly(tetramethylene oxide), PTMO, can be obtained from γ-butyrolactone[191] and THF, respectively. Another and more efficient route to P4HB is the fermentation process (see poly(hydroxyalcanoate)s, section 4). Indeed, in the case of ROP of GBL, molecular weights are limited due to the low reactivity of the five-member ring lactone. On the contrary, PTMO raised a lot of interest and more specifically as di-hydroxy-telechelic PTMO, which is widely used as polyol in the course of thermoplastic elastomers.

3. *Itaconic Acid-based Polymers*

Similarly to succinic acid, itaconic acid (IA) has raised considerable interest in the last decade. Its chemical structure composed of two carboxylic acid moieties and a double bond, makes IA a promising precursor for a large platform of chemicals (Figure 38) that are structural isomers of levulinic acid-based derivatives: for instance, 3-methyl-THF can be obtained from IA in comparison to LA-based 2-methyl-THF,

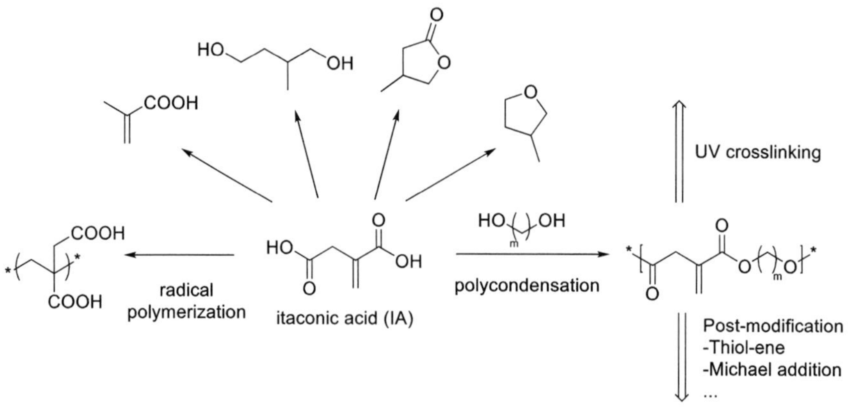

Figure 38. Main platform molecules from itaconic acid.

same for δ-valerolactone in comparison to γ-valerolactone.[5,77] The reduction of the acidic functions of IA leads to 2-methyl-1,4-butanediol, which is an interesting diol towards the synthesis of polyesters and polyurethanes.[192] Finally, the decarboxylation of itaconic acid to methyl methacrylate is of a huge interest for the industry.[193] Nowadays IA can industrially be produced with a yield of 80 g.L^{-1} *via* the fermentation of carbohydrates with fungi. However, the productivity remains at a low level, 1 g.L^{-1}h^{-1}.[194]

IA is extremely attractive to polymer chemists for the production of a wide range of polymers by providing two orthogonal reactivities *via* its two carboxylic acid functions and its activated carbon double bond. Namely, IA can either react *via* step-growth polymerization (usually radical polymerization) or chain growth polymerization (polyesterification).

a. *Chain-growth polymerization of Itaconic acid derivatives*

Due to IA similarity to acrylic acid and methacrylic acid, IA and the corresponding esters such as dimethyl itaconate, dibutyl itaconate and others have been intensively investigated as alternative (co)monomers in the production of poly(alkyl(meth)acrylate)s. The polymerization of itaconic acid and its diester derivatives has been reported since the 1970s *via* radical and anionic polymerization to produce relatively high molecular weight polymers. Their copolymerization with vinylic monomers was investigated in industry for long.[195] Polymers of the dialkyl itaconates are similar in hardness to polymers of the corresponding alkyl methacrylates. As an example, poly(di-*n*-butyl itaconate) exhibits a Tg of 25°C compared to 29°C for poly(*n*-butyl methacrylate). Homopolymer of itaconic acid is commercially available as an equivalent compound of polyacrylic acid.[194]

Furthermore, itaconamide and itaconimide were also polymerized *via* anionic and radical routes. Poly(itaconimide) exhibits excellent thermal properties due to its cyclic structure (Tg > 220°C).[196]

All these itaconate derivatives can also be synthesized by controlled radical polymerization.[197] This allows producing bio-based acrylic thermoplastic elastomers. Triblock copolymers have been prepared by the sequential block copolymerization of itaconate and itaconimide, in that

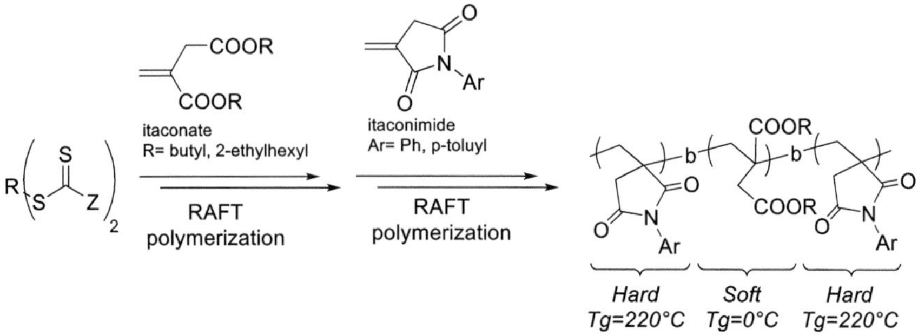

Figure 39.   Itaconic-based triblock copolymer as acrylic thermoplastic elastomer.

order, from di-functional RAFT agent, which produces chains with inner soft poly(itaconate) and outer hard poly(itaconimide) segments (Figure 39). The obtained triblock copolymers exhibit the characteristic properties of thermoplastic elastomers along with the microphase-separated morphology observed by AFM.[198]

b. *Step-growth polymerization of Itaconic acid derivatives*

Polycondensation of itaconic acid received less attention in comparison to the chain-growth polymerization.[199] The so-formed polymers are unsaturated polyesters, UPEs, with a wide range of applications thanks to their ability to undergo various post-polymerization and/or cross-linking reactions (Figure 40). The latter can be utilized in high-gloss coatings, insulating materials, drug delivery systems and biomedical applications. A wide variety of diols have been incorporated into the polyester alongside the itaconate moiety such as: 1,4-butanediol, ethylene glycol, trimethylolpropane or isosorbide. Moreover, itaconic acid can be copolymerized with other di-acids (e.g., succinic, maleic or fumaric acids) allowing the tuning of the physicochemical properties. The molecular weights of the obtained polyesters are usually relatively low. The synthesis of higher molecular weight by polycondensation has indeed been hampered by the tendency of the itaconate double bond to act itself as a cross-linker, presumably *via* Ordelt reaction. Radical inhibitors are often needed to enable

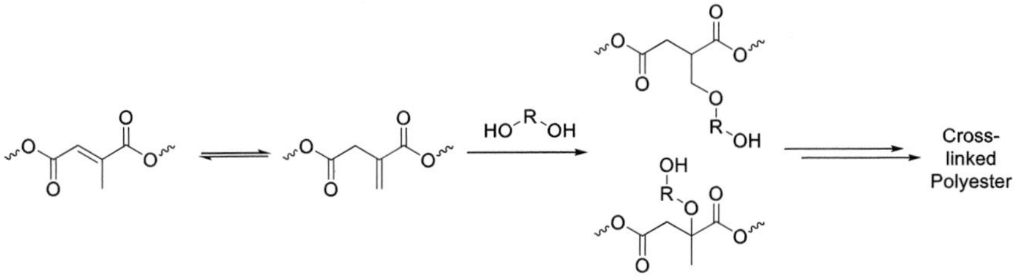

Figure 40.   Side-reaction occuring during itaconate-based polyester synthesis.

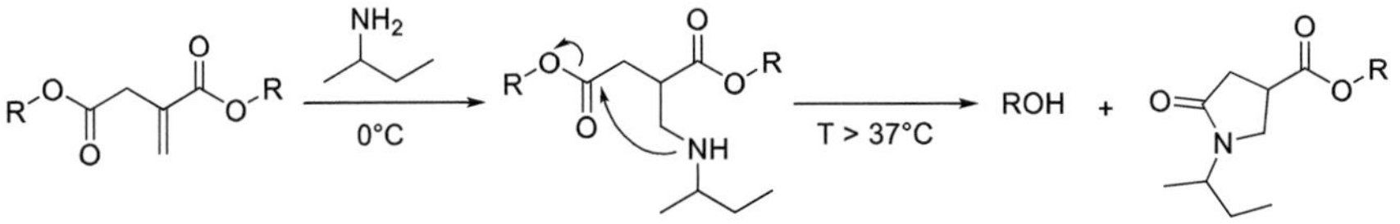

Figure 41.   Self-immolated mechanism of itaconate-based polyester.

efficient formation of even the smallest oligomers.[200] Moreover, isomerization of the double bond usually takes place leading to ill-defined polyesters.

Polyesters based on itaconic acid are frequently modified through its *exo* double bond. Because of its conjugation with the adjacent carbonyl moiety, the vinyl group acts as a Michael acceptor and allows for post-polymerization functionalization and hence property tuning of the itaconate-based polyesters. These polyesters can be subjected to Micheal additions with several sulfur- and nitrogen-based nucleophiles including protected amino acids such as proline and cysteine, for biological applications.[201] Interestingly, modification by primary amines leads to self-immolated polymers at temperature higher than 37°C *via* lactam formation. This thermally induced degradation of polyesters exhibits high potential for biomedical applications (Figure 41).

Another usually employed strategy for itaconate-based polyester is its cross-linking *via* UV-curing leading to polyester thermosets. As an example, copolyesters of succinic acid, itaconic acid and butanediol can be crosslinked, leading to thermoset with Tgs ranging from 0°C to 100°C.[202]

Finally, it is noteworthy to mention that itaconic acid can be derivatized by epichlorydrin yielding epoxy resins[202] or by Diels–Alder reaction with cyclopentadiene for ROMP of norbornene derivatives.[203]

4. *Poly(hydroxyalkanoate)s (PHAs)*

In contrast to main polymers, which are polymerized through standard routes starting from bio-based monomers, poly(hydroxyalkanoate)s, PHAs, are biogenic polyesters (Figure 42). PHAs can be naturally produced by bacteria to levels as high as 90% of the cell dry, in general cultivated on agricultural feedstock. Structurally, these polyesters are classified on the basis of the number of carbon atoms that ranges from 3 to 14 and the type of monomer units, producing homopolymers or heteropolymers. PHAs with 3–5 carbon atoms are considered as short chain length PHAs (scl-PHAs). Examples of this class include poly(3-hydroxybutyrate), P(3HB) and poly(4-hydroxybutyrate), P(4HB). Medium chain length PHAs (mcl-PHAs), contain 6–14 carbon atoms. Examples include homopolymers poly(3-hydroxyhexanoate), P(3HHx), poly(3-hydroxyoctanoate), P(3HO) and copolymers such as P(3HHx-co-3HO).[204]

Depending on the bacterial species and growth conditions, it is possible to produce homopolymers, random and block copolymers of PHA.

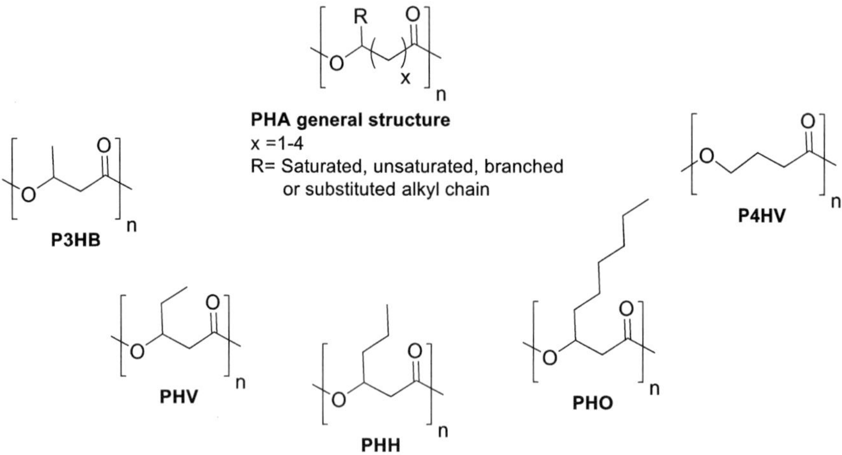

Figure 42. Synthesis and structures of main poly(hydroxyalcanoate)s, PHAs.

As a general feature, PHAs exhibit Tgs from –40°C to 4°C and melting temperatures in between 50°C to 180°C.[205] Hence PHAs characteristics range from elatomers to polypropylene mimics.

Poly(3-hydroxybutyrate), P3HB, discovered in bacteria by Lemoingne in 1923, is the most widespread and best characterized PHA. Indeed, P3HB is accumulated by a wide range of Gram-positive and negative organisms under conditions of nutrient limitation.[205] P3HB can be produced with final concentration over 200 g/L with PHB% over 75% and productivity over 3 $g.L^{-1}.h^{-1}$. Such biologically produced P3HB is a brittle semi-crystalline (elongation at break < 5%, Tm = 180°C) isotactic stereo-regular polyester with 100% of R configuration. P3HB molecular weights differ with respect to the organism, conditions of growth and method of extraction and can vary from about 50 kg/mol to over one million g/mol. This inherent brittleness of P3HB limits its most widespread practical application, such as packaging. Two approaches have been pursued to improve P3HB mechanical properties. One approach is the microbial synthesis of copolymers containing other units such as 3-hydroxyvalerate or 4-hydroxybutyrate along the polymer chain. In such a case, the physical and thermal properties of the PHAs can be regulated by varying the copolymer composition. The second approach is the blending of P3HB with other polymers, including non-biodegradable polymers (PVAc, PMMA), biodegradable polymers (PVA, PEO), bio-sourced polymers (PCL, PLA, polysaccharides).

PHAs are completely biodegradable and compostable in different environments including marine water. They are water-insoluble and relatively resistant to hydrolytic degradation. They possess good UV resistance but poor resistance to acids and bases. They are non-toxic, biocompatible and hence suitable for medical applications, the reasons of which throve considerable commercial interest. The formulation of PHAs, with tailored additives and blends improved greatly their mechanical properties as well as their processability. These advances will improve the capacity of PHAs to reach markets. Nevertheless, limitations still persist such as their elevated cost (7 €/kg). In addition, the extraction of PHAs from the culture media still represents an issue and new processes have to be developed for an efficient and economical large-scale production of PHAs.

E. Bio-Aromatic Polymers

The main resource for the synthesis of bio-based aromatic monomers is Lignin. Lignin represents 20–25% of wood in weight with a world production from pulp industry reaching 50 million tons/year. Nowadays, the main valorization of lignin is its conversion into energy. It is worth mentioning that such feedstock is not in competition with food industry, in contrast to some other renewable resources.

Lignin is a complex biopolymer composed of three different cinnamyl alcohols (Figure 43), named as p-coumaryl alcohol (a), coniferyl alcohol (b) and sinapyl alcohol (c), these three substrates differing by the number of methoxy groups linked to the phenol moiety.

The cinnamyl alcohols ratios (also called monolignols) in lignin vary from the plant species. Accordingly, the nature of the linkages in between each monolignol species varies as well (Figure 44). The most common linkage presents in lignin is the β-O-4 (aryl-ether) linkage that counts for about 50%. Such linkage is usually a main target in the industrial processes aiming at the pulp delignification. The other linkages β-5, β-1, β-β, 5-5, and 4-O-5, are generally much more difficult to cleave.

Knowing these main features, it is worth mentioning how difficult it is to precisely determine the various lignin structures. However, a hypothetical structure of lignin is generally considered as indicated in Figure 45.

Several processes have been industrially developed to remove lignin from the lignocellulosic biomass, which can be divided into two categories depending on the presence of sulfur in the raw material. Lignin, isolated from sulfite and kraft processes contains sulphur. The main

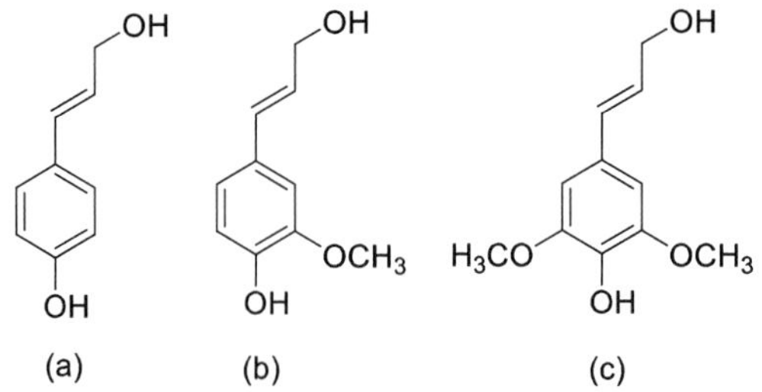

Figure 43. Monolignol monomers: (a) *p*-coumaryl alcohol, (b) coniferyl alcohol and (c) sinapyl alcohol.

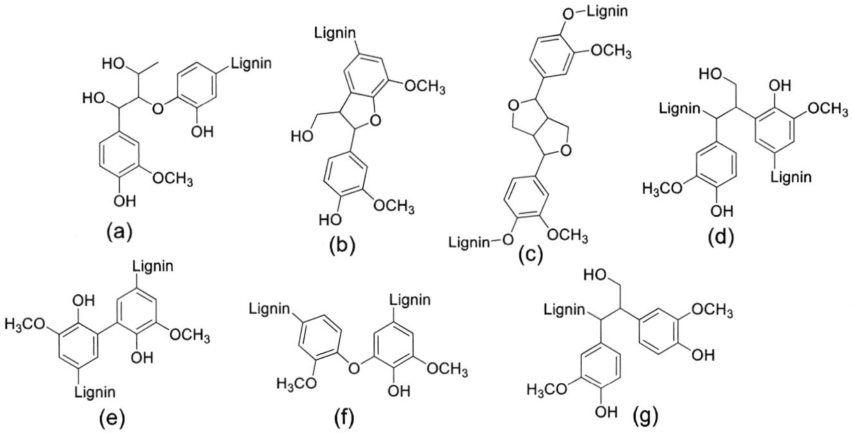

Figure 44. Main structural units between monolignols observed in lignin, (a) β-O-4, (b) β-O-5, (c) β-β, (d) β-1, (e) 5-5, (f) 4-O-5, (g) β-5.

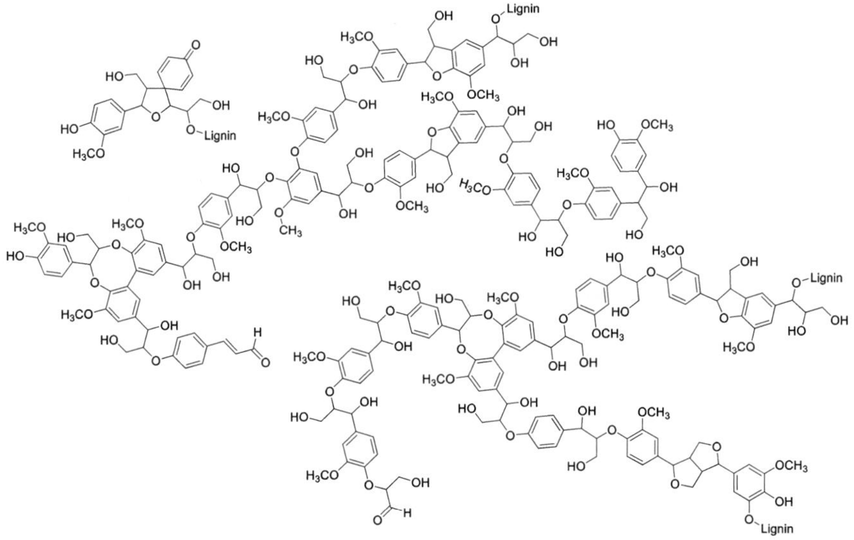

Figure 45. Hypothetical structure of lignin.

producers are Borregaard Lignotech, Rayonier and Westvaco. Lignin, which is isolated with soda or organosolv processes, does not contain sulphur atom in its structure. The major producers are Greenvalue, CIMV and Lignol Innovations. The process of isolation induces some chemical modifications of the native lignin that strongly affects the lignin structure and thus renders its structural elucidation more difficult.

1. *Lignin as a Macromononer*

A first attempt to valorize lignin as polymer precursor is to employ it as a macromonomer.[206] The main reported results concern the synthesis of polymeric materials through polyaddition and polycondensation reactions. Two important issues with respect to lignin have to be mentioned. The first one concerns the processability of lignin. Indeed, lignin is obtained as a solid, with an estimated molecular weight around 10 kg/mol., and cannot be liquefied. The second issue is finally the rather low reactivity of lignin due to some steric hindrance.

a. *Polyurethanes*

Thanks to the presence of phenolic and aliphatic hydroxyl groups, lignin was used as a polyol for the synthesis of polyurethanes. Due to its rigid structure, the lignin-based polyurethanes are stiff materials. However, to compensate its poor processability, lignin was reacted with soft aliphatic diisocyanates. Similarly and in order to enhance the reactivity of lignin as a polyol and to ease its processability, some studies reported its derivatization with poly(ethylene glycol) or poly(propylene glycol).[206–208] In such a case, the lignin structure is diluted within the final polyurethane network. As a main target of the PU industry, the synthesis of flexible and rigid foams from lignin-based polyol is a challenge that was reported but still requires strong improvements.[209]

b. *Polyepoxide networks*

Polyepoxide networks are an important class of materials used in coatings, adhesives, composites and electronic materials. The direct use of lignin as macromolecular structure in polyepoxide networks was limited for a long time to the use in blends and as ring-opening reagents.[210–213] In other routes, lignin hydroxyl functions were converted to acidic ones *via* esterification with succinic anhydride, increasing the reactivity towards

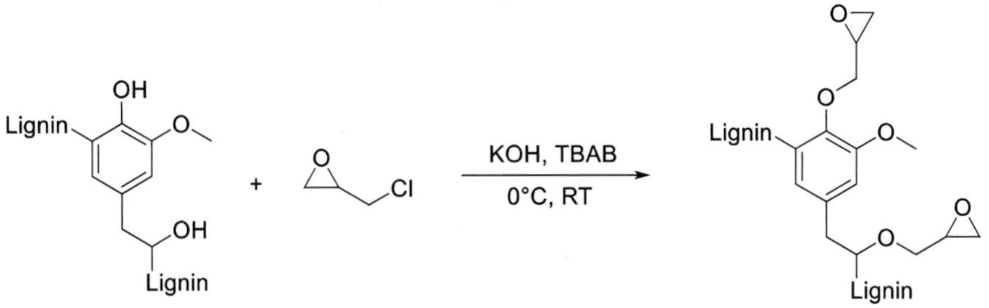

Figure 46. Glycidylation of organo-solv lignin with epichlorohydrin (idealized structure).

epoxides.[214] The interest of using lignin as epoxide substrate increased and various approaches with lignin-based epoxy thermosetting polymers were reported. A main approach consists in derivatizing lignin with epichlrohydrin to graft epoxy moiety onto lignin structure (Figure 46).[215] Generally the epoxidized lignin is mixed with other polyepoxides and then cured with various hardeners. The authors observed that the higher the lignin content, the higher the cross-linking density, resulting in a higher glass transition temperature, lower swelling ratio and increased stiffness in comparison to a reference network not embedding lignin structure.

c. *Phenolic thermosets*

Phenolic thermosets represent an important class of polymeric materials (adhesives) mainly based on two monomers, i.e. phenol and formaldehyde. Both monomers, notably formaldehyde, should be replaced because of health and environmental concerns. Indeed, lignin appeared as a good candidate as phenol precursor. Some studies reported the incorporation of lignin in phenol-formaldehyde thermosets.[216,217] Again, only a fraction of lignin was used in addition to phenol due to the lack of processability of lignin and to its uncomparable lower reactivity with respect to phenol. As a general trend, the higher the lignin content in the thermoset, the higher the thermal stability.

With respect to this polymer family, one important challenge remains the increase of phenolic functions on the lignin substrate. With this aim, the demethylation of methoxy groups from monolignol units appeared to be an efficient strategy to consider.[218–220]

232 Cramail *et al.*

2. *Depolymerization of Lignin*

As already discussed, lignin is the major source of aromatic compounds
from biomass. A big challenge is thus to selectively and efficiently depo-
lymerize lignin into well-defined, processable and low molecular weight
aromatics. Amongst such substrates, vanillin and ferulic acid appeared to
be substrates of interest for the further synthesis of (semi)aromatic polymers (Figure 47).

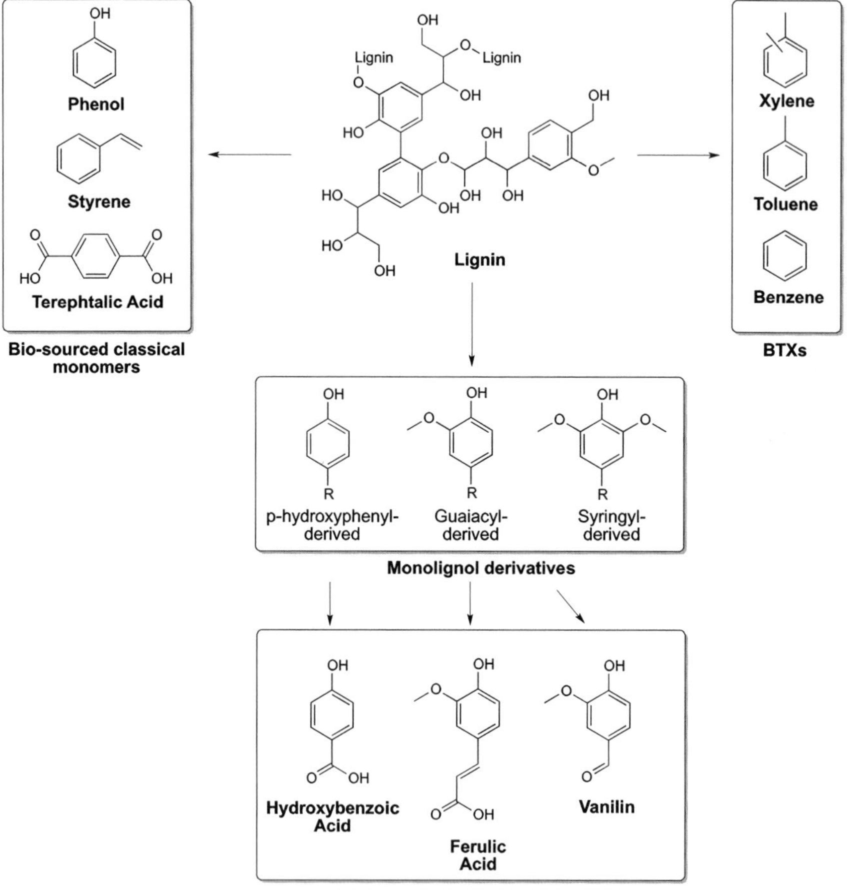

Figure 47. Targeted molecules obtained by depolymerization of lignin under
oxidative or reductive conditions.

a. *Reductive cleavage of lignin*

The depolymerization of lignin can be performed through reductive or oxidative catalyses. In the reductive process, lignin is treated at high temperature under pressurized H_2 in the presence of transition metal catalysts. This process is a two-step-procedure in which the first step is the ether bond (β-O-4, α-O-4) cleavage and the second step, a hydrodeoxygenation of the formed molecules. The main transition metal catalysts used for the reductive depolymerization of lignin are platinum, ruthenium, rhodium, and palladium.[221,222] A lot of studies focused on the reductive depolymerization of lignin models for the production of benzene, toluene and xylene (BTX products), Figure 48. To our knowledge, this method is not mature enough to be effective on crude lignin and required some improvements in terms of catalytic efficiencies and processes.

With respect to polymer synthesis, one important goal, is obviously the access through this reductive process to xylene, intermediate used for the synthesis of terephthalic acid and thus bio-poly(ethylene terephtalate), bio-PET, as already represented in Figure 5.

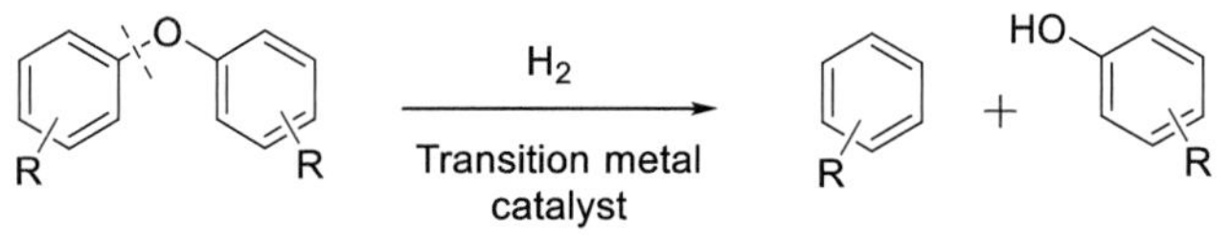

Figure 48. Model reaction of reductive cleavage of bisaromatic compounds towards the formation of BTXs.

b. *Oxidative cleavage of lignin*

Another strategy is the lignin depolymerization under oxidative conditions. Contrary to the reductive route, the oxidative cleavages of C–C and C–O bonds lead to high value functional molecules, such as ferulic acid or vanillin (Figure 47).[223] Interestingly, ferulic acid exhibits anti-oxidant, anti-ageing, anti-diabetic and anticancer properties while vanillin is widely used as aroma in agro-industry.

Since twenty years, a huge variety of oxidative process, oxidants and catalysts were tested. As for the reductive route, the main studies were reported on model molecules and not directly on lignin.

The main catalysts reported for the oxidative cleavage of lignin are organo-metalic complexes.[223,224] Polyoxometalates (POM) have also been studied as oxidative catalysts for lignin depolymerization. These metal-oxygen clusters are considered as green catalysts.[225] Zhao *et al.* performed oxidations of lignin under pressurized oxygen in a methanol/water mixture in the presence of POM, that lead to the production of acids with yields ranging from 12% to 65%.[226] Following a biomimetic approach, metalloporphyrin catalysts were also employed as analogues of peroxidases, enzymes produced by fungi capable of depolymerizing lignin. The latter catalysts are less expensive and more versatile than their enzymatic analogs.[227] Another method which employs a TEMPO derivative in a formic acid aqueous solution containing sodium formate and allowing the oxidation of lignin at 110°C during 24 h was also reported. Low molecular weight aromatic compounds were obtained in good yields (60%) through this process.[228]

Nowadays, Borregaard is the only industry producing functionalized aromatic molecules from lignin and in particular vanillin. Thanks to an oxidative process performed on lignosulfonates in the presence of copper sulfate in alkaline conditions and under pressurized oxygen, 1500 tons of vanillin are produced each year with 7% yield.

3. *Lignin-based Thermosets and Thermoplastics*

Most of the polymers produced from lignin derivatives deal with vanillin, ferulic acid or guaiacol compounds. Due to the high amount of available literature in the field, it is impossible to perform a thorough review of all the results. It is however worth mentioning the review of Cramail and colleagues on the wide scope of monomers and polymers which can originate from lignin-derivatives.[229] It is encouraged to refer to this work, should complementary information be needed.

a. *Vinyl ester and cyano ester thermosets*

Vinyl ester resins are thermoset polymers containing acrylate or methacrylate functions. The latter are used in coatings and composites materials. Wool *et al.* reported in 2012 a fully vinyl ester thermoset obtained from methacrylated vanillin derived monomer.[230]

Cyanate esters are aromatic molecules in which the phenol group is replaced by a cyanide (NCO) function. In the curing process, the trimerization of the cyanide group into triazine occurs either at high temperature or at low temperature in the presence of metal catalysts. Two studies reported the synthesis of cyanate ester thermosets from vanillin.[231] The so-formed vanillin-based thermosets exhibited similar properties than commercial cyanate ester thermosets.

b. *Epoxy thermosets*

As already discussed, the main molecule employed for the synthesis of industrial epoxy thermoset is the bisphenol-A, BPA. However, due to the toxicity of this compound, a lot of studies reported in this field aim at preparing bio-based epoxy thermosets avoiding the use of BPA. Caillol and co-workers described the synthesis of epoxy thermosets from bisepoxide of vanillin, vanillyl alcohol and vanillic acid.[232–234] These two last components are commercial compounds and can be easily obtained by reduction or oxidation of vanillin. These bisepoxides obtained by glycidylation reaction with epichlorohydrin could react with isophorone diamine to produce epoxy thermosets (Figure 49).

In order to mimic the DGEBA-based resins, the authors have also synthesized bisepoxide oligomers of vanillin (Figure 49). By playing on the bisepoxide oligomer chain length, the cross-linking of the polyepoxide network could be adjusted to target particular Tg.

c. *Polyesters*

Direct polymerization methods

The first example of polymerization of vanillic acid was described in the 50s. Indeed, vanillic acid was converted to carboxylate by etherifying the phenolic moiety with ethylene dihalides (Figure 50a). Subsequently, the carboxylate was esterified with ethylene glycol and condensed to linear semi-cristalline polyester exhibiting Tg of 80°C and Tm of 210°C.[235] Later, in the 80s, other strategies were developed to synthesize vanillic and syringic acid-based polymers.[236] In a second synthetic pathway, the

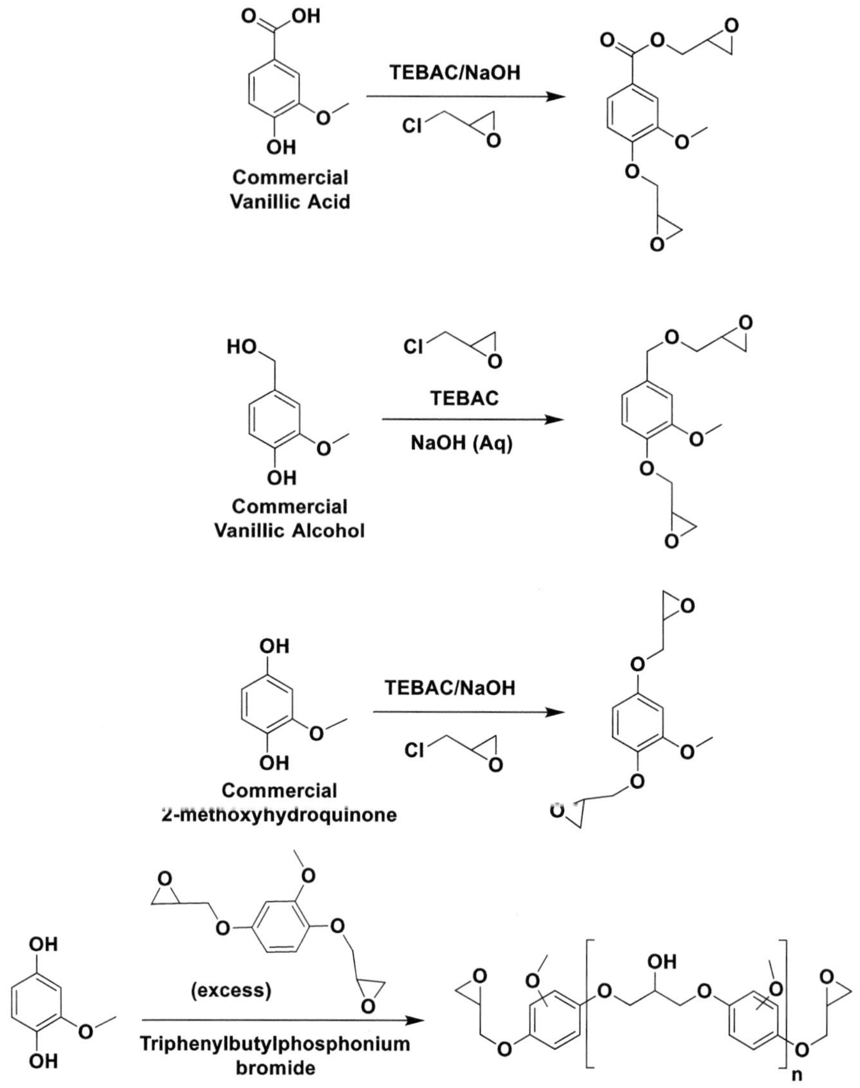

Figure 49. Synthesis of bisepoxides by chemical modification of vanillin.

phenolic moiety of vanillic acid was reacted with ethylene oxide, yielding an AB synthon able to self-polycondensate (Figure 50b). Viscometry molecular weights indicate that the first synthetic pathway (a) provides polyesters exhibiting higher molecular weight, around 50,000 g/mol both

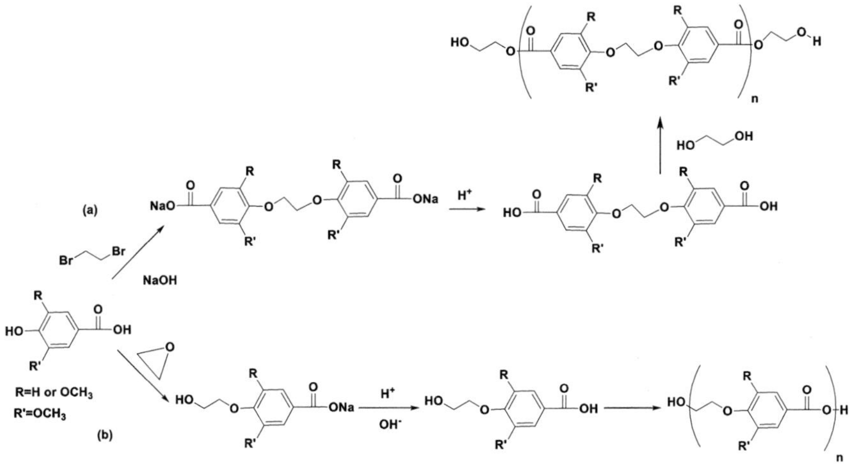

Figure 50.   Polyesters from vanillic acid (R = H) and syringic acid (R = OCH$_3$).

for vanillic acid and syringic acid in comparison to polyesters produced from the second method (b), 30,000 g/mol and 12,000 g/mol, respectively. The polyesters prepared through pathway (a) showed Tg of 69°C and Tm of 212°C in the case of vanillic acid and Tg of 58°C and Tm of 172°C for syringic acid. From the second method (b), the so-formed polyesters showed Tg of 55°C and Tm of 254°C in the case of vanillic acid and Tg of 45°C and Tm of 73°C for syringic acid. Interestingly, the polymer produced from vanillic acid exhibits thermal properties very close to PET's (Tm = 265°C, Tg = 67°C).

As fas as vanillic acid is concerned, a lot of studies were reported on thermotropic polyesters. Thermotropic polymers exhibit liquid crystal behavior in their molten form. The thermotropic polyesters described in the literature exhibit high mechanical strength, chemical resistance at elevated temperature and good flame retardancy due to the presence of aromatic rings and high order in their structures. Poly(vanillic acid) was firstly synthesized by Kricheldorf and co-workers from two different routes.[237] Based on these studies, various other polymers were synthesized by copolymerization of vanillic acid with different comonomers. Vanillic acid compared to other monomers like terephtalic acid or bisphenol-A brings to the terpolymers, faster polycondensation rate, higher molecular weights and lower melting temperatures. These vanillic acid-based thermotropic

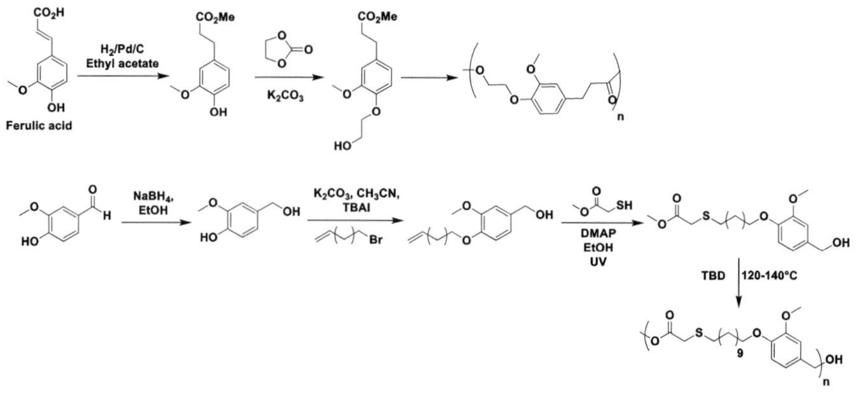

Cramail *et al.*

Figure 51. Synthesis of ferulic acid- and vanillin-based polyesters by polycondensation in a sustainable way.

polymers attracted attention for biomedical applications thanks to their good thermomechanical properties and processability. Indeed, the synthesis of bioresorbable polyesters from vanillic acid and *p*-hydroxybenzoic acid was recently reported.[238]

In the current context of "green" chemistry, 100% biobased polyesters were synthesized from vanillin, ferulic acid and vegetable oil derivatives. Meier and colleagues produced polyesters from ferulic acid by polycondensation (Figure 51).[239] The carboxylic moiety of ferulic acid was first esterified with methanol and the resulting hydrogenated methyl ferulate was then reacted with two equivalents of ethyl carbonate, yielding an AB monomer, which was polymerized in the presence of TBD as a catalyst; the polymer so-formed exhibits a molecular weight of 5400 g/mol and a semi-crystalline feature with a Tg of –27°C and a Tm of 25°C. Amorphous polyesters were also synthesized by copolymerization with fatty methylester derivatives.

The polycondensation of vanillin and fatty acid derivatives was also investigated (Figure 51).[240] The aldehyde moiety of vanillin was converted into alcohol and etherification of the phenol moiety was performed with an undecenoic acid derivative. The thiol-ene addition of methylthioglycolate on the unsaturation leads to an AB monomer able to undergo homopolymerization. The polymer so-formed is semi-crystalline with a Tg of –13°C and a Tm of 77°C. Polymers with a

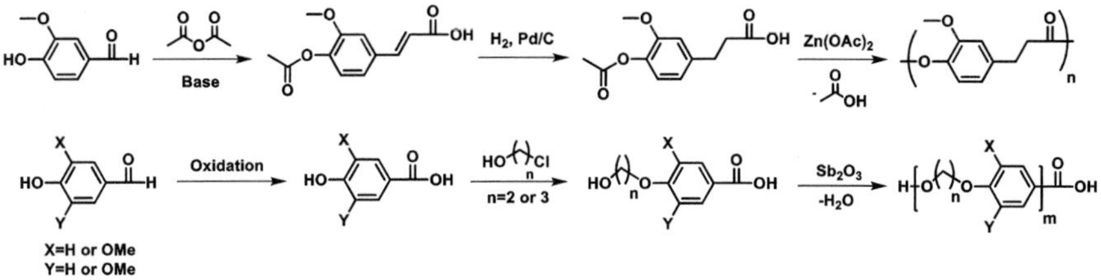

Figure 52.   Synthesis of poly(alkylenehydroxybenzoate)s from lignin derivatives.

higher crystallinity were also synthesized by copolymerization with fatty acid derived esters.

Miller and coworkers reported the synthesis of bio-renewable poly(ethylene terephthalate) mimics, derived from lignin-derived substrates (Figure 52).[241] The reaction of vanillin and acetic anhydride leads to both Perkin reaction and acetylation of the phenolic group. The resulting compound was hydrogenated and the generated acetyldihydroferulic acid was homopolymerized. ZnAc$_2$ proved to be the most efficient catalyst. The polyester so-formed exhibits a molecular weight of 17,800 g/mol, Tm of 234°C, Tg of 73°C and a 50% thermal decomposition temperature (Td$_{50\%}$) at 462°C. Such values are similar to the corresponding values of PET (Tm = 265°C, Tg = 67°C, Td$_{50\%}$ = 470°C).

The same group also synthesized poly(alkylenehydroxybenzoate)s (PAHBs) from vanillin, 4-hydroxybenzoic acid and syringic acid (Figure 52) in order to target semi-aromatic polyesters with a wide range of thermo-mechanical properties.[242] These aromatic aldehydes were oxidized into corresponding carboxylic acids and the phenol moiety was derivatized with 2-chloroethanol or 3-chloropropan-1-ol. The resulting hydroxy acid monomers (AB), were homopolymerized, in the presence of antimony oxide as catalyst. The thermo-mechanical properties of the so-formed polyesters are varying with respect to the aromatic unit substitution. Indeed, the substitution on the aromatic ring increases the free volume of the polymer chains, which decreases the Tg. In addition, the length of the aliphatic segment between the aromatic units has also an impact on the thermal properties. In this series of polymers, the glass transition temperature was tuned between 50°C and 70°C and the melting temperature between 170°C and 239°C.

240   Cramail *et al.*

Synthesis and polymerization of symmetrical difunctional monomers

An alternative access toward difunctional monomer from vanillin or lignin derivatives is based on the coupling of phenolic substrates. This coupling could occur either on the phenol yielding dicarbonyl, or on the carbonyl yielding bisphenol.[243–245] These building blocks were further used for polyester synthesis.

In order to produce difunctional symmetrical monomers, Grelier and colleagues have investigated the enzymatic coupling, with the help of laccase from *Trametes versicol*or, of several phenolic compounds potentially derived from lignin (Figure 53).[246]

The chemical modification of these building blocks enabled the synthesis of symmetrical monomers for the design of renewable (semi)aromatic polyesters (Figure 54).[247] The methylation of the phenol prevents the phenol moiety to be involved in the polycondensation. The reactivity of methylated divanillyl diol and methylated dimethyl divanillate towards polycondensation was investigated on the copolymerization with dimethyl sebacate and 1,10-decanediol. Polyester molecular weights of 65,000 g/mol were reached for methylated divanillyl diol and of 20,000 g/mol for methylated dimethyl divanillate. Considering this difference in reactivity, a series of polyesters were synthesized by transesterification of methylated divanillyl diol with several diesters of aromatic, aliphatic or cyclo-aliphatic structures (Figure 54). The polyesters displayed a Tg ranging from −5°C to

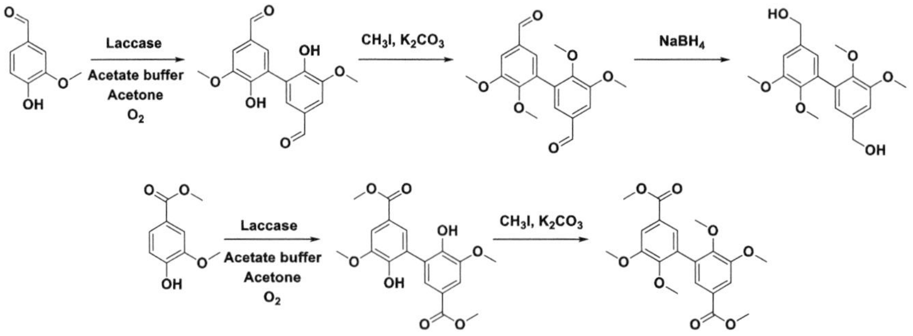

Figure 53. Synthesis of methylated divanillyl diol and methylated dimethyl divanillate from vanillin.

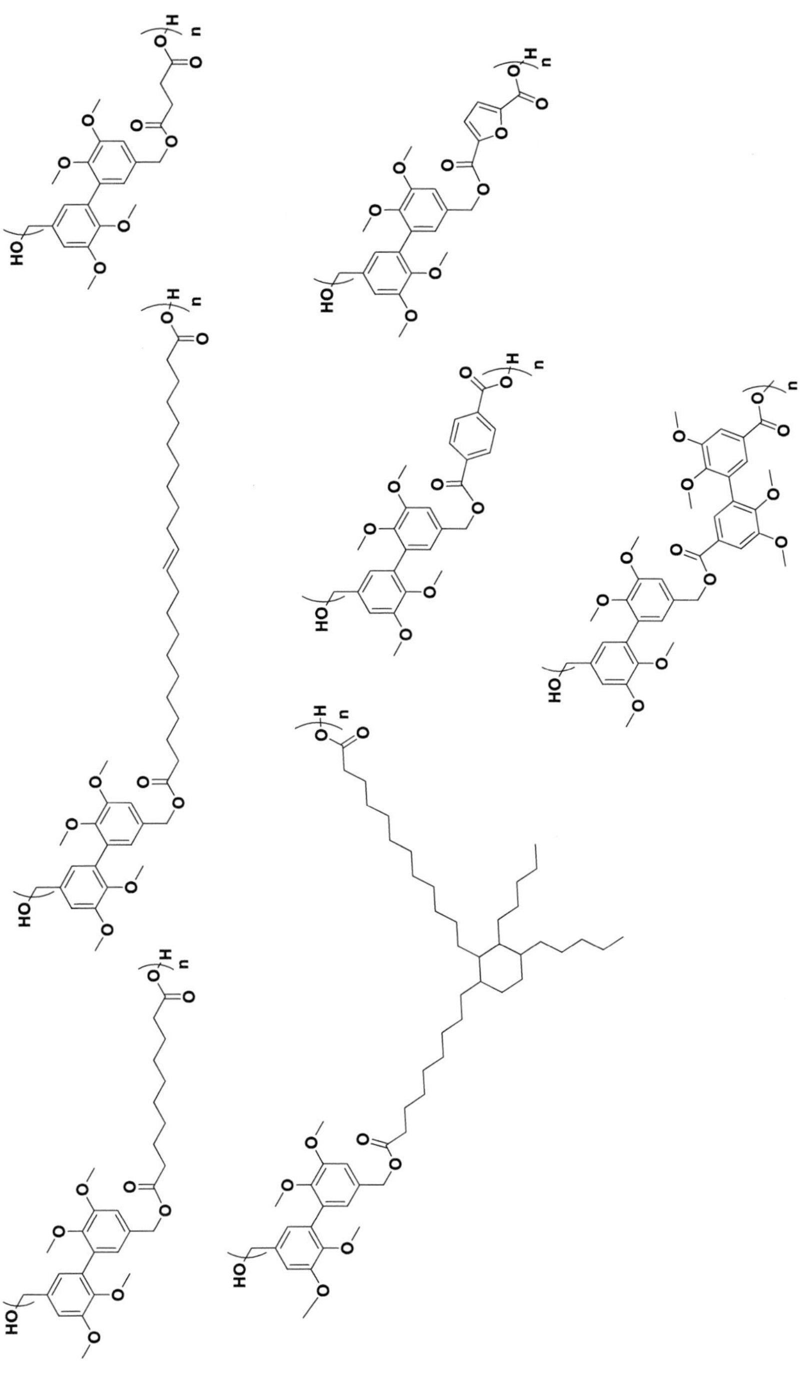

Figure 54. Structures of polyesters synthesized from methylated divanillyl diol and various diesters.

139°C, influenced by the structure of the co-monomer and 5 wt% loss temperature above 300°C.

Synthesis and polymerization of bis-unsaturated esters

Other polymerization techniques were employed to produce polyesters from lignin derivatives (Figure 55). Notably, Meier and colleagues prepared bis-unsaturated esters from ferulic acid and vanillin, which were polymerized *via* ADMET or thiol-ene methodologies.[240,248] The polymers synthesized *via* thiol-ene polyaddition were found semi-crystalline whereas the ones synthesized by ADMET were amorphous.

Allais and colleagues also modified bisphenols derived from ferulic acid to obtain bis-unsaturated esters by etherification with various bromo-alkenes in order to perform ADMET polymerization.[249] Amorphous polymers were obtained with low Tgs, from –21.6°C to 18.2°C and 5 wt.% loss temperatures between 291°C and 333°C. Our group also reported the synthesis of a bisunsaturated diester and its polymerization by ADMET methodology.[250] The latter was obtained from methylated dimethylvanillate dimer and undecenol. ADMET polymerization was performed in *Polarclean* solvent which was selected for its high boiling point, its compatibility with Grubbs catalyst and its sustainability.[251]

d. *Polyacetals*

Dialdehydes are also important monomers for the synthesis of polyacetals, a class of biodegradable polymers as acetal linkages are sensitive to hydrolysis. Miller and colleagues synthesized cyclic and polycyclic polyacetal ethers from 4-hydroxybenzaldehyde, vanillin and syringaldehyde (Figure 56).[252] The dialdehydes were prepared by reaction of lignin derivatives with 1,2-dibromoethane using sodium hydroxide and potassium iodide in water. Polycondensation of the dialdehydes with tetraol yields cyclic polyacetal ethers in the case of di-trimethylolpropane and spirocyclic acetal in the case of pentaerythritol (Figure 56). Molecular weights ranging between 10,600 and 22,200 g/mol were reached. Polymers from 4-hydroxybenzaldehyde are semi-crystalline, whereas polymers synthesized from vanillin and syringaldehyde are amorphous

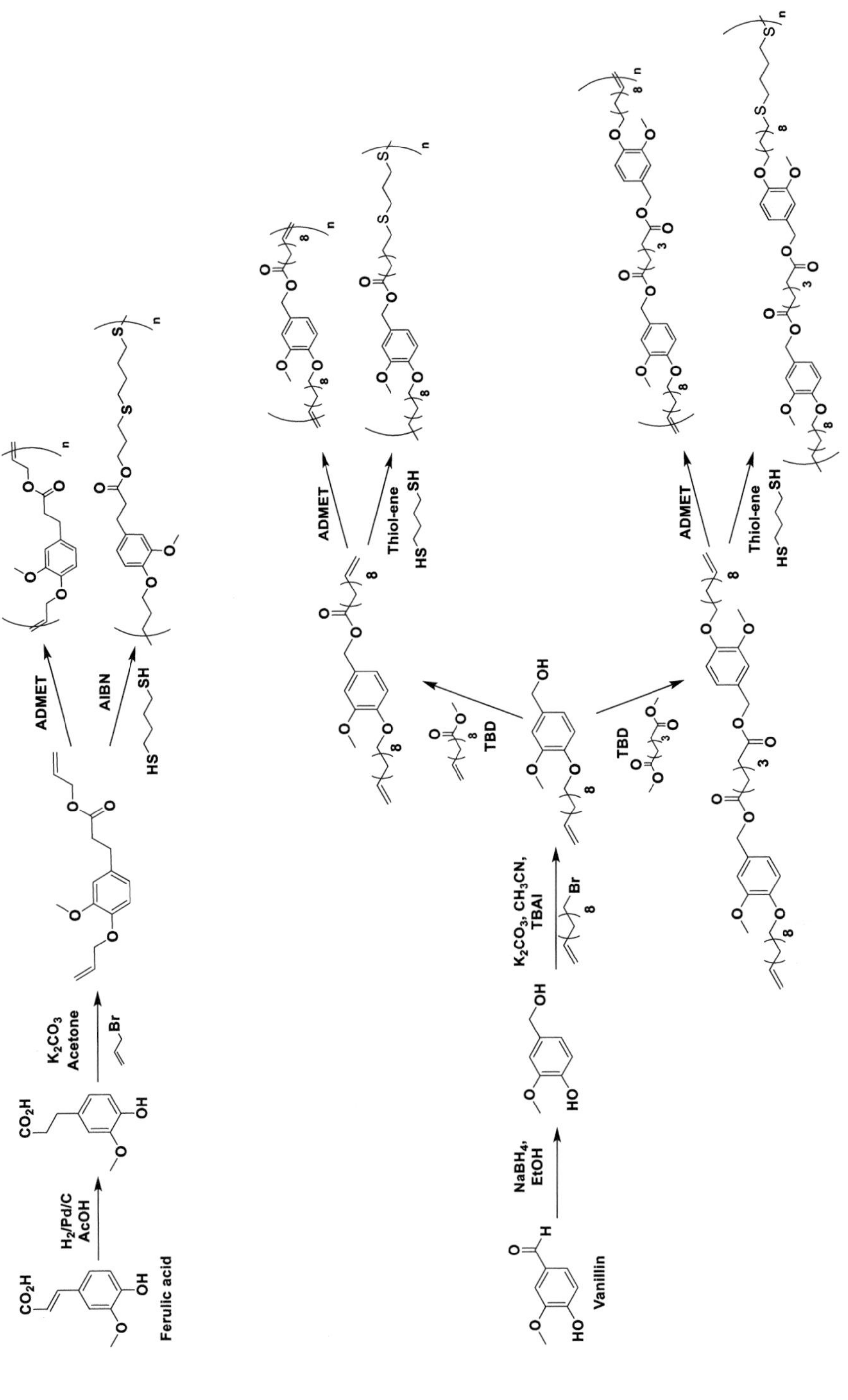

Figure 55. Synthesis of ferulic acid- and vanillin-based polyesters by ADMET polymerization and thiol-ene addition.

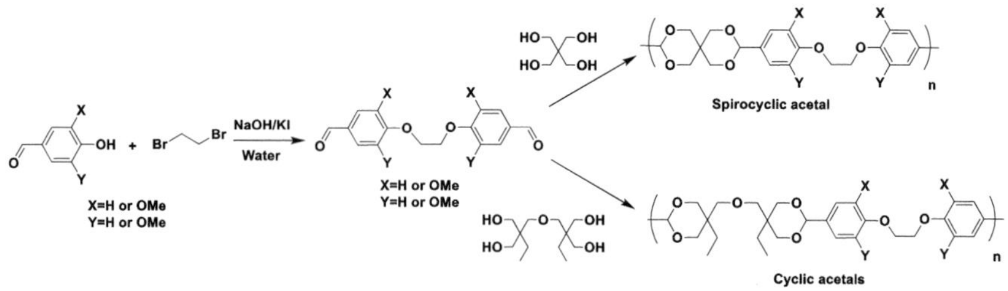

244   Cramail *et al.*

Figure 56.   Synthetic pathway of polyacetals from lignin derivatives.

due to the presence of methoxy groups. Spirocyclic polyacetals exhibit higher Tg than the cyclic equivalent ones (for vanillin: 129°C *versus* 80°C). As already reported, syringaldehyde-based polymers show higher glass transition than vanillin-based ones (152°C for the spirocyclic one against 98°C for the cyclic equivalent). All the polymers present high 5 wt% degradation temperature, between 307°C and 349°C.

e. *Polycarbonates*

Polycarbonates are a class of materials between the commodity and the engineering plastics. The latter are synthesized by transesterification of bisphenols and phosgene or carbonates. Harvey and colleagues synthesized polycarbonates by transesterification reaction between bisphenol from vanillin and diphenylcarbonate but low molecular weights were obtained.[253] Biocompatible and biodegradable poly(carbonate-amide)s were also synthesized from ferulic acid and tyrosine.[254]

f. *Radical Polymerization*

Lignin derivatives were also modified into acrylamide and acrylate derivatives and polymerized by radical polymerization. Thermoplastic polymers with lignin derivatives as side chain groups were obtained (Figure 57). Roger and coworkers synthesized the acrylamide derivative from guaiacol and vanillin.[255] The first pathway involves the Friedel-Crafts alkylation reaction of *N*-hydroxymethylacrylamide with guaiacol. The second approach is based on vanillin and requires a three-step reaction: the oxime is prepared, reduced and the resulting amine reacts with acryloyl chloride. The obtained

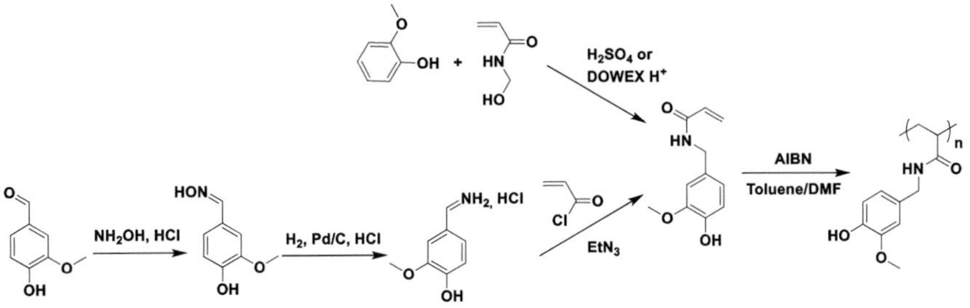

Figure 57. Synthesis of thermoplastic polymers by radical polymerization of lignin derived acrylamides.

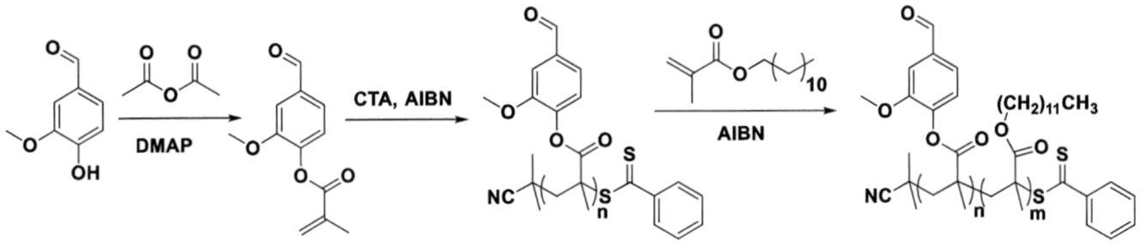

Figure 58. Synthesis of thermoplastic polymers by radical polymerization of lignin derived acrylates.

acrylamide was polymerized by free radical polymerization employing AIBN as an initiator. Despite a screening of the conditions, polymers with relatively low molecular weights, below 11,000 g/mol were obtained; branched structures and broad molar mass distributions were observed at high conversion. These observations can be attributed to the phenoxy radicals known to inhibit radical polymerization by acting as scavenger.

Controlled radical polymerization was also used to synthesize homopolymer and block copolymers from vanillin-derived acrylate.[256] An acrylate derivative was synthesized from vanillin by esterification of the phenolic group with methacrylic anhydride (Figure 58). The latter was polymerized by reversible addition-fragmentation chain transfer (RAFT) yielding a polymer with a molecular weight of 17,000 g/mol and a dispersity of 1.34. The latter polymer was employed as macro-chain transfer agent (CTA) for the polymerization of lauryl methacrylate, a fatty acid derivative, yielding a block copolymer with a molecular weight of 56,000 g/mol and a dispersity of 1.50. The vanillin homopolymer exhibits a Tg of 120°C and a degradation temperature over 300°C, similar to the values of polystyrene. The block copolymer chains self-assemble in body-entered-cubic nanospheres. Other morphologies could be reached by increasing the volume fraction of vanillin.

III. Polymers from Vegetable Oils

A. Introduction

Vegetable oils represent one of the most promising bio-resources for the synthesis of bio-based polymers. This biomass has gained attention in the past few years as the oil structures enable chemical transformation and access to new and innovative properties for some applications. Indeed, vegetable oils are a combination, *via* ester bonds, between glycerol and three fatty acids with variable carbon contents and functionalities depending on the plant species and the growing conditions (Figure 59).[257] Therefore, the nature of fatty acids present on triglycerides differentiates one vegetable oil from another. Still, one of the problematic encountered by

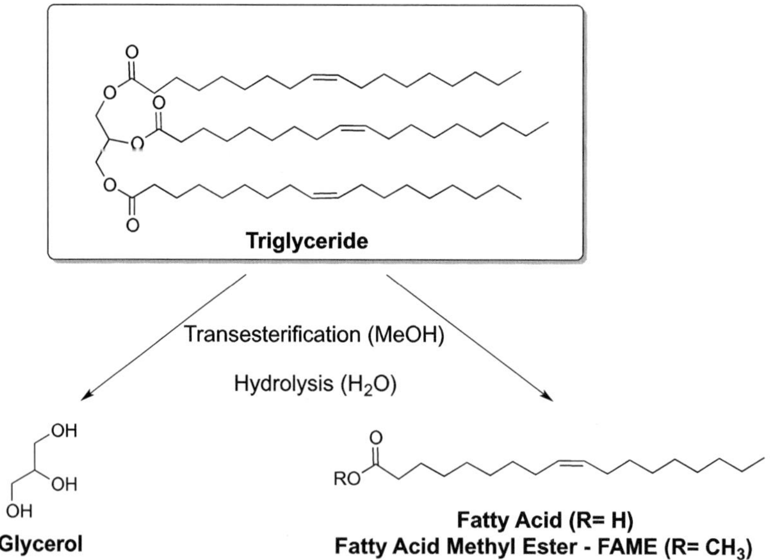

Figure 59.   Vegetable oils as a renewable resource for polymer industry.

vegetable oil-based polymers is the variable composition in triglycerides within a same kind of crop, preventing the access to materials with reproducible properties.

On the one hand, refined oil containing triglycerides can be valorized as monomers for thermosets or hyperbranched polymers, taking advantage of their plurifunctionality and their inherent biodegradability. Triglycerides can be polymerized as produced thanks to the naturally occurring functional groups present on fatty acids (esters, alcohol, insaturations, epoxide moieties) or after chemical modifications. On the other hand, glycerol and fatty acids can be extracted from the triglycerides and chemically transformed in several intermediates to prepare hyperbranched or linear thermoplastic polymers. The use of purified fatty acid derivatives enables the preparation of well-defined thermoplastic polymers with controlled structure-properties relationship. The different strategies adopted to prepare vegetable oil-based polymers will be detailed focusing mainly on renewable thermoplastic polymers.

B. Triglyceride-based Polymers

Some vegetable oils comprising castor oil or vernonia oil contain reactive functional groups such as hydroxyl or epoxide moieties that can be readily exploited for polymerization. However, the low content of available functions conferred to the final materials limited applications and weakened thermomechanical properties.[258,259] As mentioned above, triglycerides can be chemically modified to increase and diversify functional groups, leading to many well-known synthetic routes to prepare vegetable oil-based polymers.[260–262] In most cases, triglycerides are copolymerized with rigid comonomers with the aim of improving material performances. Two main families, comprising polyurethanes and polyepoxide thermosets, will be discussed by the next section. Again, the presented content is not exhaustive and aims at refer the major recent innovations in this field.

1. *Polyurethanes (Polyols) from Triglycerides*

Polyurethanes are classically prepared from polyols and polyisocyanates and cover a large range of applications in adhesives, elastomers, coatings and

248 Cramail *et al.*

foams.[261,263–265] Interestingly, only a few vegetable oil-based polyisocyanates[266] have been described in the literature in comparison with the wide range of reported renewable polyols.[266,267] In most cases, triglycerides and more specifically long aliphatic chains, play the role of soft amorphous segments, conferring to PU a high flexibility. The comonomer is often chosen as hard segment in order to balance the composition and to tune the material thermo-mechanical properties.[268]

First studies involving vegetable oil-based PUs were carried out on castor oil which presents hydroxyl moieties at the natural state and which is a non-food grade feedstock.[266] Besides, a large number of studies reported the possibilities to functionalize triglycerides for PU precursor synthesis.

Figure 60 highlights the five main routes to polyols engaging chemical modifications of triglycerides. The epoxidation of the available double bonds, followed by the ring opening of the epoxides is one of the most described pathways in the literature (Figure 60(1)).[260,269–271] The double bonds are epoxidized using hydrogen peroxide and the produced oxirane

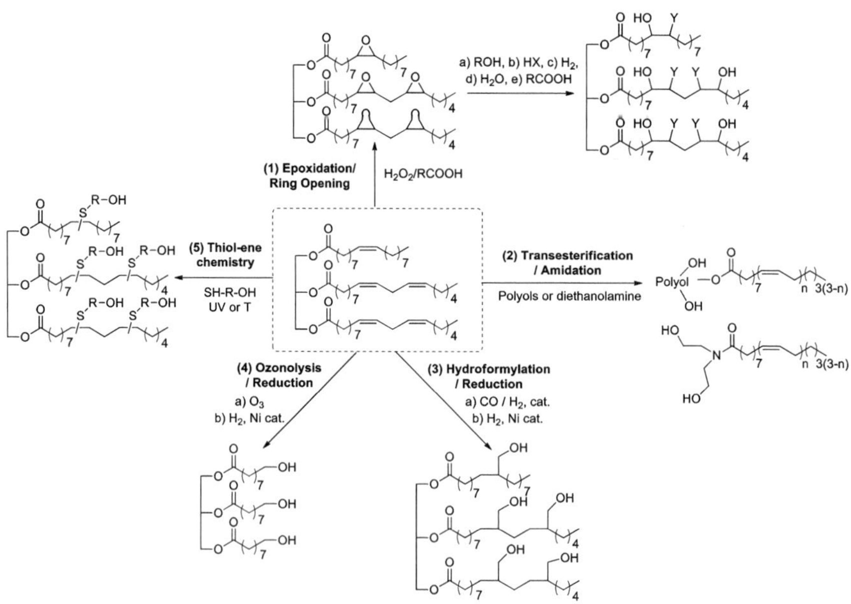

Figure 60. Strategies to prepare polyols from triglycerides.

groups are reacted with a nucleophilic ring-opening agent with or without catalyst to give the final polyol. Kessler *et al.*[267] reviewed the recent advances involving the ring opening reaction. Many nucleophiles such as acetic acid, amines methanol, glycol, 1,2-propanediol, 1,3-propanediol, diethylene glycol, poly(ethylene glycol) and even water have been used to ring-open oxiranes.[268,271–276] Such a reaction is generally catalyzed with Lewis acids or inorganic acids like phosphoric or sulfuric ones. [272,277] Vegetable oil-based polyurethanes prepared from this route exhibited properties ranged from soft elastic to hard materials. Some very interesting features such as specific optical properties, biocompatibility, antibacterial feature or shape memory could be also obtained.[278–282]

Vegetable oil-based polyols can also be synthesized from the transesterification/transamidation reaction — between ester functions of triglycerides and alcohols/amines (Figure 60(2)).[283–288] Glycerol is often used as transesterification agent to prepare such polyols.[289] Besides, the number of hydroxyl groups present on the polyol can be increased by reacting castor oil, rich in natural hydroxyl moieties, with one transesterification agent. This strategy can circumvent the presence of unreactive aliphatic chains, enabling to obtain PU with strengthen thermo-mechanical properties.[290] The PU so-formed exhibit a large palette of properties with respect to the oil, the transesterification/amidation agents and the diisocyantes used.[267]

The third route consists in the hydroformylation of double bonds in aldehydes using CO/H_2 gas under catalysis (mostly cobalt or rhodium based catalysts) that are subsequently hydrogenated under H_2 (Figure 60(3)).[291] This last step leads to the formation of primary hydroxyl groups known to be more reactive towards polyaddition with diisocyanates. Commercially available polyols were synthesized by Dow Chemical Co. and BASF [292] following the hydroformylation/reduction strategy on fatty acid methyl esters (Renuva TM polyol) and castor oil (Lupranol Balance 35) respectively. However, one study showed that this route displayed a main drawback as the PU films prepared were black.[293]

The ozonolysis of the double bounds followed by a reduction step into aldehyde, further reduced into hydroxyl functions has been investigated by several research groups in order to prepare more reactive polyols with primary hydroxyl moieties (Figure 60(4)). Besides, ozonolysis leads to fatty acid chain cleavage and hydroxyl formation at the chain end, resulting in more rigid PUs when copolymerized with diisocyanate.[294] It is noteworthy to mention that this route limits the number of hydroxyl groups to three and consequently decreases the crosslinking density.

A last method to produce polyols from vegetable oils is the use of thiol-ene click chemistry on double bonds (Figure 60(5)). Indeed, this route involves a mild synthesis displaying high reaction rate, high atom economy, high yields and low reaction times. The reaction of thiols such as 2-mercaptoethanol on vegetable oil insaturations have already been reported by several research groups to prepare polyols.[295–301] This strategy enables the simple access to other chemicals such as vegetable oil-based polyamines by reacting cysteamine on double bonds for instance.[302,303] Using thiol-ene addition, another strategy has been developed and involves the preparation of mercaptanized vegetable oils, enabling a further reaction with an unsaturated compound.[297,304]

It is noteworthy to mention than in the context of sustainable chemistry, the synthesis of non-isocyanate thermoset polyurethanes from vegetable oils has been investigated. Many works reported the preparation of poly(cyclic carbonate)s *via* the epoxidation of triglyceride carbon-carbon double bonds, followed by a carbonation under pressurized CO_2 before being polymerized with polyamines.[305] This synthetic strategy will be further discussed.

2. *Polyepoxide Networks from Triglycerides*

The scope of this chapter is limited to epoxidized vegetable oils precursors that can be cross-linked with several hardeners such as amines, acids and anhydrides to prepare epoxy networks. However, the synthesis of vegetable oil-based polyamines and their use in epoxy network syntheses is not included in this scope.

Commercially available epoxy resins are commonly prepared from bisphenol-A and epichlorohydrin whose the use is more and more restricted due to their inherent toxicity.[306] Significant efforts have been made in the last decade to substitute classical epoxy precursors by vegetable oil-based ones. The latter can naturally contain oxirane moieties as it is the case of vernonia oil, or can be epoxidized following several methodologies.

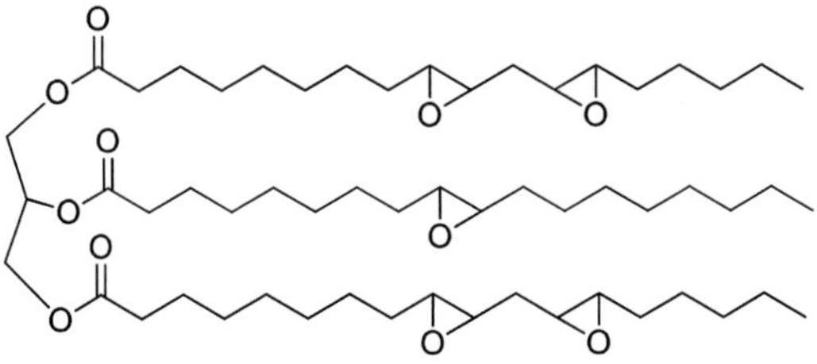

Figure 61. Epoxidized soybean oil structure.

The first route, probably the most employed, consists in the double bounds epoxidation with hydrogen peroxide as oxidative agent. This chemistry is sensitive to the double bound configuration and can lead to side reactions with subsequent low epoxide content.[306] Nonetheless, many studies have described the epoxidation of castor, soybean, grapeseed, canola or linoleic oils *via* this methodology.[307–310] Other routes using peroxides and involving metal catalysts (rhenium, oxomolybdenum, etc.)[311,312] and enzymes (ex: Novozym® 435)[313] were also investigated to increase yields and selectivity. One of the main producers of epoxidized vegetable oils is the company Arkema who has developed the production of epoxidized soybean oil (Figure 61) under the name VIKOFLEX®.[314] Also, it is important to note that epoxidized vegetable oils are used as additives or plasticizers in paints, coatings or PVC.[315]

Some research groups performed the self-polymerization of epoxidized soybean or castor oils using specific thermally latent catalysts such as *N*-benzylpyrazinium hexafluoroantimonate or $BF_3.OEt_2$.[316,317] Nonetheless, these epoxy precursors are generally copolymerized with different crosslinking agents. For instance, amines, acids and anhydrides have been used as hardener to prepare vegetable oil-based epoxy resins.

Stemmelen *et al.* described the polymerization of a linear bio-based diamine from grapeseed oil fatty acid and polyamines from the same oil with epoxidized linseed oil. The fully vegetable oil-based epoxy resins exhibited good mechanical properties but lowered chemical resistance due to the poor crosslinking density, correlated to the low monomer reactivities.[318,319]

In order to tune the epoxy network properties, epoxidized vegetable oil are commonly mixed with commercial and petroleum-based epoxy precursors. Several papers reported the impact of the epoxidized vegetable oil content in the epoxy resin formulation, containing also bisphenol-A derivative, on the network properties. These studies demonstrated that by increasing the amount of triglyceride-based epoxides, flexibility and elongation at break increased but chemical resistance, crosslinking density as well as thermo-mechanical properties were lowered.[320–322]

Various bio-based diacids were employed as crosslinking agent in the formulations. A blend of epoxidized soybean oil and diglycidyl ether of bisphenol A was cured using the fatty acid-based sebacic acid that exhibited higher efficiency than classical amine hardeners.[323] Also, another group conducted the reaction between epoxidized soybean oil and the bifunctional citric acid in order to provide fully bio-based epoxy networks.[324] Moreover, Mathary *et al.* used the diacid belonging to the Priplast® range developed by CRODA as crosslinking agent for epoxidized linseed oil. The resulting epoxy networks obtained under a base-catalyzed process displayed both good water resistance and thermal stability.[325]

Finally, other studies investigated the crosslinking of several epoxidized vegetable oils with anhydrides such as phthalic anhydride,[326] maleic anhydride,[327] methyl hexahydrophthalic anhydride,[328] methyl tetrahydrophthalic anhydride[329] and methyl nadic anhydride.[330] One interesting study highlighted the possibilities of tuning epoxy network properties by curing an epoxidized cottonseed oil with the appropriate blend of methyl nadic anhydride and dodecenyl succinic anhydride, which bring respectively rigidity and flexibility.[331]

3. *Other Triglycerides-based Polymers*

The high density of functionalities imparted by triglycerides enables a plethora of polymerization processes and subsequent materials. Some of them require prior triglyceride modification. Figure 62 summarizes some interesting polymerization routes to access cross-linked vegetable oil-based polymers.

At the industrial scale, oxypolymerization has been largely used to modify vegetable oils into "drying oils."[332–335] Indeed, after an exposure to air and oxygen, C=C double bonds composing oils (linseed, tung or soybean) generate radicals and can be oxidized to form peroxides that can cross-link oils by radicals recombinaison. These modified oils have been

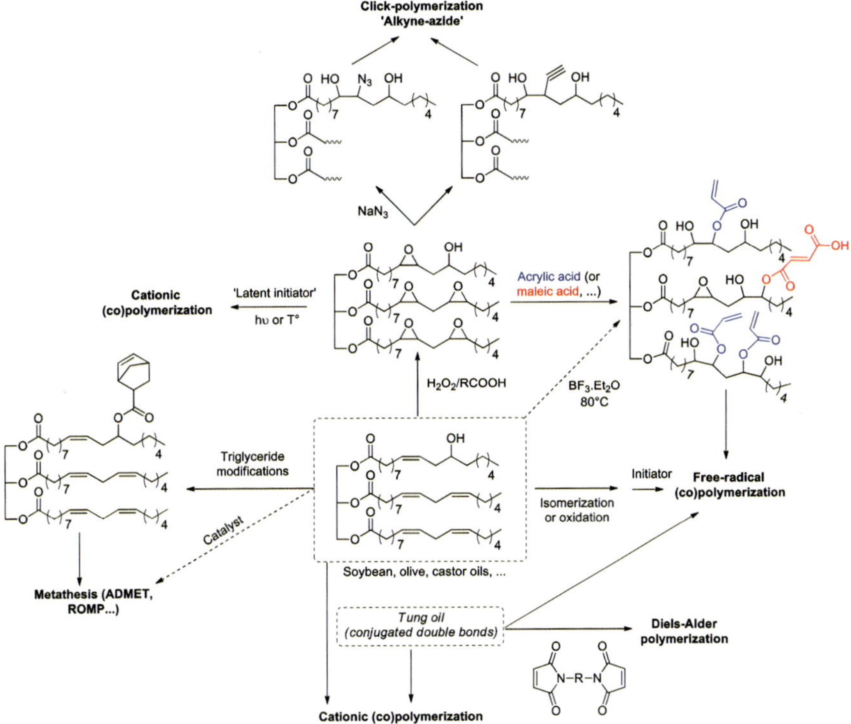

Figure 62. Triglyceride-based monomers for thermoset polymers.

widely used as free-radical macroinitiator, as linoleum floor covers or as binders in oil paints and varnishes thanks to their high viscosity.[336–340]

Besides, the majority of triglycerides are composed of unsaturated and unconjugated fatty acids with limited double bound reactivity, limiting the access to effective free radical polymerizations. One exception is made for tung oil, mainly composed of α-eleostearic acid, which comprises a naturally-occurring conjugated triene (Figure 63). This readily polymerizable oil has already been copolymerized with styrene and divinylbenzene[341,342] following a thermally-initiated free radical process.

In order to obtain more reactive conjugated monomers, other vegetable oils were first isomerized[343] before being copolymerized with various comonomers.[344,345] For instance, Larock and co-workers synthesized bio-based thermosets from the free-radical copolymerization of conjugated linseed oil with acrylonitrile and divinylbenzene.[344,345]

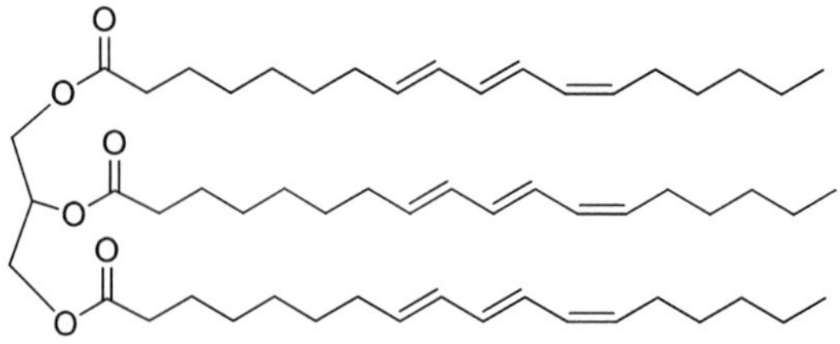

                    Cramail *et al.*

Figure 63. Hypothetical tung oil structure (~85% α-eleostearic acid).

Another strategy involving the ring-opening of epoxidized vegetable oils by acrylic acid,[346–349] maleic anhydride[350] and other (meth)acrylated compounds[351] has been widely reported in order to introduce more reactive double bonds onto triglycerides. Recently, Zhang *et al.* proposed a new route to access acrylated soybean oil from raw triglycerides in a one-step procedure using $BF_3.Et_2O$ as catalyst, avoiding the oxirane ring-opening step.[352] Generally, the reported (meth)acrylated oils are (co) polymerized following a free radical polymerization initiated by heat, UV or chemical initiators. Some of these polymers were described to be good candidates for pressure sensitive adhesives[353–355] or even for electronic materials.[351] It is noteworthy to mention that acrylated epoxidized soybean oils have been extensively used in surface coatings and have already been commercialized by UCB Chemicals Company (Ebercryl 860).[350,356]

Nonetheless, vegetable oils with less reactive carbon-carbon double bound have been copolymerized with styrene without prior modifications, in order to blend bio-based content with styrenic materials to subsequently improve the initial mechanical properties.[314]

Taking advantage of the high nucleophilicity of carbon–carbon double bounds present on triglycerides, cationic (co)polymerizations of vegetable oils have been carried out using generally protic or Lewis acids as initiators. Thus, different oils have been copolymerized *via* a cationic mechanism with divinylbenzene and styrene.[357–362] Soybean oil was also copolymerized with dicyclopentadiene to form partly bio-based thermosets.[363,364] Also, cationic polymerization of soybean oil in supercritical CO_2 has been undertaken using $BF_3.Et_2O$ as catalyst.[365] The obtained liquid soy polymer was found to be a good candidate for medical lubricants. Using the same catalyst, tung oil was found to be highly reactive towards cationic polymerization thanks to its double bond conjugation.[366]

Additionally, epoxidized vegetable oils can be cationically polymerized using "latent initiators" such as benzylpyrazinium salts under heating or photo-irradiation.[316,367–371]

Milder chemistries such as metathesis reaction, azide-alkyne Huisgen cyclo-addition or Diels–Alder reactions have been applied to triglycerides with the aim to prepare polymeric materials.

Larock's group investigated the acyclic diene metathesis (ADMET) of various vegetable oils using the first generation of Grubbs catalyst, taking advantage of the double bonds abundancy in triglycerides.[372] Further studies were conducted on the particular case of metathesized soybean oils.[373,374] Besides, vegetable oils can be modified before being polymerized *via* metathesis techniques. Indeed, rubbery thermosests could be obtained by processing the ring-opening metathesis polymerization (ROMP) of a norbornene-functionalized castor oil with cyclooctene.[375] Also, a commercially available mixture of norbornenyl-functionalized linseed oil and cyclopentadiene oligomer crosslinkers sold under the name Dilulin (Figure 64) has been polymerized by ROMP with dicyclopentadiene[376] or polycyclic norbornene-based monomers.[376]

Furthermore, Petrovic's group carried out the syntheses of azide-functionalized triglycerides and subsequently performed the "click" coupling (also called Huisgen cyclo-addition) with dialkyne or polyalkynes comprising alkynated oils.[377,378]

Lastly, an interesting polymerization, through irreversible Diels–Alder coupling, has to be mentioned. Thermosets were obtained by click coupling between tung oil conjugated double bounds and aliphatic or aromatic bismaleimides.[379,380] Contrarily to the furan-maleimide

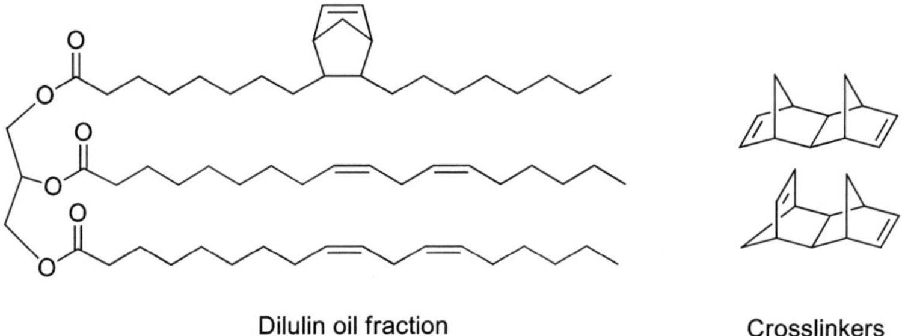

Figure 64.   Blend composition of Dilulin.

Diels–Alder reaction, the coupling was irreversible even when high temperatures were applied.

Despite the high functionality density imparted by the triglycerides, the polymerization of such chemicals results in lowered material properties. Indeed, due to the variable structure of triglycerides in a same oil sample, the relationship between structures and properties is difficult to make and the reproducibility is almost impossible to target. Therefore, building blocks-composing triglycerides, i.e. fatty acids and glycerol have been respectively used as monomers for well-defined polymer syntheses.

C. Glycerol-based Polymers

The glycerol production has been widely increased in the past few years due to the development of fatty acid methyl esters production. Indeed, this compound of interest is produced in large quantity from the triglyceride transesterification, and used as fuel additive (biodiesel). Therefore, an extensive effort has been done in the academic and industrial research in order to valorize the glycerol, which is the by-product of this reaction. Additionally, glycerol is a cheap and available intermediate that can be derivatized into numerous chemicals thanks to its multifunctional character. Figure 65 sums up the most prevalent chemicals obtained from glycerol transformations. For instance, the etherification, chlorination and hydrogenolysis of glycerol are respectively

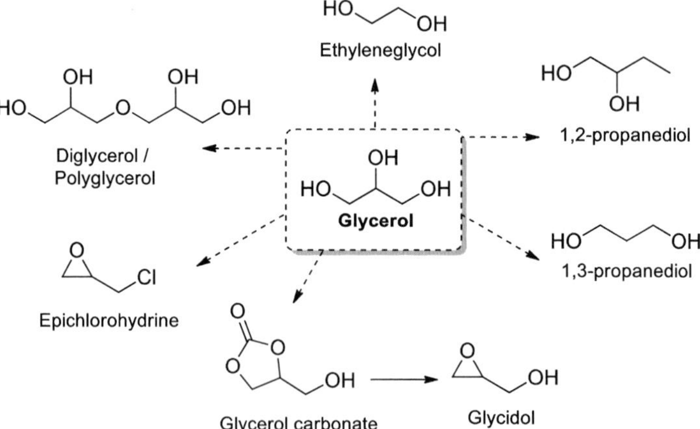

Figure 65. Most common molecules obtained from glycerol chemical transformations.

leading to oligo/polyglycerols, epichlorohydrin and 1,3-propanediol.[381] The main advantage of these intermediates is their biocompatibility suitable for biomedical applications.[382]

The syntheses of glycerol derivatives will not be detailed in this part. Special attention will be paid to the most relevant polymers obtained from glycerol or derivatives.

1. *Polyglycerol (Including Hyperbranched Polyethers)*

Glycol ethers such as ethylene glycol are commonly produced from fossil resources and are used mainly for their coalescence properties in inks, paints or solvent.[383] Due to their toxicity and CMR character, industrials tend to substitute these additives by oligo- and polyglycerols, preferentially derived from renewable resources. By varying the degree of branching or the degree of polymerization, a wide palette of oligo- and polyglycerols with tunable properties is available. For instance, some of these polymers have already been employed as plasticizers, surfactants, lubricants or in cosmetic and pharmaceutical industries.[384–394]

Lemaire and co-workers reported the different approaches to polyglycerol. They highlighted industrial routes such as the reaction between epichlorohydrin and allylic alcohol in five steps or the anionic polymerization of glycidol.[395] However, these routes represent a limitation in terms of atom economy and toxicity. Therefore, new eco-friendly etherification processes leading to polyglycerol and starting from raw glycerol or glycerol derivatives have emerged.

Glycerol can be self-polymerized following two different mechanisms producing linear or hyperbranched oligo- or polyglycerol. These two polymerizations involve respectively the primary hydroxyl functions and both primary and secondary one, depending on the catalysts employed.[396] When the etherification occurs between primary hydroxyl groups, linear polyglycerols are produced as depicted in Figure 66A, whereas hyperbranched polyglycerol schematized in Figure 66B can be obtained through the etherification of both primary and secondary hydroxyl groups. Many acidic or basic, homogeneous or heterogeneous catalysts have been tested. As reported by Martin *et al.*,[396] the catalyst efficiency was closely linked to their nature and structure. Regarding the linear polymerization of

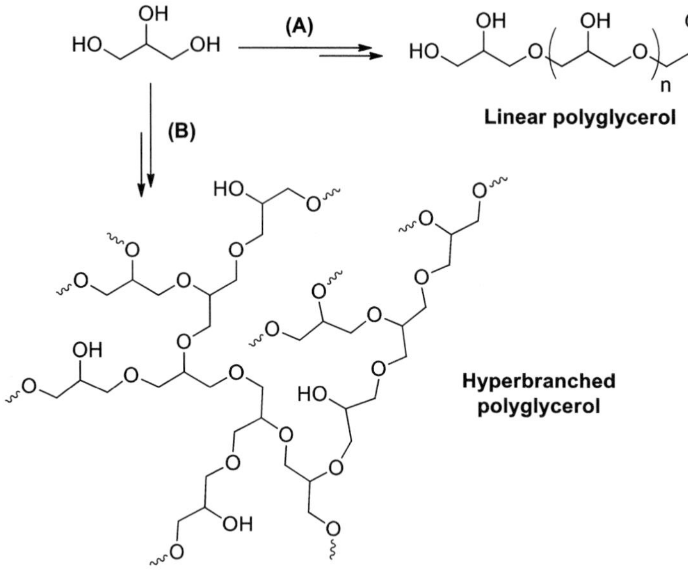

Figure 66. Homopolymerization of glycerol leading to (A) linear or (B) hyperbranched polyglycerols.

glycerol, one of the most profitable catalysts remains potassium carbonate that can be well solubilized in glycerol at high temperature, improving the etherification conversion.[397]

The main limitations of glycerol etherification remain high temperatures, high energetic costs and basic conditions employed during syntheses.[382] To tackle this issue, milder routes have been investigated to access polyethers. The preparation of polyglycerols from glycerol carbonate, itself derived from glycerol has also been carried out under microwave irradiation, enabling the *in situ* formation of glycidol by decarboxylation.[398] Another interesting monomer derived from the hydrogenolysis of glycerol is ethylene glycol that can be polymerized to obtain the widely used poly(ethylene glycol).[381]

Going further, a plethora of linear polyethers from functionalized glycidyl ethers derived from glycerol has been investigated and displayed an excellent biocompatibility.[382] Besides, mono-, oligo- and polyglycerols can be modified by transesterification with fatty acid ethyl esters in order to target applications such as cosmetic and food additives, detergents or lubricants.[399–401]

2. *Hyperbranched Polyesters (HPBEs)*

Additionally to the hyperbranched polyethers synthesized from glycerol, other types of glycerol-based hyperbranched polyesters (HBPEs) have been designed to introduce the biodegradability feature. Two main strategies have been investigated to access glycerol-based HBPEs *via* either the copolymerization of glycerol with diacids (or diesters) or the self polycondensation of a glycerol-based AB_n-type monomer.

The first route consisting in the copolymerization of glycerol with a comonomer has been widely described in the literature. Many bio-based hyperbranched copolyesters has been prepared using classical catalyst or enzymes such as the lipase *Candida antarctica* (CALB), to process the polycondensation between glycerol and various renewable diacids such as succinic, adipic, citric acids or even anhydrides.

Wyatt and Strahan[402] synthesized hyperbranched polyesters from the polycondenstaion between glycerol and various diacids of different chain lengths (succinic, glutaric and azelaic acids) with dibutyltin oxide as transesterification catalyst. The degree of branching and the diacid structure were correlated and showed that the shortest glutaric acid led to the highest degree of branching (85.6%) in comparison with the long-chain azelaic diacid (13.9%).

Another study described the copolymerization between glycerol, 1,8-octanediol and adipic acid at 70°C, under vacuum, using CALB as catalyst (Figure 67). The regioselectivity of the enzyme permitted the

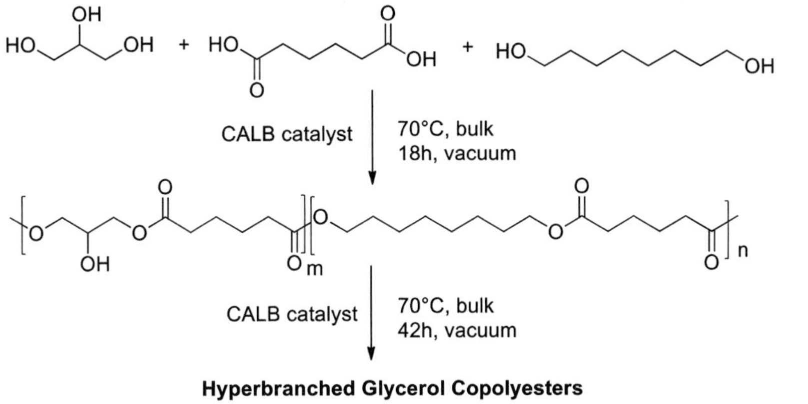

Figure 67. Enzymatic copolymerization of glycerol, adipic acid and 1,8-octanediol.[404]

preparation of linear copolyesters when reaction time was limited (18 hours). Nonetheless, as the latter was extended to 42 hours, hyperbranched copolyesters resulting from the reaction of secondary hydroxyl groups with acid moieties were obtained.[403] This work highlighted the importance of condition reactions on the copolyester final architecture.

In the same category, polyglycerol sebaçate prepared from the polycondensation between ricinoleic-based sebacic acid and glycerol has gained academic research interest. Indeed, this material is biocompatible and bioresorbable, enabling potential applications in the biomedical field.[405]

Besides, citric acid is a low cost and trifunctional monomer conferring anti-microbial and biodegradable properties when polymerized. The preparation of glycerol- and citric acid-based hyperbranched polyesters was carried out using an effective and catalyst-free $A_3 + B_3$ melt polycondensation (Figure 68).[406] The authors studied the influence of the glycerol stoichiometry on the degree of branching and the biodegradability and proved the high potential of such systems for drug delivery. Tisserat *et al.* described the same synthesis using microwave irradiation.[407] Thanks to their biocompatible characters, those bio-based hyperbranched polymers could be potential candidates for tissue engineering applications.

Another interesting example is the bulk preparation of hydrosoluble HBPEs from glycerol and maleic anhydride.[408,409] Improved water solubility

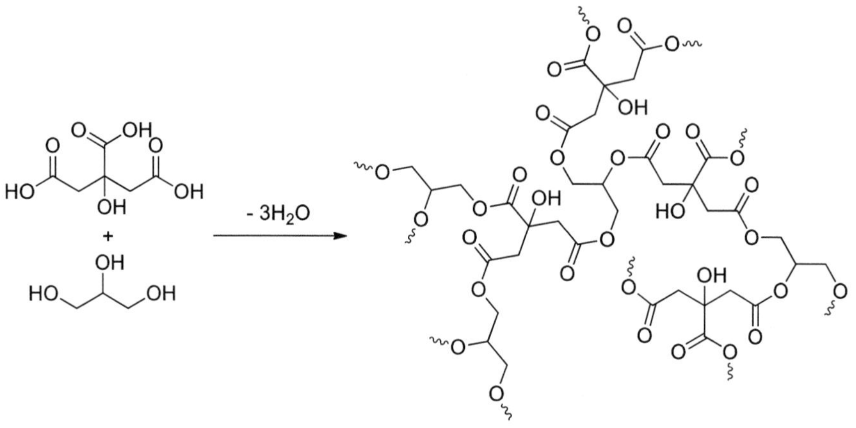

Figure 68. Hyperbranched polyester from glycerol and citric acid.

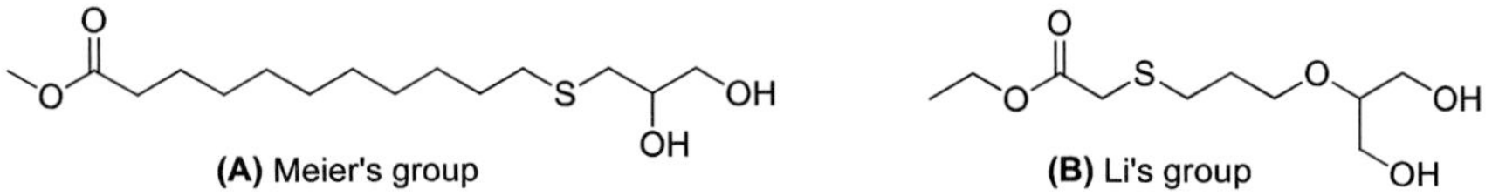

Figure 69. AB$_2$-type glycerol based monomers from (A) Meier's and (B) Li's groups.

was observed when pentaerytritol was added as a core molecule in the system. The HBPEs was then reacted with acrylic ester to form acrylic ester latex film with improved mechanical properties and minimized water adsorption. In another study, maleic anhydride was also copolymerized with glycerol and succinic acid for the preparation of HBPEs as poly(lactic acid) reinforcement.

The synthesis of HBPEs has also been carried out *via* the self-polycondensation of an AB$_n$-type monomer. Meier's and Li's groups both prepared the fatty acid- and glycerol-based AB$_2$-type monomer depicted in Figure 69A by reacting 1-thioglycerol and methyl 10-undecenoate *via* a thiol-ene addition. The first group prepared HBPEs with and without glycerol as core molecule, at 120°C using TBD as catalyst. Molecular weights were ranging between 3500 and 4400 g.mol^{-1} (Đ = 1.9–2.8). Higher HBPEs molecular weights (Mn = 11,400–60,400 g.mol^{-1}, Đ = 5.2–25.3) were obtained by the second group using metallic catalysts such as Ti(OBu)$_4$ or Zn(OAc)$_2$ and working at 160–170°C.

Parzuchowski *et al.* reported the synthesis of another AB$_2$-type monomer (Figure 69B) for the preparation of HBPEs that can be easily degraded by hydrolysis or alcoholysis, making them good candidates for recyclable materials.[408]

3. *Glycerol-based Linear Polyesters*

The examples of linear glycerol-based polyesters are very limited. One of the only specimens comes from the copolymerization between a diacid and 1,3-propanediol that is readily obtained from the selective hydrogenolysis of glycerol, using specific metallic catalysts.[381] At the industrial scale, this synthon of interest is copolymerized with terephtalic acid to produce a bio-based polyester patented under the brand SORONA® from Dupont or CORTERRA® from Shell.[410]

Besides, few papers reported the synthesis of linear poly(hydroxyester)s from the multifunctional glycerol and a diacid (or diester). Their main objective was to adjust the reaction conditions (stoichiometry, catalysts, temperature) in order to limit side reactions such as branching or crosslinking. For instance, Gross and colleagues[411] demonstrated that linear poly(hydroxyester)s with very low degree of branching could be synthesized from glycerol and oleic diacid by employing the immobilized *Candida antarctica* lipase B as transesterification catalyst. Using the same enzyme, Kobayashi and colleagues described two epoxidized aliphatic polyesters from the reaction between glycerol, fatty acids and divinyl sebaçate.[412,413] The epoxidation was carried out using hydrogen peroxide before or after copolymerization depending on the strategy adopted. The linear polyesters were then thermally cured to obtain biodegradable polyepoxide networks.

4. *Glycerol-based Polyurethanes*

Regarding the classical polyurethane chemistry, all the glycerol-based polyols including 1,3-propanediol, polyglycidol or glycerol it-self can be copolymerized with polyisocyanates to obtain many varieties of polyurethanes with tunable properties depending on the polyol structure. The numerous examples reported in the literature show a real attraction for the renewable and low-cost glycerol. Some recent studies report the preparation of such polyurethanes. For instance, Li *et al.* prepared a crude glycerol-based polyols by reacting glycerol at 150°C under vacuum for 5 hours without catalyst.[414] The subsequent foaming in presence of methylene diisocyanate, surfactant and water was performed and the resulting foams demonstrated improved cell morphologies thanks to the presence of residual fatty acid derivatives present in crude glycerol. Another research group substituted petroleum-based polyols by polyglycerol in their rigid polyurethane foam formulation and demonstrated the improved mechanical properties of the renewable materials.[415] Besides, Hu *et al.* used crude glycerol based-polyols to prepare water-bone polyurethane dispersions.[416] Glycerol can also be used as source of polyol in foam formulations.[417]

In the context of sustainable chemistry, reducing reagent toxicity and using bio-based chemicals have been two major challenges for the chemistry

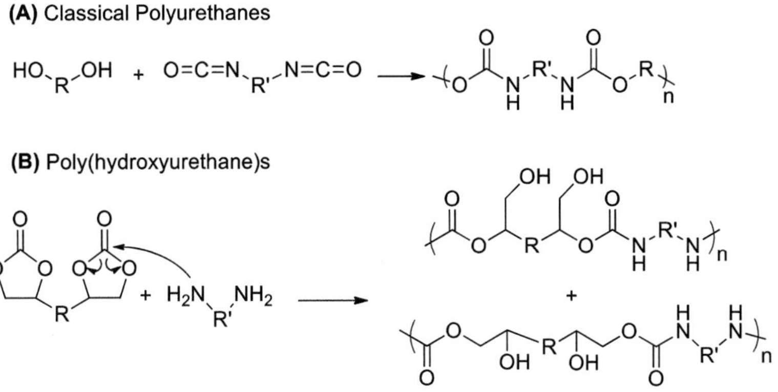

(A) Classical Polyurethanes

(B) Poly(hydroxyurethane)s

Figure 70. Synthesis of (A) classical polyurethanes and (B) poly(hydroxyurethane)s.

of the last decades. To feed the needs in the polyurethane chemistry domain, research about non-phosgene and non-isocyanate polyurethanes (NIPU) has been stepped up. The most promising route is the polyaddition between poly(cyclic carbonate)s and polyamines leading to poly(hydroxyurethane)s (Figure 70). As reviewed by Maisonneuve *et al.*, bio-based cyclic carbonates can be prepared from epichlorohydrin, glycerol carbonate or either diglycerol, making glycerol an attractive synthon.[418]

5. *Glycerol-based Polyepoxide Networks*

The great majority of epoxy monomers are produced from the reaction between epichlorohydrin and hydroxylated compounds.[419] The reaction is generally conducted in aqueous solution of sodium hydroxide, using epichlorohydrin in excess. As already discussed, the most popular remains the diglycidyl ether of bisphenol-A (DGEBA). Petroleum-based epichlorohydrin is usually produced from the chlorohydrination of allyl chloride.[420] Solvay patented the synthesis of bio-based epichlorohydrin from glycerol (Figure 71) under the brand EPICEROL®, which is produced at a level of 100 kt per year.[421] Therefore, the bio-based weight content of DGEBA reached 25% when the bisphenol-A glycidation was carried out with glycerol-based epichlorohydrin. The recent review released by Auvergne and co-workers[419] summarized the different epoxy monomers from biomass, including the ones synthesized from epichlorohydrin.

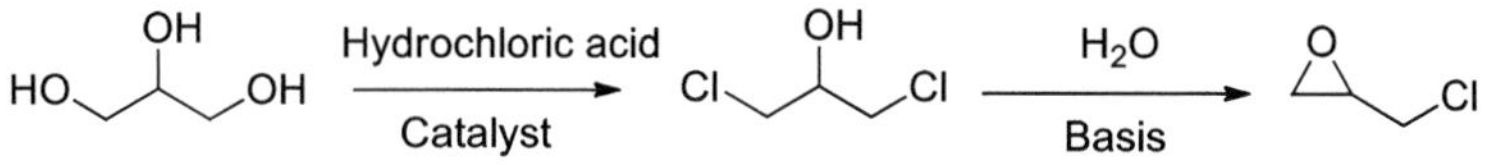

Figure 71. Glycerol chlorination to prepare bio-based epichlorohydrin (EPICEROL technology).

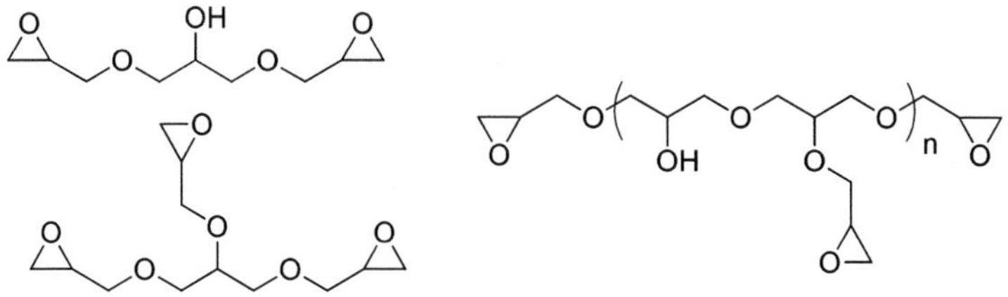

Figure 72. Glycerol-based poly(glycidyl ether)s.

One of the first glycerol-based polyepoxide network was prepared by Barua *et al.* by reacting bisphenol-A, glycerol and epichlorohydrin in aqueous basic solution.[422] Besides, poly(epoxide) prepolymers, particularly aliphatic glycidyl ether ones, can be obtained from glycerol and are commonly produced at the industrial scale. For instance, the synthesis of epoxy precursors (Figure 72) from epichlorohydrin and glycerol has been patented by Dow Chemicals. The authors have developed the range POLYPOX® glycidyl ethers used in epoxy formulations as reactive diluent.[423,424]

D. Fatty Acid-based Thermoplastic Polymers

Vegetable oils have been extensively used as intermediates for bio-based thermoset preparation, using raw oils or functionalized triglyceride monomers. However, as mentioned previously, the major problem encountered by this strategy is the heterogeneity of oil composition due to the random distribution of fatty acids within the triglyceride structure. Consequently, the correlation between the monomer structures and the polymer properties is hardly feasible. In light of this, fatty acids and their derivatives appear as the most promising building blocks for the synthesis of well-defined thermoplastic polymers with targeted and controlled properties. Indeed, fatty acids are available in a high-purity grade, enabling their use as monomers after chemical transformations. The latter are mainly conducted on the carbon-carbon double bonds along the fatty acid aliphatic chain and include oxidation, hydrogenation, hydroformylation, hydrovinylation, thiol-ene coupling, metathesis reaction, etc. Monomers, generally bifunctional, are thus designed to target a certain class of polymers with specific properties. Moreover, innovative properties can be conferred to vegetable oil-based polymers thanks to the structural difference existing between fatty acid- and petroleum-sourced chemicals.

In the following parts, the preparation of the most preponderant fatty acid-based thermoplastic polymers comprising polyamides, polyurethanes, polyesters and polycarbonates will be developed. A special attention will be given to commercially available materials and to emergent and high-potential polymers.

1. *Polyamides*

Thermoplastic polyamides (PA) represent a class of high performance materials required in many sectors involving automotive, electrical or textile industries. Indeed, due to the presence of amide linkages, polyamides exhibit unique properties such as high melting points, high impact and tensile strengths, electrical insulation, chemical, heat, and abrasion resistances. PA are commonly synthesized following three routes which are (1) the polycondensation of dicarboxylic acids and diamines, (2) the self-polycondensation of amino acids and (3) the ring-opening polymerization (ROP) of lactams (Figure 73). At the industrial scale, most aliphatic

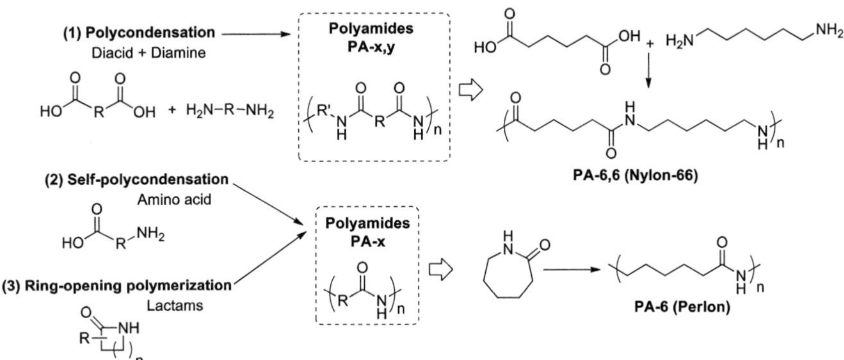

Figure 73.   Different routes towards polyamide syntheses.

polyamides are prepared by melt-phase polycondensation and reach average molecular weight around 10,000 g.mol^{-1}. [425,426]

Depending on the polymerization route employed, AABB- or AB-type polyamides are obtained resulting in different nomenclatures written respectively PA-x,y and PA-x where x and y represent the number of carbons between two nitrogen atoms. [427,428] For instance, PA-6,6 (Nylon-66) and PA-6 (Polycaprolactam, Perlon) count for 85–90% of the commercialized petroleum-sourced polyamides and are respectively synthesized by DuPont from the polycondensation of adipic acid and hexanediamine and by IG Farben from the ROP of ε-caprolactam. With regard to polyamide syntheses, it is noteworthy to mention that AABB-type polymerization is much more challenging than AB-type polyamide preparation. Indeed, according to the Carothers's law, the stoichiometry deviation between diacids and diamines can lead to low molecular weights and weakened polyamide properties. Therefore, these polycondensations are generally performed following a "nylon salt solution" [429] which consists in transforming the two monomers in their corresponding salts in order to target the exact stoichiometry before polymerizing them at elevated temperatures.

The huge potential of vegetable oils for the synthesis of bio-based polyamides comes from the available long carbon chains enabling the preparation of aliphatic diamines, diesters (or diacids) and amino esters. One of the most important sources of vegetable oil-based polyamides is the castor oil (Figure 74). On the one hand, ricinoleic acid can be extracted from the hydrolysis of triglyceride from castor oil. A subsequent pyrolysis gives access to the C11 undecenoic acid, which is an excellent building block for AB-type polyamide preparation. This synthon can be chemically modified *via* a bromination step and a treatment with ammonia in 11-aminoundecanoic acid. For instance, the AB-type PA-11 commercialized by Arkema under the trade name Rilsan® [430–432] has been produced by the self-polycondensation of 11-aminoundecanoic acid for more than 50 years. This famous fully bio-based polyamide is more resilient and thermally resistant than PA-6 and PA-6,6 and is produced at the scale of about 21,000 tons per year for high performance applications (catheters, automotive fuel lines, etc.). On the other hand, alkali fission of ricinoleic acid with NaOH or KOH can produce sebacic acid, which has been extensively used as monomer for AABB-type polyamide preparation. For

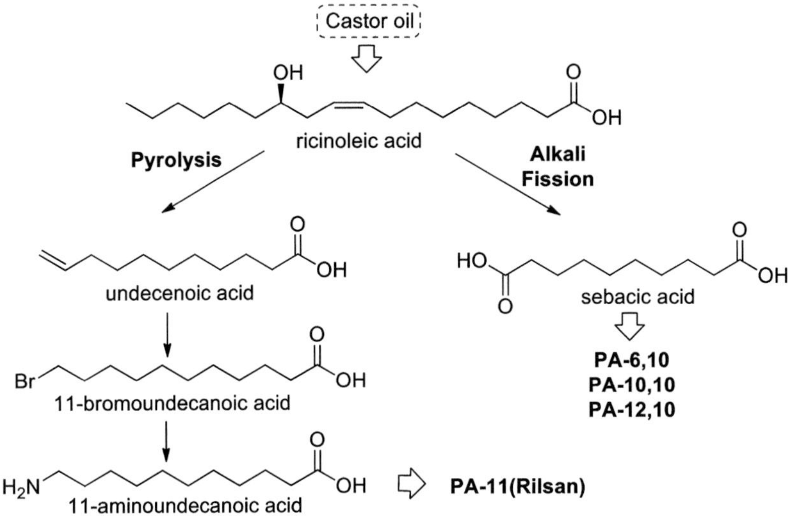

Figure 74. Castor oil as a platform for polyamide synthesis.

instance, various fatty acid-based polyamides have been synthesized from sebacic acid by different companies such as Rhodia (Technyl® eXten, PA-6,10), BASF (Ultramid® Balance, PA-6,10), Evonik (VESTAMID® Terra, fully or potentially bio-based PA-10,10 and PA-10,12) [433] and DSM (EcopaXX®, PA-4,10).[434] Due to the lack of commercial bio-based diamines, it is much more difficult to produce fully bio-based AABB-type polyamides than AB-type one. Nonetheless, sebacic acid can be either transformed in the corresponding diamine to reach fully bio-based polyamides as it is the case of PA-10,10 from Evonik.[435]

As far as material properties are concerned, polyamides coming from castor oil have been used as matrix to reinforce glass, carbon, flax and cellulosic short fibers.[436,437] The resulting composites displayed improved tensile strength, elongation at break and modulus of elasticity.

Depending on the synthesis and on the targeted application, polyamides can display various molecular weights. For example, PA-6,6 exhibits molecular weights ranging in between 25,000 and 110,000 g.mol^{-1} [438] contrarily to some fully bio-based polyamides that could only reach molecular weights up to 14,700 g.mol^{-1}.[439] More generally, melting temperatures and modulus increase with the density of amide linkages. For instance, PA-6 and PA-11 display respective melting temperatures of

218°C and 183°C demonstrating that PA-6 requires higher thermal energy to break the more prominent H-bonding between amide functions. Mecking and colleagues[181] published a complete review reporting the specific properties of long aliphatic polyamides. This review is a reference with respect to the structure-properties relationship analysis for long aliphatic polymers.

In spite of the enthusiasm for ricinoleic acid and its derivatives, other fatty acids have been derivatized and employed as building blocks for polyamide preparation. Several reactions such as enzymatic oxygenation, metathesis or thiol-ene coupling were applied to fatty acids to access polymerizable building blocks.

Regarding first the design of bifunctional monomers, several research groups were interested in erucic acid obtained from the crambeseed oil in order to prepare PA-*x*,13 (Figure 75). Indeed, the brassylic acid (a C13 diacid) can be synthesized from the erucic acid cleavage and also derivatized in 1,13-tridecane diamine using specific conditions.[440,441] PA-13,13 has been prepared from the polycondensation of both diamine and diacid from brassylic acid, resulting in polyamide with Tg of 50°C and Tm of 174°C.[442] In this line, PA-13,6 has been synthesized from 1,13-tridecane diamine and adipic acid [443] and displayed similar Tg but a higher melting temperature of 206°C than PA-13,13. This feature was explained by crystallite reorganization in this material.[444] More generally, the brassylic-based polyamides exhibited lower water uptake than PA-11 or PA-6,10.

Besides, a naturally occurring unsaturated C18 building block coming from the enzymatic oxygenation of oleic acid (using cytrochrome P450 monooxygenase enzymes) [445] and called Z-octadec-9-enedioic (Figure 76) has been copolymerized with several aliphatic and cyclo-aliphatic diamines.[446] The molecular weights of the resulting polyamides were

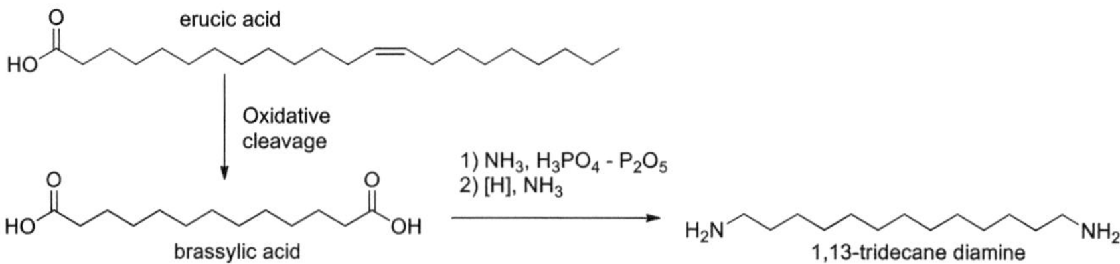

Figure 75. Synthesis of C13 dimers from erucic acid.

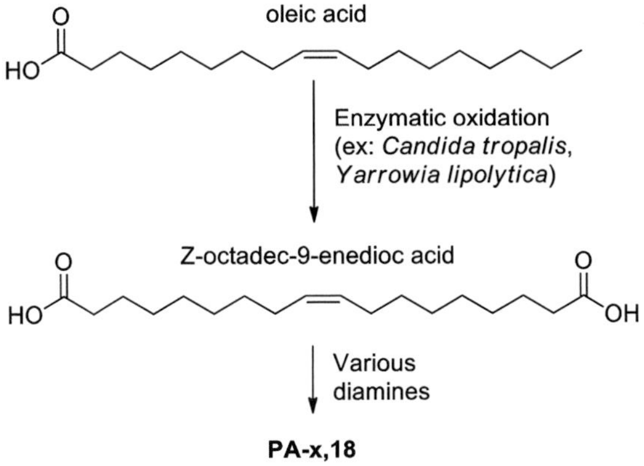

Figure 76.   Z-octadec-9-enedioc acid from oleic acid.

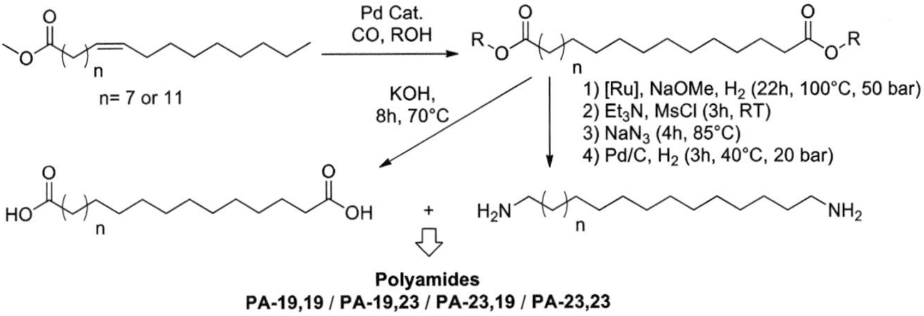

Figure 77.   Diacids and diamines from methyl erucate and methyl oleate.

close to 18,000 g.mol^{-1} and displayed lower Tg and Tm than their corresponding saturated polyamides.[447,448]

Mecking and colleagues[449] reported the synthesis of linear saturated monomers containing 19 or 23 carbons on their aliphatic chains from respectively methyl oleate and methyl erucate following an isomerizing alkoxycarbonylation and a subsequent hydrolysis. Long aliphatic α,ω-diamines can be synthesized *via* a four-step reaction including diol and diazide intermediates further reduced in the corresponding diamines (Figure 77). Interestingly, Walther *et al.*[450] also prepared the C19 diamine from the corresponding diol with an excess of ammonia at 140°C for 48 h. Both diacids and diamines coming from methyl oleate and methyl erucate were polymerized to access novel long-chain PAs with molecular weights around

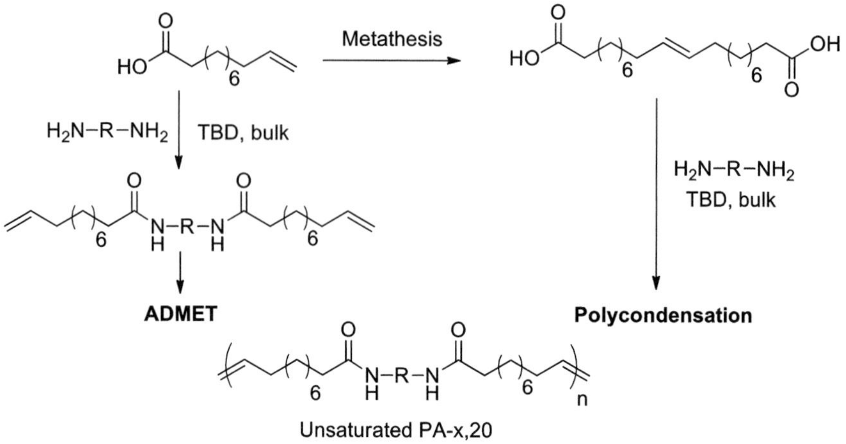

Figure 78. Preparation of fatty acid-based PA-*x*,20 *via* polycondensation or metathesis.

10,000 g.mol^{-1}. As expected, higher melting temperatures were observed for PA-11,23 and PA-12,23 than for PA-23,19 and PA-23,23.[435]

Furthermore, the preparation of unsaturated PA-*x*,20 have been carried out following two strategies. First, the monomer resulting from the self-metathesis of 10-undecenoic acid was polymerized with various diamines using TBD as catalyst (Figure 78). The second strategy involved the ADMET of the bis-unsaturated diamide obtained *via* the dimerization of methyl 10-undecenoate under TBD catalysis.[439,446] Both routes conducted to unsaturated aliphatic PA-*x*,20 with molecular weights in the range 5300–14,700 g.mol^{-1} and Tm comprised in between 180°C and 226°C. Considering the presence of unsaturation within these PAs, applications are restricted as the exposition to UV or air accelerates the material aging.

Interesting dimers originate from the dimerization between oleic and linoleic acids, well-known as *Priplast®* range commercialized by CRODA, were employed as comonomer for polyamide preparation.[451,452] The long alkyl pendant chain imparted by this dimer conferred flexibility and lowered crystallinity to the resulting material in comparison with conventional polyamides, making them good candidates for printing inks and varnishes.[453] Other C36-dimer fatty acids were synthesized by Meier and coworkers [454] *via* the thiol-ene coupling of methyl oleate with ethane-1,2-dithiol under UV

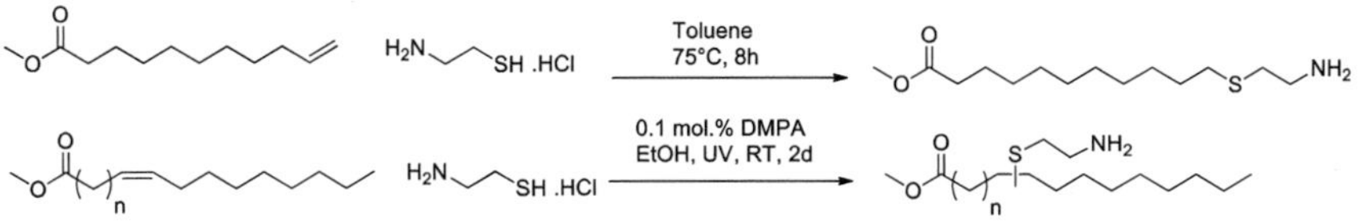

Figure 79. Thiol-ene reaction with cysteamine hydrochloride and fatty acid methyl esters to obtain AB self-condensable monomers.

irradiation using DMPA as initiator. Copolyamides were prepared by polycondensation of hexamethylenediamine, dimethyl adipate and the dimer fatty acid and displayed lower water uptake than PA-6,6 due to the long alkyl chain conferring hydrophobicity.

With respect to the synthesis of AB-type monomers, interesting strategies and chemical transformations have been applied to fatty acids to design new PA-*x* polyamides. For instance, Meier and colleagues[455] performed the thiol-ene coupling of cysteamine hydrochloride with methyl undecenoate, methyl erucate and methyl oleate unsaturations to design AB-type monomers for sustainable polyamides (Figure 79). Tm values were tuned according to the monomer polymerized. Indeed, PA obtained from methyl undecenoate-based monomer displayed a Tm of 138°C whereas the dangling chains of the methyl erucate-based one conferred to the final polyamide a Tm of 43°C.

Amino-esters were also prepared by cross-metathesis of either acrylonitrile and fatty acid methyl esters (10-undecenoate, dimethyl octadec-9-ene-1,18-dioate and methyl ricinoleate) or 10-undecenenitrile with methyl acrylate. Both nitrile derivatives were reduced in their corresponding amine resulting in AB self-condensable monomers for polyamide synthesis.[456,457]

Functionalization of methyl oleate and methyl erucate unsaturations *via* bromination, further reaction with NaN_3 and subsequent hydrogenation was also carried out by Winckler and co-workers.[458] PAs resulting from their self-polycondensation exhibited molecular weights in the range 6500–7900 g.mol^{-1} and Tm in between 40°C and 89°C. A similar strategy involving oxofunctionalization of fatty acid methyl ester unsaturations was lately engaged by the same research group.[459]

Another typical example of AB-type monomer is ω-amino lauric acid obtained from palm kernel oil *via* a biotechnological route with the aim to prepare PA-12 by self-polycondensation. This new route could enable the replacement of the petroleum-based Vestamid® L (Evonik, PA-12).

Only one example of fatty acid-based PA produced *via* ROP has been reported in the literature.[460] Indeed, the multi-step synthesis involved to obtain a polymerizable lactam is not viable at the industrial scale. Contrarily, terpenes and glucose were mentioned in a recent review[429] to be good candidates for the preparation of sustainable lactams.

2. *Polyurethanes*

Linear thermoplastic polyurethanes (TPUs) are commonly used because of their excellent properties, such as good transparency, tunable stiffness, good wear resistance, excellent biocompatibility, etc. These properties are obtained thanks to the alternating of soft and hard segments in the structure of the polymer. Soft segments are usually flexible low Tg polyether or polyester chains with molecular weights between 1000 and 4000 $g.mol^{-1}$ and can be easily obtained from fatty acid derivatives. Hard domains are usually crystalline with high melting points. The polyether polyols represent 75% of the petrochemical polyols used for PUs. Only one exemple of TPUs from polyether diol was reported from vegetable oil derivatives.[461] This polyether diol containing ester pendant groups was obtained by the acid-catalyzed ring opening polymerization of epoxidized methyl oleate and partial reduction with lithium aluminum hydride ($LiAlH_4$) (Figure 80). Polyols with various hydroxyl functionality degrees were obtained

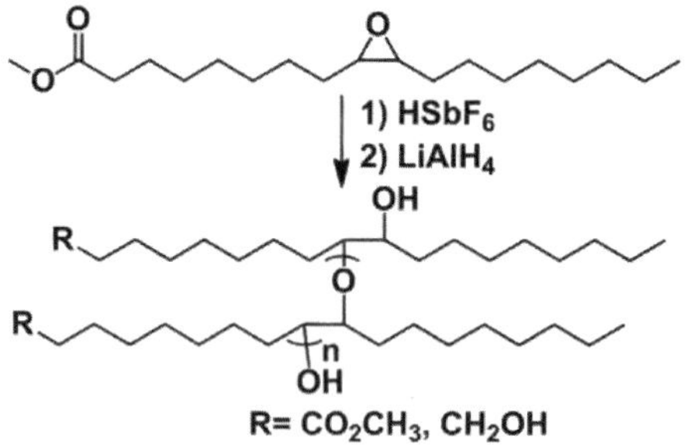

Figure 80. Polyether diol containing ester pendant groups obtained by the acid-catalyzed ring-opening polymerization of epoxidized methyl oleate and a partial reduction with lithium aluminum hydride.

depending on the conversion of the reduction step. The TPU prepared from this polyether diol and MDI presented a low Tg of –15°C and an interesting degradation temperature at 5% weight loss of 307°C.

The synthesis of diols containing ester group have been much more widely explored due to the initial presence of ester or acid function in vegetable oil derivatives. Polyester diols, diols containing ester and/or amide linkages, dimer diols, linear saturated or unsaturated diols and diols containing thioether linkages are referenced using fatty acids or fatty acid methyl esters as starting materials.[462–468] Few research groups have prepared low Tg polyester diols from vegetable oil derivatives to use them in segmented thermoplastic elastomer polyurethanes (TPUs). Petrovic and colleagues have used a poly(ricinoleic acid) diol as soft segment to prepare a series of TPUs by reacting MDI and 1,4-butanediol with various soft segment weight concentrations ranging from 40% to 70% (Figure 81). Such TPUs displayed a microphase separation and two glass transition temperatures at around –50°C and 100°C. The same group has studied the morphology of these TPUs by DMA, AFM and USAXS.[463] A co-continuous morphology with domains around 15 nm was obtained in the case of the TPUs with 50% of soft segment weight concentrations. However, in the case of TPUs with 70 wt% of soft segment, dispersed hard domains were observed in the soft matrix. The biodegradation of these TPUs has been investigated by respirometry.[467] These TPUs with poly(ricinoleic acid) (PRic) as soft segments have been found to exhibit faster degradation rates than the corresponding petrochemical poly(ester urethane)s, even if this rate is relatively slow (about 11% carbon after 30 days). The authors observed that TPUs with a co-continuous morphology demonstrated a slightly slower biodegradation than those with dispersed hard domains in the soft phase.

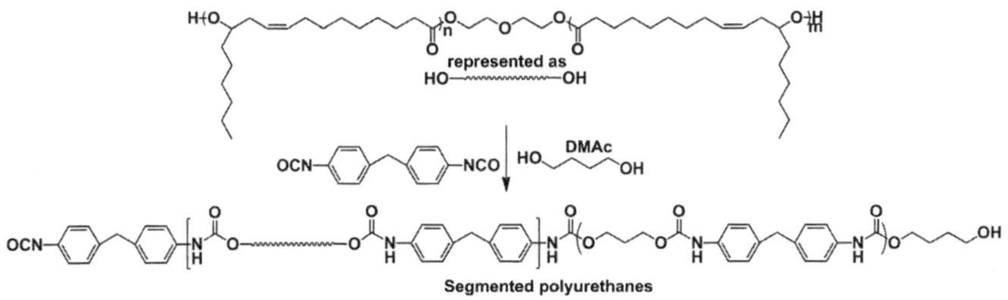

Figure 81. Thermoplastic elastomer polyurethane prepared from poly(ricinoleic acid).

An alternative to synthesize TPUs with a fatty acid based diol as soft segment, is to use the CRODA polyester diol synthesized by transesterification between dimerized fatty acids (obtained by Diels–Alder reaction) and an excess of linear diol.[469,470] By selecting the appropriate hard segment concentration, different thermo-mechanical properties were achieved and could fulfill some industrial requirements in diverse fields.

In order to obtain well-defined monomers, several research groups have synthesized accurate bis-unsaturated compounds with one or two ester functions from methyl oleate and methyl undecenoate or from the corresponding acid derivatives.[471–474] Afterwards, the hydroxyl moieties were introduced on the double bonds by thiol-ene reaction or epoxy ring-opening (Figure 82). By using the thiol-ene reaction method, asymmetric ester diols have been prepared from oleic and undecylenic acids by esterification with allyl alcohol and thiol-ene reaction with 2-mercaptoethanol. Polymerizations with MDI in the presence of tin (II) 2-ethylhexanoate as catalyst, led to TPUs with molecular weights from 50 to 70 kg.mol^{-1} and dispersity in the range 1.6–1.9. Amorphous (Tg = 8°C to 20°C) to semi-crystalline PUs (Tm = 124°C) were obtained depending on the fatty derivatives used. These TPUs revealed both good thermal and mechanical properties as well as no cytotoxic response thanks to MTT test, which make them possible candidates for biomedical purposes.

Cramail and colleagues also prepared sugar-based fatty ester polyols by selective transesterification of epoxidized methyl or ethyl oleate with unprotected methyl α-D-glucopyranoside and sucrose respectively, followed by hydrolysis of the epoxide moiety.[475] The polyols were then used in polyurethanes with IPDI in the presence of dibutyl tin dilaurate (DBTDL) as a catalyst. The authors noticed that the reactivity of the hydroxyl functions attached to the sugar and to the fatty ester chain moieties respectively could be discriminated with respect to the solvent used, enabling the synthesis of either linear or cross-linked PUs.

Vegetable oil based poly- or di-isocyanates have effectively been studied in a much less extent perhaps as a consequence of the inherent aliphatic structure of the isocyanate that could be prepared. Indeed, the most industrially used diisocyanates are aromatic namely MDI and TDI due to a higher reactivity required to produce foams. Thus, the use of aliphatic diisocyanates is limited to coatings in which the absence of unsaturation

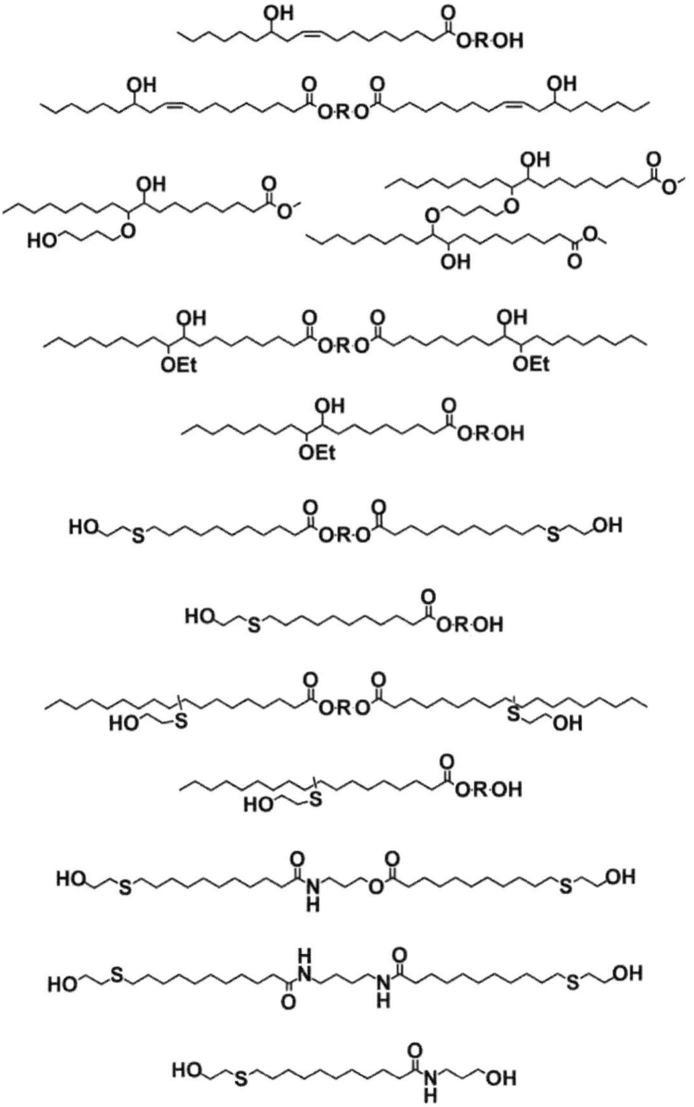

Figure 82. Range of diols (PU precursors) that can be obtained from methyl undecenoate, methyl oleate and methyl ricinoleate by transesterification, amidation, thiol-ene reaction and epoxidation/ring-opening of the epoxide.

is profitable. In industry, isocyanates are synthesized from primary amines by phosgenation. As primary amines are not easily introduced into vegetable oils, other strategies have been developed on vegetable oils and their derivatives. For instance, isocyanates can be synthesized by taking the advantage of the Curtius, Hoffman and Lossen rearrangements, which involve nitrene intermediates. Narine and colleagues used the Curtius rearrangement to prepare oleic acid-based linear diisocyanates, both saturated and unsaturated.[476,477] Two linear diacids were first synthesized then converted to acyl azides by the reaction with sodium azide and, upon heating, had decomposed to isocyanates. Cramail and colleagues also reported the synthesis of fatty acid based diisocyanates using the Curtius rearrangement through acyl hydrazide fatty acid based derivatives without the use of harmful sodium azide (Figure 83).[478] Diesters were first synthesized and then reacted with hydrazine hydrate to form diacyl hydrazides in quantitative yields. Afterwards, these diacyl hydrazides were converted into diacyl azides then into diisocyanates *via* the Curtius rearrangement. A series of partially and fully vegetable oil based TPUs were synthesized and a large range of thermo-mechanical properties were achieved. Relatively good thermal behaviors were observed with decomposition temperatures at 5% weight loss from 230°C to 280°C. For some polyurethanes, a close resemblance to HDPE was obtained in terms of solubility and thermal transitions with melting points close to 145°C.

There are few examples in the literature of fatty acid-based polyurethanes obtained by transurethanization polycondensation. The monomers, bis-alkylcarbamates and bis-hydroxyalkylcarbamates, can be obtained from dimethylcarbonate or ethylene carbonate. Fatty acids can also be modified by other means to get methylcarbamate functionalities.

Narayan and colleagues synthesized a bis-hydroxycarbamate using a dimerized fatty acid by transamidation of ethylene diamine on the carboxylic

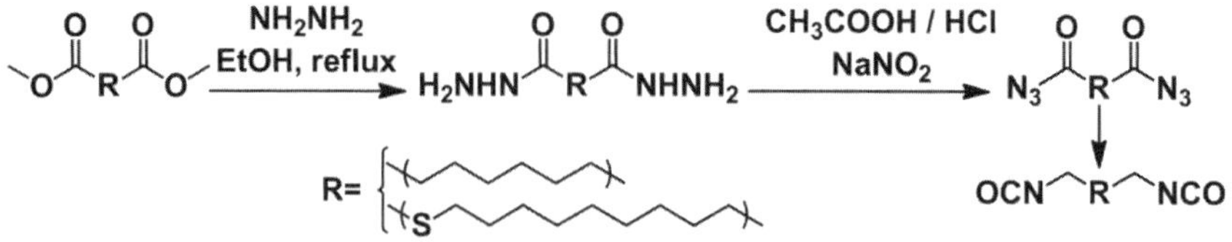

Figure 83. Fatty acid-based diisocyanates using hydrazine hydrate.

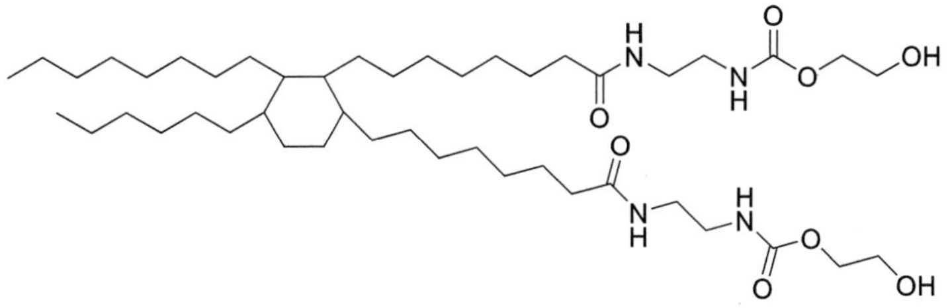

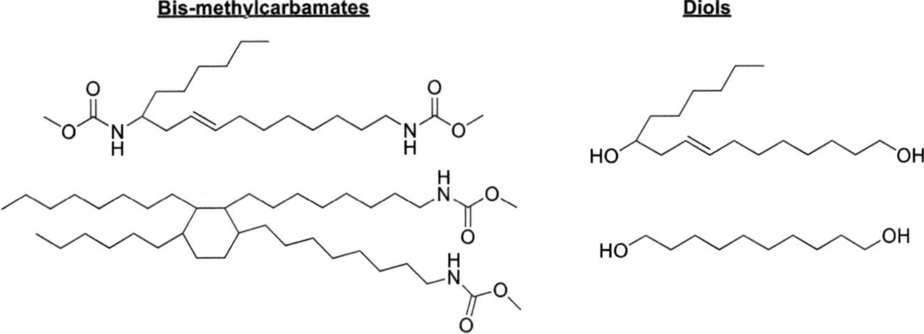

Figure 84.   Fatty acid based bis-hydroxycarbamate.

Figure 85.   Monomers derived from fatty acids for the synthesis of Pus.

acid functions and further reaction with ethylene carbonate (Figure 84). The polymerization strategy is a self-polymerization of the bis-hydroxycarbamate. The polymers obtained at 150°C under reduced pressure exhibited a $\overline{M}_n$ of 7.7 kg.mol^{-1} and $\overline{M}_w$ of 14 kg.mol^{-1}, a Tg of –10°C and a Tm of 73°C.

Burel and colleagues synthesized various bio-based PUs from bio-based diols and bis-methylcarbamates.[479] The bis-methylcarbamates were synthesized from bio-based diamines and DMC in the presence of TBD. The bis-methylcarbamates produced are presented Figure 85, the latter are derived from a fatty acid dimer and ricinoleic acid. The PUs were then synthesized by polycondensation with diols with TBD or K_2CO_3 as catalyst. Tg of the so-formed PUs are ranging from –38°C to –19°C. Most PUs are amorphous due to the presence of dangling chains brought by the fatty acid derivatives, but some present a fusion temperature around 61–64°C.

The Lossen rearrangement is usually used to transform hydroxamic acids into isocyanates. Meier and colleagues introduced a modified method to get the corresponding carbamates with a catalytic amount of an

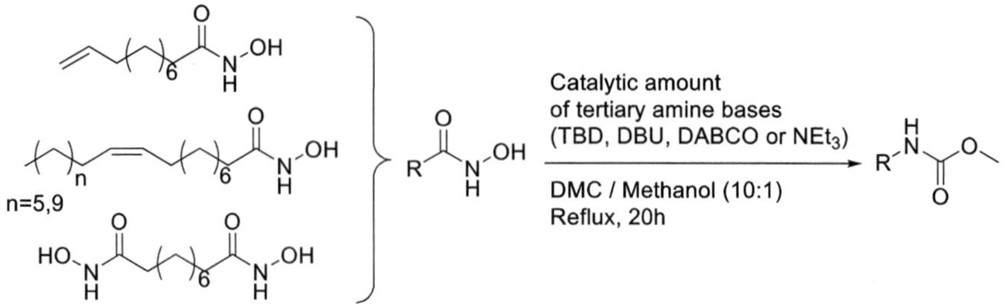

Figure 86. Catalytic Lossen rearrangement towards fatty acid-based carbamates.

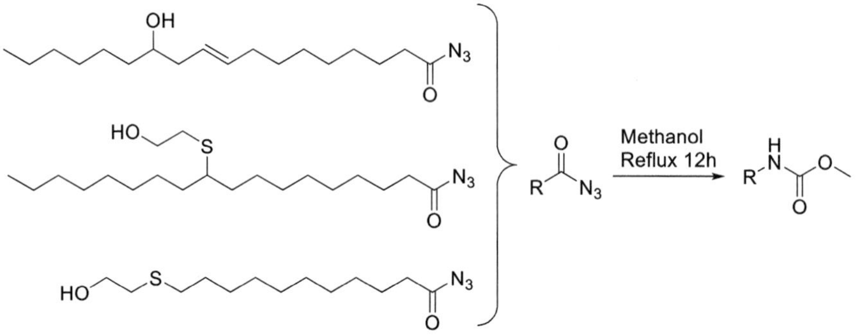

Figure 87. Fatty acids-based AB-type synthons.

organic base such as TBD in the presence of methanol.[480,481] Fatty acid methyl esters were easily turned into hydroxamic acids using hydroxylamine hydrochloride and potassium hydroxide. Bis-methylcarbamates and fatty carbamates were synthesized using this method (Figure 86).

The polymerization of such bis-methylcarbamate with bio-based aliphatic diols was performed in bulk with catalytic amount of TBD under reduced pressure by increasing the temperature from 120°C to 160°C. The PUs obtained were crystalline with melting temperatures ranging from 120°C to 145°C. The molecular weights were comprised between $\overline{M}_n = 7.1$ and 24.6 kg.mol^{-1} with dispersities of 1.2 to 2.1.

AB-type synthons were synthesized by Cramail and colleagues (Figure 87). The hydroxyl function is naturally occurring in the alkyl chain of ricinoleic acid or it was brought by thiol-ene addition on the double bond of oleic acid and derivatives. The methylcarbamate functionality was brought by transforming the carboxylic acid into acyl azide moiety. The acyl azide was then reacted with methanol to form the methylcarbamate.[482,483] Self-polymerizations were performed in the presence of $Ti(OBu)_4$ as catalyst at 130°C under N_2 atmosphere for 4 hours and then under reduced pressure for 2 hours. PUs obtained had low molecular weights with $\bar{M}_n$ from 2.1 to 6.9 kg.mol^{-1}. Tgs were ranging from –44 to –18°C. A second glass transition temperature was visible at around 25°C to 32°C for PUs synthesized from thiol-ene modified synthons, which indicated a phase-separated morphology.

Among non-isocyanate and non-phosgene routes to PUs, the approach involving the reaction of cyclic carbonate and amine functions is one of the most studied. The main route is the polyaddition of bis-five-membered cyclic carbonates and diamines leading to the formation of linear PHUs with primary or secondary alcohols, as already illustrated in Figure 70.

As far as vegetable oils are concerned, the five-membered cyclic carbonate monomers have been mainly synthesized from glycerol carbonate or by epoxidation followed by carbonation of the fatty acid chain unsaturations.

The first example relating the synthesis of thermoplastic PHUs from fatty acid derivatives was reported by Boyer et al.[484] The fatty acid-based bis 5-membered cyclic carbonates were synthesized by a three-step process composed of (1) a transesterification of fatty acid methyl esters, (2) an epoxidation of the double bonds and (3) a carbonation of the resulting epoxides. The solubility of the fatty acid or triglyceride-based mono-, bis- and poly-epoxides in supercritical CO_2 has been studied.[485] Two bis five-membered cyclic carbonates, prepared from methyl oleate and methyl undecenoate respectively, were polymerized with ethane-1,2-diamine (EDA) and isophorone diamine (IPDA) to form PHUs containing hydroxyl moieties (Figure 88).[484] The PHUs exhibited molecular weights up to 13,500 g.mol^{-1} $\left(\bar{M}_w\right)$ and relatively low Tgs ranging from –25°C to –13°C. Nevertheless, an amidation side reaction occurred between EDA and the ester linkages of the diester bis-cyclic carbonates, giving amide groups which can partly explain the low molecular weights obtained.

We further reported the synthesis of several fatty acid based bisCC using the epoxidation/carbonation strategy.[486] Two fatty acid chains were

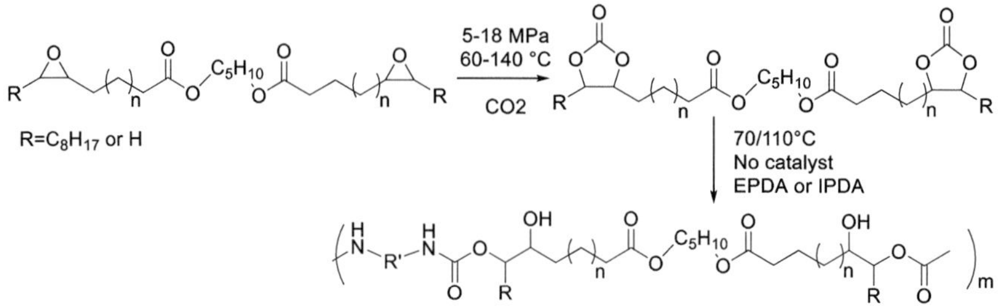

Figure 88. Cyclic carbonates and poly(hydroxyurethane)s from fatty acid or fatty acid methyl ester.

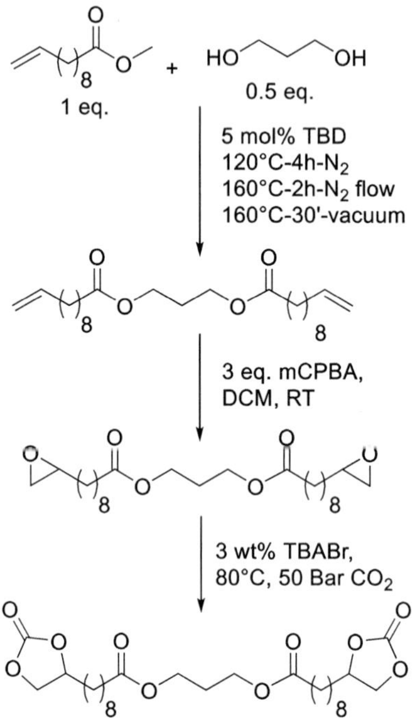

Figure 89. Strategy for the synthesis of 5-membered bis-cyclic carbonates from fatty acid derivatives.

first dimerized by amidation or transesterification reactions with butane-1,4-diamine, piperazine, *N,N′*-dimethylpropane-1,3-diamine, *N,N′*-dihexyldecane-1,10-diamine and 1,3-propanediol. The produced aliphatic bis-unsaturated substrates were epoxidized and carbonated following the procedure indicated in Figure 89. The polymerization of these dimers presenting different central blocks with various diamines (such as *Jeffamine*, isophorone diamine or diaminobutane) led to PHUs exhibiting a broad range of Tg values from –29°C to 55°C.

In order to increase the reactivity of 5CC towards aminolysis, an alternative route consists in inserting a heteroatom nearby the cyclic carbonate to improve/activate its reactivity.[487–496] Activated five-membered cyclic carbonates were prepared from glycerol, thioglycerol and fatty acid derivatives (Figure 90). Ester, ether and sulfur moieties were introduced in β position to the cyclic carbonate, in order to enhance its reactivity towards amines.[497,498] ^{1}H NMR kinetic investigation of the aminolysis of

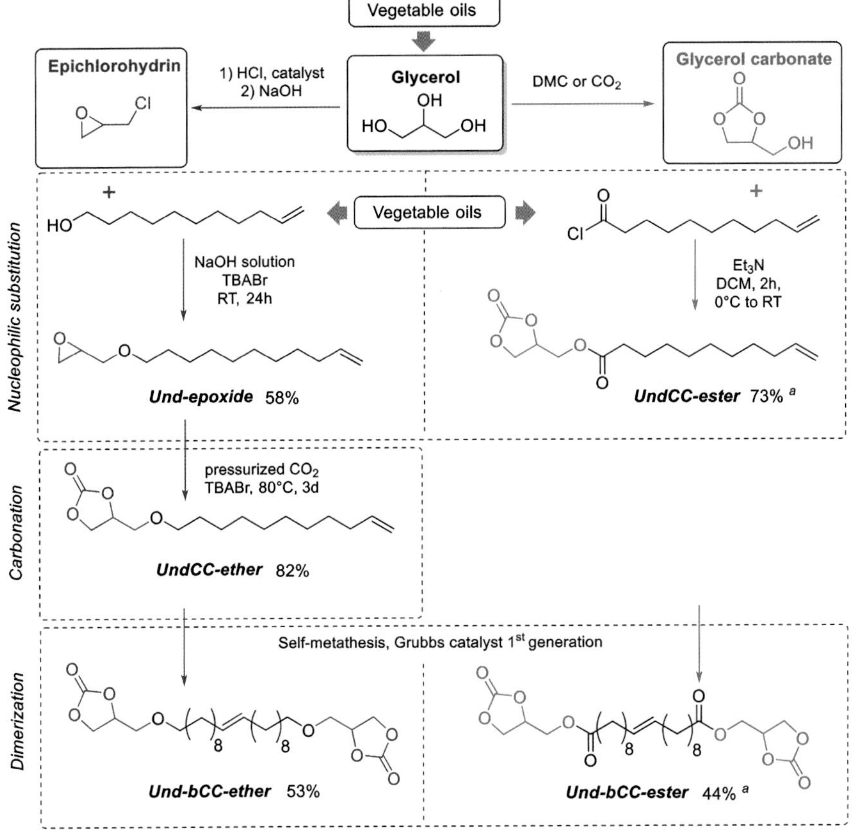

Figure 90. Strategy for the synthesis of "activated" five-membered bis-cyclic carbonates from fatty acid derivatives.

these cyclic carbonates demonstrated a higher reactivity compared to the one of alkyl substituted cyclic carbonates. In the case of ester-activated carbonates, a reactivity similar to the one of 6-membered ring cyclic carbonate was observed. Moreover, these carbonates exhibited amidation side-reactions with amines that could be however prevented by decreasing the temperature to room temperature. Poly(hydroxyurethane)s (PHUs) were then synthesized from these activated five-membered ring cyclic carbonates at 70°C in DMF and exhibited molecular weights up to 13,700 g.mol^{-1} with Tg in the range –26°C to –10°C.

Another strategy developed by Maisonneuve *et al.*[499] focused on the design of more strained and reactive six-membered cyclic carbonates. The latter were synthesized from methyl undecenoate by malonization, reduction and carbonation of the resulting diol. The mono-cyclic carbonate obtained was dimerized either by thiol-ene or metathesis reactions (Figure 91). The synthesized bifunctional six-membered cyclic carbonates were effectively used as building blocks for thermoplastic isocyanate-free PHUs in combination with dodecane-1,12-diamine as comonomer. Molecular weights up to 23,000 g.mol^{-1} (Đ = 1.7) were obtained after only one day in DMF at 50°C.

As a conclusion of this part, activated 5-CCs and 6-CCs have a comparable reactivity with respect to aminolysis. However, 5-CCs seem to be more relevant in terms of synthesis at the industrial scale. Hence, fatty acid derivatives represent a platform of bio-based building blocks for poly(hydroxyurethane) synthesis. Caillol and colleagues[500] highlighted the large possibilities offered by this platform for both linear and cross-linked polymers. The abundance of reactive hydroxyl functionalities present along the PHU backbone could enable further curing process and post-functionalization.[496]

3. *Polyesters*

As already discussed, aliphatic polyesters (PEs) combine unique properties such as film forming ability, potential biodegradability and biocompatibility.[501] Indeed, under specific conditions, PEs are prone to be degraded *via* reactions onto ester linkages, known to be hydrolyzable. Depending on the monomers and on the polymerization methods, a wide

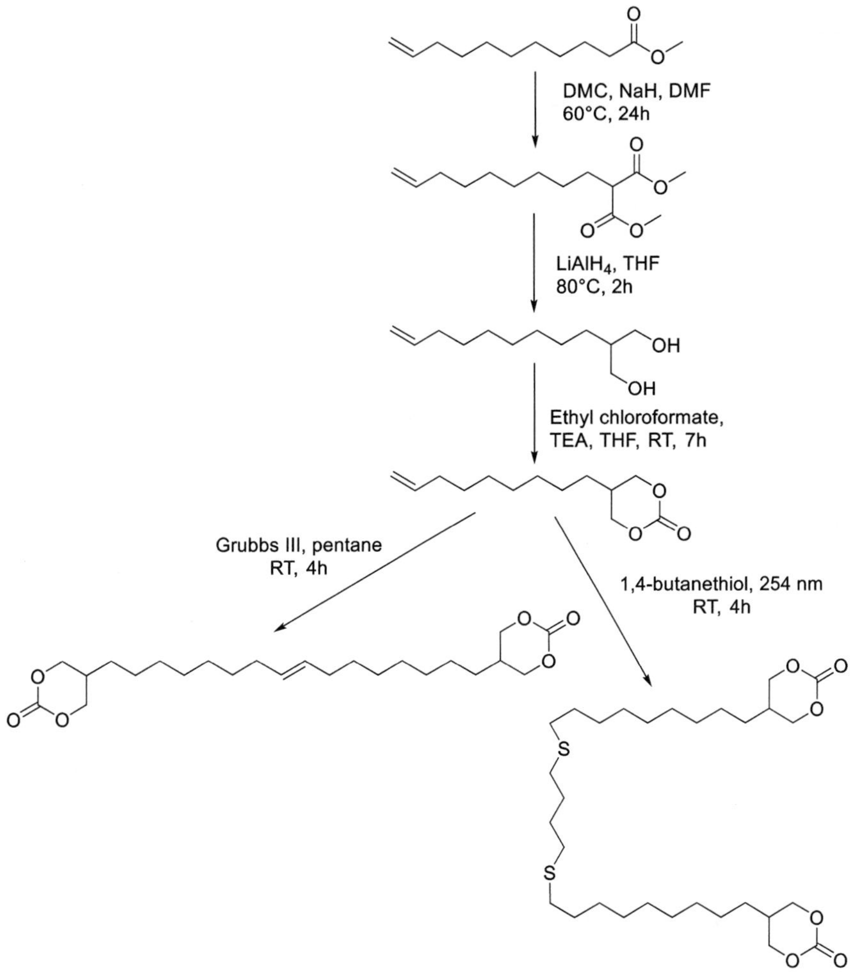

Figure 91.   Procedure for bis 6-membered cyclic carbonate synthesis from methyl undecenoate.

range of applications is covered. Indeed, polyesters can be employed in fibers, rubbers, plasticizers, surface coatings, or resins,[502,503] demonstrating the high potential and versatility of such materials. Polyesters are commonly prepared from the polycondensation by esterification between diols and diacids, or by transesterification between diols and diesters, resulting in the production of water or methanol as by product (Figure 92(1)).

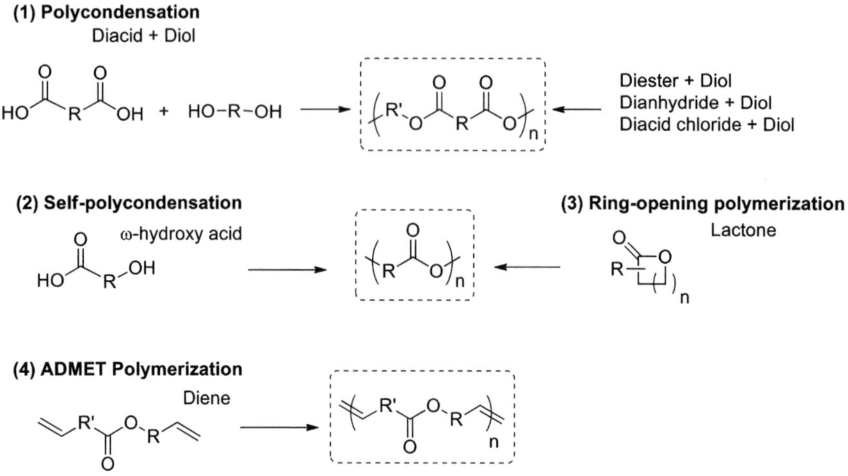

Figure 92. Different routes towards polyesters preparation.

However, the transesterification pathways require high temperature and catalysts [426,504] for the preparation of high molecular weight PEs. Another strategy is to employ "activated" monomers such as acid chlorides or anhydrides to circumvent this issue. As A_2-B_2-type polymerization requires an accurate stoichiometry (1:1) between both monomers, another strategy considering the polycondensation of ω-hydroxy acids (AB-type monomers) has been considered. Ring opening polymerization of lactones and ADMET polymerization of ester-containing dienes were also developed for the preparation of polyesters (Figures 92(2) and 92(3)).

Many reviews reported the possibilities offered to prepare bio-based polyesters from sugars, fatty acids or terpenes from which carboxylic acids are easily obtained.[404,192,505,506] Due to the extensive number of studies covering the subject, the scope of this sub-chapter will only focus on the most relevant pure fatty acid-based polyesters and their relative properties.

Fatty acids represent a promising feedstock for polyester synthesis as they provide ester, acid and sometimes hydroxyl functions. Additionally, fatty acids lead to linear aliphatic polyesters known to be degradable in a reasonable time scale, making them good candidates for drug delivery systems.[507] Fatty acid based-polyesters can also mimic polyethylene while presenting original features such as biodegradability and polarity.

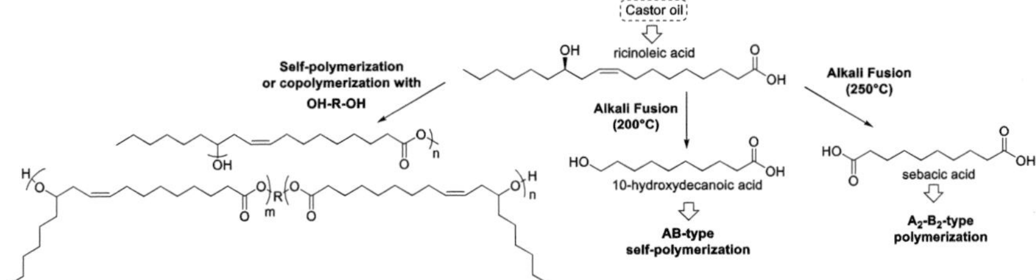

Figure 93.   Castor oil as a platform for bio-based polyesters.

Again, ricinoleic acid derived from castor oil is an interesting starting material for the synthesis of AB- and A_2-type monomers for renewable polyesters as reported by Kunduru *et al.*[508] (Figure 93). This review highlights the biocompatibility of such starting materials making them suitable for biomedical applications. The carboxylic acid and the naturally occurring hydroxyl function present on ricinoleic acid enable its self-polycondensation to yield poly(ricinoleate) also named PRic. Enzymatic polytransesterification of ricinoleic acid has been carried out using several lipases and enabled to yield polyesters with molecular weights comprised between 2000 and 100,000 g.mol^{-1}.[509] The dangling fatty acid chains conferred particular properties to the resulting polyesters; a Tg of –75°C was obtained for the highest molecular weights. Extreme temperature above 200°C and the use of tin catalysts could also permit to prepare PRic.[510] Ricinoleic acid can also be copolymerized with polyols in order to increase molecular weights and to expand the PRic range.[511] Taking advantage of the unsaturations within PRics, the latter could be cross-linked/vulcanized to yield elastomers.[509,512]

10-Hydroxydecanoic acid as well as sebacic acid can be obtained from alkali fusion leading to polymerizable molecule through an AB-type or an A_2-B_2-type process. These two molecules can also be reduced in their corresponding diol to yield another interesting A_2 monomer.[513] Regarding sebacic acid, this castor oil-based monomer has already been incorporated in copolyesters applied as biodegradable delivery systems for biomedical applications.[514,515] In a more original way, 9-hydroxynonanoic acid can be obtained from ozonolysis, reduction, transesterification and catalysis with Ti(iOPr)$_4$[516] and subsequently self-polymerized to yield a high molecular

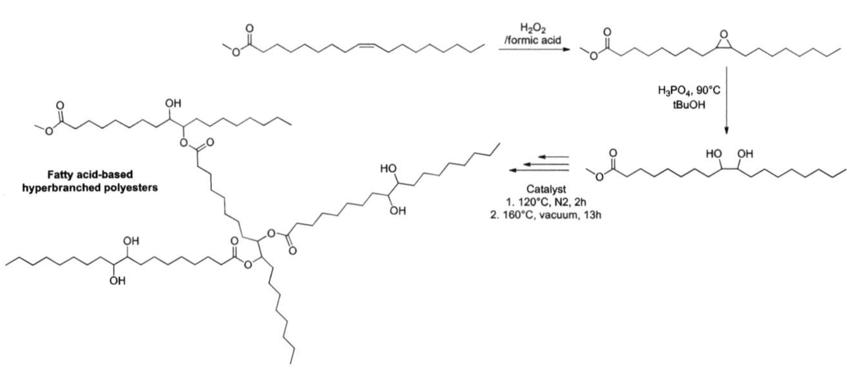

Figure 94. ABn-type monomer synthesis towards hyperbranched polyesters.

weight polymer (62,000 g.mol^{-1}) with higher Tg, Tm and thermal stability than poly(ε-caprolactone), PCL.

Regarding other fatty acid-based condensable monomers, some of them were prepared from more original techniques. As an example, Cramail and colleagues[517] published the synthesis of hyperbranched polyesters from catalyzed self-polycondensation of AB$_n$-type monomer prepared from the ring-opening of epoxidized fatty acid methyl esters in acidic conditions (Figure 94). The prepared materials exhibited molecular weights in the range 3000–10,000 g.mol^{-1} with dispersities comprised between 2 and 15.

Another example describes the biochemical conversion of unsaturated fatty acids into α,ω-dicarboxylic acids by enzymatic oxidation of terminal methyl groups into a carboxylic function using *Candida tropicalis*.[518] The unsaturated polyesters prepared *via* an enzymatic polycondensation of the prepared monomers with α,ω-diols displayed molecular weights in the range 25,000–57,000 g.mol^{-1} with relatively low Tm (23–40°C). Notably, these materials can also be post-functionalized or cross-linked through unsaturations within the backbone.

Besides, an original method to produce A$_2$-type monomers was investigated few years ago by Mecking's group[506] and consisted in the isomerizing carbonylation of unsaturated fatty acids with carbon monoxide using specific Pd catalysis. Dicarboxylic acids and diols containing 19 and 23 carbons were thus prepared and polymerized to achieve long aliphatic semicrystalline polyesters with high Tm values between 86°C and 100°C.

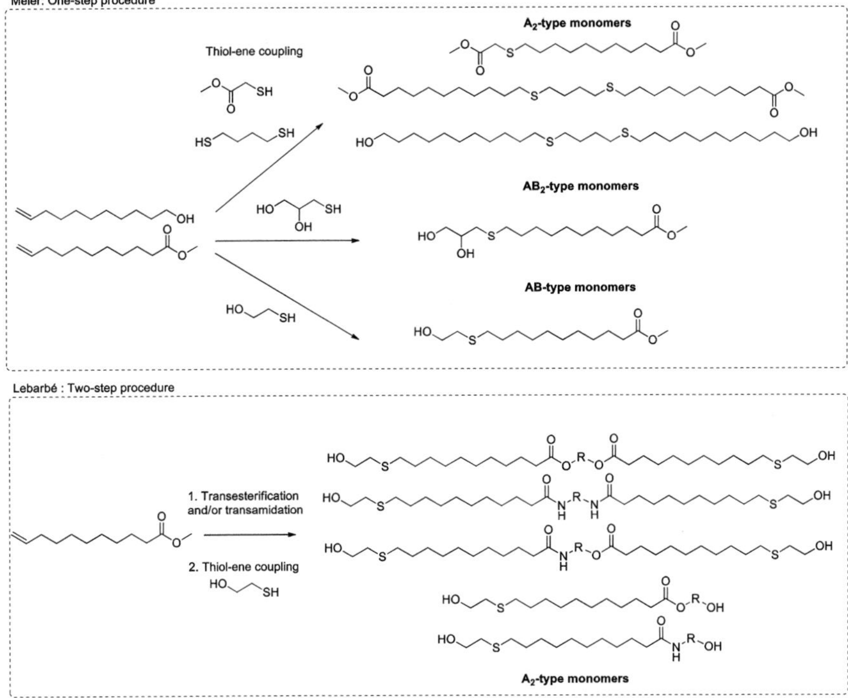

Figure 95. Monomer synthesis *via* thiol-ene click recactions.

Furthermore, ADMET and thiol-ene click chemistry have been extensively investigated for the design of AB-type, A_2-type and AB_2-type monomers as well as polymerization techniques. For instance, the group of Meier[519] carried out the preparation of different original fatty acid-based monomers from methyl castor oil derivatives (10-undecenoate, 10-undecenol) and 1,4-butanedithiol, methylthioglycolate, thioglycerol or mercaptoethanol *via* thiol-ene reactions (Figure 95). Depending on the thiol substrate used, AB-, A_2-, or AB_2-type monomers were prepared by bringing either ester or hydroxyl functions. After polycondensation, the resulting polymers displayed melting points in the range 50–71°C and thermal stability up to 300°C.

Following a similar strategy, Lebarbé *et al.*[520] synthesized symmetric and asymmetric fatty acid-based diols by associating transamidation (or transesterification) reactions and thiol-ene coupling of mercaptoethanol with fatty acid unsaturations (Figure 95). The original condensable monomers were polymerized with a diester synthesized from the self-metathesis of methyl 10-undecenoate. The resulting poly(ester-amide)s showed melting temperatures up to 127°C due to the presence of amide linkages.

Metathesis reactions were also used as a tool for the synthesis of dimers for polyester preparation. The self-metathesis of unsaturated fatty acid derivatives followed by other chemical modifications (hydrogenation, reduction of acid/ester functions into hydroxyl moieties) was conducted.[521,522] For instance, fatty diacids were synthesized by self-metathesis of respectively 10-undecenoic acid and erucic acid, followed by hydrogenation. The resulting dicarboxylic acids were polymerized with their corresponding diols (obtained after reduction) to yield PE-20,20 and PE-26,26. Besides, linear long aliphatic chain polyesters have been largely provided by ADMET polymerization[251,521,523–525] of ester-containing α,ω-dienes, themselves prepared from fatty acids (Figure 96). Most of these studies conducted the hydrogenation of the resulting unsaturated polyesters with the aim of mimicking polyethylene structure.[526] Moreover, Mecking and colleagues[521] demonstrated the strong impact of

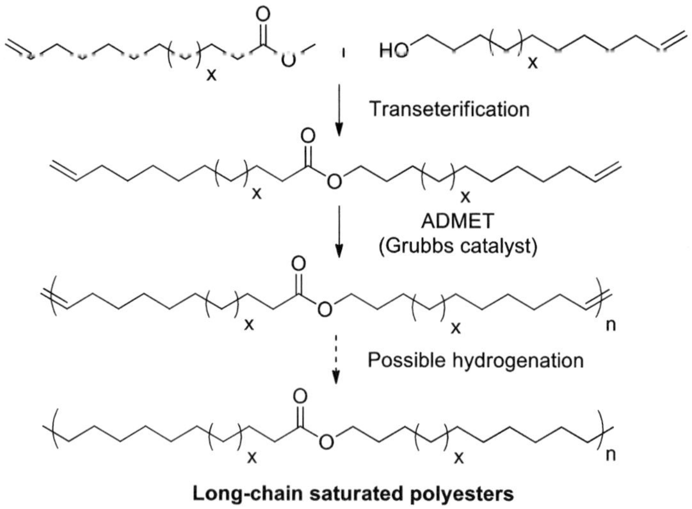

Figure 96. ADMET polymerization of fatty acid-based α,ω-dienes and possible hydrogenation towards saturated polyesters.

unsaturations within the polyesters that tends to decrease crystallinity and consequently melting temperature, justifying such hydrogenation process after polymerization.

The group of Meier[527] published a comparative study of the polymerization of fatty acid-based dienes containing ester functions *via* thiol-ene and ADMET processes. The authors demonstrated that thiol-ene click reaction was leading to higher molecular weight polyesters (11,850 g.mol^{-1}) in comparison with ADMET polymerization (9000 g.mol^{-1}) and proved the potential degradability of the synthesized polyesters in enzymatic and acidic conditions. Noteworthy, possible isomerization of the unsaturations during ADMET polymerization could create heterogeneities in the polyester structure decreasing its thermo-mechanical properties. Hopefully, the use of the first generation Grubbs catalyst and 1,4-benzoquinone would limit this side reaction.[528,529] Regarding thiol-ene coupling polymerizations, one of the major drawbacks is the presence of sulfur atoms in the resulting polymer backbone, leading to more flexible materials and to lower melting points.

The last common route to obtain polyesters is the ring opening polymerization of lactones that circumvent the stoichiometry issue encountered with AB-type polycondensation. Fatty acid-based macrolactones are less reactive than smaller size lactones and result in low molecular weight polyesters with high dispersities (around 1.4–1.5) after ROP. Additionally, these monomers are less readily available in comparison with the carboxylic acids from the same feedstock. Still, ω-pentadecalactone and ethylene brassylate are both macrolactones naturally derived from *Angelica archangelica* L. root oil[530] and musc [531] respectively (Figure 97). As very few macrolactones are naturally

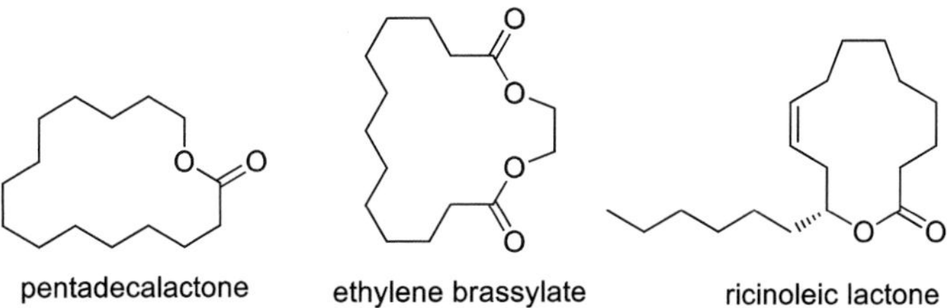

Figure 97.   Examples of vegetable oil-based macro(di)lactones.

occurring, other studies aimed at synthesizing fatty acid-based macrolactones using enzymatic or organic catalysts.[532,533] For instance, ricinoleic acid was self-transesterified using dicyclohexylcarbodimide and (dimethylamino)pyridine as catalysts in order to produce the corresponding lactone (ricinoleic lactone, Figure 97).[532] Regarding ω-pentadecalactone, numerous studies [534–539] reported the investigations on ROP catalysis of such bio-based compounds. The best results were achieved with Lipase Novozym 345 that displayed an extensive catalytic activity to ring-open this macrolactone, resulting in polyesters with molecular weight up to 470,000 g.mol^{-1} [534–536] and high elongation at break around 100–200%. Moreover, the biocompatibility and the non-toxicity of poly(pentadecalactone) was confirmed by Heise and colleagues[540] The group of Mecerreyes [531] carried out an organocatalyst screening and demonstrated that TBD was able to efficiently catalyze the ROP of ethylene brassylate. Semi-crystalline aliphatic polyesters with molecular weight up to 13,000 g.mol^{-1} and with slightly higher Tm and Tg than poly(ε-caprolactone) ones were achieved.

4. *Polycarbonates*

Aliphatic polycarbonates are very useful in a wide range of applications thanks to their biocompatibility, biodegradability and low toxicity. Only few examples of fatty acid-based polycarbonates have been reported.

The groups of Meier[541] and Cramail[542] developed a fatty acid-based carbonate-containing monomer from the transesterification between dimethyl carbonate and 10-undecenol. The resulting aliphatic diene was polymerized *via* ADMET polymerization and displayed molecular weights ranging in between 9500 and 50,800 g.mol^{-1}, depending on the reaction conditions chosen by both research teams.

Another strategy consisting in the copolymerization of dimethyl carbonate and bio-based diol was investigated. Miller and Vanderhenst[543] carried out the reaction between dimethyl carbonate (or ethyl chloroformate) and 1,10-decanediol (sebacic acid derivative), in order to prepare an α,ω-bis-carbonate. Its catalyzed self-polycondensation led to poly-carbonates with molecular weights in the range 8000–50,000 g.mol^{-1}. In the same study, the direct copolymerization of diol with dimethyl

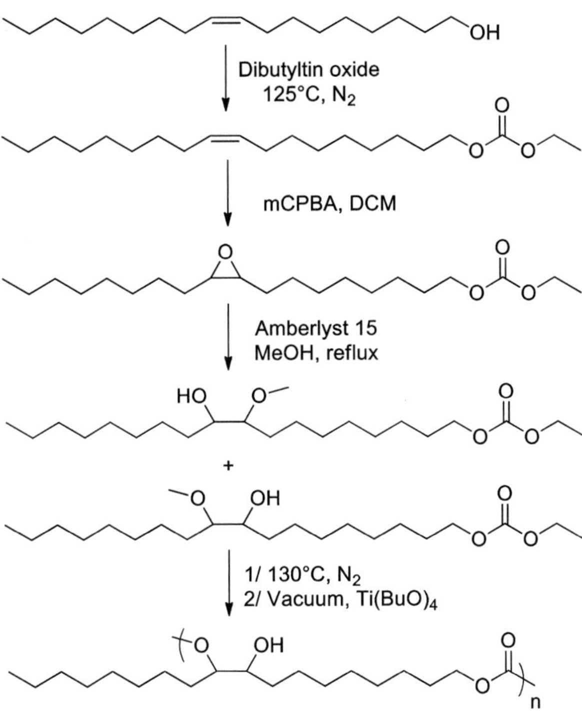

Figure 98. Multi-step synthesis of AB-type self-condensable monomer towards oleyl alcohol-based polycarbonate.

carbonate was realized and yielded polycarbonates with similar molar masses.

An original multi-step procedure involving (i) transcarbonation, (ii) epoxidation of the unsaturation and (iii) epoxide ring-opening (Figure 98) was developed by Cramail's group[542] to design AB-type monomers from ricinoleic and oleic acids. The latter were subsequently polymerized under titanium tetrabutoxide catalysis leading to polycarbonates with molecular weights up to 14,000 g.mol^{-1} and low Tg down to –60°C due to the aliphatic chain plasticizer effect.

More recently, Zhang *et al.*[544] carried out the selective copolymerization of CO_2 with epoxy methyl 10-undecenoate in the presence of a zinc-cobalt complex (Figure 99). Fatty acid-based polycarbonates were thus formed after epoxide ring-opening and displayed relatively low Tg due to

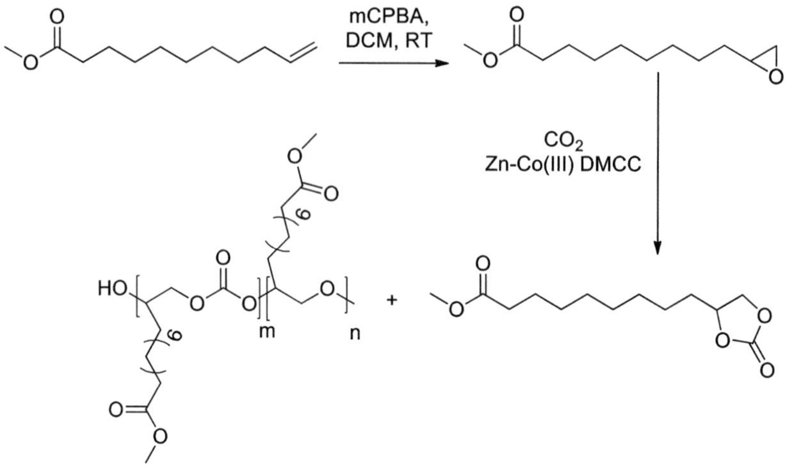

Figure 99. Copolymerization of epoxy methyl 10-undecenoate with CO_2.

internal plasticization imparted by aliphatic dangling chains and low molecular weight (2000 g.mol^{-1}). Consequently, the hydroxyl-terminated aliphatic polycarbonates were used to initiate ROP of L-lactide leading to an interesting biodegradable triblock PLLA-PC-PLLA.

IV. Polymers from Terpenic Resources

A. Introduction

Terpenes, terpenoids and rosin are compounds that are mainly synthesized from plants. They all have in common one or more isoprene unit (2-methyl-1,4-butadiene) in their chemical structure.[545] Although terpenes are mainly hydrocarbon components, terpenoids have additional constituents, for example, oxygen, containing functional groups such as carbonyl. Terpenes have been industrially important for years, particularly in the fine chemical and fragrance industries[546] because they are abundant and inexpensive. Some of them even exhibit important pharmacological functions for the treaTment of many diseases, including cancer.[547,548] In addition to these applications, terpenes are also widely used in polymer chemistry as starting materials. Here, some innovative polymeric materials based on terpene derivatives will be presented.

B. Terpenes

The most common terpenes can be isolated directly from pine trees and conifers. Indeed, the volatile fraction of resin, turpentine, is composed of a mixture of terpenes yielding directly α- and β-pinene and rosin. Turpentine is by far the main source of terpenes with a yearly terpenes production of 350,000 tons.[549] Other important terpenes as limonene or carvone can be produced by isomerization of α- or β-pinene[550,551] (Figure 100).

Thereby, only new polymers derived from α-pinene, β-pinene, limonene and two linear terpenes (myrcene and citronellol) will be considered due to the low cost and the easy isolation of such starting materials.

1. *α-Pinene-based Polymers*

Pinenes are the most abundant and easily isolated terpenes. Obtained by the steam-distillation of sap from pine or conifer trees, pinenes can undergo a multitude of chemical reactions to produce valuable monomeric compounds. Structurally, α-pinene is a bicyclic monoterpene hydrocarbon

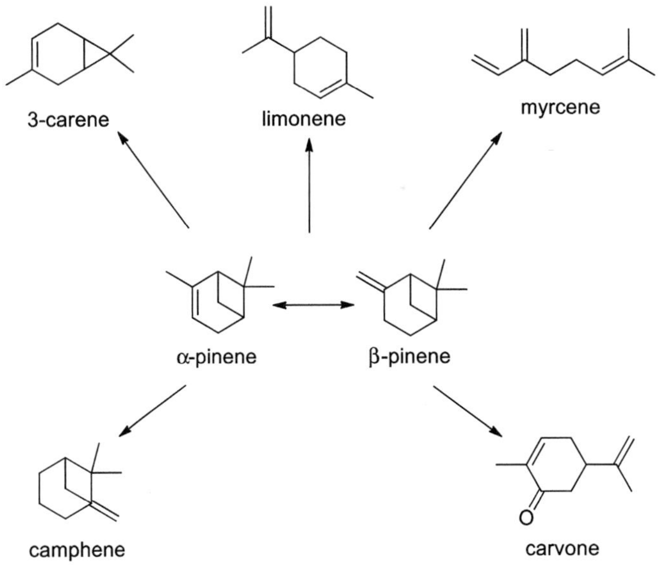

Figure 100. Various terpenes derived from α pinene and β pinene by isomerization.

consisting of four- and six-membered rings with an internal trisubstituted carbon–carbon double bond in the six-membered ring (Figure 100). However, it is noteworthy that the majority of pinene polymerizations involve the use of β-pinene and not α-pinene due to the lack of highly reactive terminal double bond in the α-pinene structure. Consequently, only oligomers of α-pinene can be obtained by direct cationic polymerization. The ring-opening metathesis polymerization (ROMP) could be a mean to access renewable materials from α-pinene. However, its highly congested trisubstituted olefin avoids initiation and propagation steps.[552] Nevertheless, Thomson and colleagues[553] investigated a novel method affording the ROMP of apopinene, an α-pinene derivative. The authors have shown, for the first time, that apopinene can act as a monomer for pinene-derived ROMP processes to yield polymers with molecular weight up to 15,600 g·mol^{-1}.

The group of Kamigaito[554] has also presented an innovative way to valorize α-pinene into pinocarvone by chemical photo-oxidation under visible-light irradiation. Pinocarvone, which bears a reactive exo methylene group, can be quantitatively polymerized in a selective ring-opening radical process to afford novel bio-based polyketone (Figure 101).

An alternative way to create new renewable polymers from α-pinene is described by Stockman and Howdle.[555] Among other

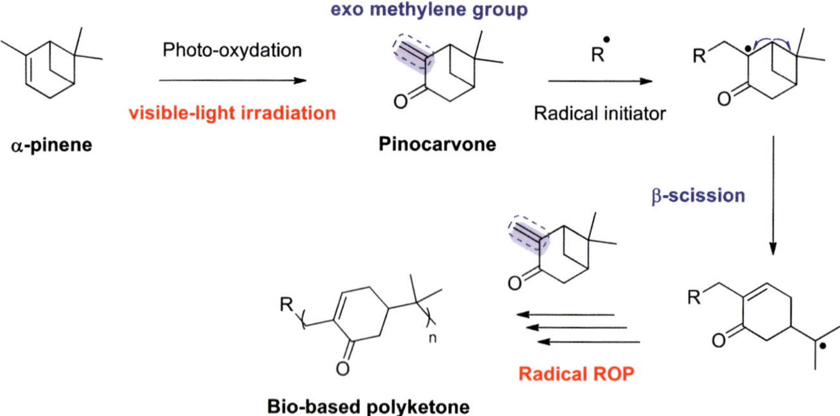

Figure 101. Structures of α-pinene and pinocarvone and polymer thereof.

terpenes, the authors transformed α-pinene into an acrylic and methacrylic monomer *via* a one-pot catalytic approach. The monomer was then polymerized into useful materials including polymer coatings.

2. *β-Pinene-based Polymers*

As previously mentioned, β-pinene is a readily suitable monomer for cationic polymerization. The mechanism is well-understood, as well as the side reactions lowering the degree of polymerization. Indeed, the cationic polymerization of β-pinene employed a Lewis acid as a catalyst. These polymerizations resulted in only low molecular weight polymers ($M_n \leq 4000$ g.mol^{-1}) which can be further modified to produce epoxy networks and polyols.[556] However, numerous studies have been done to increase the molecular weight of poly(β-pinene). Among them, Kukhta *et al.*[557] produced relatively high molecular weight poly(β-pinene) (M_n = 9000–14,000 g.mol^{-1}), with a Tg in the range 82–91°C using a complex co-initiator system comprising water, aluminum chloride and diphenyl ether at 20°C (Figure 102).

One classical method to valorize β-pinene is its copolymerization with other monomers. Various structures were tested to afford new interesting copolymers presenting a large range of properties. The copolymerization of β-pinene with methyl acrylate (MA) was investigated by Wang *et al.*[558] The feasibility of β-pinene and MA was demonstrated for the first time despite the significant difference between their reactivity ratios. The authors showed that the addition of a Lewis acid, Et$_2$AlCl, to the AIBN-initiated copolymerization enhanced the incorporation of β-pinene (up to 45 mol.%) but decreased the polymer molecular weight.

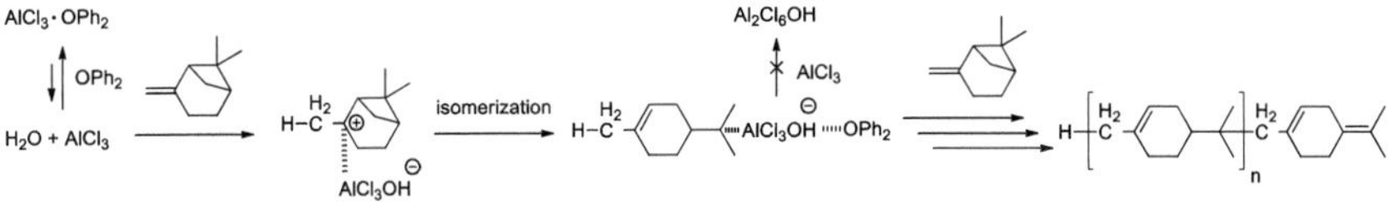

Figure 102. Proposed mechanism for β-pinene polymerization using H$_2$O/AlCl$_3$ •OPh$_2$ initiating system.

3. *Limonene-based Polymers*

Another interesting terpene is limonene. As shown in Figure 100, limonene can be obtained from the isomerization of pinene, but this chiral molecule is also naturally produced by plants. The (R)-enantiomer consists of 90–96% of citrus peel oil and its worldwide production exceeds 70,000 tons per year. Although the main use for limonene is in flavor and fragrance industry, its polymeric potential gains interest in academic research. For instance, Meier and colleagues described the synthesis of new limonene-based polyesters,[559] polyamides[560] and polyurethanes[560] *via* thiol-ene additions. For the synthesis of polyesters, ester or alcohol functionalized thiols were added regioselectively to limonene to yield monofunctional, difunctional or heterodifunctional monomers, which are interesting renewable building blocks. On the other hand, cysteamine hydrochloride was used to produce amine-functionalized monomers for polyamide or polyurethane syntheses. Thus, limonene-based polymers were prepared with mixing above-mentioned monomers with ester or alcohol functionalized fatty acids.

Limonene-based polycarbonates can be also obtained from limonene oxide. The epoxide is commercially available as a mixture of the trans and cis diastereomers. Its abundance, low cost, and structural similarity to cyclohexene oxide make (R)-limonene oxide an excellent choice as a biorenewable epoxide monomer for copolymerization with CO_2. The group of Coates[561] described the alternating copolymerization of limonene oxide and CO_2 catalyzed by a β-diiminate zinc molecule to give a new biodegradable polycarbonate, proceeding with highly selective incorporation of the trans diastereomer under mild conditions (Figure 103).

Based on this study, Greiner' group[562] managed to increase significantly the polycarbonate molecular weight from 25 kg.mol^{-1} to more than 100 kg.mol^{-1} by synthesizing the limonene oxide with very high content (>85%) of trans-isomer and by masking hydroxyl functions in monomer impurities. Greiner also demonstrated that the bio-based poly(limonene carbonate) (PLimC) is a powerful platform to create numerous new functional materials.[563] Different chemical transformations of the PLimC led to dramatic changes of its properties. For instance, a rubbery material was synthesized from the thermoplastic PLimC. Polymers with an antibacterial

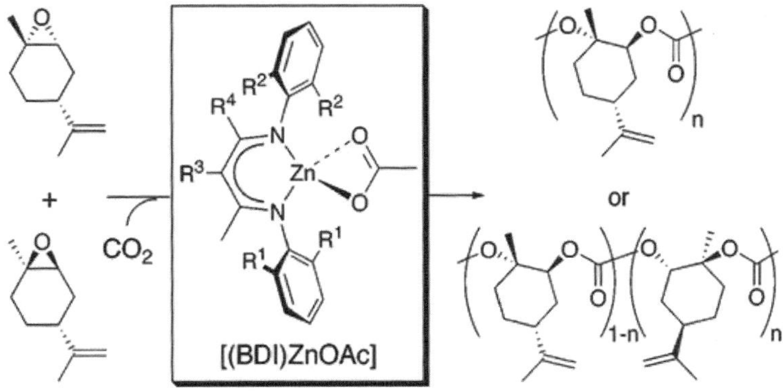

Figure 103.   Copolymerization of limonene oxide and CO_2 to produce polycarbonate.

activity, increased hydrophilicity or even water solubility were also prepared from PLimC. The heat processability can be also improved after full
hydrogenation of the polymer.

Finally, Mülhaupt[564] reported a new route to linear and thermoset
terpene-based non-isocyanate poly(hydroxyurethane)s, (NIPUs) derived
from limonene dicarbonate.

4. *Linear Terpenes-based Polymers*

As shown in Figure 100, various monoterpenes are derived from the
isomerization of pinene. One of these isomers, myrcene, has gained interest in recent years. Hillmyer *et al.*[565] prepared high-molecular weight
poly(3-methylenecyclopentene) using a combination of ring-closing
metathesis and cationic polymerization with zinc chloride (Figure 104).
This system afforded polymers with narrow molecular weight distributions. Furthermore, the molecular weight values could be tuned by changing the feed molar ratios of monomer to initiator, indicating that the
polymerization proceeds in a controlled manner.

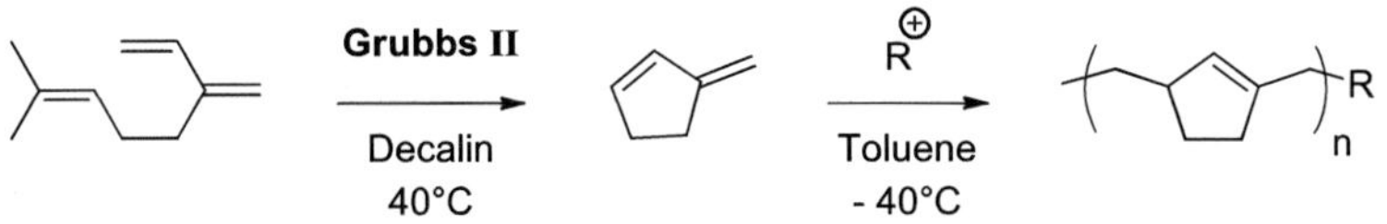

Figure 104.   Ring-closing metathesis of myrcene followed by cationic polymerization of 3-methylenecyclopentene.[565]

Another linear terpene, the citronellol, a major component of geranium and rose oils, can be directly oxidized into citronellic acid. This acid can be dimerized *via* an isomerizing alkoxycarbonylation reaction.[566] The reactive primary carboxy group formed allows the preparation of novel high molecular weight polyesters.

C. Terpenoids

A second class of terpenes, referred to as terpenoids, possess functional groups that can be converted to a cyclic ester. Carvone and menthol are two common terpenoids (Figure 105) that have been successfully derivatized and further polymerized by ROP.

Carvone is a natural product found in both Mentha spicata (spearmint) and Carum carvi (caraway) oils. The world production is estimated around 10,000 tons annually. Carvone is mainly used as a flavor in food and beverages, as well as in toothpaste. After hydrogenation, carvone leads to dihydrocarvone and carvomenthone, two ketones that can be derivatized into lactones (dihydrocarvide and carvomenthide respectively) by Baeyer–Villiger oxidation. These lactones were polymerized using the catalyst/initiating system diethyl zinc/benzyl alcohol to yield aliphatic polyesters with low Tgs and a good control of the polymer molecular weights[567] (Figure 106).

The same group oxidized dihydrocarvone to an epoxylactone on a multigram scale.[568] The resulting epoxylactone was used as a multifunctional monomer and as a cross-linker in ring-opening polymerizations leading to oligomers. However, copolymerizations of ε-caprolactone and the epoxylactone gave flexible cross-linked materials with shape mem-

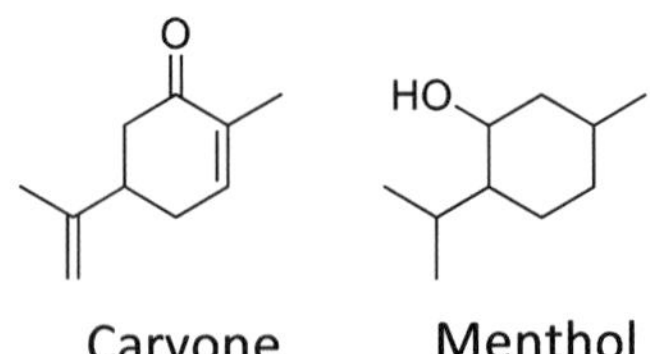

Figure 105. Two common terpenoids: carvone and menthol.

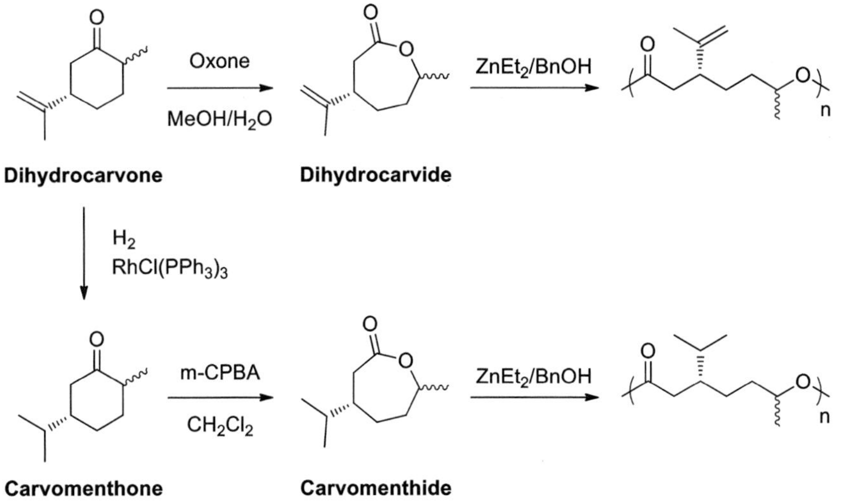

Figure 106. Baeyer–Villiger oxidation and ROP of carvone-derived monomers.

ory properties. Associating the fact that these materials are biodegradable and biocompatible, they are valuable candidates for biomedical applications.

On the other hand, menthol is found in Mentha arvenis and Mentha piperita (peppermint) oils. Its primary uses are in the flavoring industry, as well as in various medicinal applications. Menthone is readily converted to a lactone monomer, menthide, by the Baeyer–Villiger oxidation. The resulting monomer can undergo ROP with a zinc alkoxide catalyst.[569] The polymerization can take place at room temperature due to the high activity of the catalyst and the polymer molecular weight can be controlled by adjusting the monomer to catalyst ratio. The same group also worked on the synthesis of new ABA triblock copolymers prepared using the renewable monomers menthide and lactide by sequential ring-opening polymerizations.[570,571] The authors demonstrated that the properties of the new biorenewable thermoplastic elastomers can be systematically modulated by changing the stereochemistry of the polylactide end blocks. The triblock copolymers can also act as pressure-sensitive adhesive when they are combined with renewable tackifier.[572]

D. Rosin

Rosin is a general term that designates three types of rosin types. Gum rosin is the most common type of rosin extracted from pine resin. The production of rosin is more than 1 million ton per year. Gum rosin is mainly composed of abietic-type and primaric-type acids as illustrated in Figure 107 while the predominant rosin acid is abietic acid. Rosin acids can be converted to a wide range of derivatives from esters and maleic anhydride adducts to hydrogenated acids. These low cost natural products have been used in anti-fouling, adhesive and ink applications.

Several groups have developed main chain rosin-based derived polymers obtained by step-growth polymerization. The nature of the monomer can be tuned, often involving a Diels–Alder reaction between levopimaric acid and a dienophile such as acrylic acid, maleic anhydride and maleimide. The resulting monomers can react with other comonomers such as diol, diacid, diamine, etc. affording the synthesis of polyesters, polyamides, polyamideimides and polyester polyols.

Atta *et al.*[573] described the formation of polyamide and polyamideimide using respectively a diacid functionalized levopimaric acid and

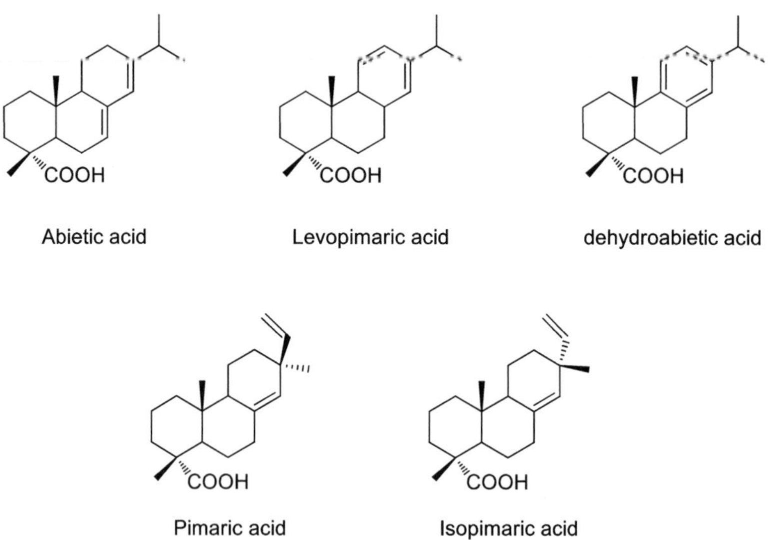

Figure 107.   Most common acids in gum rosin.

an acid-anhydride functionalized levopimaric acid. The same group prepared rosin-based unsaturated polyesters[574] using the same diacid functionalized levopimaric acid, ethylene glycol, maleic anhydride and adipic acid as the starting materials. The final unsaturated polymers were able to undergo curing in the presence of styrene to yield a cross-linked polymer with a Tg of 117°C potentially used in the field of steel coating.

Mustata group reported the synthesis of various polymers that can be obtained from this platform like water soluble poly(amide-imide)s[575] and polyamides,[576,577] polyhydroxyimides[578] and polyesters.[579] Kim *et al.*[580] have carried out studies on the formation of photoactive polyamide-imide. The polymer was made by the condensation of a maleopimaric adduct with azo-dye type diamines. Fabricated polymer films are very smooth, tough, adhesive to the substrate and optically clear.

As above mentioned, main-chain rosin-based polymers can be prepared by different step-growth polymerization methods. However, due to steric hindrance, monomer impurities and stoichiometric control, only low molecular weight polymers could be obtained. In order to access to higher molecular weights, an alternative strategy can be applied. Instead of targeting main-chain rosin-based polymers, different research group investigated the synthesis of side chain rosin-based polymers by radical polymerization. Thus, a platform of rosin-derived vinylic, allylic or acrylic monomers were developed, to be polymerized by radical polymerization.

Free radical polymerization was firstly used to obtain side chain rosin-based polymers. Only low molecular weight polymers were obtained with vinyl monomers because of the steric hindrance. To reduce the steric hindrance, allylic and acrylic monomers were synthesized. Only the latter showed promising results thanks to the higher reactivity of the double bond.[581,582,583,584] In order to obtain well-defined polymers, controlled radical polymerization (CRP) was used thereafter. This route allows the control of polymers molecular weight and polymer architectures. RAFT (Reversible Addition-Fragmentation chain Transfer) polymerization and ATRP (Atom-Transfer Radical-Polymerization) are two of the most widely used CRP methods. The group of Tang[585–588] prepared side chain rosin-based polymers using both methods. For instance, the

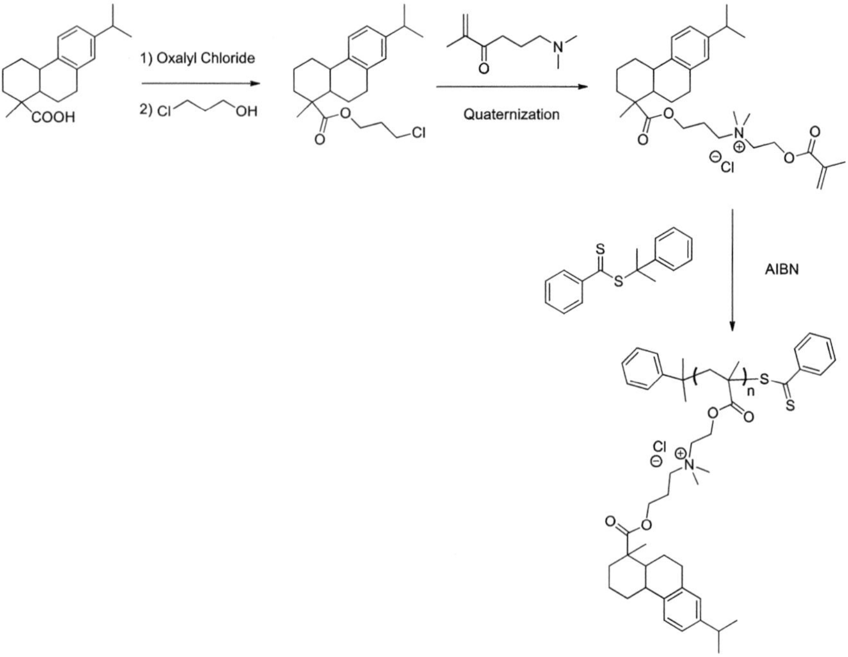

Figure 108. Cationic methacrylic polymers with natural rosin as pendant group.

authors synthesized cationic rosin-containing methacrylate polymers with controlled molar mass, potentially used as antimicrobial agents[588] (Figure 108).

The major inconvenience of using controlled radio polymerization methods is the need to purify the raw rosin to obtain a well-defined resin acid. To avoid the purification step, Tang and co-workers[589,590] combined ROP and click chemistry to prepare graft copolymers using raw rosin materials. The rosin properties have been successfully imparted to the initial polymers.

Last but not least, rosin has been widely used as curing agents to replace some of petroleum-derived ones. To do so, rosin has been derivatized to contain anhydrides, carboxyl or epoxy groups required for curing. The aromatic and cycloaliphatic structure of rosin acids affords the creation of rigid curing agents. Carlotti and co-workers[591] worked on the synthesis rosin acid oligomers as precursor for epoxy resins. Zhang and co-workers[592–595] also developed new rosin-derived rigid curing agents opening new opportunities in the field of thermosets.

V. General Conclusion

At present the development of synthons (building-blocks) stemming from the biomass is rapidly expanding, both in the academic community and industry. The production of biosourced polymers is a real challenge in the coming years to reduce the environmental impacts. If numerous ways are ongoing, some products are already commercial: from vegetable oils, as PA-11 produced by Arkema or functional intermediaries by Elevance and from sugar derivatives such as lactic acid, produced by Natureworks, or isosorbide, produced by Roquette, to mention but a few.

However, beyond the scientific questions connected to the treatment of the biomass, some challenges remain with respect to the development of the renewable resources for a sustainable chemistry.

First of all, it becomes now obvious that the mimetic way (molecules equivalent to those stemming from the petrochemistry) is not necessarily the most interesting. Indeed, most of the manufacturers choose original biosourced polymers while targeting new properties and functionalities. The only presence of renewable carbon is valuable only for very specific applications and does not justify systematically the biosourcing. Especially as the mimetic way leads to intermediaries that possess exactly the same issues as their fossil counterparts, without solving the questions of reduction of exposure levels in the dangerous compounds, for instance. On the other hand, the huge potential of the structures and the functionalities stemming from renewable resources allows to overtake certain compromises of properties, what is very sought by an innovative industry. Another major challenge consists in justifying the reduction of impact realized by the use of renewable resources. The life cycle analysis (LCA) takes here all its sense, in particular with regard to water footprint recently integrated (ISO 14046). The LCA enables to integrate, well upstream, the impacts connected to the production and to the collection of the renewable resources (inputs, transfers in fields, etc.). It emerges from these LCA that the use of renewable resources is not systematically a synonym of impact reduction. Indeed, the fossil polymers benefit from processes of polymerization (and monomer synthesis) optimized for more than 60 years and present the most reduced possible impacts. Actually, the synthesis of a kilogram of polypropylene, PP, presents at present less impact than the one of a kilogram of PLA. There are thus very wide margins of evolution for the improvement of the production processes of bio-sourced polymers. That is why the use of renewable resources has to proceed in particular of attentive choices concerning the level of crop productions, the choice and the volumes of biomass which can generate important environmental differences of impacts.

Therefore, the interest of bio-refineries takes all its sense to value at the most a given resource by integrating in particular the notion of valuation of the co-products and so to reduce the impacts and the costs. Some companies initially created at the heart of food-processing sectors as starch industry, certain stationeries, or distilleries are already organized in bio-refineries and value at the most their raw materials. The deployment of bio-refineries integrated in the territory, the judicious choice of their locations in agricultural basins and the optimization of their processes — by following the model of refineries — will guarantee the success of a large valuation of the renewable resources. This organization will permit to mutualize the costs, to make competitive the proposed intermediaries and to position with regard to hypothetical conflicts with the food use.

Also, the development of the valuation of the renewable resources — rich in atoms of oxygen in comparison to the fossil resources — generates a change of paradigm by positioning the catalytic reduction processes at the heart of the processing, which is based, in the case of fossil resources, essentially on the catalytic oxidation.

Finally, the choice of the resources and of the biosourced polymers must be also dictated by the variations of access to the oil resources. The development of the shale gas in the USA reduces the interest of the bio-sourced ethylene, but strengthens the needs for new bioresources, which can bring intermediaries with more than four carbon atoms, aromatic molecules and derivatives bearing long fatty chains. Furthermore, the use of renewable resources — globally rather available — allows to free from the business with oil-producing countries which exercise a cartelization of their production and thus permits to stabilize strategic supplies while favoring a long-lasting and sustainable agriculture, also by limiting the drift from the land and by regaining control grounds.

In conclusion, the development of polymers from renewable resources is inevitable and expects a colossal potential both for the academic research and the chemical industry in the coming years. All the world leader companies already launched into this competition by integrating chemistry and biotechnologies processes. This new research thematic is present from now in most of the laboratories and the conferences, justified by the Green Chemistry and by the current global environmental challenges. Besides, the challenges with respect to the production of monomers and biosourced polymers permit interdisciplinary crossings of interest with the experts of the biomass or the bioprocesses, which also take all their sense in a synergic approach of the biomass valuation. Finally, exceeding the hackneyed image of predation of the fossil resources, this chemistry based on the renewable resources can maybe give back to the chemistry all its sense of harmonious coexistence with the earth, foundation of its etymology: from the Egyptian *kemet*, the science and the knowledge of the earth for a sustainable future!

VI. Acknowledgments

HC most warmly thanks all the co-authors of this book chapter as well as Dr. Sylvain Caillol (Univ of Montpellier), Prof. Luc Avérous (Univ of Strasbourg), Prof. Stéphane Bruzaud (Univ of Bretagne Sud) and Prof. Stéphane Grelier (Univ. of Bordeaux) for helpful discussions.

VII. References

1. Fernando, S.; Adhikari, S.; Chandrapal, C.; Murali, N. *Energy Fuels* **2006**, *20*(4), 1727–1737.
2. Gandini, A.; Belgacem, M. N. *Prog. Polym. Sci.* **1997**, *22*(6), 1203–1379.
3. Kamm, B.; Kamm, M.; Schmidt, M.; Hirth, T.; Schulze, M. *Lignocellulose-Based Chemical Products and Product Family Trees*; **2008**; Vol. 2.
4. Rosatella, A. a.; Simeonov, S. P.; Frade, R. F. M.; Afonso, C. A. M. *Green Chem.* **2011**, *13*, 754.
5. Delidovich, I.; Hausoul, P. J. C.; Deng, L.; Pfützenreuter, R.; Rose, M.; Palkovits, R. *Chem. Rev.* **2016**, *116*(3), 1540–1599.
6. Brasholz, M.; von Känel, K.; Hornung, C. H.; Saubern, S.; Tsanaktsidis, J. *Green Chem.* **2011**, *13*(5), 1114.
7. Kohl, T. M.; Bizet, B.; Kevan, P.; Sellwood, C.; Tsanaktsidis, J.; Hornung, C. H. *React. Chem. Eng.* **2017**, *2*(4), 541–549.

8. Dutta, S.; Wu, L.; Mascal, M. *Green Chem.* **2015**, *17*(7), 3737–3739.

9. Mascal, M.; Nikitin, E. B. *Green Chem.* **2010**, *12*(3), 370.

10. Bizet, B.; Hornung, C. H.; Kohl, T. M.; Tsanaktsidis, J. *Aust. J. Chem.* **2017**, *70*(10).

11. Boisen, A.; Christensen, T. B.; Fu, W.; Gorbanev, Y. Y.; Hansen, T. S.; Jensen, J. S.; KliTgaard, S. K.; Pedersen, S.; Riisager, A.; Ståhlberg, T.; *et al. Chem. Eng. Res. Des.* **2009**, *87*(9), 1318–1327.

12. Haas, T.; Tacke, T.; Pfeffer, J. C.; Klasovsky, F.; Rimbach, M.; Volland, M.; Ortelt, M. Verfahren Zur Herstellung von 2,5-Diformylfuran Und Seiner Derivate. World Patent WO 2012004069 A1, 20120112, **2012**.

13. Ma, J.; Xu, J.; Jia, X.; Wang, M.; Shi, S.; Miao, H.; Gao, J. The Method of Preparation of 2,5-Dimethyl 2,5-Dimethyl Amino Acyl Furan Furans. Chinese Patent CN 104277018 A, 20150114, **2015**.

14. Saha, B.; Bohn, C. M.; Abu-Omar, M. M. *ChemSusChem* **2014**, *7*(11), 3095–3101.

15. Cao, Q.; Liang, W.; Guan, J.; Wang, L.; Qu, Q.; Zhang, X.; Wang, X.; Mu, X. *Appl. Catal. A Gen.* **2014**, *481*, 49–53.

16. Chatterjee, M.; Ishizaka, T.; Kawanami, H. *Green Chem.* **2014**, *16*(11), 4734–4739.

17. Chen, J.; Lu, F.; Zhang, J.; Yu, W.; Wang, F.; Gao, J.; Xu, J. *ChemCatChem* **2013**, *5*(10), 2822–2826.

18. liu, F.; Audemar, M.; De Oliveira Vigier, K.; Clacens, J.-M.; De Campo, F.; Jerome, F. *Green Chem.* **2014**, *16*(9), 4110–4114.

19. Ohyama, J.; Esaki, A.; Yamamoto, Y.; Arai, S.; Satsuma, A. *RSC Adv.* **2013**, *3*(4), 1033–1036.

20. Tamura, M.; Tokonami, K.; Nakagawa, Y.; Tomishige, K. *Chem. Commun.* **2013**, *49*(63), 7034–7036.

21. Subbiah, S.; Simeonov, S. P.; Esperanca, J. M. S. S.; Rebelo, L. P. N.; Afonso, C. A. M. *Green Chem.* **2013**, *15*(10), 2849–2853.

22. Drewitt, J. G. N.; Lincoln, J. Polyesters from Heterocyclic Components. US Patent. US 2551731 19510508, **1951**.

23. Papageorgiou, G. Z.; Papageorgiou, D. G.; Tsanaktsis, V.; Bikiaris, D. N. *Polymer (Guildf).* **2015**, *62*, 28–38.

24. Papageorgiou, G. Z.; Tsanaktsis, V.; Bikiaris, D. N. *Phys. Chem. Chem. Phys.* **2014**, *16*(17), 7946–7958.

25. Papageorgiou, G. Z.; Tsanaktsis, V.; Papageorgiou, D. G.; Exarhopoulos, S.; Papageorgiou, M.; Bikiaris, D. N. *Polymer (Guildf).* **2014**, *55*(16), 3846–3858.

26. Zhu, J.; Cai, J.; Xie, W.; Chen, P.-H.; Gazzano, M.; Scandola, M.; Gross, R. A. *Macromolecules* **2013**, *46*(3), 796–804.

27. Ma, J.; Yu, X.; Xu, J.; Pang, Y. *Polymer (Guildf).* **2012**, *53*(19), 4145–4151.

28. Papageorgiou, G. Z.; Tsanaktsis, V.; Papageorgiou, D. G.; Chrissafis, K.; Exarhopoulos, S.; Bikiaris, D. N. *Eur. Polym. J.* **2015**, *67*, 383–396.

29. Jiang, M.; Liu, Q.; Zhang, Q.; Ye, C.; Zhou, G. *J. Polym. Sci. Part A Polym. Chem.* **2012**, *50*(5), 1026–1036.

30. Gubbels, E.; Jasinska-Walc, L.; Koning, C. E. *J. Polym. Sci. Part A Polym. Chem.* **2013**, *51*(4), 890–898.

31. Carman, H. S., J.; Killman, J. I., J.; Crawford, E. D.; Jenkins, J. C. Polyester Compositions Containing Furandicarboxylic Acid or an Ester Thereof and 2,2,4,4-Tetramethyl-1,3-Cyclobutanediol. World Patent WO 2013055862 A1, 20130418, **2013**.

32. Carman, H. S. J.; Killman, J. I. J.; Crawford, E. D.; Jenkins, J. C. Polyester Compositions Containing Furandicarboxylic Acid or an Ester Thereof, Cyclobutanediol and Ethylene Glycol. U.S. Patent US 20130095272 A1, 20130418, **2013**.

33. Dong, W.; Chen, M.; Lan, D.; Yin, H.; Ni, Z.; Li, X. Preparation Method of Low-Yellowing 2,5-Furandicarboxylic Acid-Based Polyester. Patent CN 102516513 B, 20120627, **2012**.

34. Carlos Morales-Huerta, J.; Martínez de Ilarduya, A.; Muñoz-Guerra, S. *Polymer (Guildf)*. **2016**, *87*, 148–158.

35. de Jong, E.; Dam, M. A.; Sipos, L.; Gruter, G.-J. M. Furandicarboxylic acid (FDCA), a versatile building block for a very interesting class of polyesters. In *Biobased Monomers, Polymers, and Materials*; ACS Symposium Series; American Chemical Society, **2012**; Vol. 1105, pp. 1–13.

36. Burgess, S. K.; Wenz, G. B.; Kriegel, R. M.; Koros, W. J. *Polymer (Guildf)*. **2016**, *98*, 305–310.

37. Burgess, S. K.; Kriegel, R. M.; Koros, W. J. *Macromolecules* **2015**, *48*(7), 2184–2193.

38. Burgess, S. K.; Mubarak, C. R.; Kriegel, R. M.; Koros, W. J. *J. Polym. Sci. Part B Polym. Phys.* **2015**, *53*(6), 389–399.

39. Collias, D. I.; Harris, A. M.; Nagpal, V.; Cottrell, I. W.; Schultheis, M. W. *Ind. Biotechnol.* 2014.

40. Lin, Z.; Ierapetritou, M.; Nikolakis, V. *AIChE J.* **2013**, *59*(6), 2079–2087.

41. Zeng, C.; Seino, H.; Ren, J.; Yoshie, N. *ACS Appl. Mater. Interfaces* **2014**, *6*(4), 2753–2758.

42. Zeng, C.; Seino, H.; Ren, J.; Hatanaka, K.; Yoshie, N. *Polymer (Guildf)*. **2013**, *54*(20), 5351–5357.

43. Zeng, C.; Seino, H.; Ren, J.; Hatanaka, K.; Yoshie, N. *Macromolecules* **2013**, *46*(5), 1794–1802.

44. Ikezaki, T.; Matsuoka, R.; Hatanaka, K.; Yoshie, N. *J. Polym. Sci. Part A Polym. Chem.* **2014**, *52*(2), 216–222.

45. Jiang, Y.; WoorTman, A. J. J.; Alberda van Ekenstein, G. O. R.; Petrović, D. M.; Loos, K. *Biomacromolecules* **2014**, *15*(7), 2482–2493.

46. Taarning, E.; Nielsen, I. S.; Egeblad, K.; Madsen, R.; Christensen, C. H. *ChemSusChem* **2008**, *1*(1–2), 75–78.

47. Casanova, O.; Iborra, S.; Corma, A. *J. Catal.* **2009**, *265*(1), 109–116.

48. Pinna, F.; Olivo, A.; Trevisan, V.; Menegazzo, F.; Signoretto, M.; Manzoli, M.; Boccuzzi, F. *Catal. Today* **2013**, *203*, 196–201.

49. Menegazzo, F.; Fantinel, T.; Signoretto, M.; Pinna, F.; Manzoli, M. *J. Catal.* **2014**, *319*, 61–70.

50. Menegazzo, F.; Signoretto, M.; Pinna, F.; Manzoli, M.; Aina, V.; Cerrato, G.; Boccuzzi, F. *J. Catal.* **2014**, *309*, 241–247.

51. Signoretto, M.; Menegazzo, F.; Contessotto, L.; Pinna, F.; Manzoli, M.; Boccuzzi, F. *Appl. Catal. B Environ.* **2013**, *129*, 287–293.

52. Pennanen, S.; Nyman, G. Studies on the Furan Series. *Acta Chem. Scand.* **1972**, *26*, 1018–1022.

53. Khrouf, A.; Boufi, S.; Gharbi, R. El; Belgacem, N. M.; Gandini, A. *Polym. Bull.* **1996**, *37*(5), 589–596.

54. Khrouf, A.; Abid, M.; Boufi, S.; El Gharbi, R.; Gandini, A. *Macromol. Chem. Phys.* 1998.

55. Gharbi, S.; Andreolety, J.-P.; Gandini, A. *Eur. Polym. J.* **2000**, *36*(3), 463–472.

56. Kamoun, W.; Salhi, S.; Rousseau, B.; El Gharbi, R.; Fradet, A. *Macromol. Chem. Phys.* **2006**, *207*(22), 2042–2049.

57. Pentz, W. J. Polyurethanes or Isocyanurates from Alkoxylated Hydroxymethylfuran. U.S. Patent US 4,426,460 A, 19840117, **1984**.

58. Dunlop, W. R.; Pentz, W. J. Ow Fire Hazard Rigid Urethane Insulation Foam and Polyol Mixtures Used in Its Manufacture. Patent US 4,318,999 A, 19820309, **1982**.

59. The Dow Chemical Company. Describe Dow's PAPI (TM) Polymeric MDI product offerings and the applications they may be used in.

60. Cawse, J. L.; Stanford, J. L.; Still, R. H. *Die Makromol. Chemie* **1984**, *185*(4), 697–707.

61. Cawse, J. L.; Stanford, J. L.; Still, R. H. *Die Makromol. Chemie* **1984**, *185*(4), 709–723.

62. Cawse, J. L.; Stanford, J. L.; Still, R. H. *Br. Polym. J.* **1985**, *17*(2), 233–238.

63. Boufi, S.; Belgacem, M. N.; Quillerou, J.; Gandini, A. *Macromolecules* **1993**, *26*(25), 6706–6717.

64. Boufi, S.; Gandini, A.; Belgacem, M. N. *Polymer (Guildf).* **1995**, *36*(8), 1689–1696.

65. Mitiakoudis, A.; Gandini, A. *Macromolecules* **1991**, *24*(4), 830–835.

66. Ortelt, M.; Spyrou, E.; Pfeffer, J. C.; Fuchsmann, D.; Kohlstruk, B.; Haas, T. Curable Compositions Based on Epoxy Resins without Benzyl Alcohol. World Patent WO 2014037222 A2, 20140313, **2014**.

67. Amarasekara, A. S. 5-Hydroxymethylfurfural based polymers. In *Renewable Polymers: Synthesis, Processing, and Technology*; **2011**.

68. Hui, Z.; Gandini, A. *Eur. Polym. J.* **1992**, *28*(12), 1461–1469.

69. Méalares, C.; Gandini, A. *Polym. Int.* **1996**, *40*, 33–39.

70. Cooke, A. W.; Wagener, K. B. *Macromolecules* **1991**, *24*(6), 1404–1407.

71. Choi, T.-L.; Han, K.-M.; Park, J.-I.; Kim, D. H.; Park, J.-M.; Lee, S. *Macromolecules* **2010**, *43*(14), 6045–6049.

72. Kim, C. Y.; Cho, H. N.; Kim, D. Y.; Kim, Y. C.; Lee, J. Y.; Kim, J. K. Fluorene-Based Alternating Copolymers and Electroluminescence Elements Incorporating Them. Patent GB 2313127 A, 19971119, **1997**.

73. Ishida, K.; Furuhashi, Y.; Yoshie, N. *Polym. Degrad. Stab.* **2014**, *110*, 149–155.

74. Vilela, C.; Cruciani, L.; Silvestre, A. J. D.; Gandini, A. *RSC Adv.* **2012**, *2*(7), 2966–2974.

75. Vilela, C.; Silvestre, A. J. D.; Gandini, A. *J. Polym. Sci. Part A Polym. Chem.* **2013**, *51*(10), 2260–2270.

76. Banella, M.; Gioia, C.; Vannini, M.; Colonna, M.; Celli, A.; Gandini, A. *A New Approach to the Synthesis of Monomers and Polymers Incorporating Furan/maleimide Diels-Alder Adducts*; 2016; Vol. **1736**.

77. Isikgor, F. H.; Becer, C. R. *Polym. Chem.* **2015**.

78. Rose, M.; Palkovits, R. *Macromol. Rapid Commun.* **2011**.

79. The biofine process — Production of levulinic acid, furfural, and formic acid from lignocellulosic feedstocks. In *Biorefineries-Industrial Processes and Products*; Wiley-VCH Verlag GmbH, **2008**; pp 139–164.

80. Mukherjee, A.; Dumont, M. J.; Raghavan, V. *Biomass Bioenergy*. 2015.

81. The University of Maine — College of Engineering.

82. Mullen, B. D.; Scholten, M. D.; Mullen, T. J.; Leibig, C. M.; Badarinarayana, V. Polyketal Adducts, Methods of Manufacture and Uses Thereof. U.S. Patent 2012/0118201 A1, **2011**.

83. Fotinos, N.; Campo, M. A.; Popowycz, F.; Gurny, R.; Lange, N. *Photochem. Photobiol.* **2006**, *82*(4), 994–1015.

84. Bozell, J. J.; Moens, L.; Elliott, D. C.; Wang, Y.; Neuenscwander, G. G.; Fitzpatrick, S. W.; Bilski, R. J.; Jarnefeld, J. L. *Resour. Conserv. Recycl.* **2000**, *28*(3), 227–239.

85. Serrano-Ruiz, J. C.; West, R. M.; Dumesic, J. A. *Annu. Rev. Chem. Biomol. Eng.* **2010**, *1*(1), 79–100.

86. Al-Shaal, M. G.; Wright, W. R. H.; Palkovits, R. *Green Chem.* **2012**, *14*(5), 1260–1263.

87. Wright, W. R. H.; Palkovits, R. *ChemSusChem* **2012**, *5*(9), 1657–1667.

88. Manzer, L. E. *Appl. Catal. A Gen.* **2004**, *272*(1), 249–256.

89. Manzer, L.; Hutchenson, K. Process for the Production of Y-Methyl-a-Methylene-Y-Butyrolactone from Reaction of Levulinic Acid and Hydrogen Followed by Reaction of Crude Y-Valerolactone and Formaldehyde, Both Reactions Being Carried out in the Supercritical or near-Critical Fluid Ph. U.S. Patent US 200600100450 A1, **2006**.

90. Marvel, C. S.; Levesque, C. L. *J. Am. Chem. Soc.* **1939**, *61*(7), 1682–1684.

91. Leonard, R. H. Method of Converting Levulinic Acid into Alpha Angelica Lactone. U.S. Patent US 2,809,203, **1957**.

92. Ertl, J.; Cerri, E.; Rizzuto, M.; Caretti, D. *AIP Conf. Proc.* **2014**, *1599*(1), 326–329.

93. Liu, H.-F.; Zeng, F.-X.; Deng, L.; Liao, B.; Pang, H.; Guo, Q.-X. *Green Chem.* **2013**, *15*(1), 81–84.

94. Arasa, M.; Pethrick, R. A.; Mantecón, A.; Serra, A. *Eur. Polym. J.* **2010**, *46*(1), 5–13.

95. Lee, C. W.; Urakawa, R.; Kimura, Y. *Eur. Polym. J.* **1998**, *34*(1), 117–122.

96. Chalid, M.; Heeres, H. J.; Broekhuis, A. A. *Procedia Chem.* **2012**, *4*, 260–267.

97. Alonso, D. M.; Bond, J. Q.; Dumesic, J. A. *Green Chem.* **2010**, *12*(9), 1493–1513.

98. Rothen-Weinhold, A.; Schwach-Abdellaoui, K.; Barr, J.; Ng, S. Y.; Shen, H.-R.; Gurny, R.; Heller, J. *J. Control. Release* **2001**, *71*(1), 31–37.

99. Lange, J.-P.; Vestering, J. Z.; Haan, R. J. *Chem. Commun.* **2007**, No. 33, 3488–3490.

100. Tarabanko, V. E.; Kaygorodov, N. K. L. *Chem. Sustain. Develop.***2010**,*18*, 321–328.

101. Chen, T.; Qin, Z.; Qi, Y.; Deng, T.; Ge, X.; Wang, J.; Hou, X. *Polym. Chem.* **2011**, *2*(5), 1190–1194.

102. Qi, G.; Nolan, M.; Schork, F. J.; Jones, C. W. *J. Polym. Sci. Part A Polym. Chem.* **2008**, *46*(17), 5929–5944.

103. Miyake, G. M.; Zhang, Y.; Chen, E. Y.-X. *Macromolecules* **2010**, *43*(11), 4902–4908.

104. Gowda, R. R.; Chen, E. Y.-X. *Dalt. Trans.* **2013**, *42*(25), 9263–9273.

105. Hu, Y.; Xu, X.; Zhang, Y.; Chen, Y.; Chen, E. Y.-X. *Macromolecules* **2010**, *43*(22), 9328–9336.

106. Hu, Y.; Miyake, G. M.; Wang, B.; Cui, D.; Chen, E. Y.-X. *Chem. — A Eur. J.* **2012**, *18*(11), 3345–3354.

107. Schmitt, M.; Falivene, L.; Caporaso, L.; Cavallo, L.; Chen, E. Y.-X. *Polym. Chem.* **2014**, *5*(9), 3261–3270.

108. Zhang, Y.; Miyake, G. M.; Chen, E. Y.-X. *Angew. Chemie Int. Ed.* **2010**, *49*(52), 10158–10162.

109. Xu, T.; Chen, E. Y.-X. *J. Am. Chem. Soc.* **2014**, *136*(5), 1774–1777.

110. Chen, X.; Caporaso, L.; Cavallo, L.; Chen, E. Y.-X. *J. Am. Chem. Soc.* **2012**, *134*(17), 7278–7281.

111. Hu, Y.; Gustafson, L. O.; Zhu, H.; Chen, E. Y.-X. *J. Polym. Sci. Part A Polym. Chem.* **2011**, *49*(9), 2008–2017.

112. Tagle, L. H.; Diaz, F. R.; Donoso, A. *J. Macromol. Sci. Part A* **1996**, *33*(11), 1643–1651.

113. Moore, J. A.; Tannahill, T. *High Perform. Polym.* **2001**.

114. Zhang, R.; Moore, J. A. *Macromol. Symp.* **2003**, *199*(1), 375–390.

115. Zhou, S.; Kim, D. *Electrochim. Acta* **2012**, *63*, 238–244.

116. Maiorana, A.; Spinella, S.; Gross, R. A. *Biomacromolecules* **2015**, *16*(3), 1021–1031.

117. Patel, A.; Maiorana, A.; Yue, L.; Gross, R. A.; Manas-Zloczower, I. *Macromolecules* **2016**, *49*(15), 5315–5324.

118. Chu, F.; Hawker, C. J.; Pomery, P. J.; Hill, D. J. T. *J. Polym. Sci., Part A Polym. Chem.* **1997**, *35*, 1627–1633.

119. Kambouris, P.; Hawker, C. J. *J. Chem. Soc. Perkin Trans. 1* **1993**, No. 22, 2717–2721.

120. Foix, D.; Ramis, X.; Sangermano, M.; Serra, A. *J. Polym. Sci. Part A Polym. Chem.* **2012**, *50*(6), 1133–1142.

121. Galbis, J. A.; García-Martín, M. D. G.; De Paz, M. V.; Galbis, E. *Chem. Rev.* 2016.

122. Fenouillot, F.; Rousseau, A.; Colomines, G.; Saint-Loup, R.; Pascault, J. P. Polymers from Renewable 1,4:3,6-Dianhydrohexitols (Isosorbide, Isomannide and Isoidide): A Review. *Progress in Polymer Science (Oxford)*. **2010**.

123. Caouthar, A.; Roger, P.; Tessier, M.; Chatti, S.; Blais, J. C.; Bortolussi, M. *Eur. Polym. J.* **2007**, *43*(1), 220–230.

124. Thiem, J.; Lüders, H. *Starch — Stärke* **1984**, *36*(5), 170–176.

125. Thiem, J.; Lüders, H. *Die Makromol. Chemie* **1986**, *187*(12), 2775–2785.

126. Stoss, P.; Hemmer, R. 1,4:3,6-Dianhydrohexitols. In *Advances in Carbohydrate Chemistry and Biochemistry*; Horton, D., Ed.; Academic Press, **1991**; Vol. 49, pp. 93–173.

127. Xi, J.; Zhang, Y.; Ding, D.; Xia, Q.; Wang, J.; Liu, X.; Lu, G.; Wang, Y. *Appl. Catal. A Gen.* **2014**, *469*, 108–115.

128. Bocqué, M.; Voirin, C.; Lapinte, V.; Caillol, S.; Robin, J.-J. *J. Polym. Sci. Part A Polym. Chem.* **2016**, *54*(1), 11–33.

129. Braun, D.; Bergmann, M. *J. für Prakt. Chemie/Chemiker-Zeitung* **1992**, *334*(4), 298–310.

130. Okada, M.; Okada, Y.; Tao, A.; Aoi, K. *J. Appl. Polym. Sci.* **1996**, *59*(1199–1202).

131. Okada, M.; Okada, Y.; Aoi, K. *J. Polym. Sci. Part A Polym. Chem.* **1995**, *33*(16), 2813–2820.

132. Storbeck, R.; Rehahn, M.; Ballauff, M. *Die Makromol. Chemie* **1993**, *194*(1), 53–64.

133. Storbeck, R.; Ballauff, M. *J. Appl. Polym. Sci.* **1996**, *59*, 1199–1202.

134. Storbeck, R.; Ballauff, M. *Polymer (Guildf).* **1993**, *34*(23), 5003–5006.

135. Chatti, S.; Schwarz, G.; Kricheldorf, H. R. *Macromolecules* **2006**, *39*(26), 9064–9070.

136. Chatti, S.; Kricheldorf, H. R.; Schwarz, G. *J. Polym. Sci. Part A Polym. Chem.* **2006**, *44*(11), 3616–3628.

137. Bachmann, F.; Reimer, J.; Ruppenstein, M.; Thiem, J. *Macromol. Chem. Phys.* **2001**.

138. Sangeetha, V. H.; Deka, H.; Varghese, T. O.; Nayak, S. K. *Polym. Compos.* **2018**, *39*(1), 81–101.

139. Hofvendahl, K.; Hahn–Hägerdal, B. *Enzyme Microb. Technol.* **2000**, *26*(2), 87–107.

140. Vink, E. T. H.; Rábago, K. R.; Glassner, D. A.; Gruber, P. R. *Polym. Degrad. Stab.* **2003**, *80*(3), 403–419.

141. Madhavan Nampoothiri, K.; Nair, N. R.; John, R. P. *Bioresour. Technol.* **2010**, *101*(22), 8493–8501.

142. John, R. P.; Nampoothiri, K. M.; Pandey, A. *Appl. Microbiol. Biotechnol.* **2007**, *74*(3), 524–534.

143. Datta, R.; Henry, M. *J. Chem. Technol. Biotechnol.* **2006**, *81*(7), 1119–1129.

144. Reddy, G.; Altaf, M.; Naveena, B. J.; Venkateshwar, M.; Kumar, E. V. *Biotechnol. Adv.* **2008**, *26*(1), 22–34.

145. Södergård, A.; Stolt, M. *Prog. Polym. Sci.* **2002**, *27*(6), 1123–1163.

146. Garlotta, D. *J. Polym. Environ.* **2001**, *9*(2), 63–84.

147. Lim, L.-T.; Auras, R.; Rubino, M. *Prog. Polym. Sci.* **2008**, *33*(8), 820–852.

148. Hiltunen, K.; Seppälä, J. V.; Härkönen, M. L. *J. Appl. Polym. Sci.* **1997**, *63*, 1091–1100.

149. Tuominen, J.; Seppälä, J. V. *Macromolecules* **2000**, *33*(10), 3530–3535.

150. Moon, S.-I.; Lee, C.-W.; Taniguchi, I.; Miyamoto, M.; Kimura, Y. *Polymer (Guildf)*. **2001**, *42*(11), 5059–5062.

151. Penczek, S.; Duda, A.; Szymanski, R.; Biela, T. *Macromol. Symp.* **2000**, *153*, 1–15.

152. Dove, A. P. *ACS Macro Lett.* **2012**, *1*(12), 1409–1412.

153. Albertsson, A.-C.; Srivastava, R. K. *Adv. Drug Deliv. Rev.* **2008**, *60*(9), 1077–1093.

154. Thomas, C. M. *Chem. Soc. Rev.* **2010**, *39*(1), 165–173.

155. Dechy-Cabaret, O.; Martin-Vaca, B.; Bourissou, D. *Chem. Rev.* **2004**, *104*(12), 6147–6176.

156. Jérôme, C.; Lecomte, P. *Adv. Drug Deliv. Rev.* **2008**, *60*(9), 1056–1076.

157. Kamber, N. E.; Jeong, W.; Waymouth, R. M.; Pratt, R. C.; Lohmeijer, B. G. G.; Hedrick, J. L. *Chem. Rev.* **2007**, *107*(12), 5813–5840.

158. Nijenhuis, A. J.; Grijpma, D. W.; Pennings, A. J. *Macromolecules* **1992**, *25*(24), 6419–6424.

159. Rasal, R. M.; Janorkar, A. V; Hirt, D. E. P. *Prog. Polym. Sci.* **2010**, *35*(3), 338–356.

160. Stanford, M. J.; Dove, A. P. *Chem. Soc. Rev.* **2010**, *39*(2), 486–494.

161. Vasanthakumari, R.; Pennings, A. J. *Polymer (Guildf)*. **1983**, *24*(2), 175–178.

162. Kalb, B.; Pennings, A. J. *Polymer (Guildf)*. **1980**, *21*(6), 607–612.

163. Ikada, Y.; Jamshidi, K.; Tsuji, H.; Hyon, S. H. *Macromolecules* **1987**, *20*(4), 904–906.

164. Tsuji, H.; Ikada, Y. *Macromolecules* **1993**, *26*(25), 6918–6926.

165. Tsuji, H. *Macromol. Biosci.* **2005**, *5*(7), 569–597.

166. Hoogsteen, W.; Postema, A. R.; Pennings, A. J.; Ten Brinke, G.; Zugenmaier, P. *Macromolecules* **1990**, *23*(2), 634–642.

167. Puiggali, J.; Ikada, Y.; Tsuji, H.; Cartier, L.; Okihara, T.; Lotz, B. *Polymer (Guildf)*. **2000**, *41*(25), 8921–8930.

168. Cartier, L.; Okihara, T.; Ikada, Y.; Tsuji, H.; Puiggali, J.; Lotz, B. *Polymer (Guildf)*. **2000**, *41*(25), 8909–8919.

169. Pan, P.; Inoue, Y. *Prog. Polym. Sci.* **2009**, *34*(7), 605–640.

170. Perego, G.; Cella, G. D.; Bastioli, C. *J. Appl. Polym. Sci.* **1996**, *59*, 37–43.

171. Auras, R.; Harte, B.; Selke, S. *Macromol. Biosci.* **2004**, *4*(9), 835–864.

172. Taubner, V.; Shishoo, R. *J. Appl. Polym. Sci.* **2001**, *79*, 2128–2135.

173. Kale, G.; Auras, R.; Singh, S. P.; Narayan, R. *Polym. Test.* **2007**, *26*(8), 1049–1061.

174. Kale, G.; Kijchavengkul, T.; Auras, R.; Rubino, M.; Selke, S. E.; Singh, S. P. *Macromol. Biosci.* **2007**, *7*(3), 255–277.

175. Park, K. I.; Xanthos, M. A *Polym. Degrad. Stab.* **2009**, *94*(5), 834–844.

176. Bordes, P.; Pollet, E.; Avérous, L. *Prog. Polym. Sci.* **2009**, *34*(2), 125–155.

177. Muzumdar, A. V; Sawant, S. B.; Pangarkar, V. G. *Org. Process Res. Dev.* **2004**, *8*(4), 685–688.

178. Choudhary, H.; Nishimura, S.; Ebitani, K. *Appl. Catal. A Gen.* **2013**, *458*, 55–62.

179. Song, H.; Lee, S. Y. *Enzyme Microb. Technol.* **2006**, *39*(3), 352–361.

180. Lee, P. C.; Lee, W. G.; Lee, S. Y.; Chang, H. N. *Biotechnol. Bioeng.* **2001**, *72*(1), 41–48.

181. Stempfle, F.; OrTmann, P.; Mecking, S. *Chem. Rev.* **2016**, *116*(7), 4597–4641.

182. Zhu, C.; Zhang, Z.; Liu, Q.; Wang, Z.; Jin, J. *J. Appl. Polym. Sci.* **2003**, *90*(4), 982–990.

183. Nikolic, M. S.; Djonlagic, J. *Polym. Degrad. Stab.* **2001**, *74*(2), 263–270.

184. Bikiaris, D. N.; Papageorgiou, G. Z.; Achilias, D. S. *Polym. Degrad. Stab.* **2006**, *91*(1), 31–43.

185. Fujimaki, T. *Polym. Degrad. Stab.* **1998**, *59*(1), 209–214.

186. Shirahama, H.; Kawaguchi, Y.; Aludin, M. S.; Yasuda, H. *J. Appl. Polym. Sci.* **2001**, *80*, 340–347.

187. Wang, Q.; Shao, Z.; Yu, T. *Polym. Bull.* **1996**, *36*(6), 659–665.

188. Gaymans, R. J.; Venkatraman, V. S.; Schuijer, J. *J. Polym. Sci. Polym. Chem. Ed.* **1984**, *22*(6), 1373–1382.

189. Tokiwa, Y.; Suzuki, T.; Ando, T. *J. Appl. Polym. Sci.* **1979**, *24*(7), 1701–1711.

190. Grigat, E.; Koch, R.; Timmermann, R. *Polym. Degrad. Stab.* **1998**, *59*(1), 223–226.

191. Moore, T.; Adhikari, R.; Gunatillake, P. *Biomaterials* **2005**, *26*(18), 3771–3782.

192. Maisonneuve, L.; Lebarbe, T.; Grau, E.; Cramail, H. *Polym. Chem.* **2013**, *4*(22), 5472–5517.

193. Nagai, K. *Appl. Catal. A Gen.* **2001**, *221*(1), 367–377.

194. Willke, T.; Vorlop, K.-D. *Appl. Microbiol. Biotechnol.* **2001**, *56*(3), 289–295.

195. Ishida, S.; Saito, S. *J. Polym. Sci. Part A-1 Polym. Chem.* **1967**, *5*(4), 689–705.

196. Watanabe, H.; Matsumoto, A.; Otsu, T. *J. Polym. Sci. Part A Polym. Chem.* **1994**, *32*(11), 2073–2083.

197. Satoh, K. *Polym. J.* **2015**, *47*, 527.

198. Satoh, K.; Lee, D.-H.; Nagai, K.; Kamigaito, M. *Macromol. Rapid Commun.* **2014**, *35*(2), 161–167.

199. Robert, T.; Friebel, S. *Green Chem.* **2016**, *18*(10), 2922–2934.

200. Farmer, T. J.; Castle, R. L.; Clark, J. H.; Macquarrie, D. J. *Int. J. Mol. Sci.* **2015**, *16*, 14912–14932.

201. Chanda, S.; Ramakrishnan, S. *Polym. Chem.* **2015**, *6*(11), 2108–2114.

202. Ma, S.; Liu, X.; Jiang, Y.; Tang, Z.; Zhang, C.; Zhu, J. *Green Chem.* **2013**, *15*(1), 245–254.

203. Winkler, M.; Lacerda, T. M.; Mack, F.; Meier, M. A. R. *Macromolecules* **2015**, *48*(5), 1398–1403.

204. Bugnicourt, E.; Cinelli, P.; Lazzeri, A.; Alvarez, V. *Express Polym. Lett.* **2014**, *8*(11), 791–808.

205. Laycock, B.; Halley, P.; Pratt, S.; Werker, A.; Lant, P. *Prog. Polym. Sci.* **2013**, *38*(3), 536–583.

206. Laurichesse, S.; Avérous, L. *Prog. Polym. Sci.* **2014**, *39*(7), 1266–1290.

207. Silva, E. A. B. da; Zabkova, M.; Araújo, J. D.; Cateto, C. A.; Barreiro, M. F.; Belgacem, M. N.; Rodrigues, A. E. *Chem. Eng. Res. Des.* **2009**, *87*(9), 1276–1292.

208. Hatakeyama, H.; Hirogaki, A.; Matsumura, H.; Hatakeyama, T. *J. Therm. Anal. Calorim.* **2013**, *114*(3), 1075–1082.

209. Cinelli, P.; Anguillesi, I.; Lazzeri, A. *Eur. Polym. J.* **2013**, *49*(6), 1174–1184.

210. Simionescu, C. I.; Rusan, V.; Macoveanu, M. M.; Cazacu, G.; Lipsa, R.; Vasile, C.; Stoleriu, A.; Ioanid, A. *Compos. Sci. Technol.* **1993**, *48*(1), 317–323.

211. Feldman, D.; Banu, D.; Natansohn, A.; Wang, J. *J. Appl. Polym. Sci.* **1991**, *42*(6), 1537–1550.

212. Feldman, D.; Banu, D.; Luchian, C.; Wang, J. *J. Appl. Polym. Sci.* **1991**, *42*(5), 1307–1318.

213. Delmas, G.-H.; Benjelloun-Mlayah, B.; Bigot, Y. Le; Delmas, M. *J. Appl. Polym. Sci.* **2013**, *127*(3), 1863–1872.

214. Hirose, S.; Hatakeyama, T.; Hatakeyama, H. *Macromol. Symp.* **2003**, *197*(1), 157–170.

215. Over, L. C.; Grau, E.; Grelier, S.; Meier, M. A. R.; Cramail, H. *Macromol. Chem. Phys.* **2017**, *218*(4), 1600411–n/a.

216. Muller, P. C.; Kelley, S. S.; Glasser, W. G. *J. Adhes.* **1984**, *17*(3), 185–206.

217. Wang, M.; Leitch, M.; (Charles) Xu, C. *Eur. Polym. J.* **2009**, *45*(12), 3380–3388.

218. Li, J.; Zhang, J.; Zhang, S.; Gao, Q.; Li, J.; Zhang, W. *Polymers (Basel).* **2017**.

219. Sawamura, K.; Tobimatsu, Y.; Kamitakahara, H.; Takano, T. *ACS Sustain. Chem. Eng.* **2017**, *5*(6), 5424–5431.

220. Ferhan, M.; Yan, N.; Sain, M. *J. Chem. Eng. Process. Technol.* **2013**, *4*, 160.

221. Zakzeski, J.; Jongerius, A. L.; Bruijnincx, P. C. A.; Weckhuysen, B. M. *ChemSusChem* **2012**, *5*(8), 1602–1609.

222. Chen, L.; Xin, J.; Ni, L.; Dong, H.; Yan, D.; Lu, X.; Zhang, S. *Green Chem.* **2016**, *18*(8), 2341–2352.

223. Lange, H.; Decina, S.; Crestini, C. *Eur. Polym. J.* **2013**, *49*(6), 1151–1173.

224. Harms, R. G.; Markovits, I. I. E.; Drees, M.; Herrmann, H. C. Mult. W. A.; Cokoja, M.; Kühn, F. E. *ChemSusChem* **2014**, *7*(2), 429–434.

225. Sik Kim, Y.; Chang, H.; Kadla, J. F. *J. Wood Chem. Technol.* **2007**, *27*(3–4), 225–241.

226. Zhao, Y.; Xu, Q.; Pan, T.; Zuo, Y.; Fu, Y.; Guo, Q.-X. *Appl. Catal. A Gen.* **2013**, *467*, 504–508.

227. Crestini, C.; Saladino, R.; Tagliatesta, P.; Boschi, T. *Bioorg. Med. Chem.* **1999**, *7*(9), 1897–1905.

228. Rahimi, A.; Ulbrich, A.; Coon, J. J.; Stahl, S. S. *Nature* **2014**, *515*, 249.

229. Llevot, A.; Grau, E.; Carlotti, S.; Grelier, S.; Cramail, H. *Macromol. Rapid Commun.* **2016**, *37*(1), 9–28.

230. Stanzione III, J. F.; Sadler, J. M.; La Scala, J. J.; Reno, K. H.; Wool, R. P. *Green Chem.* **2012**, *14*(8), 2346–2352.

231. Meylemans, H. A.; Harvey, B. G.; Reams, J. T.; Guenthner, A. J.; Cambrea, L. R.; Groshens, T. J.; Baldwin, L. C.; Garrison, M. D.; Mabry, J. M. *Biomacromolecules* **2013**, *14*(3), 771–780.

232. Fache, M.; Darroman, E.; Besse, V.; Auvergne, R.; Caillol, S.; Boutevin, B. *Green Chem.* **2014**, *16*(4), 1987–1998.

233. Fache, M.; Auvergne, R.; Boutevin, B.; Caillol, S. *Eur. Polym. J.* **2015**, *67*, 527–538.

234. Fache, M.; Viola, A.; Auvergne, R.; Boutevin, B.; Caillol, S. *Eur. Polym. J.* **2015**, *68*, 526–535.

235. Bock, L. H.; Anderson, J. K. *J. Polym. Sci.* **1955**, *17*(86), 553–558.

236. Lange, W.; Kordsachia, O. *Holz als Roh- und Werkst.* **1981**, *39*(3), 107–112.

237. Kricheldorf, H. R.; Löhden, G. Whisker 11. *Polymer (Guildf).* **1995**, *36*(8), 1697–1705.

238. Montes de Oca, H.; Wilson, J. E.; Penrose, A.; Langton, D. M.; Dagger, A. C.; Anderson, M.; Farrar, D. F.; Lovell, C. S.; Ries, M. E.; Ward, I. M.; *et al. Biomaterials* **2010**, *31*(30), 7599–7605.

239. Kreye, O.; Oelmann, S.; Meier, M. A. R. *Macromol. Chem. Phys.* **2013**, *214*(13), 1452–1464.

240. Firdaus, M.; Meier, M. A. R. *Eur. Polym. J.* **2013**, *49*(1), 156–166.

241. Mialon, L.; Pemba, A. G.; Miller, S. A. *Green Chem.* **2010**, *12*(10), 1704–1706.

242. Mialon, L.; Vanderhenst, R.; Pemba, A. G.; Miller, S. A. *Macromol. Rapid Commun.* **2011**, *32*(17), 1386–1392.

243. Pang, C.; Zhang, J.; Wu, G.; Wang, Y.; Gao, H.; Ma, J. *Polym. Chem.* **2014**, *5*(8), 2843–2853.

244. Pion, F.; Reano, A. F.; Ducrot, P.-H.; Allais, F. *RSC Adv.* **2013**, *3*(23), 8988–8997.

245. Pion, F.; Ducrot, P.-H.; Allais, F. *Macromol. Chem. Phys.* **2014**, *215*(5), 431–439.

246. Llevot, A.; Grau, E.; Carlotti, S.; Grelier, S.; Cramail, H. *J. Mol. Catal. B Enzym.* **2016**, *125*, 34–41.

247. Llevot, A.; Grau, E.; Carlotti, S.; Grelier, S.; Cramail, H. *Polym. Chem.* **2015**, *6*(33), 6058–6066.

248. Kreye, O.; Tóth, T.; Meier, M. A. R. *Eur. Polym. J.* **2011**, *47*(9), 1804–1816.

249. Barbara, I.; Flourat, A. L.; Allais, F. *Eur. Polym. J.* **2015**, *62*, 236–243.

250. Llevot, A.; Grau, E.; Carlotti, S.; Grelier, S.; *Polym. Chem.* **2015**, *6*(44), 7693–7700.

251. Lebarbé, T.; More, A. S.; Sane, P. S.; Grau, E.; Alfos, C.; Cramail, H. *Macromol. Rapid Commun.* **2014**, *35*(4), 479–483.

252. Pemba, A. G.; Rostagno, M.; Lee, T. A.; Miller, S. A. *Polym. Chem.* **2014**, *5*(9), 3214–3221.

253. Harvey, B. G.; Guenthner, A. J.; Meylemans, H. A.; Haines, S. R. L.; Lamison, K. R.; Groshens, T. J.; Cambrea, L. R.; Davis, M. C.; Lai, W. W. *Green Chem.* **2015**, *17*(2), 1249–1258.

254. Noel, A.; Borguet, Y. P.; Raymond, J. E.; Wooley, K. L. *Macromolecules* **2014**, *47*(9), 2974–2983.

255. Liu, H.; Lepoittevin, B.; Roddier, C.; Guerineau, V.; Bech, L.; Herry, J.-M.; Bellon-Fontaine, M.-N.; Roger, P. *Polymer (Guildf).* **2011**, *52*(9), 1908–1916.

256. Holmberg, A. L.; Stanzione, J. F.; Wool, R. P.; Epps, T. H. *ACS Sustain. Chem. Eng.* **2014**, *2*(4), 569–573.

257. Pfister, D. P.; Xia, Y.; Larock, R. C. *ChemSusChem* **2011**, *4*(6), 703–717.

258. Seniha Güner, F.; Yağcı, Y.; Tuncer Erciyes, A. *Prog. Polym. Sci.* **2006**, *31*(7), 633–670.

259. Miao, S.; Wang, P.; Su, Z.; Zhang, S. *Acta Biomater.* **2014**, *10*(4), 1692–1704.

260. Petrovic, Z. *Polym. Rev.* **2008**, *48*(1), 109–155.

261. Desroches, M.; Escouvois, M.; Auvergne, R.; Caillol, S.; Boutevin, B. *Polym. Rev.* **2012**, *52*(1), 38–79.

262. Mosiewicki, M. A.; Aranguren, M. I. *Eur. Polym. J.* **2013**, *49*(6), 1243–1256.

263. De Espinosa, L.; Meier, M. A. R. *Eur. Polym. J.* **2011**, *47*(5), 837–852.

264. Lligadas, G.; Ronda, J. C.; Galià, M.; Cadiz, V. *Mater. Today* **2013**, *16*(9), 337–343.

265. Engels, H. W.; Pirkl, H. G.; Albers, R.; Albach, R. W.; Krause, J.; Hoffmann, A.; Casselmann, H.; Dormish, J. *Angew. Chemie — Int. Ed.* **2013**, *52*(36), 9422–9441.

266. Maisonneuve, L.; Chollet, G.; Grau, E.; Cramail, H. *OCL* **2016**, *23*(5), D508.

267. Zhang, C.; Garrison, T. F.; Madbouly, S. A.; Kessler, M. R. *Prog. Polym. Sci.* 2017.

268. Datta, J.; Głowińska, E. *J. Elastomers Plast.* **2014**, *46*(1), 33–42.

269. Alam, M.; Akram, D.; Sharmin, E.; Zafar, F.; Ahmad, S. *Arab. J. Chem.* **2014**, *7*(4), 469–479.

270. Petrovic, Z. S.; Guo, A.; Zhang, W. *J. Polym. Sci. Part A Polym. Chem.* **2000**, *38*(22), 4062–4069.

271. Wang, C.-S.; Yang, L.-T.; Ni, B.-L.; Shi, G. *J. Appl. Polym. Sci.* **2009**, *114*(1), 125–131.

272. Zaher, F. A.; El-Mallah, M. H.; El-Hefnawy, M. M. *J. Am. Oil Chem. Soc.* **1989**, *66*(5), 698–700.

273. Datta, J.; Głowińska, E. *Ind. Crops Prod.* **2014**, *61*, 84–91.

274. Zieleniewska, M.; Auguścik, M.; Prociak, A.; Rojek, P.; Ryszkowska, J. *Polym. Degrad. Stab.* **2014**, *108*, 241–249.

275. Gaikwad, M. S.; Gite, V. V.; Mahulikar, P. P.; Hundiwale, D. G.; Yemul, O. S. *Prog. Org. Coatings* **2015**, *86*, 164–172.

276. Miao, S.; Zhang, S.; Su, Z.; Wang, P. *J. Appl. Polym. Sci.* **2013**, *127*(3), 1929–1936.

277. Guo, R.; Ma, C.; Sun, S.; Ma, Y. *J. Am. Oil Chem. Soc.* **2011**, *88*(4), 517–521.

278. Tathe, D. S.; Jagtap, R. N. *J. Am. Oil Chem. Soc.* **2013**, *90*(9), 1405–1413.

279. Miao, S.; Callow, N.; Wang, P.; Liu, Y.; Su, Z.; Zhang, S. *J. Am. Oil Chem. Soc.* **2013**, *90*(9), 1415–1421.

280. Bakhshi, H.; Yeganeh, H.; Mehdipour-Ataei, S.; Solouk, A.; Irani, S. *Macromolecules* **2013**, *46*(19), 7777–7788.

281. Bakhshi, H.; Yeganeh, H.; Mehdipour-Ataei, S. *J. Biomed. Mater. Res. Part A* **2013**, *101A* (6), 1599–1611.

282. Yari, A.; Yeganeh, H.; Bakhshi, H.; Gharibi, R. *J. Biomed. Mater. Res. Part A* **2014**, *102*(1), 84–96.

283. Kong, X.; Liu, G.; Curtis, J. M. *Eur. Polym. J.* **2012**, *48*(12), 2097–2106.

284. Kong, X.; Liu, G.; Qi, H.; Curtis, J. M. *Prog. Org. Coatings* **2013**, *76*(9), 1151–1160.

285. Petrović, Z. S.; Cvetković, I.; Hong, D.; Wan, X.; Zhang, W.; Abraham, T. W.; Malsam, J. *Eur. J. Lipid Sci. Technol.* **2010**, *112*(1), 97–102.

286. Chaudhari, A.; Kulkarni, R.; Mahulikar, P.; Sohn, D.; Gite, V. *J. Am. Oil Chem. Soc.* **2015**, *92*(5), 733–741.

287. Stirna, U.; Fridrihsone, A.; Lazdiņa, B.; Misāne, M.; Vilsone, D. *J. Polym. Environ.* **2013**, *21*(4), 952–962.

288. Stirna, U.; Fridrihsone-Girone, A.; Yakushin, V.; Vilsone, D. *J. Coatings Technol. Res.* **2014**, *11*(3), 409–420.

289. Das, B.; Konwar, U.; Mandal, M.; Karak, N. *Ind. Crops Prod.* **2013**, *44*, 396–404.

290. Das, S.; Pandey, P.; Mohanty, S.; Nayak, S. K. I*Mater. Express* **2015**, *5*(5), 377–389.

291. Guo, A.; Demydov, D.; Zhang, W.; Petrovic, Z. S. *J. Polym. Environ.* **2002**, *10*(1/2), 49–52.

292. Vanbesien, T.; Hapiot, F.; Monflier, E. *Lipid Technol.* **2013**, *25*(8), 175–178.

293. Petrović, Z. S.; Guo, A.; Javni, I.; Cvetković, I.; Hong, D. P. *Polym. Int.* **2008**, *57*(2), 275–281.

294. Zhang, C.; Ding, R.; Kessler, M. R. *Macromol. Rapid Commun.* **2014**, *35*(11), 1068–1074.

295. Bantchev, G. B.; Kenar, J. A.; Biresaw, G.; Han, M. G. *J. Agric. Food Chem.* **2009**, *57*(4), 1282–1290.

296. Hoyle, C. E.; Bowman, C. N. *Angew. Chemie Int. Ed.* **2010**, *49*(9), 1540–1573.

297. Ionescu, M.; Radojčić, D.; Wan, X.; Petrović, Z. S.; Upshaw, T. A. *Eur. Polym. J.* **2015**, *67*, 439–448.

298. Desroches, M.; Caillol, S.; Lapinte, V.; Auvergne, R.; Boutevin, B. *Macromolecules* **2011**, *44*(8), 2489–2500.

299. Caillol, S.; Desroches, M.; Carlotti, S.; Auvergne, R.; Boutevin, B. *Green Mater.* **2013**, *1*(1), 16–26.

300. Lowe, A. *Polym. Chem.* **2010**, *1*, 17–36.

301. Alagi, P.; Choi, Y. J.; Hong, S. C. *Eur. Polym. J.* **2016**, *78*, 46–60.

302. Darroman, E.; Auvergne, R.; Boutevin, B.; Caillol, S. *Eur. J. Lipid Sci. Technol.* **2014**, *116*(2), 178–189.

303. Darroman, E.; Bonnot, L.; Auvergne, R.; Boutevin, B.; Caillol, S. *Eur. J. Lipid Sci. Technol.* **2015**, *117*(2), 178–189.

304. Ionescu, M.; Radojčić, D.; Wan, X.; Shrestha, M. L.; Petrović, Z. S.; Upshaw, T. A. *Eur. Polym. J.* **2016**, *84*, 736–749.

305. Poussard, L.; Mariage, J.; Grignard, B.; Detrembleur, C.; Jérôme, C.; Calberg, C.; Heinrichs, B.; De Winter, J.; Gerbaux, P.; Raquez, J.-M.; *et al. Macromolecules* **2016**, *49*(6), 2162–2171.

306. Raquez, J.-M.; Deléglise, M.; Lacrampe, M.-F.; Krawczak, P. *Prog. Polym. Sci.* **2010**, *35*(4), 487–509.

307. Zhang, C.; Madbouly, S. A.; Kessler, M. R. *ACS Appl. Mater. Interfaces* **2015**, *7*(2), 1226–1233.

308. Can, E.; Wool, R. P.; Küsefoğlu, S. *J. Appl. Polym. Sci.* **2006**, *102*(2), 1497–1504.

309. Samper, M. D.; Fombuena, V.; Boronat, T.; García-Sanoguera, D.; Balart, R. *J. Am. Oil Chem. Soc.* **2012**.

310. Espinoza-Perez, J. D.; Nerenz, B. A.; Haagenson, D. M.; Chen, Z.; Ulven, C. A.; Wiesenborn, D. P. *Polym. Compos.* **2011**, *32*(11), 1806–1816.

311. Farias, M.; Martinelli, M.; Bottega, D. P. *Appl. Catal. A Gen.* **2010**, *384*(1–2), 213–219.

312. Farias, M.; Martinelli, M.; Rolim, G. K. *Appl. Catal. A Gen.* **2011**, *403*(1–2), 119–127.

313. Rüsch gen. Klaas, M.; Warwel, S. *Ind. Crops Prod.* **1999**, *9*(2), 125–132.

314. Seniha Güner, F.; Yağcı, Y.; Tuncer Erciyes, A. *Prog. Polym. Sci.* **2006**, *31*(7), 633–670.

315. Metzger, J. O. *Eur. J. Lipid Sci. Technol.* **2009**, *111*(9), 865–876.

316. Park, S.-J.; Jin, F.-L.; Lee, J.-R. *Macromol. Rapid Commun.* **2004**, *25*(6), 724–727.

317. Liu, Z.; Erhan, S. Z. *J. Am. Oil Chem. Soc.* **2010**, *87*(4), 437–444.

318. Stemmelen, M.; Lapinte, V.; Habas, J.; Robin, J. **2015**, *68*, 536–545.

319. Stemmelen, M.; Pessel, F.; Lapinte, V.; Caillol, S.; Habas, J.-P.; Robin, J.-J. *J. Polym. Sci. Part A Polym. Chem.* **2011**, *49*(11), 2434–2444.

320. Czub, Pi.; Franek, I. *Polimery* **2013**, *58*(2), 135–139.

321. Czub, P. *Macromol. Symp.* **2006**, *242*(1), 60–64.

322. Yang, G.; Rohde, B. J.; Robertson, M. L. *Green Mater.* **2013**, *1*(2), 125–134.

323. Mashouf Roudsari, G.; Mohanty, A. K.; Misra, M. *ACS Sustain. Chem. Eng.* **2014**, *2*(9), 2111–2116.

324. Altuna, F. I.; Pettarin, V.; Williams, R. J. J. *Green Chem.* **2013**, *15*(12), 3360.

325. Supanchaiyamat, N.; Shuttleworth, P. S.; Hunt, A. J.; Clark, J. H.; Matharu, A. S. *Green Chem.* **2012**, *14*(6), 1759.

326. Omonov, T. S.; Curtis, J. M. *J. Appl. Polym. Sci.* **2014**, *131*(8), n/a-n/a.

327. Omonov, T.; Curtis, J. Aldehyde Free Thermoset Bioresins and Biocomposites. WO2014075182A1, 2014.

328. Tan, S. G.; Ahmad, Z.; Chow, W. S. *Ind. Crops Prod.* **2013**, *43*, 378–385.

329. Altuna, F. I.; Espósito, L. H.; Ruseckaite, R. A.; Stefani, P. M. *J. Appl. Polym. Sci.* **2011**, *120*(2), 789–798.

330. Chen, Y.; Yang, L.; Wu, J.; Ma, L.; Finlow, D. E.; Lin, S.; Song, K. *J. Therm. Anal. Calorim.* **2013**, *113*(2), 939–945.

331. Carbonell-Verdu, A.; Bernardi, L.; Garcia-Garcia, D.; Sanchez-Nacher, L.; Balart, R. *Eur. Polym. J.* **2015**, *63*, 1–10.

332. Mallégol, J.; Lemaire, J.; Gardette, J.-L. *Prog. Org. Coatings* **2000**, *39*(2), 107–113.

333. Hess, P. S.; O'Hare, G. A. *Ind. Eng. Chem.* **1950**, *42*(7), 1424–1431.

334. Tallman, K. A.; Roschek, B.; Porter, N. A. *J. Am. Chem. Soc.* **2004**, *126*(30), 9240–9247.

335. Taylor, W. L. *J. Am. Oil Chem. Soc.* **1950**, *27*(11), 472–476.

336. Çakmaklı, B.; Hazer, B.; Tekin, İ. Ö.; Cömert, F. B. *Biomacromolecules* **2005**, *6*(3), 1750–1758.

337. Keles, E.; Hazer, B. *Macromol. Symp.* **2008**, *269*(1), 154–160.

338. Allı, A.; Hazer, B. *Eur. Polym. J.* **2008**, *44*(6), 1701–1713.

339. Çakmaklı, B.; Hazer, B.; Açıkgöz, Ş.; Can, M.; Cömert, F. B. *J. Appl. Polym. Sci.* **2007**, *105*(6), 3448–3457.

340. Cakmakli, B.; Hazer, B.; Tekin, I. O.; Kizgut, S.; Koksal, M.; Menceloglu, Y. *Macromol. Biosci.* **2004**, *4*(7), 649–655.

341. Fengkui, L.; Larock, R. C. *Biomacromolecules* 2003.

342. Gabel, A. R.; Stoesser, S. M. Styrene-Tung Oil Copolymer, US2190906 A. **1939**.

343. Krompiec, S.; Suwiński, J.; Majewski, J.; Grobelny, J. *Polish J. Appl. Chem.* **1998**, *42*(1), 43–48.

344. Henna, P. H.; Andjelkovic, D. D.; Kundu, P. P.; Larock, R. C. *J. Appl. Polym. Sci.* **2007**, *104*(2), 979–985.

345. Valverde, M.; Andjelkovic, D.; Kundu, P. P.; Larock, R. C. *J. Appl. Polym. Sci.* **2008**, *107*(1), 423–430.

346. Pelletier, H.; Belgacem, N.; Gandini, A. *J. Appl. Polym. Sci.* **2006**, *99*(6), 3218–3221.

347. La Scala, J.; Wool, R. P. *Polymer (Guildf).* **2005**, *46*(1), 61–69.

348. Luo, Q.; Liu, M.; Xu, Y. (Frank); Ionescu, M.; Petrović, Z. S. T*Macromolecules* **2011**, *44*(18), 7149–7157.

349. Bunker, S. P.; Wool, R. P. *J. Polym. Sci. Part A Polym. Chem.* **2002**, *40*(4), 451–458.

350. Lu, J.; Khot, S.; Wool, R. P. *Polymer (Guildf).* **2005**, *46*(1), 71–80.

351. Zhan, M.; Wool, R. P. *J. Appl. Polym. Sci.* **2010**, *118*(6), 3274–3283.

352. Zhang, P.; Zhang, J. *One-Step Acrylation of Soybean Oil (SO) for the Preparation of SO-Based Macromonomers*; **2013**; Vol. 15.

353. Ahn, B. K.; Sung, J.; Rahmani, N.; Wang, G.; Kim, N.; Lease, K.; Sun, X. S. *J. Adhes.* **2013**, *89*(4), 323–338.

354. Li, Y.; Sun, X. S. *RSC Adv.* **2015**, *5*(55), 44009–44017.

355. Çolak, S.; Küsefoğlu, S. H. *J. Appl. Polym. Sci.* **2007**, *104*(4), 2244–2253.

356. Pashley, R. M.; Senden, T. J.; Morris, R. A.; Guthrie, J. T.; He, W. D. Polymerizable Porphyrins. US 5360880A, 1994.

357. Andjelkovic, D. D.; Valverde, M.; Henna, P.; Li, F.; Larock, R. C. *Polymer (Guildf).* **2005**, *46*(23), 9674–9685.

358. Li, F.; Larock, R. C. *J. Appl. Polym. Sci.* **2001**, *80*(4), 658–670.

359. Li, F.; Larock, R. C. *Polym. Adv. Technol.* **2002**, *13*(6), 436–449.

360. Li, F.; Larock, R. C. *J. Polym. Sci. Part B Polym. Phys.* **2000**, *38*(21), 2721–2738.

361. Li, F.; Larock, R. C. *J. Polym. Sci. Part B Polym. Phys.* **2001**, *39*(1), 60–77.

362. Lu, Y.; Larock, R. C. *Macromol. Mater. Eng.* **2007**, *292*(10–11), 1085–1094.

363. Andjelkovic, D. D.; Larock, R. C. Novel Rubbers from Cationic Copolymerization of Soybean Oils and Dicyclopentadiene. 1. Synthesis and Characterization. **2006**.

364. Andjelkovic, D. D.; Lu, Y.; Kessler, M. R.; Larock, R. C. *Macromol. Mater. Eng.* **2009**, *294*(8), 472–483.

365. Liu, Z.; Sharma, B. K.; Erhan, S. Z. *Biomacromolecules* **2007**, *8*(1), 233–239.

366. Li, F.; Larock, R. C. *J. Appl. Polym. Sci.* **2000**, *78*(5), 1044–1056.

367. Jin, F.-L.; Park, S.-J. *Polym. Int.* **2008**, *57*(4), 577–583.

368. Jin, F.-L.; Park, S.-J. *Mater. Sci. Eng. A* **2008**, *478*(1–2), 402–405.

369. Park, S.-J.; Jin, F.-L.; Lee, J.-R.; Shin, J.-S. *Eur. Polym. J.* **2005**, *41*(2), 231–237.

370. Park, S.-J.; Jin, F.-L.; Lee, J.-R. *Macromol. Chem. Phys.* **2004**, *205*(15), 2048–2054.

371. Kim, M. S.; Lee, K. W.; Endo, T.; Lee, S. B. *Macromolecules* **2004**, *37*(15), 5830–5834.

372. Refvik, M. D.; Larock, R. C.; Tian, Q. *J. Am. Oil Chem. Soc.* **1999**, *76*(1), 93–98.

373. Refvik, M. D.; Larock, R. C. *J. Am. Oil Chem. Soc.* **1999**, *76*(1), 99–102.

374. Tian, Q.; Larock, R. C. *J. Am. Oil Chem. Soc.* **2002**, *79*(5), 479–488.

375. Henna, P. H.; Larock, R. C. *Macromol. Mater. Eng.* **2007**, *292*(12), 1201–1209.

376. Mauldin, T. C.; Haman, K.; Sheng, X.; Henna, P.; Larock, R. C.; Kessler, M. R. *J. Polym. Sci. Part A Polym. Chem.* **2008**, *46*(20), 6851–6860.

377. Hong, J.; Luo, Q.; Wan, X.; Petrović, Z. S.; Shah, B. K. *Biomacromolecules* **2012**, *13*(1), 261–266.

378. Hong, J.; Shah, B. K.; Petrović, Z. S. *Eur. J. Lipid Sci. Technol.* **2013**, *115*(1), 55–60.

379. Shibata, M.; Teramoto, N.; Nakamura, Y. *J. Appl. Polym. Sci.* **2011**, *119*(2), 896–901.

380. Lacerda, T. M.; Carvalho, A. J. F.; Gandini, A. *RSC Adv.* **2014**, *4*(51), 26829.

381. Gandini, A.; Lacerda, T. M. *Prog. Polym. Sci.* **2015**.

382. Gandini, A.; Lacerda, T. M.; Carvalho, A. J. F.; Trovatti, E. *Chem. Rev.* **2016**.

383. Weissermel, K.; Arpe, H.-J. *Industrial Organic Chemistry, Fourth Edition*; WILEY-VCH: Weinheim, Ed.; Germany, 2013.

384. Haouet, A.; Sepulchre, M.; Spassky, *Eur. Polym. J.* **1983**, *19*(12), 1089–1098.

385. Sunder, A.; Hanselmann, R.; Frey, H.; Mülhaupt, R. *Macromolecules* **1999**, *32*(13), 4240–4246.

386. Sunder, A.; Krämer, M.; Hanselmann, R.; Mülhaupt, R.; Frey, H. *Angew. Chemie Int. Ed.* **1999**, *38*, 3552–3555.

387. Wilms, D.; Stiriba, S.-E.; Frey, H. *Acc. Chem. Res.* **2010**, *43*(1), 129–141.

388. Cesteros, L. C. *Green Chem.* **2011**, *13*(1), 197–206.

389. Cauwet, D.; Dubief, C. Cosmetic Composition Containing a Surfactant such as an Alkylpolyglycoside And/or Polyglycerol and an Urethanpolyether. Patent EP 0555155, **1993**.

390. Kunieda, H.; Akahane, A.; Jin-Feng; Ishitobi, M. *J. Colloid Interface Sci.* **2002**, *245*(2), 365–370.

391. Morlat, S.; Cezard, N.; Loubinoux, B.; Philippart, J.-L.; Gardette, J.-L. *Polym. Degrad. Stab.* **2001**, *72*(2), 199–208.

392. Charles, G.; Clacens, J.-M.; Pouilloux, Y.; Barrault, J. *OCL* **2003**, *10*(1), 74–82.

393. Calderón, M.; Quadir, M. A.; Sharma, S. K.; Haag, R. *Adv. Mater.* **2010**, 22(2), 190–218.

394. Frey, H.; Haag, R. *Rev. Mol. Biotechnol.* **2002**, *90*(3), 257–267.

395. Sutter, M.; Da Silva, E.; Duguet, N.; Raoul, Y.; Métay, E.; Lemaire, M. *Chem. Rev.* **2015**.

396. Martin, A.; Richter, M. *Eur. J. Lipid Sci. Technol.* **2011**, *113*(1), 100–117.

397. Garti, N.; Aserin, A.; Zaidman, B. *J. Am. Oil Chem. Soc.* **1981**, *58*(9), 878–883.

398. Iaych, K.; Dumarçay, S.; Fredon, E.; Gérardin, C.; Lemor, A.; Gérardin, P. *J. Appl. Polym. Sci.* **2011**, *120*(4), 2354–2360.

399. Macierzanka, A.; Szelag, H. *Ind. Eng. Chem. Res.* **2004**, *43*(24), 7744–7753.

400. Detergents and Detergency. In *Bailey's Industrial Oil and Fat Products*; John Wiley & Sons, Inc., **2005**.

401. Johnson, D. T.; Taconi, K. A. *Environ. Prog.* **2007**, *26*(4), 338–348.

402. Wyatt, V. T.; Strahan, G. D. *Polymers (Basel).* **2012**, *4*(1), 396–407.

403. Kulshrestha, A. S.; Gao, W.; Gross, R. A. *Macromolecules* **2005**, *38*(8), 3193–3204.

404. Vilela, C.; Sousa, A. F.; Fonseca, A. C.; Serra, A. C.; Coelho, J. F. J.; Freire, C. S. R.; Silvestre, A. J. D.; Heise, A.; Benthem, R. A. T. M. van; Péter, F.; *et al. Polym. Chem.* **2014**, *5*(9), 3119–3141.

405. Rai, R.; Tallawi, M.; Grigore, A.; Boccaccini, A. R. *Prog. Polym. Sci.* **2012**, *37*(8), 1051–1078.

406. Halpern, J. M.; Urbanski, R.; Weinstock, A. K.; Iwig, D. F.; Mathers, R. T.; von Recum, H. A. *J. Biomed. Mater. Res. Part A* **2014**, *102*(5), 1467–1477.

407. Tisserat, B.; O'kuru, R. H.; Hwang, H.; Mohamed, A. A.; Holser, R. *J. Appl. Polym. Sci.* **2012**, *125*(5), 3429–3437.

408. Parzuchowski, P. G.; Grabowska, M.; Jaroch, M.; Kusznerczuk, M. *J. Polym. Sci. Part A Polym. Chem.* **2009**, *47*(15), 3860–3868.

409. Zhao, X.; Liu, L.; Dai, H.; Ma, C.; Tan, X.; Yu, R. *J. Appl. Polym. Sci.* **2009**, *113*(5), 3376–3381.

410. Zhou, C.-H. C.; Beltramini, J. N.; Fan, Y.-X.; Lu, G. Q. M. *Chem. Soc. Rev.* **2008**, *37*(3), 527–549.

411. Yang, Y.; Lu, W.; Cai, J.; Hou, Y.; Ouyang, S.; Xie, W.; Gross, R. A. *Macromolecules* **2011**, *44*(7), 1977–1985.

412. Uyama, H.; Kuwabara, M.; Tsujimoto, T.; Kobayashi, S. *Biomacromolecules* **2003**, *4*(2), 211–215.

413. Tsujimoto, T.; Uyama, H.; Kobayashi, S. *Biomacromolecules* **2001**, *2*(1), 29–31.

414. Li, C.; Luo, X.; Li, T.; Tong, X.; Li, Y. *Polym. (United Kingdom)* **2014**.

415. Piszczyk, Ł.; Strankowski, M.; Danowska, M.; Hejna, A.; Haponiuk, J. T. *Eur. Polym. J.* **2014**.

416. Hu, S.; Luo, X.; Li, Y. *J. Appl. Polym. Sci.* **2015**.

417. Gama, N. V.; Silva, R.; Costa, M.; Barros-Timmons, A.; Ferreira, *Polym. Test.* **2016**.

418. Maisonneuve, L.; Lamarzelle, O.; Rix, E.; Grau, E.; Cramail, H. *Chem. Rev.* **2015**, *115*(22), 12407–12439.

419. Auvergne, R.; Caillol, S.; David, G.; Boutevin, B.; Pascault, J. P. *Chem. Rev.*. **2014**.

420. Grunchard, F. Process for the Manufacture of Epichlorohydrin. EP 561441, **1993**.

421. Tuck, C. O.; Pérez, E.; Horvath, I. T.; Sheldon, R. A.; Poliakoff, M. *Science* **2012**, *337*(6095), 695.

422. Barua, S.; Dutta, G.; Karak, N. *Chem. Eng. Sci.* **2013**, *95*, 138–147.

423. Shibata, M.; Nakai, K. *J. Polym. Sci. Part B Polym. Phys.* **2010**, *48*(4), 425–433.

424. Takada, Y.; Shinbo, K.; Someya, Y.; Shibata, M. *J. Appl. Polym. Sci.* **2009**, *113*(1), 479–484.

425. Herzog, B.; Kohan, M.; Mestemacher, S.; Pagilagan, R. *Polyamides. Ullmann's Encyclopedia of Industrial Chemistry*, Wiley-VCH:; **2000**.

426. Rogers, M. E.; Long, T. E. *Synthetic Methods in Step-Growth Polymers*; Wiley-Interscience, **2003**.

427. Marchildon, K. *Macromol. React. Eng.* **2011**, *5*(1), 22–54.

428. Schmitz, K.; Schepers, U. *Angew. Chemie Int. Ed.* **2004**, *43*(19), 2472–2475.

429. Winnacker, M.; Rieger, B. *Macromol. Rapid Commun.* **2016**, *37*(17), 1391–1413.

430. Genas, M. *Angew. Chemie* **1962**, *74*(15), 535–540.

431. Baumann, H.; Bühler, M.; Fochem, H.; Hirsinger, F.; Zoebelein, H.; Falbe, J. *Angew. Chemie Int. Ed. English* **1988**, *27*(1), 41–62.

432. Arkema. *Arkema Puteeaux, Fr.* **2009**.

433. Häger, H. *Evonik Degussa GmbH, Marl, Ger.* **2011**.

434. DSM. EcoPaXX : The Green Performer. *DSM Eng. Plast. Galeen, Netherlands* **2011**.

435. Wang, M. S.; Huang, J. C. *J. Polym. Eng.* **1994**, *13*(2), 155–174.

436. Kuciel, S.; Kuniar, P.; Liber-Knec, A. *Polimery* **2012**, No. 9, 627.

437. Feldmann, M.; Bledzki, A. K. *Compos. Sci. Technol.* **2014**, *100*, 113–120.

438. Samanta, S. R. *J. Appl. Polym. Sci.* **1992**, *45*(9), 1635–1640.

439. Mutlu, H.; Meier, M. A. R. *Macromol. Chem. Phys.* **2009**, *210*(12), 1019–1025.

440. Greene, J. L.; Huffman, E. L.; Burks, R. E.; Sheehan, W. C.; Wolff, I. A. *J. Polym. Sci. Part A-1 Polym. Chem.* **1967**, *5*(2), 391–394.

441. Nieschlag, H. J.; Rothfus, J. A.; Sohns, V. E.; Perkins, R. B. *Ind. Eng. Chem. Prod. Res. Dev.* **1977**, *16*(1), 101–107.

442. Wang, Y.; Liu, M.; Wang, Z.; Li, X.; Zhao, Q.; Fu, P.-F. *J. Appl. Polym. Sci.* **2007**, *104*(3), 1415–1422.

443. Samanta, S.; He, J.; Selvakumar, S.; Lattimer, J.; Ulven, C.; Sibi, M.; Bahr, J.; Chisholm, B. J. *Polymer (Guildf).* **2013**, *54*(3), 1141–1149.

444. He, J.; Samanta, S.; Selvakumar, S.; Lattimer, J.; Ulven, C.; Sibi, M.; Bahr, J.; Chisholm, B. J. *Green Mater.* **2013**, *1*(2), 114–124.

445. Eschenfeldt, W. H.; Zhang, Y.; Samaha, H.; Eirich, L. D.; Wilson, C. R.; Mark, I.; Stols, L.; Donnelly, M. I. *Appl. Environ. Microbiol.* **2003**, *69*(10), 5992–5999.

446. Pardal, F.; Salhi, S.; Rousseau, B.; Tessier, M.; Claude, S.; Fradet, A. *Macromol. Chem. Phys.* **2008**, *209*(1), 64–74.

447. Bennett, C.; Mathias, L. J. *J. Polym. Sci. Part A Polym. Chem.* **2005**, *43*(5), 936–945.

448. Cui, X.; Li, W.; Yan, D.; Yuan, C.; Di Silvestro, G. *J. Appl. Polym. Sci.* **2005**, *98*(4), 1565–1571.

449. Stempfle, F.; Quinzler, D.; Heckler, I.; Mecking, S. *Macromolecules* **2011**, *44*(11), 4159–4166.

450. Walther, G.; Deutsch, J.; Martin, A.; Baumann, F.-E.; Fridag, D.; Franke, R.; Köckritz, A. *ChemSusChem* **2011**, *4*(8), 1052–1054.

451. Çavuş, S.; Gürkaynak, M. A. *Polym. Adv. Technol.* **2006**, *17*(1), 30–36.

452. Heidarian, J.; Ghasem, N.; Daud, W. *Chem. Eng. J.* **2004**, *100*(1–3), 85–93.

453. Chen, X.; Zhong, H.; Jia, L.; Ning, J.; Tang, R.; Qiao, J.; Zhang, Z. *Int. J. Adhes. Adhes.* **2002**, *22*(1), 75–79.

454. Unverferth, M.; Meier, M. A. R. *Eur. J. Lipid Sci. Technol.* **2016**, *118*(10), 1470–1474.

455. Turunc, O.; Firdaus, M.; Klein, G.; Meier, M. A. R. *Green Chem.* **2012**, *14*(9), 2577–2583.

456. Miao, X.; Malacea, R.; Fischmeister, C.; Bruneau, C.; Dixneuf, P. H. *Green Chem.* **2011**, *13*(10), 2911–2919.

457. Miao, X.; Fischmeister, C.; Dixneuf, P. H.; Bruneau, C.; Dubois, J. L.; Couturier, J. L. *Green Chem.* **2012**, *14*(8), 2179–2183.

458. Winkler, M.; Steinbiß, M.; Meier, M. A. R. *Eur. J. Lipid Sci. Technol.* **2014**, *116*(1), 44–51.

459. Winkler, M.; Meier, M. A. R. *Green Chem.* **2014**, *16*(4), 1784–1788.

460. Oelmann, S.; Meier, M. A. R. *Macromol. Chem. Phys.* **2015**, *216*(19), 1972–1981.

461. Lligadas, G.; Ronda, J. C.; Galià, M.; Biermann, U.; Metzger, J. O. *J. Polym. Sci. Part A Polym. Chem.* **2006**, *44*(1), 634–645.

462. Xu, Y.; Petrovic, Z.; Das, S.; Wilkes, G. L. *Polymer (Guildf).* **2008**, *49*(19), 4248–4258.

463. Petrović, Z. S.; Hong, D.; Javni, I.; Erina, N.; Zhang, F.; Ilavský, J. *Polymer (Guildf).* **2013**, *54*(1), 372–380.

464. Saralegi, A.; Rueda, L.; Fernández-d'Arlas, B.; Mondragon, I.; Eceiza, A.; Corcuera, M. A. *Polym. Int.* **2013**, *62*(1), 106–115.

465. Hojabri, L.; Jose, J.; Leao, A. L.; Bouzidi, L.; Narine, S. S. *Polymer (Guildf).* **2012**, *53*(17), 3762–3771.

466. Lluch, C.; Ronda, J. C.; Galià, M.; Lligadas, G.; Cádiz, V. *Biomacromolecules* **2010**, *11*(6), 1646–1653.

467. Petrović, Z. S.; Xu, Y.; Milić, J.; Glenn, G.; Klamczynski, A. *J. Polym. Environ.* **2010**, *18*(2), 94–97.

468. Tang, D.; Noordover, B. A. J.; Sablong, R. J.; Koning, C. E. *J. Polym. Sci. Part A Polym. Chem.* **2011**, *49*(13), 2959–2968.

469. Bueno-Ferrer, C.; Hablot, E.; Perrin-Sarazin, F.; Garrigós, M. C.; Jiménez, A.; Averous, L. *Macromol. Mater. Eng.* **2012**, *297*(8), 777–784.

470. Bueno-Ferrer, C.; Hablot, E.; Garrigós, M. del, C.; Bocchini, S.; Averous, L.; Jiménez, A. *Polym. Degrad. Stab.* **2012**, *97*(10), 1964–1969.

471. González-Paz, R. J.; Lluch, C.; Lligadas, G.; Ronda, J. C.; Galià, M.; Cádiz, V. *J. Polym. Sci. Part A Polym. Chem.* **2011**, *49*(11), 2407–2416.

472. Desroches, M.; Caillol, S.; Auvergne, R.; Boutevin, B. *Eur. J. Lipid Sci. Technol.* **2012**, *114*(1), 84–91.

473. Palaskar, D. V.; Boyer, A.; Cloutet, E.; Le Meins, J.-F.; Gadenne, B.; Alfos, C.; Farcet, C.; Cramail, H. *J. Polym. Sci. Part A Polym. Chem.* **2012**, *50*(9), 1766–1782.

474. Maisonneuve, L.; Lebarbe, T.; Nguyen, T. H. N.; Cloutet, E.; Gadenne, B.; Alfos, C.; Cramail, H. *Polym. Chem.* **2012**, *3*(9), 2583–2595.

475. Boyer, A.; Lingome, C. E.; Condassamy, O.; Schappacher, M.; Moebs-Sanchez, S.; Queneau, Y.; Gadenne, B.; Alfos, C.; Cramail, H. *Polym. Chem.* **2013**, *4*(2), 296–306.

476. Hojabri, L.; Kong, X.; Narine, S. S. *Biomacromolecules* **2009**, *10*(4), 884–891.

477. Hojabri, L.; Kong, X.; Narine, S. S. *J. Polym. Sci. Part A Polym. Chem.* **2010**, *48*(15), 3302–3310.

478. More, A. S.; Lebarbé, T.; Maisonneuve, L.; Gadenne, B.; Alfos, C.; Cramail, H. *Eur. Polym. J.* **2013**, *49*(4), 823–833.

479. Duval, C.; Kébir, N.; Charvet, A.; Martin, A.; Burel, F. *J. Polym. Sci. Part A Polym. Chem.* **2015**, *53*(11), 1351–1359.

480. Unverferth, M.; Kreye, O.; Prohammer, A.; Meier, M. A. R. *Macromol. Rapid Commun.* **2013**, *34*(19), 1569–1574.

481. Kreye, O.; Wald, S.; Meier, M. A. R. *Adv. Synth. Catal.* **2013**, *355*(1), 81–86.

482. More, A. S.; Gadenne, B.; Alfos, C.; Cramail, H. *Polym. Chem.* **2012**, *3*(6), 1594–1605.

483. Palaskar, D. V.; Boyer, A.; Cloutet, E.; Alfos, C.; Cramail, H. *Biomacromolecules* **2010**, *11*(5), 1202–1211.

484. Boyer, A.; Cloutet, E.; Tassaing, T.; Gadenne, B.; Alfos, C.; Cramail, H. *Green Chem.* **2010**, *12*(12), 2205–2213.

485. Foltran, S.; Maisonneuve, L.; Cloutet, E.; Gadenne, B.; Alfos, C.; Tassaing, T.; Cramail, H. *Polym. Chem.* **2012**, *3*(2), 525–532.

486. Maisonneuve, L.; More, A. S.; Foltran, S.; Alfos, C.; Robert, F.; Landais, Y.; Tassaing, T.; Grau, E.; Cramail, H. *RSC Adv.* **2014**, *4*(49), 25795–25803.

487. Tomita, H.; Sanda, F.; Endo, T. *J. Polym. Sci. Part A Polym. Chem.* **2000**, *39*(1), 162–168.

488. Tomita, H.; Sanda, F.; Endo, T. *J. Polym. Sci. Part A Polym. Chem.* **2001**, *39*(6), 860–867.

489. Tomita, H.; Sanda, F.; Endo, T. *J. Polym. Sci. Part A Polym. Chem.* **2001**, *39*(23), 4091–4100.

490. Besse, V.; Foyer, G.; Auvergne, R.; Caillol, S.; Boutevin, B. *J. Polym. Sci. Part A Polym. Chem.* **2013**, *51*(15), 3284–3296.

491. Kihara, N.; Kushida, Y.; Endo, T. *J. Polym. Sci. Part A Polym. Chem.* **1996**, *34*(11), 2173–2179.

492. Steblyanko, A.; Choi, W.; Sanda, F.; Endo, T. *J. Polym. Sci. Part A Polym. Chem.* **2000**, *38*(13), 2375–2380.

493. Kihara, N.; Endo, T. *J. Polym. Sci. Part A Polym. Chem.* **1993**, *31*(11), 2765–2773.

494. Fleischer, M.; BlatTmann, H.; Mulhaupt, R. *Green Chem.* **2013**, *15*(4), 934–942.

495. Keul, H.; Mommer, S.; Möller, M. *Eur. Polym. J.* **2013**, *49*(4), 853–864.

496. van Velthoven, J. L. J.; Gootjes, L.; van Es, D. S.; Noordover, B. A. J.; Meuldijk, J. *Eur. Polym. J.* **2015**, *70*, 125–135.

497. Lamarzelle, O.; Durand, P.-L.; Wirotius, A.-L.; Chollet, G.; Grau, E.; Cramail, H. *Polym. Chem.* **2016**, *7*, 1439–1451.

498. Lamarzelle, O.; Hibert, G.; Lecommandoux, S.; Grau, E.; Cramail, H. *Polym. Chem.* **2017**, *8*(22), 3438–3447.

499. Maisonneuve, L.; Wirotius, A.-L.; Alfos, C.; Grau, E.; Cramail, H. *Polym. Chem.* **2014**, *5*, 6142–6147.

500. Desroches, M.; Benyahya, S.; Besse, V.; Auvergne, R.; Boutevin, B.; Caillol, S. *Lipid Technol.* **2014**, *26*(2), 35–38.

501. Edlund, U.; Albertsson, A.-C. *Adv. Drug Deliv. Rev.* **2003**, *55*(4), 585–609.

502. Zhang, J.; Li, J.; Tang, Y.; Lin, L.; Long, M. *Carbohydr. Polym.* **2015**, *130*, 420–428.

503. Bakare, I. O.; Pavithran, C.; Okieimen, F. E.; Pillai, C. K. S. *J. Appl. Polym. Sci.* **2006**, *100*(5), 3748–3755.

504. Fakirov, S.; Wiley InterScience (Online service). *Transreactions in Condensation Polymers*; Wiley-VCH, **1999**.

505. Zia, K. M.; Noreen, A.; Zuber, M.; Tabasum, S.; Mujahid, M. *Int. J. Biol. Macromol.* **2016**, *82*, 1028–1040.

506. Quinzler, D.; Mecking, S. *Angew. Chemie Int. Ed.* **2010**, *49*(25), 4306–4308.

507. Ravi Kumar, M. N. V. *Handbook of Polyester Drug Delivery Systems.*

508. Kunduru, K. R.; Basu, A.; Haim Zada, M.; Domb, A. J. *Biomacromolecules* **2015**, *16*(9), 2572–2587.

509. Ebata, H.; Toshima, K.; Matsumura, S. *Macromol. Biosci.* **2007**, *7*(6), 798–803.

510. Totaro, G.; Cruciani, L.; Vannini, M.; Mazzola, G.; Di Gioia, D.; Celli, A.; Sisti, L. *Eur. Polym. J.* **2014**, *56*, 174–184.

511. Kelly, A. R.; Hayes, D. G. *J. Appl. Polym. Sci.* **2006**, *101*(3), 1646–1656.

512. Ebata, H.; Yasuda, M.; Toshima, K.; Matsumura, S. *J. Oleo Sci.* **2008**, *57*(6), 315–320.

513. Mutlu, H.; Meier, M. A. R. *Eur. J. Lipid Sci. Technol.* **2010**, *112*(1), 10–30.

514. Hiremath, J. G.; Kusum Devi, V.; Devi, K.; Domb, A. J. *J. Appl. Polym. Sci.* **2008**, *107*(5), 2745–2754.

515. Shikanov, A.; Vaisman, B.; Shikanov, S.; Domb, A. J. *J. Biomed. Mater. Res. Part A* **2009**, *9999A* (4), NA-NA.

516. Petrović, Z. S.; Milić, J.; Xu, Y.; Cvetković, I. *Macromolecules* **2010**, *43*(9), 4120–4125.

517. Testud, B.; Pintori, D.; Grau, E.; Taton, D.; Cramail, H. *Green Chem.* **2017**, 6–11.

518. Yang, Y.; Lu, W.; Zhang, X.; Xie, W.; Cai, M.; Gross, R. A. *Biomacromolecules* **2010**, *11*(1), 259–268.

519. Türünç, O.; Meier, M. A. R. *Macromol. Rapid Commun.* **2010**, *31*(20), 1822–1826.

520. Lebarbé, T.; Maisonneuve, L.; Nga Nguyen, T. H.; Gadenne, B.; Alfos, C.; Cramail, H.; Gross, R. A.; Cramail, H. *Polym. Chem.* **2012**, *3*(10), 2842.

521. Trzaskowski, J.; Quinzler, D.; Bährle, C.; Mecking, S. A. *Macromol. Rapid Commun.* **2011**, *32*(17), 1352–1356.

522. Vilela, C.; Silvestre, A. J. D.; Meier, M. A. R. P. *Macromol. Chem. Phys.* **2012**, *213*(21), 2220–2227.

523. Warwel, S.; Tillack, J.; Demes, C.; Kunz, M. *Macromol. Chem. Phys.* **2001**, *202*(7), 1114–1121.

524. Bauch, C. G.; Wagener, K. B.; Boncella, J. M. *Die Makromol. Chemie, Rapid Commun.* **1991**, *12*(7), 413–417.

525. Hall, A. J.; Hodge, P.; Kamau, S. D.; Ben-Haida, A. Acyclic Diene Metathesis (ADMET) Polymerization of Allyl Undec-10-Enoate and Some Related Esters. *J. Organomet. Chem.* **2006**, *691*(24–25), 5431–5437.

526. Stempfle, F.; OrTmann, P.; Mecking, S. *Macromol. Rapid Commun.* **2013**, *34*(1), 47–50.

527. Türünç, O.; Meier, M. A. R. *Green Chem.* **2011**, *13*(2), 314.

528. Fokou, P. A.; Meier, M. A. R. *J. Am. Chem. Soc.* **2009**, *131*(5), 1664–1665.

529. Fokou, P. A.; Meier, M. A. R. *Macromol. Rapid Commun.* **2010**, *31*(4), 368–373.

530. Nivinskiene, O.; Butkiene, R.; Mockute, D. *Chemija* **2003**, *14*(1), 52–56.

531. Pascual, A.; Sardon, H.; Veloso, A.; Ruipérez, F.; Mecerreyes, D. *ACS Macro Lett.* **2014**, *3*(9), 849–853.

532. Raia Slivniak; Abraham J. Domb*. Macrolactones and Polyesters from Ricinoleic Acid. **2005**.

533. Gargouri, M.; Drouet, P.; Legoy, M.-D. *J. Biotechnol.* **2002**, *92*(3), 259–266.

534. de Geus, M.; van der Meulen, I.; Goderis, B.; van Hecke, K.; Dorschu, M.; van der Werff, H.; Koning, C. E.; Heise, A. *Polym. Chem.* **2010**, *1*(4), 525.

535. Kırpal S. Bisht; Lori A. Henderson, and; Gross*, R. A.; Kaplan, D. L.; Swift, G. Enzyme-Catalyzed Ring-Opening Polymerization of ω-Pentadecalactone†. **1997**.

536. Cai, J.; Liu, C.; Cai, M.; Zhu, J.; Zuo, F.; Hsiao, B. S.; Gross, R. A. *Polymer (Guildf).* **2010**, *51*(5), 1088–1099.

537. Nakayama, Y.; Watanabe, N.; Kusaba, K.; Sasaki, K.; Cai, Z.; Shiono, T.; Tsutsumi, C. *J. Appl. Polym. Sci.* **2011**, *121*(4), 2098–2103.

538. Zhong, Z.; Dijkstra, P. J.; Feijen, J. *Macromol. Chem. Phys.* **2000**, *201*(12), 1329–1333.

539. Bouyahyi, M.; Pepels, M. P. F.; Heise, A.; Duchateau, R. *Macromolecules* **2012**, *45*(8), 3356–3366.

540. van der Meulen, I.; de Geus, M.; Antheunis, H.; Deumens, R.; Joosten, E. A. J.; Koning, C. E.; Heise, A. *Biomacromolecules* **2008**, *9*(12), 3404–3410.

541. Mutlu, H.; Ruiz, J.; Solleder, S. C.; Meier, M. A. R.; Meier, M. A. R.; Cadiz, V.; Xiao, F.; Wei, W.; Sun, Y. *Green Chem.* **2012**, *14*(6), 1728.

542. More, A. S.; Palaskar, D. V.; Cloutet, E.; Gadenne, B.; Alfos, C.; Cramail, H.; Wang, F. *Polym. Chem.* **2011**, *2*(12), 2796.

543. Vanderhenst, R.; Miller, S. A. *Green Mater.* **2013**, *1*(2), 64–78.

544. Zhang, Y.-Y.; Zhang, X.-H.; Wei, R.-J.; Du, B.-Y.; Fan, Z.-Q.; Qi, G.-R.; Qi, G. R.; Qi, G. R. *RSC Adv.* **2014**, *4*(68), 36183.

545. Winnacker, M.; Rieger, B. *ChemSusChem* **2015**, *8*(15), 2455–2471.

546. BreiTmaier, E. *Terpenes: Flavors, Fragrances, Pharmaca, Pheromones*; Wiley-VCH, Ed.; **2006**.

547. Sun, I.; Wang, H.; Kashiwada, Y.; Shen, J.; Cosentino, L. M.; Chen, C.; Yang, L.; Lee, K. *J. Med. Chem.* **1998**, *41*, 4648–4657.

548. Cichewicz, R. H.; Kouzi, S. A. *Med. Res. Rev.* **2004**, *24*(1), 90–114.

549. Gandini, A. *Green Chem.* **2011**, *13*(5), 1061.

550. Corma Canos, A.; Iborra, S.; Velty, A. *Chem. Rev.* **2007**, *107*(6), 2411–2502.

551. Belgacem, M.; Gandini, A. *Monomers, Polymers and Composites from Renewable Resources*; **2008**.

552. Fomine, S.; Tlenkopatchev, M. A. *J. Organomet. Chem.* **2012**, *701*, 68–74.

553. Strick, B. F.; Delferro, M.; Geiger, F. M.; Thomson, R. J. *ACS Sustain. Chem. Eng.* **2015**, *3*(7), 1278–1281.

554. Miyaji, H.; Satoh, K.; Kamigaito, M. *Angew. Chemie — Int. Ed.* **2016**, *55*(4), 1372–1376.

555. Sainz, M. F.; Souto, J. A.; Regentova, D.; Johansson, M. K. G.; Timhagen, S. T.; Irvine, D. J.; Buijsen, P.; Koning, C. E.; Stockman, R. A.; Howdle, S. M. *Polym. Chem.* **2016**, *7*(16), 2882–2887.

556. Wu, G.; Kong, Z.; Huang, H.; Chen, J.; Chu, F. *J. Appl. Polym. Sci.* **2009**, *113*, 2894–2901.

557. Kukhta, N. A.; Vasilenko, I. V.; Kostjuk, S. V. **2011**, 2362–2364.

558. Wang, Y.; Li, A.-L.; Liang, H.; Lu, J. *Eur. Polym. J.* **2006**, *42*(10), 2695–2702.

559. Firdaus, M.; Montero De Espinosa, L.; Meier, M. A. R. *Macromolecules* **2011**, *44*(18), 7253–7262.

560. Firdaus, M.; Meier, M. A. R. *Green Chem.* **2013**, *15*(2), 370.

561. Byrne, C. M.; Allen, S. D.; Lobkovsky, E. B.; Coates, G. W. *J. Am. Chem. Soc.* **2004**, *126*(37), 11404–11405.

562. Hauenstein, O.; Reiter, M.; Agarwal, S.; Rieger, B.; Greiner, *Green Chem.* **2016**, *18*(3), 760–770.

563. Hauenstein, O.; Agarwal, S.; Greiner, A. *Nat. Commun.* **2016**, *7*(May), 1–7.

564. Bähr, M.; Bitto, A.; Mülhaupt, R. *Green Chem.* **2012**, *14*(5), 1447.

565. Kobayashi, S.; Lu, C.; Hoye, T. R.; Hillmyer, M. A. *J. Am. Chem. Soc.* **2009**, *131*(23), 7960–7961.

566. Busch, H.; Stempfle, F.; Heß, S.; Grau, E.; Mecking, S. *Green Chem.* **2014**, *16*(10), 4541–4545.

567. Lowe, J. R.; Martello, M. T.; Tolman, W. B.; Hillmyer, M. A. *Polym. Chem.* **2011**, *2*(3), 702–708.

568. Lowe, J. R.; Tolman, W. B.; Hillmyer, M. A. *Biomacromolecules* **2009**, *10*, 2003.

569. Zhang, D.; Hillmyer, M. A.; Tolman, W. B. *Biomacromolecules* **2005**, *6*(4), 2091–2095.

570. Wanamaker, C. L.; Bluemle, M. J.; Pitet, L. M.; O'Leary, L. E.; Tolman, W. B.; Hillmyer, M. A. *Biomacromolecules* **2009**, *10*(10), 2904–2911.

571. Wanamaker, C. L.; O'Leary, L. E.; Lynd, N. A.; Hillmeyer, M. A.; Tolman, W. B. *Biomacromolecules* **2007**, *8*(11), 3634–3640.

572. Shin, J.; Martello, M. T.; Shrestha, M.; Wissinger, J. E.; Tolman, W. B.; Hillmyer, M. A. *Macromolecules* **2011**, *44*(1), 87–94.

573. Atta, A. M.; Mansour, R.; Abdou, M. I.; Sayed, A. M. *Polym. Adv. Technol.* **2004**, *15*(9), 514–522.

574. Atta, A. M.; Elsaeed, A. M.; Farag, R. K.; El-Saeed, S. M. *React. Funct. Polym.* **2007**, *67*(6), 549–563.

575. Bicu, I.; Mustata, F. *Macromol. Mater. Eng.* **2000**, *280–281*(1), 47–53.

576. Bicu, I.; Mustata, F. *J. Polym. Sci. Part A Polym. Chem.* **2005**, *43*(24), 6308–6322.

577. Bicu, I.; Mustata, F. *J. Polym. Sci. Part A Polym. Chem.* **2007**, *45*(24), 5979–5990.

578. Bicu, I.; Mustata, F. *Polimery* **2000**, *45*(4), 258.

579. Mustata, F.; Bicu, I. *Eur. Polym. J.* **2010**, *46*(6), 1316–1327.

580. Kim, S. J.; Kim, B. J.; Jang, D. W.; Kim, S. H.; Park, S. Y.; Lee, J. H.; Lee, S. D.; Choi, D. H. *J. Appl. Polym. Sci.* **2001**, *79*(4), 687–695.

581. Atta, A. M.; El-Saeed, S. M.; Farag, R. K. *React. Funct. Polym.* **2006**, *66*(12), 1596–1608.

582. Lee, J. S.; Hong, S. Il. *Eur. Polym. J.* **2002**, *38*(2), 387–392.

583. Kim, H.-J. *J. Appl. Polym. Sci.* **2009**, *111*, 1172–1176.

584. Chu, F. *J. Appl. Polym. Sci.* **2009**, *113*, 3757–3765.

585. Wilbon, P. A.; Zheng, Y.; Yao, K.; Tang, C. *Macromolecules* **2010**, *43*(21), 8747–8754.

586. Zheng, Y.; Yao, K.; Lee, J.; Chandler, D.; Wang, J.; Wang, C.; Chu, F.; Tang, C. *Macromolecules* **2010**, *43*(14), 5922–5924.

587. Wang, J.; Yao, K.; Korich, A. L.; Li, S.; Ma, S.; Ploehn, H. J.; Iovine, P. M.; Wang, C.; Chu, F.; Tang, C. *J. Polym. Sci. Part A Polym. Chem.* **2011**, *49*(17), 3728–3738.

588. Chen, Y.; Wilbon, P. A.; Chen, Y. P.; Zhou, J.; Nagarkatti, M.; Wang, C.; Chu, F.; Decho, A. W.; Tang, C. A. *RSC Adv.* **2012**, *2*(27), 10275.

589. Yao, K.; Wang, J.; Zhang, W.; Lee, J. S.; Wang, C.; Chu, F.; He, X.; Tang, C. *Biomacromolecules* **2011**, *12*, 2171–2177.

590. Wang, J.; Chen, Y. P.; Yao, K.; Wilbon, P. A.; Zhang, W.; Ren, L.; Zhou, J.; Nagarkatti, M.; Wang, C.; Chu, F.; *et al. Chem. Commun.* **2012**, *48*(6), 916.

591. Mantzaridis, C.; Brocas, A.-L.; Llevot, A.; Cendejas, G.; Auvergne, R.; Caillol, S.; Carlotti, S.; Cramail, H. *Green Chem.* **2013**, *15*, 3091–3098.

592. Liu, X.; Xin, W.; Zhang, J. *Green Chem.* **2009**, *11*(7), 1018.

593. Wang, H.; Liu, X.; Liu, B.; Zhang, J.; Xian, M. *Polym. Int.* **2009**, *58*(12), 1435–1441.

594. Wang, H.; Liu, B.; Liu, X.; Zhang, J.; Xian, M. *Green Chem.* **2008**, *10*(11), 1190.

595. Liu, X.; Xin, W.; Zhang, J. *Bioresour. Technol.* **2010**, *101*(7), 2520–2524.

6 Insight into Continental Organic Matter: A Chemist View

Katell Quénéa,[*,‡] Sylvie Derenne[*,§] and Marc F. Benedetti[†]

*METIS, Sorbonne université-CNRS-EPHE, UMR 7619,
Paris, France
[†]IPGP-Univ.Paris Diderot-USPC-CNRS UMR 7154, Paris, France
[‡]katell.quenea@upmc.fr
[§]sylvie.derenne@upmc.fr

Table of Contents

List of Abbreviations

(ATR)-FTIR	attenuated total reflectance fourier transform infrared
A4F	asymmetrical flow field flow fractionation
AFM	atomic force microscopy
a_λ	Napierian absorption coeffcicient (m^{-1})
BC	black carbon
BIX	biological index
CFS	chemical force spectroscopy
CLE-AdSV	competitive ligand exchange adsorptive cathodic stripping voltametry
CP/MAS	cross polarization magic angle spinning
DGT	diffusion gradient in thin films
DIN	dissolved inorganic nitrogen
DLT	diffusion layer thickness
DMT	donnan membrane technique
DOC	dissolved organic carbon
DOM	dissolved organic matter

DON	dissolved organic nitrogen
DRIFT	diffuse reflectance FTIR
DSC	differential scanning calorimetry
ED	electrodialysis
EEM	excitation-emission matrix
ESI-FTICR-MS	electrospray ionization coupled to fourier transform ion cyclotron resonance mass sprectrometry
F-D	force displacement
FA	fulvic acid
FFF	field flow fractionation
FTICR-MS	Fourrier transform ion cyclotron resonance mass sprectrometry
FTIR	fourier transform infrared
GC-C-irMS	gas chromatography coupled to carbon isotopic ratio mass spectrometry
GC/MS-C-irMS	gas chromatrography coupled to mass spectrometry and carbon isotopic ratio mass spectrometry
GIME	gel immobilize micro electrode
HA	humic acid
HIX	humification index
HP-SEC	high pressure steric exclusion chromatography
HS	humic subtances
I	ionic strength
IR	infrared
irMS	isotopic ratio mass spectrometry
K	binding affinity constant
K_{av}	median value for the binding affinity constant K
L	ligand
LC-C-irMS	liquid chromatography couple to carbon isotopic ratio mass spectrometry
LIFS	laser-induced fluorescence spectroscopy
M	metal ion
MIRS	mid infrared spectroscopy
MRT	mean residence time
MWEL	milled wall enzymatic lignin
NanoSIMS	nano secondary ions mass spectrometry

NEXAFS	near edge X-ray adsoprtion fine structure
NICA	non-ideal competitive adsorption model
NMR	nuclear magnetic resonance
NOM	natural organic matter
OM	organic matter
PAH	polyaromatic hydrocarbons
PARAFAC	parallel factor analysis
PLS	partial least square
POM	particulate organic matter
PPHA	purified peat extracted humic acid
ppm	part per million
Py-FIMS	pyrolysis coupled to field ionization mass spectrometry
Py-GC/MS	pyrolysis coupled to gas chromatography and mass sprectrometry
RO	reverse osmosis
S	spectral slope (nm^{-1})
SEC	steric exclusion chromatography
SEM	scanning electron microscopy
SOM	soil organic matter
SPE	solid phase extraction
SR	ratio of spectral slopes
STXM	scanning transmission X-ray microscopy
TEM	transmission electron microscopy
TG	thermogravimetry
TMAH	tetramethylammonium hydroxide
TN	total nitrogen
UF	ultrafiltration
UV	ultraviolet
UV/Vis	ultraviolet visible
XAD®	hydrophobic copolymer of styrene-divinylbenzene resin
XANES	X-ray absorption near-edge structure
XPS	X-ray photonelectron spectroscopy

I. Introduction

Continental organic matter (OM) stands for OM occurring in environments encompassing sediments, soil and aquatic systems. The study of natural OM is mainly driven by a better understanding of the C geochemical cycle. Indeed, continental OM comprises the most reactive pool of C and the most sensitive to anthropic pressure. As a result, it is prone to interact with trace elements.

Approximately 90% of organic C on Earth, accounting for 12–15 10^6 Gt, occurs within sedimentary rocks.[1] Among these, 1000 Gt are considered to occur in surface sediments. Although continental margins only account for 20% of the global ocean area, they represent the major repository of C with 80–90% of the accumulated OM[2,3]. This reflects the large primary productivity ($\approx$ 2.2 Gt C yr^{-1}) in these coastal areas[2] but also the input through river discharge ($\approx$ 0.4 Gt C yr^{-1})[4–6] including 0.2 Gt C yr^{-1} as dissolved organic carbon (DOC).[7] Moreover, 0.6 Gt C.yr^{-1} are buried in freshwater sediments[6]. Due to the great volume of the oceans, although occurring in low concentration, DOC accounts for $\approx$ 670 Gt C,[8] in the same range as the atmospheric C pool. Most of this C ($\approx$ 70%) occurs in the bathypelagic layer, i.e. below 1000 m.[9] Terrestrial OM mainly results from the decay of vascular plants, which represent the largest pool of living biomass on Earth, accounting for 570 Gt of C. Taken together, soils are the largest active pool of organic C. Indeed, they contain 1550 Gt of C in the top meter and 2350 Gt in the top three meters.[10] This C pool thus represents 3.3 times the atmospheric one and 4.5 times the biotic one.[11] In the present chapter, we will restrict the sediments to the surficial ones and the aquatic environments to the freshwater (riverine and lacustrine) and estuarine ones.

Due to differences in deposition environments and associated source organisms, different geochemical behaviors are expected between sediment and soil OM as emphasized by Hedges and Oades (1997).[5]

Sedimentary OM was deposited in lacustrine or marine environments and therefore it is mostly of phytoplanktonic origin, along with microorganisms including fungi, bacteria and archaea. These microorganisms were involved both in the decomposition of the biomass after the death of the source organisms and in the contribution to the sedimentary OM through their own biomass.

Terrestrial vascular plant biomass can be either locally deposited to be transformed into coal beds or transported to aquatic systems to contribute to sediments.[12] Although it is commonly admitted that the OM incorporated in sediments accounts for less than 1% of the living biomass of the source organisms,[13] it is difficult to quantify each contribution to the sedimentary OM.[14] However, based on the lacustrine, marine or terrestrial origin of the OM, three types of kerogens (defined as the insoluble OM dispersed in sedimentary rocks, Durand, 1980) can be distinguished. They are characterized by differences in elemental composition, mostly H/C and O/C atomic ratios (commonly reported in the so-called van Krevelen diagram, i.e. H/C = f(O/C))[15] which reflect differences in the chemical composition of the source organism biomass.

Vascular plants are the main source of terrestrial OM where they undergo biological and physico-chemical degradation to various extent.[16] The terrestrial OM stock therefore reflects the balance between the input of OM from plant and organisms and degradation of this OM. This equilibrium is mainly controlled by biological processes such as mineralization. In consequence, when considering the evolution of soil organic C, time scales from years to decades have to be considered. The C sequestration potential of soil OM was estimated between 0.6 and 1.2 Gt C/year[17] and could represent a way to mitigate atmospheric CO_2 increase through long term storage of C in soil.[11] If the main input of soil OM is considered to be plant residues at different decay stage and exudates, the contribution of microbial biomass was reappraised around 2007 and microbial contribution seems to be far more important than previously reported.[18–22] It was even considered to reach 80% of soil organic carbon,[23] the large majority being composed of bacterial necromass.[19] This important bacterial OM contribution to soil OM and its potential stabilization in soil was tentatively explained in a framework proposed by Cotrufo and Wallenstein (2013).[24]

The behavior (fate and reactivity) of natural OM is driven by its chemical composition, which is directly linked to its source organisms. The turnover of continental OM depends on its biodegradability. Moreover, different recalcitrance and preservation potential occur within natural OM. Whatever its terrestrial or aquatic origin, the biomass comprises several biopolymers such as polysaccharides or proteins which result in distinct

OM composition. However, additional macromolecules are specific for a given type of source organisms and others result from secondary reactions associated with transformations of the initial biomass constituents. In the nature, all these different constituents exhibit variable recalcitrance toward degradation which in addition, depends on the environmental conditions. They will contribute at different degradation stages to the natural OM, thus resulting in a large heterogeneity and diversity, hence a high complexity of the OM. This diversity is also observed at the molecular scale in the chemical functions and characteristic molecular moieties of the natural OM.[16,25,26] As a result, so as to minimize bias, a combination of analytical approaches, encompassing spectroscopic methods, chemical and thermal degradation, nanoscale observations and analyses, is required to achieve a complete chemical characterization of natural OM.[27] However, prior to any characterization, samples must often be subjected to pretreatment, as detailed in Section II, for several reasons such as to reduce sample volume and make transport easier, or to improve sample conservation, or to isolate OM fractions, which may be operational or scientific.

In natural environments, OM is tightly associated with minerals. Due to the high reactivity of OM, organo-mineral associations and complexation occur. They impact OM stabilization and trace element transportation and therefore their biodisponibility and potential toxicity, as discussed in Section III.B. This reciprocal interaction illustrates the coupling between the biogeochemical cycles of the trace elements and of carbon. Although several analytical techniques have been developed to investigate the interactions between OM and trace elements, modelling these interactions is still required to describe the fate of trace elements.

II. Sample Pretreatment in OM Characterization

OM occurs at different concentrations in the natural environment but it is often dispersed within a mineral matrix or diluted in water, sometimes further associated with salts. When the organic content is too low, a concentration step is a prerequisite for OM detailed analysis. Sample fractionation may also be used to separate OM in pools with homogeneous properties. Physical and chemical fractionations have thus been developed.

Density and particle-size fractionation are commonly achieved on soil samples. Several types of fractionation, based on molecular size, are increasingly performed on aquatic OM to investigate its diversity. Similarly, chemical fractionation is often used to provide homogeneous fractions in terms of physico-chemical properties, biochemical composition or turnover. These different points are detailed below.

A. Concentration of OM

Depending on the nature (sediment/soil or aquatic) of the sample, different procedures are used for OM concentration.

1. *OM from Sediment and Soil*

In sediments, kerogens are commonly isolated through the classical HF/HCl treatment which aims at destroying most of the mineral.[14,28] HCl is first used to eliminate carbonates, followed by HF to remove silicates. The role of a final HCl treatment is mainly to prevent the formation of fluorides upon HF/HCl treatment, some of them like CaF_2 being highly insoluble. The removal of specific minerals is also required for some further analyses; for example, it has been shown that the presence of clay minerals may bias the amount and nature of pyrolysis products through selective adsorption of some effluents or catalyzed reaction.[29-31] However, it must be noted that some minerals such as pyrite and heavy oxides can survive this treatment. In 2007, a treatment with a ferric nitrate solution was proposed to remove most of pyrite in coals.[32] HNO_3 treatment was also shown to be efficient to eliminate pyrite from coals but it induced nitration of the OM.[33] Similarly, to avoid the use of HCl considered as a too strong acid, a method based on the *in situ* production of BF_3 was developed to dissolve the neofluorides and convert them into soluble fluoroborates.[34] Methods to produce ultra-clean coal through chemical treatment were reviewed in 2015.[35]

Although the classical HF/HCl treatment was shown not to induce any substantial chemical modification in the sedimentary OM, micro-structural alteration due to collapse of kerogen-hosted pores was reported and an improved method involving critical point drying was proposed to overcome this problem.[36] Moreover, highly immature or recent sediments still contain functional groups, which are eliminated during diagenesis in more mature samples, at least for the most reactive ones. These groups are potentially sensitive to hydrolysis, and caution must be exercised in the isolation of the corresponding kerogen. The use of less concentrated acids was therefore recommended and a procedure to recover the OM solubilized into HF was developed leading to a carbon concentration factor reaching up to 20%.[37] A similar approach was used to remove minerals from soils but it was often associated with C losses. An "optimized" demineralization technique using dilute HCl to dissolve the neoformed fluorides after HF treatment and recovery of the soil OM fraction in the HF solution was proposed to investigate the H isotope composition of the bulk soil OM.[38] The potential effect of demineralization on soil OM chemical structure was therefore assessed through several methods including ^{13}C nuclear magnetic resonance (NMR),[39–44] pyrolysis coupled to gas chromatography and mass spectrometry (GC-MS)[43,45] or Rock-Eval pyrolysis.[45,46] The influence of the HF treatment on the nature of specific constituents such as lignin, the second most abundant macromolecule in plants (as detailed in Section C1.2.2), protein or sugars[42,43] or on the surface area of black carbon, which encompass carbonization residues (as detailed in Section C1.2.1.1)[47] was also investigated as well as on the ^{14}C age.[44]

2. *Aquatic OM*

Streams, rivers, lakes, and other inland waters play major role in the coupling of the organic carbon cycle between continents, atmosphere, and oceans. Particulate organic matter (POM) and dissolved organic matter (DOM) biogeochemical transformations are followed by various techniques (see Section C) that sometimes require a preliminary concentration. POM and DOM are operationally defined from size fractionation with a cutoff at 0.7 μm, POM being retained upon filtration contrary to DOM. It must be noted that a cutoff at 0.45 μm is favored for DOM from soil groundwater. Concentration can be achieved through rotary evaporation and freeze drying.

However, when samples with high ionic strength are considered, such as estuarine ones, salts and OM are simultaneously concentrated. Due to the large amount of salts (up to 30 g.L^{-1} to be compared with a few mg.L^{-1} of DOC), extraction procedures dedicated to OM concentration had to be developed as reviewed in 2015 by Sandron *et al.*[48]

The reverse osmosis (RO) was carried out since the 1990s to concentrate natural OM[49,50] but it also concentrates salts, requiring further elimination of salts through electrodialysis (ED) when dealing with high ionic strength samples. Portable RO devices are increasingly used for isolating freshwater DOM thanks to their high recovery yield (> 80%).[51] The coupling of RO/ED was proven to be efficient even for saline waters, first in improving the recovery yield of OM and second in reducing the relative abundance of salts versus organic matter by two orders of magnitude.[52–55] The OM obtained after RO/ED could thus be analyzed using various spectroscopic methods.[53,56,57] An intercomparison published in 2014 suggested coupling solid phase extraction (SPE) using a functionalized styrene divinylbenzene PPL resin with RO/ED to quantitatively isolate oceanic DOC.[58]

It must be noted that most of these studies on DOM isolation are focused on DOC and relatively little is known about the effect of these preparation steps on dissolved organic nitrogen (DON). However, due to the cumulative errors in the quantification of DON as the result of the difference of total nitrogen (TN) content and of the various dissolved inorganic forms of N (DIN), DIN removal is a prerequisite for DON quantification. The coupling between ED and size exclusion chromatography (SEC) was shown to end with a DIN/TN ratio below 0.6 and to induce limited DOC losses (6%).[59] In their review of methods for DON analysis, Chen *et al.* (2015) point that the potential of ED/RO coupling has not yet been evaluated and that it should be further investigated.[60]

B. Fractionation of OM

Sample fractionation is performed to separate OM pools with distinct properties. When dealing with soils, there were numerous attempts to separate fractions with homogeneous chemical composition or turnover, with the ultimate goal of feeding models. To this end, physical and chemical fractionations were developed as reviewed in von Lützow *et al.* (2007).[61] The main two types of physical fractionation are based on density or particle size differences.[62] Chemical fractionations aims at isolating fractions from soils or sediments that survive to various treatment such as extraction, hydrolysis or oxidation. Aquatic OM physical fractionation commonly provides fractions of different molecular size whereas chemical fractionation mostly involves solid phase extractions. Fractionation techniques are summarized in Table 1.

1. *Density Fractionation*

Density fractionation is based on differences in the nature of the minerals, and hence their density, associated with the OM. The initial concept underlying this fractionation is that the less degraded OM occurs as free OM (light fraction in sodium polytungstate solution) whereas the most degraded one is associated with minerals (heavy fraction). Moreover, the extent of aggregation is also a key factor in density differences. As a result, a first step of aggregate disruption is recommended prior to fractionation. This dispersion step, commonly achieved through sonication, leads to an additional light fraction, considered as the occluded or intra-aggregate light fraction.[63] Taken together, the main parameters in such fractionation are therefore the density cut-off and the sonication intensity.[64,65] Although variable cut-off was used, 1.6 g.cm^{-3} was recommended to allow most organic material to float and to prevent the suspension of organo-mineral particles.[64,66] Most attention was paid to the light fractions as they are enriched in C but the heavy fractions could be further separated using successive fractionation with sodium polytungstate solutions of density of up to 2.6 g.cm^{-3}, thus allowing further discrimination between the mineral phases and associated OM.[67–70] In their literature review, Wagai *et al.* (2009) showed that the occluded light fraction exhibits a strongly variable chemical composition (as reflected by the atomic C/N ratio and ^{13}C NMR) and ^{14}C-derived mean residence time, hence the necessity to improve the models underlying density fractionation.[71]

The optimum sonication energy, i.e. that which allows aggregate disruption without dispersion of organo-mineral associations, must be determined

Table 1. Main OM fractionation techniques.

OM Compartment	Fractionation			Isolation of pools with different turnover	Isolation of chemically homogeneous pool
Soil	Physical	Size	Sieving	X	
		Density	Inorganic Salts	X	
	Chemical		Oxidation		
			Extraction	X	X
			Hydrolysis		
Sediment	Chemical		Oxidation		
			Extraction		X
			Hydrolysis		
Dissolved	Physical Chemical	Size	Steric exclusion		
			Chromatography		
			Electrodialysis		X
			Ultranltration		
			SPE		X

for each studied soil, due to the variability in aggregate stability.[64] Such optimization led Sohi *et al.* (2001) to use a combination of 1.8 g.cm^{-3} cutoff and 1500 J.g^{-1} soil sonication intensity.[72] Unfortunately, such preliminary experiments are barely conducted and a wide range of sonication energy is used, making comparison between studies difficult. Moreover, it was highlighted in 2014 that not only ultrasonic energy but also ultrasonic power needs to be controlled, differences in power with the same final energy leading to significant differences in aggregate dispersion.[73] Two aggregate dispersion methods (sonication and shaking) were however shown not to affect the elemental (C/N ratio) and isotopic composition (^{13}C, ^{14}C) of the density separated fractions.[70,74]

Significant differences, according to ^{13}C NMR and Fourier transform infrared (FTIR), observed between free and intra-aggregate fractions, made Sohi *et al.* (2001) confident for further use of their separation protocol in defining model pools.[72] Similarly, contrasted composition between these fractions was revealed through a combination of techniques including ^{13}C NMR, FTIR, Py-GC/MS and Scanning transmission X-ray microscopy (STXM)/ near edge X-ray absorption fine structure spectroscopy (NEXAFS)[75,76] and an enrichment in pyrogenic matter was observed in the occluded fraction thanks to ^{13}C NMR and Py-GC/MS.[77] OM from free and intra-aggregates (see Section III.B.2) were shown to mainly derive from plant residues whereas mineral-associated OM was predominantly of microbial origin.[78] Boeni *et al.* (2014) interpreted the high abundance of labile O-alkyl C in the heavy fraction by the strong OM-mineral interactions which reduce microbial attack.[79] The decrease in C/N with density suggested a preferential association of organic N with minerals,[80,81] most of which being of peptidic nature according to X-ray photoelectron spectroscopy (XPS) and NEXAFS spectroscopy,[82] thus providing evidence for the role of minerals in the transformation and stabilization of soil organic N.

Although designed for soil, density fractionation was also applied to estuarine sediments revealing the importance of organo–clay associations in a shelf sediment[83] and of the selective degradation and flocculation processes during cross-shelf transport.[84] The light fractions were generally dominated by terrestrial OM, whereas the marine OM was predominant in the heavy fractions.[85–87]

2. *Size Fractionation*

In soil, the main other type of physical fractionation is based on particle-size differences. It is generally performed after ultrasonic dispersion in water although the smallest aggregates ($< 2\ \mu m$) were shown to be able to survive this ultrasonic treatment.[88] It combines sieving (wet or dry) to separate the sand fraction ($< 2000\ \mu m$), with sedimentation and decantation to recover the clay and silt-size fractions (usual cutoff at 50 or 63 μm).[61] An additional flocculation step with $CaCl_2$ is sometimes included. Particle-size fractionation is based on the difference of sorption ability between sand (coarse fraction), on the one hand and silt and clay (fine fractions), on the other hand. Soil fine fractions are commonly considered to be enriched in OM, and the results from physical fractionations led to three levels of structural and functional complexity in the turnover of organic matter in soil.[89] Comparison between the two physical fractionation procedures showed that particle-size fractions differed less in chemical composition than density-separated ones.[72] Finally, both types of fractionation are commonly associated with (i) wet sieving or decantation being further performed on the heavy fraction[90–93] or conversely, (ii) density fractionation being carried out on previously size-separated fractions.[94]

Another size fractionation cutoff distinguishes macroaggregates ($> 250\ \mu m$) from microaggregates ($50–250\ \mu m$), leading to the aggregate hierarchy theory, first proposed by Tisdall and Oades (1982)[95] and now widely accepted.[96,97] Microaggregates, which are supposed to retain the stabilized OM as supported by strong differences in mean residence time,[98] form within macroaggregates[99,100] (Six *et al.*, 2000, 2004) as detailed in Section III.B.2. Moreover, the microaggregates occurring within macroaggregates were proposed in 2014 as a key parameter to reflect management-induced changes and to be introduced in models.[101]

Ultrafiltration (UF) leads to size fractionation with a higher recovery yield than solid phase extraction (SPE)[102–104] but as with SPE, extracted OM cannot be considered as representative of the bulk OM.[105] The coupling between the two techniques, i.e. SPE extraction on previously separated UF fractions was also used to investigate the nature of the hydrophobic DOM.[106]

Although the study of natural OM from aquatic system frequently implies to isolate different fractions, its characteristics (size, molecular weight, and composition, etc.) can also be directly followed by coupling chromatographic techniques with on line detection working at low DOM concentrations (< 2 mg/L). For instance, the steric exclusion chromatography (SEC) has the advantage of being non-destructive and relatively fast. It requires a small amount of sample and does not need any pre-treatment.[107] It is a technique of chromatography where the separation of the molecules is based on the assumption that the phenomenon of exclusion is the only one responsible for the separation of chemical molecules. The separation is based on the differences in permeability and size within a mixture of molecules in a porous matrix. So, larger molecules will less interact with the matrix than smaller ones and therefore they will be less retained in the column whereas the smaller molecules will be eluted later. The new generation of SEC techniques, such as high-pressure steric exclusion chromatography, have been commonly used for, the study of the distribution of the molecular size of the DOM in aquatic and terrestrial ecosystems,[108–113] water swamp or wetland,[114,115] lakes,[116–118] estuaries.[110,119]

Another approach is to separate the organic molecules by applying field-flow-fractionation (FFF). The FFF family of chromatographic-like elution techniques was developed for the first time by Giddings at the end of the 1970s as a method of separation of macromolecules and colloids.[120] The separation is based on a physical principle that takes place in a small channel of separation (with a thickness ranging from 300 to 500 nm) under the effect of a field force perpendicular to the stream of molecules moving in the channel. The field force can have different nature (electrical, thermal, magnetic, flow force of mobile phase) depending on the desired information (size, density, molar mass, etc.)[120] and the studied object. It has the advantage of being a versatile, robust technique able to split a wide range of molecules ranging from 0.1 nm to 100 μm.[120] Typically, this technique can be coupled with UV/Vis spectrophotometer, spectrofluorimeter, diffusion of light (multi-angle light scattering) photometer, a refractometer or an inductively coupled plasma-mass spectrometer. Many parameters can influence the elution of the colloids, such as the composition of the mobile separation phase (pH, nature of ions and ionic strength), the flow (carrier flow, focusing time), the size of the channel, the cut-off threshold of the membrane and the making of the membrane. In all cases, the separation mechanism is born from differences between particles under the forces of the field, in equilibrium with the forces of diffusion; a parabolic laminar-flow-velocity profile in the channel determines the velocity of a particular particle, based on its equilibrium position from the wall of the channel. Within the FFF family, AF4 (asymmetrical flow field flow fractionation) is remarkable because the field flow is applied through an asymmetrical channel and there is only one permeable wall. As a result, only the carrier liquid constantly exiting through the semi-permeable wall on the bottom of the channel causes the cross-flow. Splitting by AF4 is based on the particle diffusion coefficient. Separation by asymmetrical flow allows a faster split of the molecules and reduces the dilution of the samples, often problematic for environmental studies at low concentration. This AF4 technique has turned out to be a good choice for the study of the DOM in natural aquatic environments and has been used in the case of rivers,[108,121–126] lakes[127–130] and estuarine waters.[123,131,132] These studies report changes in molecular weight of the DOM as well as modifications of the polydispersity of the DOM due to changes in river conditions (pH, salinity), inputs of new sources or *in situ* bio-geochemical processes modifying DOM properties.

3. *Chemical Fractionation*

Chemical fractionation is often performed on NOM to separate homogenous fractions either in terms of turnover/stability or of biochemical composition (Table 1).

a. *Isolation of turnover fractions*

Chemical treatments, mainly involving hydrolyses or oxidation with various reagents, were developed to isolate stable OM, stability which is often assessed from ^{14}C age. A more stable fraction than the bulk soil was thus successfully isolated after $Na_2S_2O_8$,[133] H_2O_2,[134] NaClO treatment,[135,136] HCl 6M hydrolysis.[137,138] Moreover, Zimmermann *et al.* (2007) found the NaClO oxidation resistant residue to be more stable than that having survived HCl 6M hydrolysis, which might be explained by the higher recalcitrance of some fresh biopolymers (such as lignin) towards hydrolysis than oxidation.[139] However, as previously mentioned for HF treatment, the chemical fractionations which aim at isolating a given OM may induce some transformations in the chemical structure of the left-over OM, making more difficult its interpretation.[140]

A combination of physical and chemical methods was even set up to try isolating the most stable OM.[141,142] The thus obtained OM pool was shown to be of similar size as that directly recovered through H_2O_2 oxidation and the inconsistency of the size of this pool made this method questionable.[141] Moreover, the comparison between the non-hydrolysable (HCl 6 M) OM recovered from the previously isolated heavy fraction and OM fractions separated through sequential density fractionation pointed to a role of the chemical recalcitrance rather than the nature of associated minerals in the turnover rate of the slowest cycling carbon.[74]

b. *Biochemical fractionation*

Models of OM dynamics are based on OM pools with different mean residence time (MRT). This MRT was sometimes considered as dependent on the biochemical composition of OM from plant, compost, sewage sludge, or other organic amendment.[143–146]

Thus, biochemical fractionations based on solubility in acid or neutral detergents, often referred to as Van Soest fractionation[147,148] were used to isolate OM fractions with different biodegradability. This was first implemented in fodder,[147] then in compost[149] and organic amendment,[150] sometimes before their incorporation in soil, to thereafter, follow their influence on soil carbon content. This fractionation method provides four fractions that are sequentially obtained: a soluble fraction, obtained after extraction with a neutral detergent (sodium dodecyl sulfate), a "hemicellulosic" fraction released upon hydrolysis of the insoluble residue with an acid detergent (hexadecyl bromure trimethyl ammonium), a "cellulosic" fraction released after H_2SO_4 (72%) treatment on the previously obtained residue and a final insoluble fraction supposed to contain lignin and cutin. However, chemical analyses of these different fractions indicated that they do not correspond to biochemically homogeneous fractions.[143,151,152] For ^{18}O isotope measurement, holocellulose was isolated from tree rings through extractions and $NaClO_2$ treatment aiming at removing lignin.[153]

The isolation of compounds such as lipids or lignin or carbohydrates could also be performed by solvent extraction or acid hydrolyses. Total lipid extract obtained by extraction in organic solvent is used to characterize lipids from sediment as well as soils to investigate OM source or degradation.[154–161]

Lignin was often isolated as the remaining residue after acid hydrolysis, the two main methods being the aforementioned Van Soest fractionation, also termed acid detergent lignin, and the so-called Klason lignin.[162] Both methods use H_2SO_4, the main difference being the order of the two steps, high acid concentration at low temperature and low concentration at high temperature. A so-called milled wall enzymatic lignin (MWEL) fraction was isolated from soil[163] using a method combining ball-milling, cellulolysis and solvent extraction previously developed for plants.[164] Interestingly, when dealing with soil, MWEL was shown to contain less associated polysaccharides, likely due to alteration upon soil formation.

c. *Humic substance fractionation*

The study of OM in the environment was confronted to the intimate mixing of OM with minerals and sometimes to the low concentration of OM. A procedure was developed to extract OM based on its solubility in base and acid solutions, leading to the definition of the so-called humic substances (HS), which encompass the fulvic acids (FA) (soluble in both base and acid), humic acids (HA) (soluble in base solutions, insoluble in acids) and humin (insoluble).[165] The idea behind this operational definition was that the gradient in OM solubility would parallel that of stability. However, this concept of humic substances has been largely debated and it is now more and more accepted that these base and acid treatment can alter molecular structures.[166,167] The characterization of soil OM based on humic substances may therefore have induced some misinterpretations. In their paper, Kleber and Johnson (2010) showed the evolution of the concept of molecular structure of soil OM.[167] They discussed the different points of the concept and finally refuted the traditional soil OM structure based on humic substances and thus their recalcitrance. The chemical structure of soil OM itself is not consensual as underlined by Lehmann and Kleber (2015).[168] The macromolecular structure of SOM has been discussed and a model based on the aggregation of low molecular weight compounds has been proposed instead of the macromolecular structure.[169] Indeed, an array of analytical techniques supports the low molecular weight of SOM. The analysis by gel permeation chromatography and high-pressure size-exclusion chromatography performed by Piccolo *et al.* (1999) evidenced self-assembled aggregates[169] and this structure was supported by Simpson (2002) using diffusion ordered spectroscopy.[170] Electrospray ionization revealed a molecular mass distribution of humic cations ranging from 1000 to 2000 Da whereas laser desorption ionization provided mass/charge distribution with an average close to 500 Da. However, the fractionated humic substances still constitute key materials to investigate and model physico-chemical properties of NOM.

d. *Dissolved OM fractionation*

The study of natural OM from aquatic system frequently implies to isolate different DOM fractions that are more homogeneous, thus allowing improvement of their characterization. SPE has been largely used and is still the most common way to fractionate water samples through various types of cartridges. Its principle is that a fraction of the bulk material is firstly retained on the solid phase of the cartridge and it is released through a subsequent elution using an appropriate solvent.[171] The composition of the SPE-extracted OM will inherently depend on its affinity with the solid phase and on the nature of the eluting mixtures.[172,173] This was also demonstrated in 2016 in a high resolution mass spectrometry study using Fourier transform ion cyclotron resonance mass spectrometry (FTICR-MS), the authors recommending to only compare fractions extracted with the same solid phase.[174] XAD resins have been mainly used for long, often through the XAD-8/XAD-4 tandem,[175] before being progressively replaced by silica columns bonded to a C_{18} hydrocarbon sorbent[176,177] and by functionalized styrene divinylbenzene phases[178,179] as reviewed by Liška (2000).[180] This evolution is associated with an increase in DOM recovery[181,182] but the question of how representative the extracted material is, remains open.[183] Other types of designs (disks instead of cartridges) were also tested to improve the separation performance.[179,184] The efficiency and the duration of this extraction step were also improved thanks to the use of a combination of stacked cartridges.[185,186] In addition to being used for OM concentration, SPE was used to develop a rapid method of characterization of natural OM, the so-called polarity rapid assessment method based on UV absorbance of SPE-eluted fractions using different types of solid phase.[187,188]

III. Fate of NOM

It is commonly considered that the fate of NOM is influenced by its chemical composition, at an early stage, due to differential biodegradability of compounds, and in the longer term, by its role on organo-mineral interactions. The predominance of one mechanism involved in stabilization, i.e. its persistence in the environment, over the others is influenced by environmental conditions.[189] The main obstacle to assess the contribution of the different mechanisms is the ability to clearly distinguish one mechanism from the others.

A. Differential Degradation

Organic C constitutes a highly reactive pool of the biogeochemical C cycle. This cycle is controlled by a balance between the inputs of new OM (mainly primary production) and the outputs, either by exportation or by microorganism mineralization. As stated in the introduction, the main source organisms of natural OM are phytoplankton, vascular plants and microorganisms such as bacteria and fungi. Depending on the considered environment, the relative abundance of these different types of organisms varies. However, from the living biomass to soil or sediment OM, several transformations take place. Among these transformations, the most prominent is the mineralization of the most labile constituents of the biomass, which are converted into CO_2. However, it is commonly considered that 0.1 to 1 wt% of the initial biomass escapes this mineralization and survives as sedimentary and soil OM.[13,190]

The identification of the mechanisms involved in this stabilization of OM is still challenging[191] especially in soils. Thus, some gaps have to be filled by integrating the different mechanisms implied in the balance between carbon input and mineralization of OM. The most frequently identified controlling factors of OM degradation are: (i) environmental factors, especially pH, temperature and availability in oxygen and water and their subsequent interactions, (ii) accessibility of microorganisms to nutrients including OM. However, in similar environmental conditions, the mineralization rate by the microorganisms could also be influenced by OM chemical composition as their metabolic capacity could be more or less efficient in OM degradation. Indeed, the chemical composition and structure of soil OM must not be the only factors influencing its turnover rate. As a result, the intrinsic recalcitrance (resistance to biodegradation due to OM chemical composition) of OM or of some OM compounds does not seem relevant, as inert OM has not been identified.[192,193] Thus, different processes involved in OM stabilization should take place simultaneously in the environment with different relative importance depending on considered time and space.

1. *Characterization of Bulk Chemical Composition to Assess NOM Stability*

The chemical composition and structure of soil OM were proposed to play a role only in the short to medium term.[25,194] Differential degradation or selective preservation leading to persistence of OM due to its chemical composition, would result in OM pools with contrasted MRT and thus of stability (i.e. persistence). Selective preservation of OM was demonstrated in sediments in the 80's[195] through morphological and chemical comparison between extant microorganisms and their fossil counterparts, but it was only considered later in soils.[196] In addition, the characterization by fluorescence spectroscopy and FTICR-MS of DOM from lake has revealed that the main factor controlling DOM was its elemental composition, whereas in soil, organo-mineral interactions could dominate.[197] The hypothesis of a relative preservation of some OM components has arisen from the monitoring of litter degradation, or analyses of OM from different soil depths or different soil fractions (chemical, density or size fractionations) as detailed in Section B. Indeed, the evolution of OM from different soil depths was frequently used as representative of soil OM of different extent of degradation.[151,198,199] The potential role of chemical composition on

OM persistence in soils was put forward through the characterization of OM having persisted in soil for decades.[200]

The most abundant biopolymers of the biomass, polysaccharides and proteins are known to be extensively degraded during the first steps of the biomass transformation when not physically protected. The thus released monomers may either be further altered and mineralized or, for a small part, they can escape this mineralization through the Maillard reaction to form the so-called melanoidins (for more details, see melanoidin section below). This mechanism, termed degradation-recondensation, was for long considered as the only one to form kerogens.[13] Despite the later recognition of additional fossilization pathways, the degradation-recondensation is still shown to be responsible for the formation of sedimentary OM in various environments.[201,202] A similar pathway, although involving phenolic monomers derived from lignin rather than amino acids (lignin being more abundant than proteins in vascular plants), was proposed as soil OM formation pathway.[203] This abiotic condensation would have formed high molecular weight highly aromatic compounds of increasing recalcitrance termed FA, HA and humin.[183] However, as mentioned in the previous section, this is presently debated, and Sutton and Sposito (2005) suggested a new molecular structure of soil OM which led them to recommend an examination of biogeochemical pathways of humic substances formation and "to redefine the concept of humification."[204] Although humification was supposed to go along with aromatization, the ^{13}C NMR analysis of soil horizons indicated no increase in aromaticity with increasing depth.[199] In consequence, the increase in aromaticity of HA has been also reconsidered. Moreover, the contribution of aromatic compounds could also reflect the contribution of black carbon (BC, defined as the residue of incomplete combustion, see black carbon section below) to soil.[205] Hempfling *et al.* (1987) followed litter degradation in different layers by IR, solid state ^{13}C NMR and pyrolysis coupled to field ionization mass spectrometry (Py-FIMS).[206] The molecular chemical information obtained by Py-FIMS on OM composition during degradation did not support the occurrence of condensation reactions of the macromolecules constitutive of plants. The humification concept was re-evaluated by Lehmann and Kleber in 2015.[186] They exposed the different models of formation of SOM and finally proposed a soil continuum model combining residue degradation, interaction of compounds with minerals and aggregation. In this model, the authors discarded humic substances considering that they are not present in soils but a biased result of chemical extraction of SOM. In addition, Kelleher *et al.* (2006) compared two-dimensional NMR spectra of biopolymers (protein, lignin, cutin) and spectra of HA.[207] Their results showed neither indication of abundant polyphenolic polymers, often present in HA models, nor increasing cross-linking in HA compared to the pristine biopolymers of the plant. As the stability of humic substances was attributed to their high complexity, with their aromaticity and their high molecular weight, the reconsideration of the humic model has important consequences on the evaluation of the fate of OM.

Several analytical tools have been developed and applied to assess NOM stability from its bulk features. They include thermal techniques, UV absorbance spectrophotometry, several spectroscopic methods such as fluorescence, infrared and NMR spectroscopy, isotope measurements at the bulk or molecular level as well as high resolution mass spectrometry.

a. *Thermal techniques*

As the recalcitrance of OM to biodegradation was postulated to be linked to OM thermal stability, thermal techniques, including the so-called "Rock-Eval" pyrolysis are often used in natural OM analysis. Rock-Eval pyrolysis was initially designed to evaluate the oil potential of sedimentary rocks, through a simple and fast technique without any sample preparation.[208,209] The maximum temperature of effluent release (S2 curve) is used to assess the refractory nature of the studied sample (Figure 1). Since these pioneer studies, these parameters were used to characterize the OM in recent sediments, peats and soil. However, in such samples, the S2 curve shows a multimodal shape reflecting the co-occurrence of biopolymers with different thermal stability.[210] The deconvolution of the S2 curve led to the definition of new parameters which were shown to evolve along a soil depth profile[210] and the resulting sub-peaks were assigned to some biochemical classes such as little resistant fresh plant material, lignin and cellulose, humic substances and "mature geomacromolecules" encompassing BC and OM stabilized by organo-mineral interactions.[200–212] The possibility of

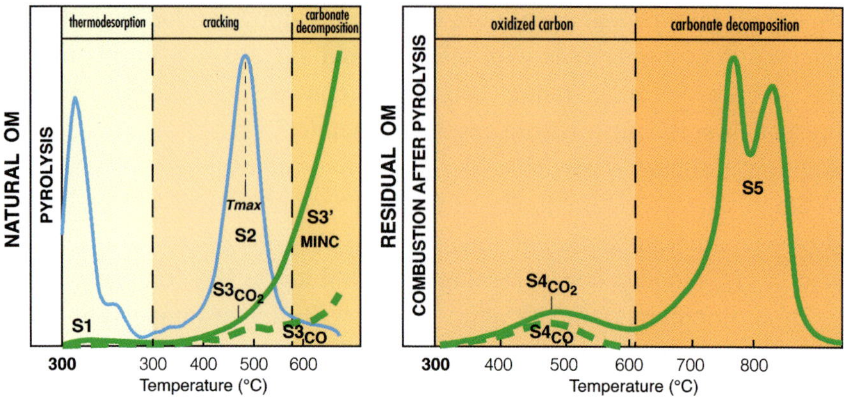

S1 : Labile compounds **S2** : Oil potential **S3** : CO & CO_2 **S4** : Residual C_{org} **S3'+S5** : Inorganic carbon

Figure 1. Typical curves recorded upon Rock-Eval pyrolysis showing different thermal stability pools.

determining stable pool of OM by "Rock-Eval pyrolysis" to survey soil at a large space scale was argued.[213,214] Barré *et al.* (2016) performed analytical thermal analyses and characterization of OM from five soils supposed to be enriched in old OM as they do not receive any fresh OM input.[200] The Rock-Eval analyses indicated an increased in thermal stability with persistence of OM and an energetic signature for persistent OM was proposed despite variable chemical composition of the OM. These features are notably related to the relative contributions of labile and resistant biological constituents and to the OM degradation stage but they cannot be used to reflect the molecular structure of the OM.[215]

The main thermal analysis techniques applied in soil science are thermogravimetry (TG), differential thermal analysis (DTA), differential scanning calorimetry (DSC) and pyrolysis which could provide information on soil sample properties as described by Plante *et al.* (2009).[216] They allow fast analysis, require little sample preparation and are rather cheap, which constitute important advantages in soil screening. They were also applied to monitor BC.[217]

TG was used to evaluate OM stability. It measures a weight loss in a furnace with increasing temperature after calibration with standard reference materials. This technique indicates modification of thermal stability as a function of the chemical composition or organo-mineral interaction. Three main weight losses determined three different pools of soil OM from labile to refractory.[218] TG is often coupled to DSC to characterize OM. DSC measures differences in heat flow between a sample and a reference as a function of temperature during a controlled temperature program. DSC was also applied on the heavy fraction, which OM is supposed to be more stable, from different soils in combination with NEXAFS and compared with previous ^{14}C measurements. The oldest OM, surprisingly, appeared as the most thermally labile and it exhibited an important contribution of O-alkyl C (attributed to microbial compounds) compared to aromatic C.[219] The authors concluded that not only OM composition controls its turnover. TG was also coupled to mass spectrometry to investigate changes in OM such as those occurring in soil OM after fire,[220] and even to isotope ratio mass spectrometry to monitor wheat straw degradation by fungi.[221] Lopez-Capel *et al.* (2008) coupled thermal analysis to mass spectrometry to identify the cause of the weight loss during TG analysis from some clay in an Australian soil after C4 to C3 vegetation change.[222] They also investigated OM stability with this thermal analysis coupled to quadrupole mass spectrometry-isotope ratio mass spectrometry and revealed important input of new OM on the ten first cm of the soil.

b. *Ultraviolet absorbance spectrophotometry*

Ultraviolet (UV) absorbance spectrophotometry is often used to overcome sample limitations (volume, concentration) by providing estimations of properties (aromaticity, molecular weight, hydrophobic C content, degree of humification, biodegradability) of DOM during its transformation (biodegradation or interactions with mineral phases). UV absorbance spectrophotometry is fast and requires little preparation.

The biodegradability is supposed to be controlled by intrinsic properties (i.e. lignin-derived phenols, etc.), site-specific parameters and surrounding conditions.[223] A trend, between the DOM mineralization at 90 days and the absorbance at 280 nm, was reported by Kalbitz *et al.* (2003).[224] Similarly, Strauss and Lamberti (2002) showed that nitrification rates were correlated with the specific absorbance of OM at 254 nm.[225]

The aromatic carbon content drives the UV properties of the DOM since the π–π* transition absorbs the light corresponding to the UV spectral range. These aromatic structures pre-exist or originate from various OM degradation pathways.[226] The UV absorbance spectrum does not show any significant features and is monotonous. However for specific wavelengths, various empirical relationships are found between UV absorbance and several DOM properties. The first DOM property that can also be used to follow DOM degradation or transformation is concentration (mg OM/L). Several attempts have been made to apply UV spectrophotometry to predict total DOM in aquatic systems or soil solutions. For instance, relationship between the absorbance at 250 nm and 350 nm[227] or at 254 nm[228] or at 280 nm[229] and the DOC content were proposed. The fate of organic matter can also be affected by its hydrophobicity that may be controlled by UV-chromophore aromatic compounds, considering that hydrophilic organic matter does not absorb UV light. This was supported by good relationship between the absorbance at 260 nm or 280 nm and the hydrophobicity of OM.[224,230] The spectral slope (S, nm^{-1}) has been derived from DOM absorption spectra by fitting the absorption data to the equation:

$$a_\lambda = a_{\lambda_{ref}} e^{-S(\lambda - \lambda_{ref})}$$

where a_λ is a Napierian absorption coefficient (m–1), λ is the wavelength (nm), and λ_{ref} is the reference wavelength (nm).[231]

The spectral slope is considered to be a powerful indicator for the characterization of the DOM, although the derived values are highly dependent on the spectral area chosen for the calculation.[232,233] The spectral slope calculated between wavelengths 275 and 295 nm (S275–295) was used to characterize the DOM in terms of molecular size, aromaticity and as a tracer of the terrigenous DOM inputs.[233,234] The ratio of the spectral slopes (SR) is calculated as the ratio of the spectral slope for the short wavelengths (S275–295) on the spectral slope of longer wavelengths (S350–400). The SR is negatively correlated with the size of the DOM[233] and gives information about its origin. Thus, SR is greater than 1 for the waters under marine influence and less than 1 for the DOM of terrestrial origin.[233]

c. *Fluorescence spectroscopy*

Like UV absorbance, fluorescence spectroscopy is used to explore the properties or characteristics of DOM and their link to DOM function in the ecosystem or evolution and degradation.[235–238] It is a quantitative and sensitive technique that requires small quantities of sample, with little sample preparation and offers a spectral fingerprint for any type of water body. The fluorescence signal can be recorded as excitation–emission matrix, generated by scanning both excitation and emission wavelengths.[235] Thus, analysing the excitation/emission wavelength pairs can identify the fluorescence maxima. Coble (1996) and Parlanti *et al.* (2000) established major fluorescence band nomenclature for water OM (Figure 2).[235,239] It was however pointed out by Murphy *et al.* (2010) that the standardisation of data treatment procedure is critical in order to be able to compare results from different laboratories and sites.[240] The fluorophores detected in aquatic systems are microbial- and terrestrial-derived compounds. Amino acid-like components, microbial- or algal-derived, represent the labile fraction of OM showing the bacterial or algal activity in a water system.[238] Uncontaminated waters exhibit low amino acid-like fluorescence. The terrestrial-derived fluorescence is given by humic-like substances produced through biological and chemical degradation of plant material present in the ecosystem. Another more simple way to use fluorescence spectroscopy is to determine the humification index (HIX) by dividing the area of fluorescence intensity between 435 and 480 nm by that between 300 and 345 nm.[241,242] An increase in the HIX is associated with an increase in emission wavelength due to increased aromatization (lower H/C ratios). A new fluorescence index (BIX) associated to OM characteristic of autochthonous biological activity in water samples was proposed by Huguet *et al.* (2009).[243] Its calculation is based on the broadening of the emission fluorescence spectrum due to the presence of the M fluorophore (Figure 2). The origin and degree of transformation of DOM can be assessed through the calculation of these indices. Huguet *et al.* (2010), studying DOM distribution and characteristics in the Gironde and Seine estuaries, demonstrated that the most humified OM was associated with different size fractions, depending on water sample origin and that DOM distribution and characteristics differ between the two estuaries.[244]

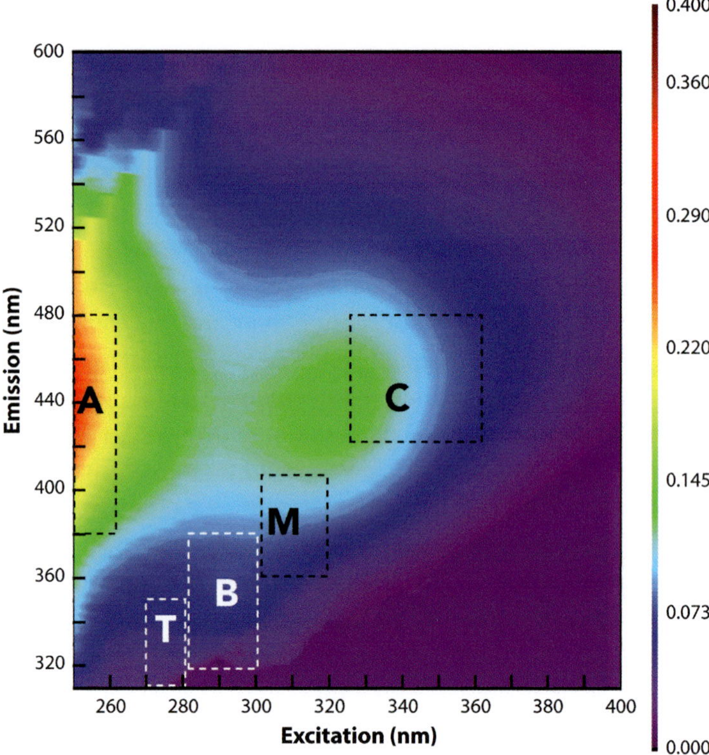

Figure 2. Contour excitation–emission matrix (EEM) plot of EEM fluorescence spectra of a river sample from Peru (South America). Fluorescence intensities are expressed in Raman units (nm^{-1}). The letters A, C, M, T and B correspond to the Coble (1996) fluorophore nomenclature.[235] The dashed line boxes in the graph correspond to the domain of the fluorophores as defined by Parlanti *et al.* (2000).[239] Boxes A and C corresponding to humic-like substances of allochthonous origin, Box M to organic molecules from bacterial or algal origin, Box T to proteins and markers of bacterial activities of autochthonous origin, Box B to proteins and phenolic compounds of autochthonous origin.

In the early 2000s, the application of parallel factor analysis (PARAFAC)[245] to fluorescence excitation-emission matrices allows the connection of fluorescence properties with DOM functions such as a source of nutrients and energy,[246–248] a metal-binding agent.[249,250] Moreover the technique can provide a fingerprint for DOM origins and end members[251–254] or a tracer for the production and degradation of autochthonous DOM.[251] For Cuss and Gueguen (2015), the correlations between optical properties and molecular weight were revealed by PARAFAC, with higher protein/polyphenol-like fluorescence (peaks B and T) and humic-like fluorescence (Peaks A and C) being observed for the lower (<0.5 kDa) and larger (> 1 kDa) molecular weights, respectively.[255] They concluded that the structure of aged DOM arises through supramolecular assembly. Helton *et al.* (2015) showed with PARAFAC analysis that DOC lability increases with water residence time in the alluvial aquifer of a river floodplain ecosystem.[256] Total OM, fluorescent humic-like components, and specific UV absorbance decreased with water residence time suggesting sorption to sediments. Conversely, labile DOC (both concentration and fraction) increased with water residence time, pointing to a competing influx or production of labile DOC.

Although mostly developed for DOM, fluorescence techniques have been then applied to soil and humin characterization. The laser-induced fluorescence spectroscopy (LIFS) is based on the fluorescence emission of aromatic components present in the soil and can be implemented without any sample pre-treatment. Using LIFS, it is possible to determine the humification index of a sample, which is related to the concentration of rigid structures, like quinone groups and aromatic rings.[257] The humification index is calculated as the ratio between the area of the fluorescence emission spectrum (excitation at 465 nm, signal integrated from 450 to 700 nm) and the total organic carbon amount (g kg^{-1}) in the soil sample.[257] This index increased along a depth profile and showed that the humin fraction could be a larger fraction of SOM than previously foreseen (i.e. 80–93%) in some Amazon soils.[258]

d. *Infrared spectroscopy*

Modification of litter composition occurring with degradation was evidenced by Fourier transform infrared (FTIR) and diffuse reflectance FTIR (DRIFT) spectroscopy. These techniques enable a rapid and easy distinction between litter horizons and organo-mineral horizon as well as an estimation of the degradation extent of organic horizon.[259] FTIR spectroscopy represents a rapid, cheap and accessible analytical method. Mid infrared spectroscopy (MIRS) is often used to provide information on the nature of the chemical functions in humic substances[260] or OM from soil fractions,[261,262] as detailed in Ellerbrock and Gerke (2013).[263] DRIFT appeared as an attractive method through direct analysis of samples without preparation such as inclusion in KBr pellets. However, samples should be ground and sometimes mixed with KBr. Partial least squares (PLS) associated to DRIFT calibrated with NMR data were shown to provide satisfactory results on quantifying some OM functions,[264] leading to the conclusion that MIRS could predict NMR characteristic functions (aromatic C, carbonyl-C, O-alkyl-C and alkyl-C).[265] Characterization of POM from river using attenuated total reflectance (ATR)-FTIR and PLS treatment evidenced differential degradation between OM functions; aromatic and carboxylic functions would degrade more slowly than amino acids and carbohydrates.[266] In addition to provide information on OM chemical structure, near-infrared and mid-infrared spectroscopies have been recognized as valuable tools to predict C content in soil and litter after calibration.[265,267,268]

e. *Nuclear magnetic resonance*

^{13}C solid-state NMR is commonly used to characterize NOM,[269–274] by providing the relative contribution of the different functional groups. Due the insoluble character of most of the sedimentary and soil OM, solid state NMR has to be used (Figure 3).

In sediments, solid state ^{13}C NMR was mainly used to identify OM sources, as reviewed by Cao *et al.* (2013).[274] In aquatic environment, solid state ^{13}C CP/MAS (cross polarization magic angle spinning) NMR allowed monitoring OM evolution. To this aim, dissolved OM and separated fractions were characterized.[275,276] A correlation was established between the aromaticity determined from NMR and from specific UV absorbance at 254 nm.[277] Solid state ^{13}C CP/MAS NMR was used to provide constraints on data obtained from mass spectrometry on dissolved OM from Lake Drummond.[278] It also revealed a substantial contribution of terpenoids in dissolved OM from several continental environments.[279,280] In 2016, high resolution ^{1}H NMR and two-dimensional NMR were used to provide evidence for wetland-specific molecular assemblies.[281] Such approaches could be used to assess the role of aquatic OM composition on its stability.

In soil, the use of ^{13}C CP/MAS NMR to follow the chemical structure of plant residues and OM with time, either in litter bags in soil or through

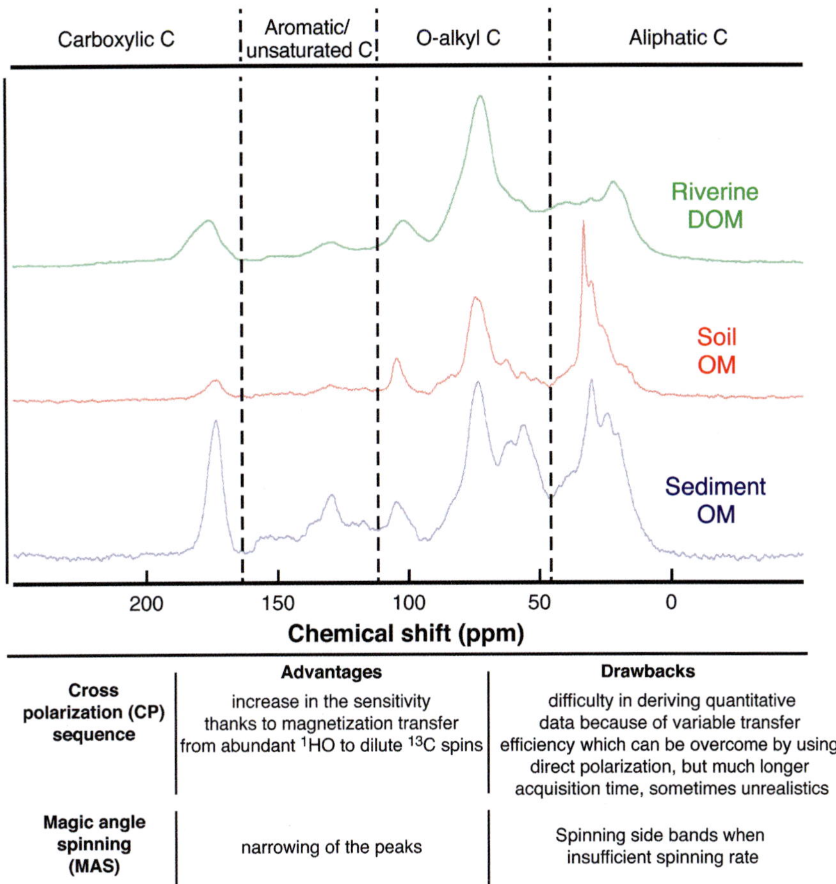

	Advantages	Drawbacks
Cross polarization (CP) sequence	increase in the sensitivity thanks to magnetization transfer from abundant ^1HO to dilute ^{13}C spins	difficulty in deriving quantitative data because of variable transfer efficiency which can be overcome by using direct polarization, but much longer acquisition time, sometimes unrealistics
Magic angle spinning (MAS)	narrowing of the peaks	Spinning side bands when insufficient spinning rate

Figure 3. Solid state ^{13}C CP/MAS NMR spectra of riverine DOM, Soil OM and sediment OM illustrating the differences in their functional groups.

the comparison of OM from different soil horizons, provides information on the evolution of NOM during biodegradation.[206,271,282,283] The origin of the alkyl C signal is commonly attributed to aliphatic compounds (lipids, cutin and suberin monomers, etc.) whereas O-alkyl signal is assigned to polysaccharides. During biodegradation, a general trend of decrease in O-alkyl C and increase in alkyl C, carboxylic C and signals specifically assigned to lignin is observed in parallel to the C/N decrease.[271] These changes have led to hypothesize a slower degradation of the aliphatic components and lignin. In this context, high resolution magic-angle spinning NMR, a recently developed technique to analyze kerogen or whole soil after swelling,[284,285] was also used and it confirmed the aforementioned observations.[207] Note however, that this technique can only be applied to semi-solid samples. ^{13}C and ^{15}N NMR analyses were also associated with isotopic analyses of the soil and a correlation between δ^{13}C[269,271] and δ^{15}N,[286] and the extent of biodegradation of plant residues and soil horizons could be evidenced.

f. *Isotopic ratio mass spectrometry analyses (irMS)*

The measurement of the isotope composition *via* irMS is largely applied to OM studies for various purposes such as determination of OM origin, paleoenvironmental reconstruction or evaluation of OM turnover. Natural or artificial ^{13}C labelling was notably used to trace OM and to estimate its MRT. The use of OM δ^{13}C measurements in bulk soil, as well as in soil fractions, was extensively reviewed by Glaser.[287] The MRT of OM from soils was generally estimated by means of isotopic measurements, especially δ^{13}C and ^{14}C[288,289] and reviewed in Amelung *et al.*[290] Indeed, natural ^{13}C abundance in plants differs according to their metabolic pathways. Plants present δ^{13}C values around –27‰ and –12‰ for C3 and C4 metabolisms, respectively.[291–293] In the case of a vegetation change implying C3 and C4 plants, the δ^{13}C value of soil C progressively shifts toward the new vegetation δ^{13}C value. The progressive incorporation of new labeled carbon can then be calculated and the MRT deduced.[294] In consequence, the differences in δ^{13}C values were extensively used to monitor OM in bulk soil[295,296] or in soil fractions.[288,297,298] Differences in turnover of C were evidenced in soil particle-size fractions with values of less than 1 year and 63 years in the > 2000 μm and 0–50 μm fractions, respectively.[299] An increased age (^{14}C measurement) and an increase in ^{13}C and ^{15}N content were also reported in fractions with increasing density.[18] Such an increase is commonly observed during biodegradation as well as with particle-size decrease and with soil depth pointing to higher MRT in fine fractions or in deeper horizons.[300,301] Several hypotheses have been put forward to explain the 1–3‰ relative enrichment in ^{13}C with depth and/or degradation. Biological factors could influence the OM δ^{13}C through preferential accumulation of ^{13}C depleted compounds such as lignin,[302] isotopic fractionation by microorganism respiration[301] or accumulation of ^{13}C enriched microorganisms compared to plant residues.[303] In the same way, $\delta^{15}N$ increases with depth because of OM mineralization and plant nutrition. In addition, the decrease in $\delta^{13}C$ of atmospheric CO_2 induced by combustion of fossil fuels depleted in ^{13}C may also account for the ^{13}C enrichment with depth.[303]

Labeling, especially with ^{13}C and ^{15}N, turned out to be a powerful way to monitor OM. Incubation of artificially ^{13}C and/or ^{15}N-enriched organic matter is becoming increasingly favored to unravel biogeochemical processes in ecological studies.[304–311]

g. Electrospray ionization coupled to Fourier transform ion cyclotron resonance mass spectrometry (ESI-FTICR-MS)

ESI-FTICR-MS is increasingly used to characterize natural OM at the molecular level.[113,312,313] FTICR-MS is a technique employed to determine mass to charge ratio (m/z), of ionized compounds of molecular weight ranging from 100 to 1000 Da. Compounds are either positively or negatively charged in the ion source. The ultra-high resolving power is capable of detecting mass to charge ratio within ±0.000005 amu or 5 ppm of thousands of individual compounds present in a sample.[314] This precision also allows the detection of individual carbon isotopic counterparts, ^{12}C and ^{13}C peaks. Results are commonly displayed in a van Krevelen diagram in which areas corresponding to the source macromolecules are defined.[115] However, due to the ionization method, its application is restricted to solubilized samples and it was therefore mainly applied to humic acids[315,316] and dissolved OM from numerous environments such as rivers, lakes as well as soil or sediment pore waters[9,281,314,317–321] (Figure 4). In soils, it has been used to track the changes in water soluble organic fractions upon litter transformation, showing an increase in lignin and carbohydrates and a decrease in proteins.[322] It also revealed an increase in condensed aromatics in HA when humification increased (assessed by NMR from aromaticity)[322,323] and the condensed aromatics were proposed to result, at least partly, from reactions between lignin-derived compounds and hydroxyl or other organic radicals.[324] The occurrence of such photochemical reactions is supported by experimental photo irradiation of DOM[56] and was proposed to play a role in coal formation.[325]

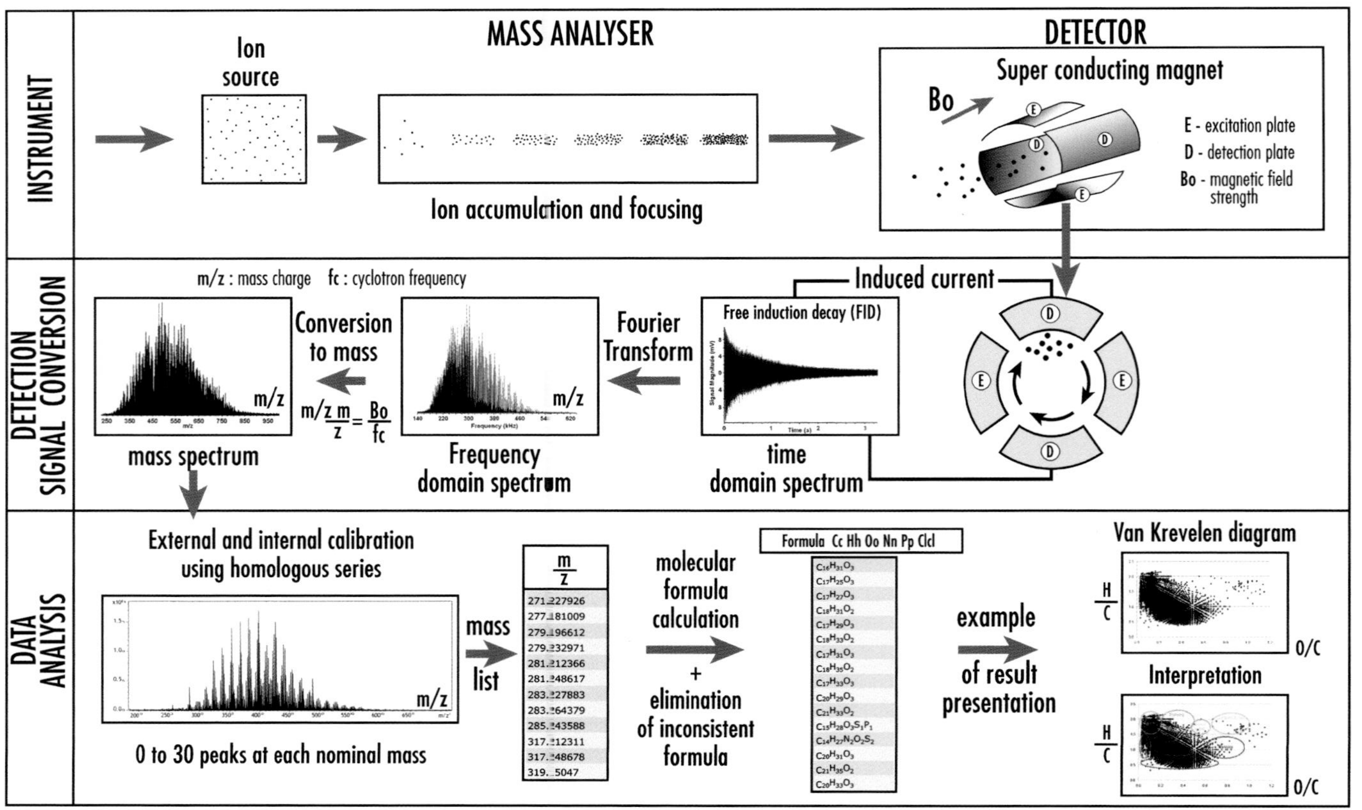

Figure 4. Principle of FT-ICR-MS and example of data interpretation.

Using the van Krevelen representation, a "molecular lability boundary" was determined, leading to a scale of lability depending on the environment (glacial > marine > freshwater).[326]

2. *Molecular Analysis*

In addition to the aforementioned techniques that provide insight into the chemical functions or the bulk elemental composition, the analysis of bulk OM, or of OM fractions at the molecular level, shed a new light on OM knowledge. Indeed, among the different constituents of OM, MRT could vary largely depending on the considered organic compounds, some being commonly reported to have high MRT while others are supposed highly labile. Thus, some specific compounds can be used as markers to follow the evolution of their source organisms, such as lignin or cutin and suberin for vascular plants, or algaenan for microalgae. In addition to these constituents of the source biomass, some components of natural OM result from transformation during the first stages of soil (humification) or sediment (diagenesis) formation, resulting in an induced recalcitrance. Among them, one can mention melanoidin-like formation, partial combustion leading to black carbon and sulphur incorporation in OM.

The molecular analyses have also revealed some unexpected results on compound turnovers and the biodegradability of various OM constituents has been questioned.[273,327] Indeed, the development of artificial plant labelling and of techniques combining irMS to other analytical methods, such as gas chromatography or liquid chromatography followed by on-line combustion (GC-C-irMS and LC-C-irMS), or pyrolysis GC/MS-C-irMS, has promoted new elements in OM understanding of individual compound fate.[290,328–333] In consequence, recalcitrance and lability of molecules have been questioned.

a. *Induced recalcitrance*

Melanoidins

Melanoidins are the products of the Maillard reaction between amino acids and carbohydrates.[334] Although this reaction takes place in the laboratory in hot acid or base medium, it has been suggested to also occur in the natural environment, although at a different time scale.[201,335,336] Based on the ubiquitous occurrence of their precursor, melanoidins are expected to be widespread in the natural environment. Moreover, in sediments, they were considered for long as the most refractory constituents, as the products of the degradation-recondensation mechanism.[13] This was further supported by comparison between their chemical structure and that of natural OM.[179,202,337] Synthetic melanoidins are therefore used as model material to investigate natural OM properties such as adsorption capacity.[338,339]

Black carbon

Black carbon (BC) stands for all residues from incomplete combustion of plant biomass and fossil fuels.[340,341] It encompasses a continuum from partly charred material to graphite and soot particles.[342] BC is ubiquitous in soils and sediments and was also shown to significantly contribute to the global riverine flux of dissolved OM.[343,344] It plays a role in xenobiotic transportation and it can be used as a tracer of fire.[341,345–347] Biochar which is part of BC[348] is added to soil to improve fertility. However the efficiency of this amendment is debated[349] and likely depends on the origin of the biochar.[350] The continuum of charred OM, and the highly aromatic structure, made BC analysis challenging. Due to its chemical recalcitrance, several methods have been developed for its quantification[351] assuming that BC was the material left over after various types of degradations: chemothermal oxidation at 375°C,[37,352] Cr_2O_7 oxidation,[353] NaClO oxidation,[354] UV photooxidation.[355] However, some materials, such as melanoidins or natural OM from kerogens or coals were shown to interfere with BC leading to false positive results or overestimation of BC content.[356] It was therefore recommended to characterize the remaining material after oxidation.[357] An alternative was to reductively remove the labile OM, i.e. OM easily biodegradable, thanks to hydropyrolysis.[358] On the other side, difficulties in such quantification are heightened by the continuum nature of BC,[359] which can be evidenced by oxidative scanning calorimetry.[217] Consequently, some methods are too aggressive towards poorly charred and thus lead to underestimation of BC in such samples.[356] Instead of comparing the sample weight before and after oxidation, products of BC oxidation, namely benzenepolycarboxylic acids were quantified and their amount converted into BC quantity[360–363] but caution should be exercised when applying this method to organic C-rich (> 5 mg C_{org}) samples.[364] Similarly, an attempt was made to use the levoglucosan level as a proxy for BC quantification; although results exhibited a too large variability, this compound is able to probe the less charred materials of the continuum.[365] Characterization of BC was achieved by Py GC-MS and tetramethylammonium hydroxide (TMAH) thermochemolysis (Figure 5) and high abundance of benzene, toluene, benzonitrile and PAHs were identified as markers of BC.[366] A direct quantification method was expected to be achieved through spectroscopic methods, such as NMR[366,367] or C 1s near-edge X-ray absorption fine edge structure (XANES). However, the chemical structure of BC, based on fused aromatic C results in a low observability of the C by NMR.[368] A large range of uncertainty was observed with XANES even through the use of deconvolution.[369]

BC is frequently found in soils in low quantity although more than 30% of BC was reported in some soils.[370] Charring, by inducing molecular structure alteration of the initial OM (reviewed by Knicker *et al.*, 2007b),[371] was reported to stabilize OM in soil. Indeed, MRT of BC in soil and sediments was assessed to range between hundred and more than 10,000 years based on ^{14}C age measurements.[340,372,373] This high stability was attributed to the high aromaticity of BC. XPS and STXM coupled with NEXAFS spectroscopy were applied to BC from soils and indicated that more than 60% of the C was found in aromatic C and the interior region of the particle have more C in un-oxidized form than the exterior.[367] The high MRT of BC has led to consider the thermal stabilization of part of OM as biochar in soil[348] and BC stability in soil was reviewed by Spokas (2010).[374]

Although most studies deal with BC, this concept is likely oversimplified as such combustion residues also comprise heteroelements, especially N[375] and the dissolved black nitrogen was suggested to play a role in the global nitrogen cycle.[376]

Compounds formed upon natural sulphurisation

Although not preexisting, organo sulphur compounds, which acquired recalcitrance through sulphur incorporation, occur in various environments,

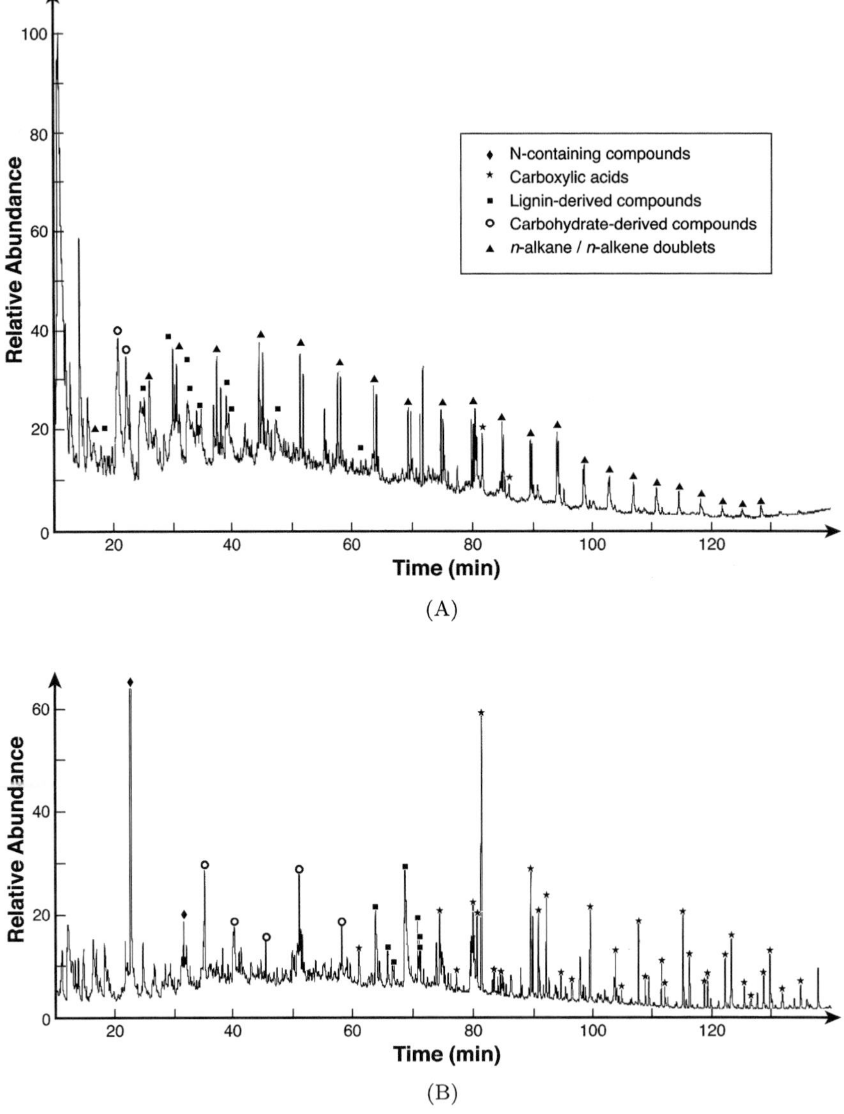

Figure 5. Chromatograms of soil OM obtained upon pyrolysis (A) and TMAH thermochemolysis (B) coupled with GC-MS. TMAH as a base and methylating agent enhances bond cleavage and induces *in situ* methylation of the pyrolysis products allowing the detection of the polar moieties.

especially in sediments. In sedimentary rocks, sulphur occurs in OM and in associated minerals such as pyrite, which can survive the HF/HCl treatment used for kerogen isolation. S is incorporated in kerogens through intra- or intermolecular reactions of sulphides produced by bacterial reduction of sulphates.[377] The intermolecular reactions result in cross-linking between some lipids or some carbohydrates and thus provide an increased preservation, the initial molecules being no longer recognized by enzymes. The precursors can be identified after thermal or chemical degradation of the kerogen.[378–380] Pyrolysis products commonly consist in a complex mixture of organo-sulphur compounds which has to be reduced with Raney Nickel prior to GC-MS identification.[381] Using RuO_4 oxidation, a series of lipids typical for sulphate-reducing bacteria was shown to be incorporated in immature sulphur-rich kerogens.[382,383] The extent of S incorporation in kerogens is assessed by the Sorg/C atomic ratio, and kerogens with S/C > 0.04 are considered as S-rich. They mainly fall in the type II (marine origin) and led to define a type II-S.[384] In contrast, only rare occurrences of lacustrine S-rich kerogens were reported.[385,386] However, it must be noted that, thanks to the development of FTICR-MS, the contribution of sulphur compounds in lacustrine[318,387] and riverine[388] DOM is increased, this technique having allowed the recognition of a wide range of sulphur compounds, hence a possible implication of the sulphurization mechanism upon diagenesis in these environments. Due to the higher thermal lability of aliphatic sulphur bonds, sulphur-rich kerogens are considered to produce oil at lower maturation level than other kerogens.[384]

b. *Differential intrinsic degradability of molecules*

Aliphatic compounds

Aliphatic macromolecules were suggested to be more resistant to biodegradation than bulk OM. This was derived from molecular analysis,[389,390] ^{13}C NMR analysis,[283,391] or through characterization of chemically isolated soil fractions reported as refractory.[154,392] In algal-derived sediments, the high contribution of aliphatic material led to the concept of selective preservation in which some aliphatic biopolymers survive the first steps of fossilization whereas the other biomass constituents are degraded.[391] Indeed, as for aquatic microorganisms, some microalgae were shown to contain in their outer walls an aliphatic biopolymer, termed algaenan.[195,393–396] These algae mainly belong to three classes and encompass marine and freshwater species.[397,398] Several procedures were proposed to isolate algaenan from the algal biomass so as to avoid the artefactual formation of insoluble OM (melanoidin) upon reaction between amino acids and carbohydrates.[399–403] The chemical structure of algaenans was investigated through a large range of analytical methods including spectroscopic analyses and chemical and thermal degradations.[393,404,405]

Comparative studies performed on extant microalgae and their fossil counterparts revealed striking morphological (scanning and transmission electron microscopy, SEM and TEM) and chemical (distribution of pyrolysis products) similarities between algaenans and kerogens and provided chemical evidence for the involvement of the selective preservation pathway in kerogen formation.[195,396,397,406–410] Due to the highly aliphatic character of the algaenans, this mechanism yields kerogens with a high oil potential as reflected by their high H/C atomic ratio and position in the van Krevelen diagram. Contrary to the classical degradation-recondensation pathway, it can account for the presence of organic microfossils in kerogens such as those derived from *Botryococcus braunii* and *Tetraedron minimum*.[406,411,412] When the outer walls of the source microalgae are thinner (10–30 nm), the isolated algaenan and the derived kerogens appear as ultralaminae observed by TEM and this morphological similarity is strengthened by the common production of a series of alkylnitriles upon pyrolysis.[407] A study published in 2015 reported the isolation of an aliphatic biopolymer in cyanobacteria[413] whereas the isolation of a melanoidin-like material similar to that obtained from mycobacterial biomass was previously suspected.[414,415]

In terrestrial plants, cutin and suberin are amorphous biopolyesters acting as protective barriers against chemical, biological and physical attacks and involved in the regulation of plant water content and gas exchanges.[416–418] Cutin is embedded within intracuticular waxes and covered with epicuticular waxes to form the cuticle which covers the aerial parts of vascular plants.[389,419,420] Suberin mostly occurs in roots, bark, and some suberized tissues of shoots.[421–427] Although both biopolyesters are

Cutin		Suberin
Major monomers		**Major monomers**
C_{16}	C_{18}	
$CH_3(CH_2)_{14}COOH$	$CH_3(CH_2)_7CH = CH(CH_2)_7 COOH$	$CH_3(CH_2)_m COOH$
$\underset{OH}{CH_2(CH_2)_{14}COOH}$	$\underset{OH}{CH_2(CH_2)_7CH} = CH(CH_2)_7 COOH$	$CH_3(CH_2)_m CH_2OH$
$\underset{OH\quad\;\; OH}{CH_2(CH_2)_x CH(CH_2)_y COOH}$	$\underset{OH\qquad\quad O}{CH_2(CH_2)_7 CH - CH(CH_2)_7COOH}$	$\underset{OH}{CH_2(CH_2)_n COOH}$
$y = 8, 7, 6\ or\ 5;$	$\underset{OH\qquad OH\;\; OH}{CH_2(CH_2)_7 CH - CH(CH_2)_7COOH}$	$HOOC(CH_2)_n COOH$
$x + y = 13$		$(m = 18\text{-}30; n = 14\text{ -}24)$
		phenolic compounds

Figure 6. Cutin and suberin major monomers, adapted from Kögel-Knabner *et al.* (2002).[16]

mainly composed of *n*-carboxylic acids, hydroxyacids, diacids and epoxy-acids esterified to each other and to glycerol, they exhibit differences in their chemical structure.[428] Indeed, whereas the chain length of most cutin monomers is C_{16} and C_{18}, that of suberin ones maximizes at C_{24} and suberin comprises an aromatic domain in addition to its aliphatic units[429] (Figure 6).

Their monomers are commonly used to discriminate below ground and above ground OM input in soils.[430] Based on their chemical structure, cutin and suberin are thought to contribute to the stable aliphatic component of soil OM[282,389,431,432] and therefore to the ^{13}C NMR aliphatic peak.[433] Their C–O bonds also lead to a signal in NMR in the 62–74 ppm region,[434,435] interfering with peaks commonly assigned to carbohydrates.[436] As polyesters, they are also partly responsible for the carboxylic C signal.[434,435] Although saponification was logically the first method used for the depolymerization of these polyesters,[437,438] transesterification with BF_3,[389,439] CuO oxidation previously developed for lignin characterization[420] and the aforementioned thermochemolysis in the presence of TMAH[440,441] were also used for their characterization. In a comparative study, saponification appeared the most appropriate method to quantify cutin monomers in soils.[430]

In some cuticles and suberized tissues, cutin and suberin are respectively associated with non-hydrolysable macromolecules termed cutan[442] and suberan.[443] Contrary to cutin and suberin, the chemical structure of cutan and suberan is still debated. It is commonly considered to be based on long polymethylenic chains[443–447] although some aromatic units were evidenced in cutan by NMR.[448] Cutan was proposed to be related to drought-adapted plants[445] or to provide defense against pathogens in tea leaves.[449] Based on their non-hydrolysable character, these aliphatic biopolymers were initially proposed to survive upon fossilization of plant cuticles.[450] This was supported by comparative studies performed on cutan from vascular plants including angiosperms and gymnosperms and fossil cuticles from Pliocene to Carboniferous deposits.[442,450–454] However, this was further debated and *in situ* polycondensation of aliphatic lipids and cutin monomers was rather proposed to account for the preservation of cuticle during diagenesis.[455–457]

By analogy with sediment, the potential role of the selective preservation pathway was investigated in soil[282,458] where it was proposed to account for the presence of aliphatic chains in the OM.[196] Moreover, fractions from different subsoils with the oldest ^{14}C ages were shown to be enriched in alkyl C suggesting that aliphatic compounds were selectively preserved.[459]

When dealing with smaller molecules, such as those analyzed by GC-MS in lipid extracts, aliphatic suberin-derived compounds were shown to exhibit a low level of degradation.[460] The increase in these compounds upon OM evolution and the accumulation of ester bound lipids from suberin in sandy soils were reported by Nierop *et al.*[461,462] They hypothesized that cutin and suberin could have formed stable cross-linkages resistant to decomposition. The isolated aliphatic compounds were identified by GC-MS after saponification and by Py GC-MS in the presence of TMAH. In a study of a pristine grassland, suberin and cutin degradation was observed with soil depth, but the corresponding aliphatic monomers were more preserved than the aromatic lignin monomers in the deepest horizon.[463]

In contrast, GC-C-irMS of *n*-alkanes isolated from different size fractions of a soil after conversion from forest to maize cultivation revealed an important degradation of these aliphatic compounds[179] and the turnover of *n*-alkanes was reported to be faster than that of bulk soil.[464] Moreover, in an experiment investigating changes in DOM during biodegradation, an increase in aromaticity was observed whereas lipids decreased. In consequence, the chemical stability of alkyl compounds in DOM was debated.[224]

In addition, estimated turnover rate of hydrolysable lipids was shorter than expected (despite longer turnover than bulk soil).[465] Therefore, the persistence of aliphatic compounds was proposed to be better explained by specific interaction with minerals or hydrophobicity than by their chemical structure[466] as detailed in Section III.B. Similarly, the higher residence time of root compared to shoot was also attributed to their different chemical composition, roots having higher lignin and suberin content than shoots.[467]

Lignin

Lignin is the second most abundant constituent of vascular plants after polysaccharides, accounting for up to 30 wt% in wood. It allows the rigidity of the plant and increases the water transport efficiency in specialized cells. This biopolymer is commonly used as a marker of terrestrial OM. It is based on three types of phenylpropanoid units, namely p-coumaryl alcohol, coniferyl alcohol, and sinapyl alcohol (Figure 7). The proportion of the three monomers varies between plant families (herbs, angiosperms and gymnosperms) but also within individual plant.[420,468] Based on its chemical structure, lignin contributes to the aromatic peak in NMR of natural OM and is also responsible for the methoxy C peak at 56 ppm allowing its recognition (Figure 3).

Lignin degradation is mainly achieved by fungi with demethylation during the first steps.[469] The relatively high aromaticity of lignin and the presence of non-hydrolysable C–O–C and C–C bonds between the monomeric units were often put forward to support its relative recalcitrance to degradation in soil.[470] For analytical purposes, several techniques of depolymerisation of lignin were developed, especially CuO oxidation which releases the acid counterparts of the monomers[471,472]

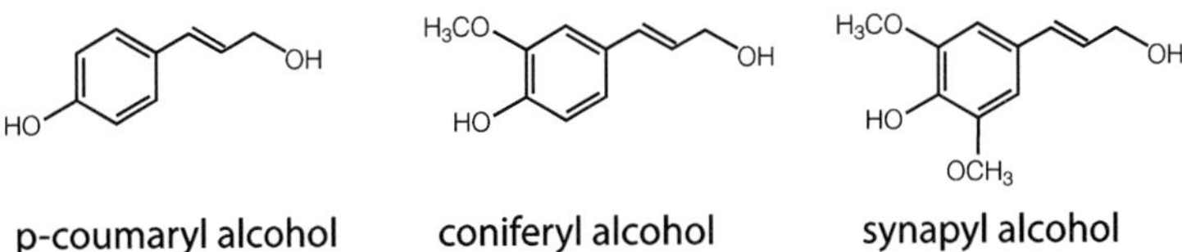

Figure 7. Lignin monomers.

and thermochemolysis in the presence of TMAH (Figure 5) which induces the cleavage of the β-O-4 linkages and *in situ* methylation of the released phenolic monomers.[473–477] GC-MS analysis of the released compounds is then used to monitor lignin degradation. Indeed, a decrease in the sum of these CuO products occurs during biodegradation.[478] In addition, as these lignin-derived CuO oxidation products could bear acid or aldehyde functions, the acid to aldehyde ratio was frequently used to estimate the lignin degradation rate.[479–482] A lower ratio was observed in fine soil fractions[483,484] indicating a lower degradation stage in these fractions which are commonly considered to contain old OM. However, since the last two decades, several authors have pointed out an overestimation of lignin recalcitrance, with lignin turnover being faster than bulk SOM.[333,485] Indeed, no increase in lignin was observed in subsoil horizon of soils where part of the OM is supposed to be stabilized[199] and Kiem and Kögel-Knabner have shown that lignin was not accumulated in the refractory C pool of their studied soils.[486] Similarly, the lignin contribution to OM fractions enriched in old C was shown to be very low, both by NMR and quantification after CuO oxidation, hence a low preservation of lignin in subsoils.[154] The lignin content was also reported to decrease with depth as well as in fine fractions and exhibited an increase of acid to aldehyde ratio indicative of higher level of degradation.[486] In addition, the turnover of lignin was assessed using compound specific ^{13}C analysis by GC-C-irMS in wheat to maize conversion chronosequence[333] and a model with two pools was developed. The first pool, representing more than 90% of the lignin, was composed by undecomposed plant residues and exhibited turnover rate of less than 1 year whereas that of the second pool, which comprised lignin fragments being partially protected by interaction with minerals was 0.05 years^{-1}.[487] Thus, conversely to the initial assumption, lignin does not present an intrinsic recalcitrance in soils (reviewed by Thevenot *et al.*[488]) and its persistence in soil could be considered at most, only in the medium term. Indeed, in a 27 month experiment using litter bags to follow the degradation of lignin in different samples of needles and leaves, thermochemolysis with ^{13}C labeled TMAH and CuO oxidation revealed a progressive oxidation of the propyl side chains indicative of white-rot decay rather than brown rot decay which is associated with lignin demethylation.[489]

Carbohydrates

Carbohydrates principally consist in cellulose, composed of glucose units, and were generally considered as belonging to a relatively labile pool. However, some studies have revealed that carbohydrates could occur in fractions of soils belonging to the recalcitrant C pool.[486] Derrien *et al.* found carbohydrates with a mean age comprised between 60 and 100 years thanks to natural labeling of C_3 and C_4 plants and GC-C-irMS analysis.[490] The stabilization of carbohydrates was attributed to their interaction with minerals, occlusion in aggregates or incorporation of the carbon from microbial biomass. However, carbohydrates present in deep horizons or in fine fractions could originate from different sources and not directly from plant residues. In the coarse fractions, carbohydrates were mainly derived from plants whereas in fine fractions, carbohydrates show a high contribution of microbial polysaccharides.[483,486]

Nitrogen compounds

Although nitrogen is the 4th most abundant element in the bio- and geosphere, the fate of organic nitrogen in natural environment is far from being elucidated at the molecular level. N-containing compounds are commonly considered as labile, with the C/N atomic ratio being often used to follow the extent of OM degradation. However, organic compounds comprising amino sugars or amino acid derivatives, including proteinaceous material, were reported in natural environment, pointing to some preservation potential. Py-GC -C-IRMS of soil after C3 to C4 plant conversion have highlighted the high residence time of N-containing molecules (around 50 years).[491,492]

Although they are often considered as labile, amino sugars account for 1–5% of soil organic C, i.e. a much higher concentration than in the soil microbial biomass. This shows their stabilization in soil where they are considered as markers of microbial necromass.[493] Amino sugars are reliable microbial molecular markers as they are not found in plants. Galactosamine and muramic acid are indicative of bacteria, glucosamine being indicative of fungal residues,[494] although it also occurs in arthropod cuticles, which contribution to NOM is minor. Muramic acid was thus used as a biomarker of detrital biomass in aquatic environments such as Amazon

POM and DOM[495] and the presence of glucosamine, together with the absence of muramic acid, suggested a much higher fungal than bacterial contribution in the hydrophilic fraction of a soil.[496] This stabilization was assumed to result from the polymeric nature of the amino sugar source (peptidoglycan). However, in an incubation experiment on a ^{13}C and ^{15}N double-labeled other aminosugar polymer, fungal chitin, Zeglin and Myrold observed a rapid mineralization of the macromolecule.[497] Sorption on mineral matrices has been put forward to account for the survival of amino sugars[82,498] as detailed in Section III.B. However, the individual monomer turnovers were still difficult to estimate. Indeed, the determination of individual δ^{13}C values in amino sugars by GC-C-irMS was associated with a large error due to the need of preliminary derivatization.[499] LC-C-irMS[500] and ion chromatography-oxidation-isotope ratio mass spectrometry[501] were developed to overcome these limitations by avoiding derivatization. It would be interesting now to apply these methods to natural ecosystems. Chitin was also shown to survive in sediments where its recalcitrance is assumed to be provided by its complexation with proteins within the cuticle.[502] D-glucosamine was identified in the pyrolysates of fossil insect cuticles from deposits as old as 25 million years.[503,504] Chitin was also reported to be preserved in 200-million-year-old gastropod egg capsules.[505]

Proteins are the main biological sources of N and they are reported to be labile compounds as they are easily depolymerized into single amino acids through enzymatic hydrolysis. They account for a large fraction (ca. 30%) of soil SOM[506,507] where they are the dominant (70–90%) form of N.[508,509] The preservation of proteinaceous material was first suggested by ^{15}N NMR, the abundant amide signal being assigned to proteins[270] and hydrolysis of soils was later shown to release amino acids which could be analyzed by GC-MS using a chiral capillary column to distinguish amino acid enantiomers.[510] The preservation of proteinaceous moieties in soils was confirmed by on line thermochemolysis in the presence of TMAH combined with GC-MS[511,512] and these moieties were proposed to originate from microbial biomass.[19] Similarly, proteinaceous moieties were shown to survive in sediments[513,514] and encapsulation in organic biopolymers was put forward to account for this preservation.[511–514] A similar mechanism was suggested to occur in soil.[515] Encapsulation of intact proteins by humic substances was further demonstrated through the use of a combination of two *in situ* surface adsorption techniques[516] and covalent linkages were evidenced by two-dimensional NMR.[517] Finally, the formation of complexes between proteins and tannins was invoked to account for protein preservation although no evidence could be brought.[507]

The extent to which soil N could be assigned to amides was questioned as most of the heterocyclic N appeared to be poorly detected by ^{15}N CP/MAS NMR[518] and the presence of pyridinic N was revealed by N K-edge XANES.[519] Amides were thus often overestimated at the expense of heterocyclic compounds. Heterocyclic nitrogen is commonly considered as a marker of fire. Nitrogen heterocyclic polyaromatic rings probably formed as a consequence of the thermal-induced dehydration of aromatic amides or the ring scission and subsequent rearrangement of heterocyclic N.[520,521]

Degradation of NOM is not only dependent on its molecular composition or structure but it can also be regulated by organo-mineral interactions such as sorption, aggregation or occlusion.

B. Protection due to Organo-Mineral Interactions

OM and minerals coexist in the environment and organo-mineral interactions occur in aquatic systems, in soils as well as in sediments.[522–525] The potential transport and susceptibility to degradation of NOM is largely determined by the organic molecule partitions to mineral surfaces or persistence as dissolved OC (Figure 8).

Many studies have evidenced the prominent role of organo-mineral interactions on OM protection and thus stabilization in soils and sediments.[100,184,498,522,526–528] Indeed, in sediments, OM content and surface area were correlated[522] and sediment microstructure was recognized as important in OM diagenetic degradation.[5] In soil, due to biological, chemical and physical processes, the organic and inorganic compartments are intimately associated through organo-mineral interactions (reviewed by Baldock and Skjemstad (2000), von Lützow *et al.* (2006), Kögel-Knabner *et al.* (2008)).[194,539,540] Thus, lower OM turnover has been reported for a long time in clay and silt sized aggregates.[541–544] The identification of old OM in these fine fractions, and in particular in clay fraction[81,193] has

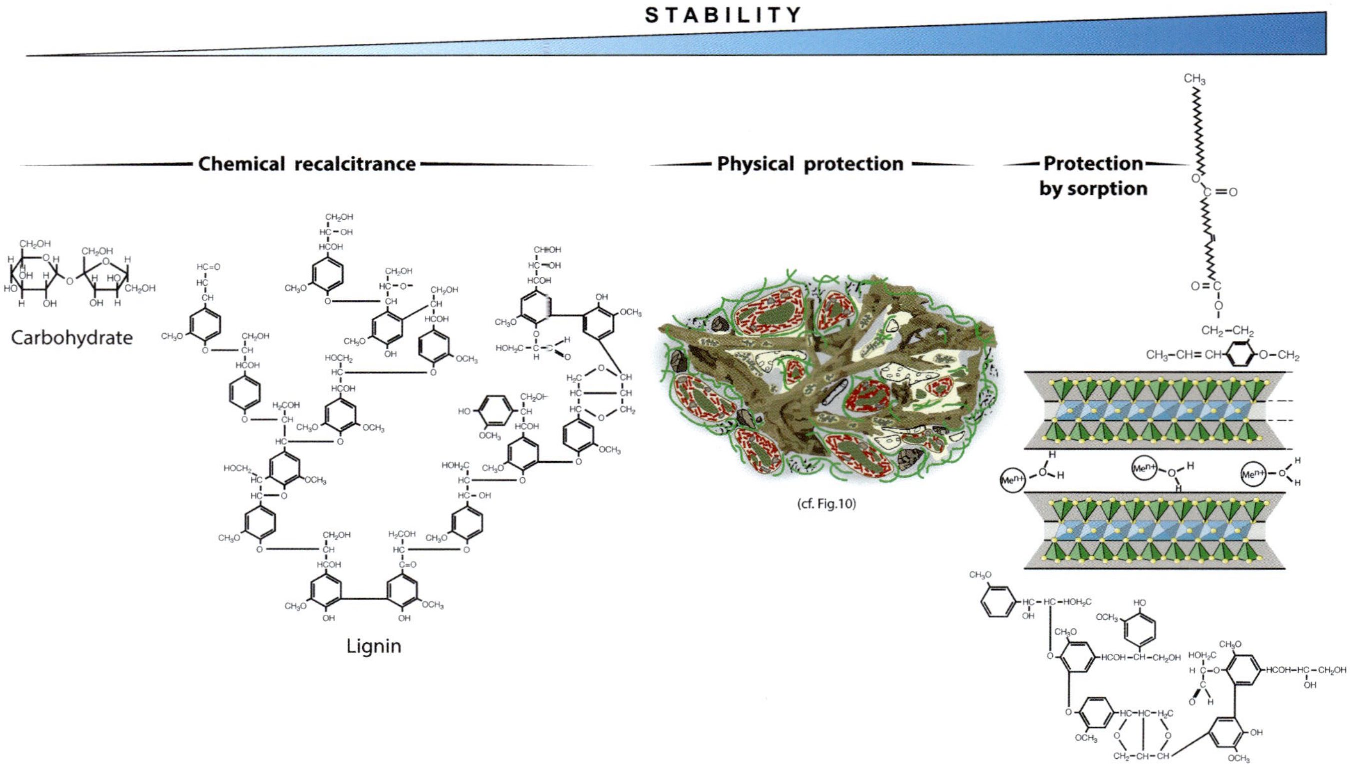

Figure 8. Examples illustrating the different OM stabilization processes leading to different mean residence times. (A) Chemical recalcitrance related to chemical composition, (B) physical protection (see Figure 10) and (C) protection by sorption onto clay sheets. Me^{n+}: metal ions.

promoted the role of the organo-mineral interactions in OM stabilization at the century time scale.[44,151] Moreover, radiocarbon activity confirmed preferential stabilization of OM in occluded clay fractions compared with free clay fractions,[545] but also in high density fractions encompassing OM associated to minerals.[18,81,546] As the clay-OM interactions were generally accepted, several authors have used the clay content as a proxy to the potential of organic C sequestration[547,548] even though recent studies have evidenced that this correlation was more complex.[549]

Plante *et al.* demonstrated that chemically resistant OM was more resistant to thermal oxidation, as thermal techniques relied on differential energy of interaction between soil organic molecules and between soil minerals.[550] In consequence, OM stabilization through organo-mineral interactions was suggested to induce thermal stability, as chemical bonding of OM with mineral was shown to increase the pyrolysis temperature of volatilization of compounds when compared with the light soil fraction, where OM is supposed to be free.[546] Few studies have explored how the thermal stability of OM sorbed on soil minerals changed with carbon loading. Feng *et al.* focused on testing the hypothesis that adsorption and variations in loading alter the intrinsic thermochemical stability of NOM.[547] They showed that OM thermal stability decreased along with OM loading onto goethite. Mineral surface interactions at larger loadings involved a smaller fraction of the adsorbed OM, thus increasing thermal stability toward that of its unbound counterpart. However, there is still a knowledge gap concerning the mechanisms affecting the thermal stability of OM bound to mineral surfaces.

Organo-mineral interactions stabilize OM but this stabilization could be the result of the inaccessibility of OM within aggregates or adsorption on minerals. Ekschmitt *et al.* (2005) have referred to two distinct types of "refuge" which could constitute OM "partial" and "absolute" protection from biodegradation and two different kinds of bacteria strategy to explain the persistence of biodegradable OM in soil.[192] They estimate that more than 80% of SOM were temporally stabilized due to their location on partial refuge.

Whatever this inaccessibility could be provided by sorption or protection in aggregates, the relative importance of sorption versus aggregation is difficult to estimate and could influence OM persistence in soil. To disentangle sorbed OM from that occluded in small aggregates, distinct soil fractions were isolated after hydrolysis of minerals with HF or granulometric fractionation.[545] It appeared that the implied process was dependent on the considered horizons and soils.[545] In addition, ^{14}C age measurement of carbon from different density fractions highlighted the different contribution of aggregation and sorption onto minerals in C protection according to soil depth.[548] These two types of organo-mineral interactions are successively developed in the following sections.

1. *Sorption onto Minerals*

According to Keil and Mayer, OM preservation by minerals could be influenced by the mineralogy but organo-mineral interactions could also play a role on the types of OM preserved.[549] This mechanism was put forward to explain the persistence of DOM in deep mineral horizon. Indeed, Kalbitz and Kaiser attributed this persistence to sorption and co-precipitation of DOM with mineral and Al.[550] The chemical composition of OM could be a factor influencing organo-mineral interactions, as well as mineralogy and surface characteristics (e.g. chemistry, specific surface area, roughness) of minerals.[498,551] Sorption onto minerals is therefore influenced by both the properties of soil minerals and the nature of sorbing organic species

a. *Role of minerals*

Hassink *et al.* linked the proportion of OM protected in soil to the proportions of silt and clay in soil.[548] The negative correlation between the total organic carbon content and the size of soil or sediment fractions was often explained by adsorption of OM onto the minerals.[552] Among the different stabilization pathways based on organo-mineral interactions, ligand exchange between Fe oxides and OM is often put forward. To differentiate the OM fraction concerned by this mechanism, a so-called high gradient magnetic separation method has been developed. It is based on differences in magnetic susceptibilities at different field strengths.[553] The use of this technique evolved in 2014 towards decontamination of soil polluted with heavy metals.[554] However, minerals from fine fractions in soil encompass phyllosilicates as well as metallic oxides and hydroxides or carbonates. Due to their high specific surface area and their surface reactivity, phyllosilicates and metallic oxides and hydroxydes are prone to interact with OM.[151,555,556] The role of different phyllosilicate mineralogy on their mechanisms of interaction with OM has been reviewed by Barré et al.[557]

The adsorption strength of OM on minerals results in a higher affinity than the affinity of the enzyme active site[327] making the OM not biodegradable by microorganisms. Thus, the surface specific area of minerals, but also the mineralogy, plays an important role in OM sorption onto minerals.[529,541,558–560] The soil OM sorbed onto minerals was proposed to constitute the OM pool with the highest MRT.[548,561] The protection of OM by minerals was also illustrated by adsorption of previously extracted DOM onto mineral soils. Mikutta et al. evidenced an effect of the mineral nature and that desorption of OM controlled its biodegradation, bound OM being less mineralized than the soluble OM.[459] In a similar experiment, Kalbitz et al. showed that after one year, DOM was less mineralized when sorbed.[561]

To evaluate the role of OM sorption on minerals, Eusterheus et al. have performed radiocarbon analysis after HF treatment on soil.[151] HF allows the dissolution of minerals and OM sorbed on these minerals are released in the solution. The negative correlation between ^{14}C activity and the HF-soluble organic carbon was interpreted as indicative of carbon stabilization in soil *via* sorption onto minerals. The relative importance of this stabilization pathway seemed to increase from top soil to subsoils.[151] Using a combination of NaClO and HF treatment, Mikutta et al. isolated a chemically resistant fraction from subsoils, with radiocarbon age 900 years older than bulk OM.[154] 70% of this OM was shown to be protected by organomineral interactions and to comprise polysaccharides and proteins which have thus been stabilized.

Galindo et al. reported the fractionation of aquatic FA and terrestrial HA upon sorption onto aluminium oxide.[562,563] They demonstrated that molecule acidity and hydrophobicity controlled the fractionation of FA and HA on the oxide. Three common Fe oxyhydroxides (poorly crystalline ferrihydrite, crystalline goethite and lepidocrocite) were selected by Lv et al. to investigate the molecular fractionation of OM at Fe oxyhydroxide/water interfaces using ESI-FTICR-MS.[564] Higher affinity for OM and molecular fractionation of OM were recorded for ferrihydrite than goethite and lepidocrocite. In another experiment, the turnover of the OM from a volcanic ash soil containing mainly poorly crystalline aluminosilicates was estimated as millennia by isotopic measurements (^{14}C and ^{13}C).[565]

In addition, OM fractions isolated by pyrophosphate and hydroxide treatment from soils dominated either by kaolinite clays or smectite clays, were analyzed by solid-state ^{13}C NMR and Py-GC/MS. Despite no difference in C content was observed between the clay size fractions of the soils, in kaolinitic soil, polysaccharides have been more abundant than in smectitic soil, in which aromatic and phenolic compounds were more abundant. Thus, the type of clay influences mechanisms involved in sorption of OM and consequently also the nature of the sorbed and eventually protected OM.[566]

b. *Role of OM nature*

Experiments comparing the sorption of hydrophilic and hydrophobic DOM revealed that the latter was more sorbed than hydrophilic DOM.[567] OM composition and distribution could contribute to an increased hydrophobicity[568] and this hydrophobic OM could exhibit lower degradation rate.[466] This is consistent with previous investigations which suggested that the OM fractions of higher molecular weight, higher aromatic and carboxyl functional group content, and more hydrophobicity show preferential adsorption onto mineral oxides and hydroxydes.[555,564,569–571] In turn, because of its heterogeneous nature, the composition and chemical properties of humic substance-like OM are modified during adsorption onto mineral oxides.[551,572–577] Sorption experiments of different dissolved OM also revealed that aromatic OM is more sorbed onto minerals and hence more stabilized.[459,561] Along with the study of Lv *et al.*,[564] this support results Galindo *et al.*, stating that the double bond equivalence and the number of oxygen atoms are valuable parameters indicating the selective fractionation of OM at mineral and water interfaces.[563]

Carboxyl functional groups play an important role in sorption and their role can be investigated at the macroscopic scale by potentiometric titration. Janot *et al.* used spectrophotometric titrations at different ionic strengths to examine the deprotonation of purified OM before and after sorption onto alumina.[578,579] They showed a preferential sorption of aromatic OM but it could only be evidenced for OM/mineral ratio below 20 mg OM/g mineral. Size-exclusion chromatography coupled with organic carbon detection measurements also showed a preferential adsorption of higher size fractions.[579] The amount of low proton-affinity type of sites (i.e. carboxyl functional groups) and the value of their median affinity constant decrease after adsorption. These changes in reactivity could explain the difficulty to model the behavior of ternary systems composed of pollutants/HS/mineral since additivity is not respected.[579] The role of acidic groups in OM sorption was also put forward when comparing amino acid and amino sugar sorption onto various mineral phases.[498]

Organo-mineral interactions were also considered to provide an efficient protection to organic N[24,568] especially as proteins and nucleic acids are known to be efficiently adsorbed onto clays and Fe-oxides.[580–582] The role of carboxylic functional groups (pKa value, functional groups) for the sorption and retention of these carboxylic and amino acids on Fe oxides (and kaolinite) and of amino group on 2:1 phyllosilicates is important. Yeasmin *et al.* used the attenuated total reflectance infrared (ATR)-FTIR method that is highly sensitive to small structural changes at the OM-mineral interface.[583] They examined the role of the nature of OM compounds and mineral properties on the OM sorption and desorption capacity using batch experiments where different ^{14}C labeled organic compounds (carboxylic acids, amino acids with varying side chains) are mixed with pure minerals (from oxides to clays). Cationic lysine was shown to exhibit a stronger affinity for permanently charged phyllosilicates than Fe oxides, while alanine and phenylalanine were poorly adsorbed on all minerals. Thanks to the development of *in situ* analytical methods such as STXM in combination with XANES and Nano Secondary Ion Mass Spectrometry (NanoSIMS), direct observation of the organo-mineral associations could be achieved. This led to the hypothesis that amide groups from fungal cell walls were assimilated by associated bacteria, yielding amide groups concentrated onto Fe (hydr)oxide surfaces.[584]

Contrasted results, obtained after HF experiments to isolate mineral bound OM fractions led Eusterhues *et al.* to consider different kinds of interaction depending on the soil environment and depth.[44] Indeed, the investigation at the organo-mineral interface remains challenging as highlighted by Totsche *et al.*, even though development of techniques is ongoing.[585] Indeed, chemical force spectroscopy (CFS) with functionalized atomic force microscope (AFM) tips can directly measure the interaction of OM with mineral surfaces[586–588] by determining the force between the tip and the sample surface as a function of separation distance. The resulting force displacement (F–D) curve reflects the strength of the binding interaction between the functionalized surface of the probe and the mineral surface substrate. Aubry *et al.* prepared OM-coated colloidal AFM probes using iron oxide as an intermediate layer. In addition, the interaction forces between mica and four well-characterized OM samples of different physicochemical properties and origins were studied using AFM.[589] They demonstrated that OM adhesion behavior on mica was consistent with its physicochemical characteristics. Humic-like substances poorly adsorbed on hydrophilic mica due to their high content of ionized carboxyl groups and aromatic/hydrophobic character contrary to polysaccharide-like substances. Hydrogen-bonding between hydroxyl groups on polysaccharide-like substances and highly electronegative elements on mica was suggested as the main adsorption mechanism, where the adhesion force decreased with increasing ionic strength. Moreover, OM samples were shown to retain their characteristics after the coating procedure. Chassé *et al.* used molecular-scale information provided by the combination of CFS and FTICR-MS in order to increase our understanding of the processes driving OM sequestration.[590] The abundance of condensed aromatic, aromatic, and N-containing aliphatic molecules of OM was correlated with the adhesion force between the OM and iron (oxy) hydroxide functionalized AFM tips. These results show that the force of OM binding to the mineral surfaces, at least at the initial stages of adsorption, decreases with increasing molecular mass.

Taken together, molecular weight, aromaticity and carboxylic group abundance seem to play a prominent role in OM sorption onto minerals.

c. *Pathway of organo-mineral interactions*

Several pathways were put forward to account for the sorption of OM onto mineral surfaces, including electrostatic/hydrophobic interactions, ligand exchange-surface complexation, entropic effect, hydrogen bonding and cation bridging.[194,549,591–593] However, due to the complex nature of OM, different combinations of interaction between OM and surfaces may arise and have not been fully elucidated. Different models were then proposed such as the mono-layer or multilayer coating, the uniform or patchy distribution of OM and the carbon saturation concept, as presented below.

Oxides and phyllosilicates present at their surfaces hydroxyl groups which could be ionized, especially at low pH. Moreover, phyllosilicates bear negative charges due to isomorphic cation substitution, and hydrophobic surfaces. On the other hand, OM contains reactive polar and charged moieties through some chemical functions (comprising carboxyl, hydroxyl) which could be ionized at high pH. In addition, the alkyl and aromatic functions in OM constitute hydrophobic regions. Mikutta *et al.* assessed the stabilizing effect of the different binding mechanisms and ordered these mechanisms: "van der Waals forces" "Ca^{2+} bridging" "ligand exchange".[459] However, depending on the soil pH, some reactions prevail. For example, at low pH, hydrophobic interactions and ligand exchange are dominant. OM can precipitate with polyvalent cations and the proportion of precipitated OM depends on pH, especially with Al[594] and precipitated OM would present lower turnover than sorbed one.[459,550]

FTICR-MS can be used to assess the effect of functional group homologue series on reactivity, providing further information on structure–reactivity relationships.[595] It was thus shown that carboxyl-rich aromatic and N-containing aliphatic molecules are involved in the most stable bonds with the mineral surface and that two carboxyl groups are involved in the inner-sphere bond formation to natural organic matter. This work suggests that changes in OM composition due to the environmental processes generating OM (e.g., forest cover change or land use change) may affect stabilization of OM at a molecular level with landscape-scale consequences for carbon sequestration and loss.[590]

Most of the information on mineral organic interactions discussed above was derived from indirect methods and/or macroscopic observations. Therefore, there is a need to relate these laboratory results to processes and to directly investigate the associations of OM with specific minerals using high spatial resolution and state-of-the-art spectroscopic techniques. However, this is challenging because there are few techniques that can simultaneously characterize OM and minerals. STXM, combined with NEXAFS, is a powerful method for analyses of C functional group distribution at the nanoscale.[596–599] However, most synchrotron-based soft X-ray spectroscopic analyses do not provide enough flux at higher energies to quantify Al and Si, which are the major elements in clay minerals. In 2007, Wan *et al.* showed the spatial association of C with Fe, Al, Si, Ca and K in soils but the associations between OM and mineral cations were obscured by the presence of very thick particles. Further research was performed by Chen *et al.* in 2014 demonstrating that C was associated with Ca, Fe, Al and Si without any separate phase in soil clay particles. In soil clay particles, the pervasive C forms were aromatic C, carboxyl C and polysaccharides with the relative abundance of carboxyl C and polysaccharides varying spatially at the submicrometer scale. Aliphatic C, generally from cell wall lipids, was localized in limited regions in the soil clay particles. This study suggests the importance of Ca in organo-(metal)-mineral interactions and/or organo-metal cation complexation.

Based on the relationship between mineral specific surface area and organic carbon concentration, the adsorption of a monolayer of OM onto the mineral surfaces was proposed to account for the protection of potentially labile OM,[523] but samples from various environments fall out of this correlation. Indeed, a preferential localization of the OM originating from the litter on surfaces already covered with OM was underlined[81,568,600] and specific surface area measurements provided results which suggested the adsorption of OM by patches or even by multilayers.[528]

Kleber *et al.* developed a model of organo-mineral interaction with OM being constituted of amphiphilic molecules with different ranges of polarity and charge.[568] This model proposed the existence of layered zones in which different mechanisms of organo-mineral interactions would be successively implied. This layered mode of OM accumulation on mineral surfaces was indirectly evidenced using density fractionation.[18] In addition, OM analyses from density fractions suggested an "onion layering model"[81] and is supported by results from Chassé *et al.*[590] Intimate associations between OM and clay minerals were revealed by electron microscopy and electron energy-loss spectroscopy in recent estuarine sediments where OM was shown to be tightly attached to the surfaces of individual clay plates or structurally incorporated into clay crystals.[601] The alternation of organic and clay nanolayers was also evidenced by TEM[527] and XPS analyses of density fractions indicated that OM was located in discrete spots rather than uniformly distributed onto surfaces.[602] Moreover, direct microscopic observations performed by SEM or TEM confirmed the patchy distribution of OM on minerals.[19,88,600] Miltner *et al.* observed soil particles by SEM and they attributed the patchy fragment rich in carbon to microbial cell wall.[19] In the early 2000s, the NanoSIMS was applied to biogeochemistry and was of high interest in soil studies at a micro- and nano-metre scale (reviewed by Herrman *et al.*,[603] especially to investigate the elemental and isotopic composition of soils (Figure 9). Mueller *et al.* pointed out the opportunity of NanoSIMS to supply information on SOM and mineral interactions for long term protection by tracing dissolved ^{13}C and ^{15}N labeled amino acid at the submicrometer scale.[604] Thus, the patchy distribution of OM was evidenced by Vogel *et al.* in an incubation of ^{15}N and ^{13}C labelled litter by combining SEM with NanoSIMS.[599,600] Remusat *et al.* have performed NanoSIMS analysis on forest soil density fractions where ^{15}N labeled

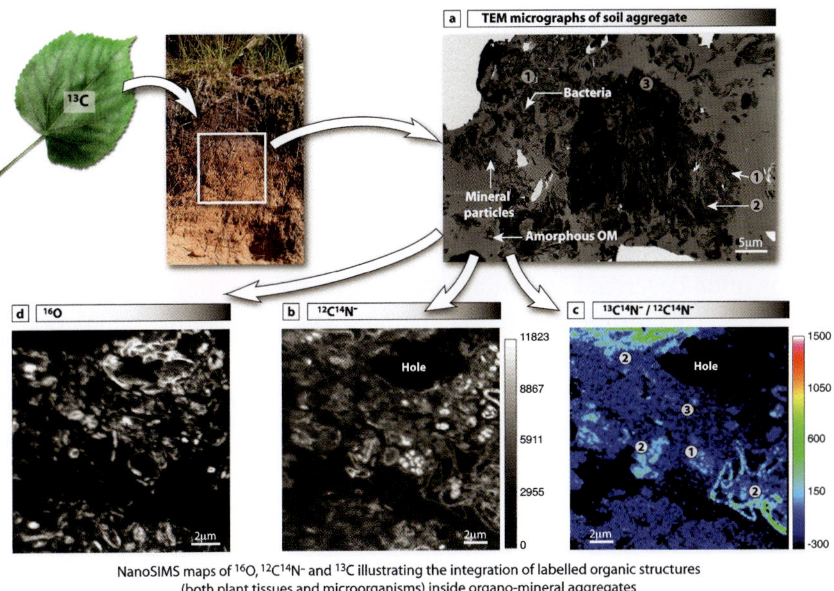

NanoSIMS maps of ^{16}O, $^{12}C^{14}N^-$ and ^{13}C illustrating the integration of labelled organic structures (both plant tissues and microorganisms) inside organo-mineral aggregates
① Bacteria   ② Vegetal cell wall residues   ③ Organo-mineral aggregate

Figure 9.   Illustration of the potential of ^{13}C labeling to trace OM at the microscale.

leaves were deposited 12 years before sampling.[605] The distribution of nitrogen and carbon were different with nitrogen located around the particles whereas carbon occurred as patches. They revealed that ^{15}N labeling was still present on the mineral surfaces after 12 years and, thanks to the characterization of the OM by STXM/NEXAFS, located by NanoSIMS, they suggested that the ^{15}N-rich areas would correspond to microbial metabolites. Whether the nature of the organo-mineral interaction influences the duration of stabilization of OM and could vary with time,[606] it could also be assumed that the different OM layers adsorbed on the mineral surfaces could present different residence time with outer layers with shorter residence time than inner layer.[568,607]

Experiments of DOM sorption evidenced limited sorption of DOM onto mineral and led to propose a concept of availability of mineral sorption sites.[561] This sorption capacity was detailed by Guggenberger and Kaiser.[607] From this assumption arose a concept of carbon saturation in soil (reviewed by Six *et al.*, 2002 and Stewart *et al.*, 2007).[558,608] Thereafter, based on a set of data, Stewart *et al.* proposed a model including a C saturation level which better reflects C content evolution with time in soil.[608] However, considering different pools of C defined by their size fractions (macro-, microaggregates and silt plus clay size), this saturation concept has been proposed to be hierarchical: the finest fraction potentially stabilizing less carbon than the largest one.[609]

2. *Occlusion in Aggregates*

Soil is often described as a hierarchical system exhibiting organo mineral interactions occurring at different spatial scales. Indeed, the interaction of minerals and OM could occur through adsorption but could also form aggregates by rearrangement of particles through biological and/or abiotic processes. These processes depend on pedoclimatic conditions (e.g. dry-wetting, occurrence of clays and/or oxides, of divalent cations, calcium especially) and on the occurrence and activity of microorganisms, roots and fauna in the soil. The aggregates constituting the soil structure were separated according to their size: micro aggregates (< 250 μm) and macroaggregates (> 250 μm). However, the actual structure is more complex as microaggregates can be included into macroaggregates (Figure 10).

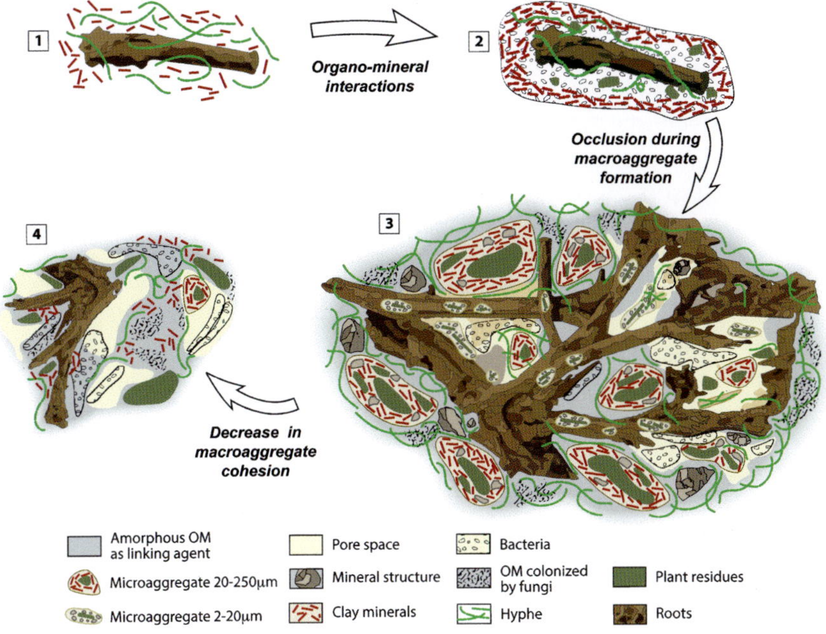

Figure 10.   Schematic representation of root occlusion within aggregates, showing the evolution of the organo-mineral structure with time. (1) Initial dead root fragment in soil, (2) microaggregate formed through organo-mineral interactions, (3) macroaggregate including microaggregates, (4) macroaggregate showing a decreased cohesion due to OM mineralization.

Soil aggregation was involved in OM protection mainly through inaccessibility of microorganisms (and/or enzymes) to OM[529–611] and/or limited activity due to pore size exclusion and creation of microenvironment unsuitable to microorganism development in aggregates, e.g. limited oxygen diffusion in aggregates.[612,613]

The OM protection depends on its localization, outside aggregates (as particulate OM) with a high turnover[61] or within macro- or microaggregates, further depending on whether the microaggregates are occluded within macroaggregate or free. The incorporation of POM into small aggregates (microaggregates or silt and clay-size aggregates) could represent an important process of protection of the easily biodegradable (labile) OM as evidenced by Paradelo *et al.* in a bare fallow experiment.[614] However, aggregates have limited life span due to a decrease in their cohesion caused by OM mineralization (Figure 10). In a review on aggregates and aggregation, Six *et al.* report the evolution of models of aggregate formation and turnover.[100] SOM within microaggregates is stabilized for longer time than within macroaggregates.[88,95,97,99,305,543,558,615–618] Jastrow *et al.* calculated a turnover of OM of 140 years in macroaggregates and of 400 years in microaggregates by using $\delta^{13}C$ isotopic measurement of OM after conversion of C4 cropland to C3 grass pasture.[619] This higher persistence of OM in microaggregates is partly due to the higher stability of microaggregates.[615] Thus the protection of SOM depends on aggregate stability and microaggregates were proposed as target for potential OC sequestration.[620] Indeed, with time, macroaggregate stability decreases, as microorganisms mineralize SOM, until its disaggregation. When their disintegration occurs, SOM is released and can be degraded by microorganisms again[615] and microaggregates previously occluded in macroaggregates are also released. Similarly POM previously included in macroaggregates undergoes an important degradation.[616,621,622]

In addition, silt-size (2–50 μm)[623] and clay-size (< 2 μm) aggregates[88] were also evidenced in soils and have been pointed in long term protection of OM either by adsorption onto clay or entrapment of OM particles. In addition, clay-size fractions include OM sorbed on clay but also microaggregates of 2 μm. Chenu and Plante have isolated the clay-size fraction and performed a density fractionation of this fraction.[88] The obtained fractions displayed a continuum with free OM particles, OM sorbed onto primary particles and microaggregates. The clay-size fraction could constitute a protection of OM on the very long term (century to millennia).[624] This heterogeneity in the clay-size fraction was put forward to explain important differences in the turnover of OM in the clay fraction depending on the authors.

Corollary to hierarchical arrangement of aggregates, a pore hierarchy self-organizes.[625] The size and tortuosity of pores appear to influence soil organism activity,[613] by constraining the location and influence area of the organisms, and also the production of oxygen or exo-enzyme by bacteria diffusion. Kaiser and Guggenberger inferred the possible protection of OM preferentially sorbed at mouth of micropores.[528] The recognition of stable and strong sorption onto these specific sites was highlighted by experimental sorption of OM on soils and subsequent analysis of specific surface area by multiple N_2 adsorption–desorption steps. According to Strong *et al.*, SOM located in large pore (> 60 μm) would decompose more slowly because of limited mobility of bacteria, limited nutrient availability due to limited solute diffusion than OM in pore < 4 μm which would be protected.[626] Indeed, the authors observed an accumulation of C in the pore small size class. Likewise, in sediment, small pores (< 10 nm) would afford a higher protection to OM as most of the surface area was shown to be provided by such pores.[523,524] Depending on the soil texture, 15 to 52% of sandy and clayey soils, respectively, are not accessible to bacteria due to size of the pores.[611] In addition, microaggregation could also create specific environment in which gas, water and nutrient circulation could be limited and, in consequence, where degradation rate would be decreased,[611] core microaggregates being potentially anaerobic.[612] Therefore, the pore size distribution and connectivity appear as key elements of biological processes.

The examination by Kinyangi *et al.* of microaggregates by STXM/ NEXAFS has indicated a differential oxidation state in OM functional groups between and within the microaggregates[627] and supports previous results indicating discrepancies between OM quantity[305] and composition[63] from macro and microaggregates. They inferred that the occurrence of highly processed OM present at the exterior of the microaggregate compared to the interior could be indicative of stabilization of C in microaggregates.

IV. Organic Matter and Trace Element Dynamics

OM is found in soils and sediments at concentrations of 0.1–10% by weight. Despite this relatively low abundance, OM may play key role on mineral fate, such as alteration or complexation. OM binds a variety of trace elements and typically contains 2–10 eq kg^{-1} of ionizable groups. The binding of protons and metal ions to OM is important for the speciation, transport, and toxicity of many trace metals, but it has proven difficult to find equations that describe this binding over a wide range of conditions. The direct measurement of their speciation in natural matrices is challenging since the analytical methods used should be sensitive for low concentration, and highly specific to the element and compartment of interest.[628] In order to circumvent these potential analytical limitations, various speciation models are developed and included in software assuming a thermodynamic equilibrium.

A. Effect of Organic Ligands on Mineral Alteration

The long-term biogeochemical cycles of carbon and trace elements can be influenced by rock weathering. Although the effect of pH and pCO_2 on rock dissolution is rather well studied and quantified[629] the role of OM is still a matter of debate as contradictory effects are reported.[630] In the following, we will therefore focus on the OM role. The main obstacle in quantifying the effect of OM on rock mineral weathering consists in disentangling the influence of pH and that of the organic ligands. The effects of extracellular acid and neutral polysaccharides, uronic acids (galacturonic, guloronic), peptides, aminoacids, low molecular weight organic carboxylic acids (oxalic, citric, tartaric, gluconic, lactic) on aluminosilicate (for example $NaAlSi_3O_8$) dissolution or on iron mobilization from silicates (for example $(Mg^{+2}, Fe^{+2})_2SiO_4$) have been widely described leading to the general conclusion that dissolution of minerals is affected by the presence of OM.[631] The interaction of OM with Ca or Mg-bearing silicates ($CaMgSi_2O_6$) was also reviewed and results demonstrate that very high concentrations (0.01–0.1 M) of organic ligands, whatever they originated from enzymatic degradation of OM, or bacterial metabolic activity, are needed to substantially affect dissolution.[632] Thus, the effect of natural organic ligands on the weathering of Ca or Mg-bearing silicates in soil environment is likely to be weak. The reactivity of secondary silicate clays except kaolinite[633] is also poorly known. Similarly, the effect of natural organic ligands on the dissolution kinetics of smectitic minerals $((Na,Ca)_{0.3}(Al, Mg)_2Si_4O_{10}(OH)_2 \cdot nH_2O$ in soil environment is likely to be weak.[630] Among silicates, the weathering of phyllosilicates will also influence geochemical cycles due to their abundance, large specific surface areas, and high ion-exchange capacities[634] and among phyllosilicates, biotite ($K(Mg,Fe)_3(OH,F)_2(Si_3AlO_{10})$) is the key source of primary (K) and secondary (Fe, Mg, Al) nutrients in soils and freshly exposed rocks.[635] In 2015, Bray *et al.* showed that the presence of organic ligands, at low concentrations, enhanced the release of di- and tri-valent metals from both octahedral and tetrahedral layers of biotite, by the formation of metal–organic ligand complexes (i.e. Al and Fe OM complexes).[636] However, this increased release of metals did not significantly influence the biotite dissolution rate, based on Si release. OM could also play a role in the precipitation of minerals during weathering processes; however this effect has not been properly documented yet. Future studies should address this issue and test in the field whether the lack or very weak effects of organic ligands observed in laboratory experiments are confirmed or not.

B. Measurements and Models for Trace Element Speciation in the Presence of OM

Several methods have been developed to determine trace element speciation in the presence of dissolved OM (Table 2).

Trace element complexation with soil or aquatic OM is strongly pH-dependent and a function of the element affinity constant for a given binding site and ionic strength (Figure 11).[638,653]

OM, as seen in Section C, is highly complex and comprises a large diversity of components, with various functional groups and therefore different types of binding sites. They can undergo homo-aggregation or heteroaggregation generating a wealth of distinct physicochemical species. Through binding, trace elements can form chemically heterogeneous complexes with OM. In fact, due to the physical size, polyfunctional and polyelectrolytic nature of the OM, the interactions cover a broad range of free energies (Figure 12).[654]

Chemical speciation is a key issue when assessing the bioavailability and eco-toxicity of metals and the distribution of trace elements in the bio-geosphere. Each fraction of OM in soils or aquatic systems will have different reactivity and binding capacities towards trace elements. Thus, the understanding of their speciation requires not only the knowledge of trace elements binding equilibrium parameters but, also the diffusion and/or kinetic fluxes of the various species in solution, both depending on the time-scale considered.[656] Several analytical techniques investigating trace element speciation are needed to collect equilibrium and dynamic parameters[657] (Figure 13). Two families of techniques can be distinguished (Figure 13).

Table 2. Analytical methods for trace element speciation in the presence of dissolved OM

Non-Electrochemical methods		References
Ion-selective electrode (ISE).	An ISE is a sensor converting the activity of a specific ion in a solution into an electrical potential. The voltage is theoretically dependent on the logarithm of the ionic activity, according to the Nernst equation. Potentiometry using ISEs is a well-established technique for free ion quantification without disturbing the equilibrium and the sample composition. Electrodes are available only for a limited number of elements (Ca, Cd, Cu, Pb…) usually with detection limits above 10^{-9} M. The development of potentiometric sensors using polymeric membranes has increased the ISE detection limit to values as low as 10^{-11} M.	637, 638, 639
Donnan membrane technique (DMT).	The device consists of a continuous flow system with a donor cell and an acceptor cell, separated by a semipermeable, negatively charged, cation exchange membrane. The negative electrostatic potential generated in the membrane by the deprotonated sulfonic acid groups allows only cations to pass through the membrane. The Donnan equilibrium at the two interfaces can be reached without any perturbation of the chemistry of the solutions in both compartments. Compared to ISEs, the DMT has the advantage to allow simultaneous measurement of several cations in a multicomponent sample solution, the lowest detection limits (10^{-11} M) being reached when metal ions in the acceptor side are detected by inductively coupled plasma mass spectrometry.	640, 641
Diffusive gradients in thin films (DGT).	It is a passive sampler that houses a binding gel, diffusive gel and membrane filter. The element or compound passes through the membrane filter and diffusive gel and is assimilated by the binding gel in a rate-controlled manner. The binding gel is a metal ion binding resin at the back of the diffusive layer formed by the diffusive gel and the membrane. A concentration gradient is formed due to the sorption of the free metal ion on the binding resign. DGT is based on the application of Fick's law. The mass of the accumulated metal after a given accumulation time is measured providing a flux that is related to the labile metal concentration in the sample solution. Once the mass of an analyte has been determined, the time-averaged concentration of the analyte in the bulk, can be determined	642, 643

Electrochemical methods		
Absence of gradients and Nernstian equilibrium stripping technique (AGNES).	It utilizes a bulk electrolysis step to pre-concentrate a substance from solution into the small volume of a mercury electrode (a hanging mercury drop or a thin film) or onto the surface of an electrode. It consists of two conceptual steps: (i) application of a potential program (e.g. a step at a fixed potential) generating a known concentration gain between the outer and inner concentrations of the metal at the mercury electrode surface together with null gradients of the concentration profiles (inside and outside the mercury electrode) and (ii) determination of the concentration of reduced metal inside the amalgam in a stripping step	644, 645
Scanned Stripping Chrono-Potentiometry (SSCP).	It is two-step technique: the deposition step is the same as stripping voltammetry, but quantification of the accumulated metal its achieved by application of a constant oxidizing current and the analytical signal it's the time taken for reoxidazion (the transition time, t). SSCP curves are constructed from a series of individual SCP measurements made over an interval of deposition potentials, E_d. This range of E_d covers the situations where no element is reduced at the electrode, passing through the Nerrstian regime to the region where all the elements reaching the electrode are reduced. Each SCP comprises two steps: (i) the deposition (accumulation) step. Where ions are reduced on the working electrode at the chosen E_d for a fixed period of time (t_d), followed by (ii) the stripping step, where the accumulated ion is reoxidized through the application of a low constant oxidizing current. The analytical signal is the electrolysis time required for reoxidation, τ (transition time). The representation of τ as a function of E_d results in a sigmoidal-shape curve	646, 647, 648
Competitive ligand exchange adsorptive cathodic stripping voltammetry (CIE-AdCSV).	Technique is characterized by a ligand exchange reaction. A well-characterized ligand (Lad) of known competition strength is added to the sample solution and equilibrium is set up between it and the different species of the target metal M. Measurement by cathodic stripping voltammetry. Lad needs to have well-defined complexation characteristics with respect to M, and MLad, must have a strong affinity for adsorption at the electrode surface	649, 650
Gel-integrated microelectrode (GIME) voltammetry.	An integrated microanalytical system coupling separation with voltammetric detection in a small volume. GIME measurements consist of two successive steps: equilibration of the agarose gel with the sample solution, and voltammetric analysis inside the gel	651, 652

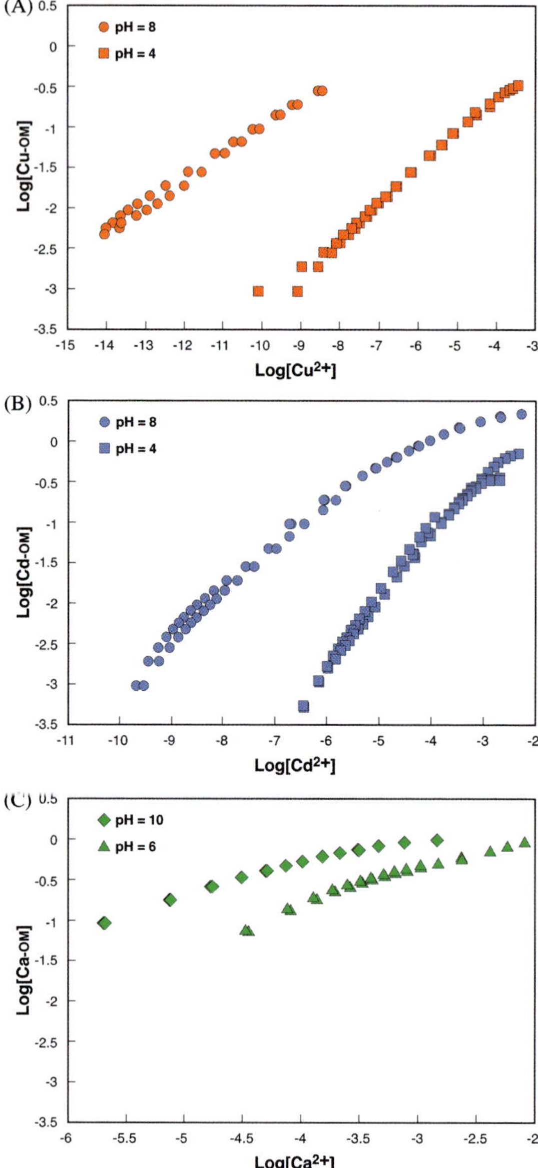

Figure 11. Cu, Cd and Ca binding (in mol $_{metal\ ion}$/kg $_{PPHA}$) to a purified peat extracted humic acid (PPHA) as function of pH. The slope of the binding curves reflects the heterogeneity of the binding process caused by the position of the binding moieties on the organic matter molecules as well as the type of binding for the metal ion (i.e. mono, bi or polydendate). The smaller the slope, the stronger is the heterogeneity. Here the slope is close to 0.4, 0.7 and 0.8 for Cu, Cd and Ca, respectively. The binding affinity is increasing in the following order Ca < Cd < Cu.

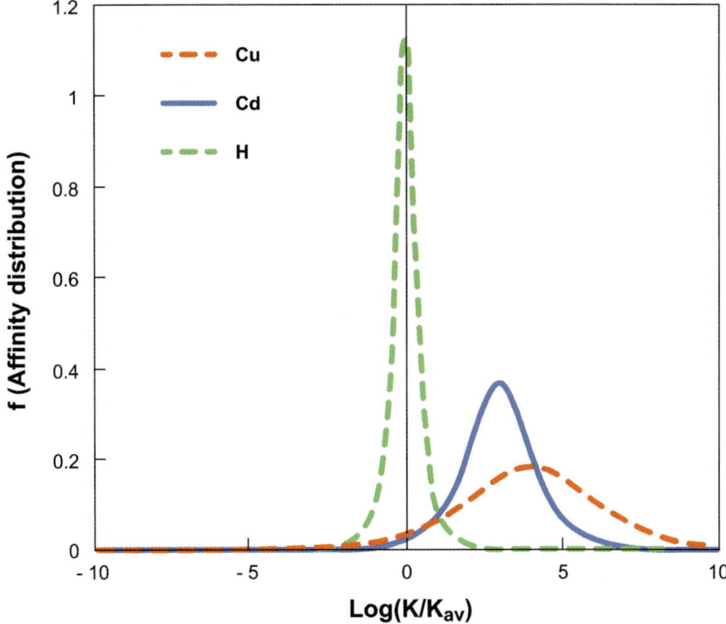

Figure 12. Calculated H, Cd and Cu affinity distributions' combining a binding Langmuir Freundlich isotherm and a Sips distribution for K values.[655] The distribution of log(K/K$_{av}$) represents the range of the binding energy to be considered for a given element when interacting with OM (i.e. binding chemical heterogeneity). K$_{av}$ is the median value for the binding affinity constant K. The ratio K/K$_{av}$ shows the partially correlated affinity distribution between H, Cd and Cu. Each element also experiences a different type of heterogeneity when binding to OM which explains the different width of their respective affinity distribution.

The electrochemical ones can give information on both dynamic and equilibrium speciation of the trace elements.[658] However, the obtained signal is sometimes difficult to interpret due to the variable nature of the so called "labile pool" for some elements (i.e. the concentration of the labile pool can be pH-and/or ligand- dependent). The second family corresponds to a range of techniques that will mostly give information on the free metal ion in solution. However, these approaches can suffer from poor detection limits outside the laboratory or are prone to interferences (i.e. Ca interferences on Cd ion selective electrode response).[637] Moreover, they do not account for the labile fraction of the trace element speciation. Each technique gives the fraction of free or labile element when in contact

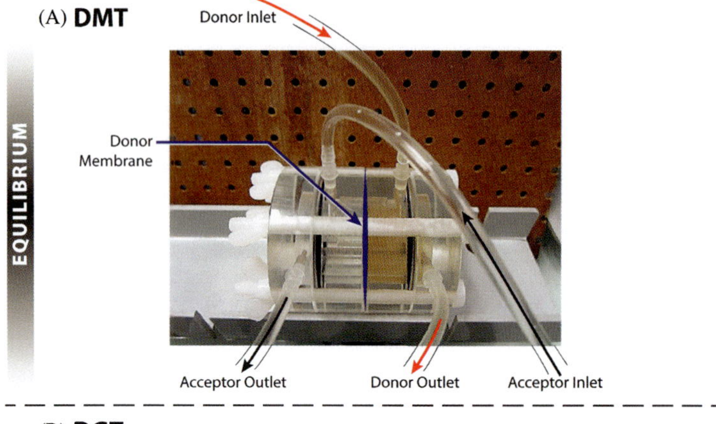

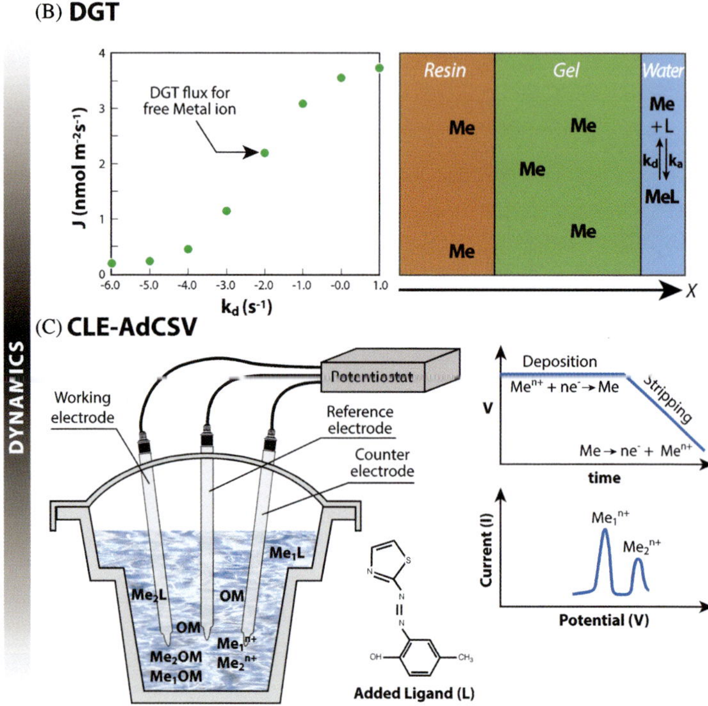

Figure 13. Examples of equilibrium (A) and dynamic (B, C) techniques to investigate trace element speciation. (A) Donnan membrane technique (DMT), (B) diffusive gradients in thin films (DGT), (C) competitive ligand exchange adsorptive cathodic stripping voltammetry (CLE-AdCSV). For further details on these techniques, see Table 2.

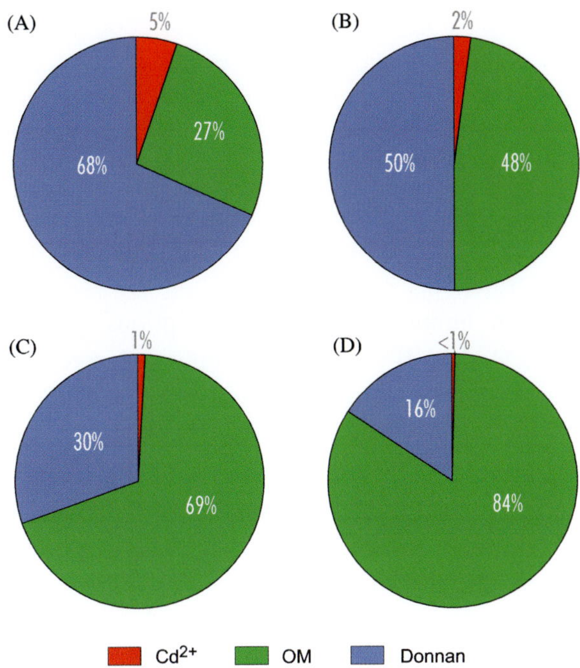

Figure 14. Example of Cd speciation when interacting with OM at pH 5 (A), 6 (B), 7 (C) and 8 (D), total [Cd] equal to 1 μM, ionic strength I = 0.001 and [OM] is 20 mg/L. Donnan: corresponding to Cd nonspecific electrostatic binding, OM: to Cd specifically bound to reactive groups of the OM (i.e. carboxylic and phenolic type groups), Cd^{2+} corresponding to free Cd ions in solution.

with OM in solution, and the exact amount of the element bound to the OM can be calculated by difference between the free or labile and the total amount of the trace element. Equilibrium models can, for instance, be used to characterize the nature on the binding (covalent versus ionic) (Figure 14).[640]

1. *Equilibrium Models and Measurements*

Speciation models are developed using intrinsic thermodynamic parameters to describe the interactions between trace elements and OM in the soil, sediment and water. They are based on the concentrations of elements or densities of reactive sites and the associated set of affinity constants. In addition, most of these models include a description of the electrostatic potential developed by organic macromolecules usually charged in solution. The addition of OM in speciation codes is usually considered as a daunting task,[659] and constitutes one of the latest evolutions in modelling.[660] Actually, binding of trace elements to humic substances (i.e. the mostly used OM surrogate for environmental metal binding studies despite its flaws), is complicated by their relative heterogeneity in charge, site densities and distribution.[661]

The two main humic substances descriptions are NICA-Donnan model,[638,662] and Model VII[663] and are incorporated in geochemical codes such as Visual Minteq (http://vminteq.lwr.kth.se), ECOSAT[664] and ORCHESTRA (www.orchestra.meeussen.nl). These two approaches are equilibrium-based models, and calibrated with the same published database for proton and metal binding to HA and FA, obtained from a broad spectrum of freshwaters and soils. Being calibrated with the same set of experimental data, the two model results are extremely close. However, some differences have been reported such as database distinctions for inorganic complexes.[665]

NOM is highly heterogeneous, for instance for its hydrophobicity degree, and having a generic description of NOM for all environments seems unrealistic.[666,667] As a result, adjustments to existing databases are made for a good description of the NOM reactivity in various environments. For instance, the proportion of HA over FA should be refined using field experiments.[640,668,669] Once fixed, these parameters allow robust modelling within the same system. Meylan *et al.* compared the main models for metal ion complexation to HA and FA using competitive ligand exchange adsorptive cathodic stripping voltammetry (CLE-AdSV) measurements (Figure 13) (Cu and Zn).[670] Interestingly, predictions were similar for the models with a good agreement for free zinc ion and labile concentrations, and an overestimation of free copper ion. Modeling of Cu interactions with humic substances is a demanding task and requires to consider the competitive effect of Fe(III) or Al(III) binding. However, in complex environmental matrices, Groenenberg *et al.* claimed that uncertainties in predicted speciation for Cu and Pb are mainly due to the natural variation in binding properties of FA, while uncertainties for lower affinity metals such as Cd result from inaccuracies in the reactive nature of the fulvic fraction in DOM and the maximum site density of the FA.[671] Indeed, Ren *et al.* showed, with DMT (Figure 13), that for a whole range of soil solutions, only 26.2% of DOM is reactive and consists mainly of FA, concluding that to make accurate modelling of metal speciation in soil solutions, the knowledge of DOM composition is the crucial information, especially for Cu.[640] Unsworth *et al.* used the DMT to measure the free metal ion activities, while the concentrations of the labile and mobile dynamic species have been estimated using DGT and GIME (Figure 13).[672] The authors concluded that a correct agreement was obtained between measured and modelled concentrations using Model VII and NICA Donnan models for the dynamic species, while the free ion concentrations were systematically underestimated by 3–5 orders of magnitude for Pb and Cu. Mueller *et al.* found that it was necessary to refine Model VII predictions by using spectroscopic determinations of the active fraction of DOM involved in Cd, Cu, Zn and Ni complexation in several Canadian Shield lakes to obtain good comparisons between measured and modelled metal speciation.[604] This led to suggest that the DOM binding sites active in Cd, Cu and Zn complexation are different from those involved in Ni complexation. Han *et al.* compared DGT and Model VII model in Japanese river waters.[673] It appeared that the model did not equally describe all the divalent cations investigated. Assumptions on precipitated Fe(III) and Al(III) resulted in reasonably predicted dynamic concentrations of Ni, Cu and Zn, while the model was not able to determine Pb dynamic concentrations. To account for the small differences between modelled and measured Ni or Cu, the authors also invoked a kinetic control on the slow dissociation rate of humic complexes. Similarly, Vega and Weng were able to describe correctly the speciation of Zn, Cu, Ni, and Cd in the Rhine River by estimating the free fraction using the ECOSAT computer program and corresponding database[664] with a NICA-Donnan description of the NOM metal ion interactions.[674] The results were then compared to direct free concentration measurements obtained by DMT analysis. However, as in Han *et al.*,[673] Pb was found to be poorly described, and the authors invoked an overestimation of the DOM complexation in the model, combined to an underestimation of Pb sorption onto iron hydroxides.

In addition, humic substances are not the only constituents of NOM, and some other organic ligands of natural or anthropic origin can bind metal ions in natural systems. For instance, siderophores are natural complexing agents, present in various environments such as the rhizosphere[675] or the Ocean.[676] These organic molecules are produced by microorganisms, usually as a strategy to enhance Fe(III) or other metal availability in solution, or as recently shown as a detoxification strategy by bacteria.[677] They exhibit high affinities for metals with reported concentrations ranging from tens of micromoles to a few millimoles per litre in soil solutions.[678] Given their potential role as a control on metal cycling and metal bioavailability, studies focused on the siderophore impact on mineral stabilities during weathering.[679] Despite their possible important role, detailed information concerning the influence of siderophores on heavy metal mobility in environmental systems is still missing in speciation software. Another type of organic ligands produced by microbial cells is exopolymeric substances. These molecules are mainly composed of polysaccharides and proteins.[680] Literature and codes are not well-documented regarding the different types, functionalities and thermodynamic constants associated with these complexants. However, attempts to take into account their role in natural environments showed that their contribution to metal binding can be significant, although being less than 20% for Cd binding, under most environmental conditions.[681]

These modeling approaches postulate the studied system to be at thermodynamic equilibrium. They should thus consider reactions that are sufficiently fast and fully reversible. However, natural aquatic systems are frequently subject to changing conditions and practically never reach chemical equilibrium.[682] For instance, dissolution/precipitation reactions can be associated to slow kinetics when regarding nanoparticle stability in natural waters,[683,684] and recent studies reported different kinetics for metal exchange rate between different bearing phases in ultramafic soil solution systems for Ni(II).[685]

2. *Kinetics versus Thermodynamics*

OM exhibits a wide distribution of complexation affinity for metals (Figures 11 and 12), and consequently a broad distribution of metal complex dissociation kinetics that can vary over orders of magnitude. As a result, some metal complexes will be labile, while in the same solution, other metal complexes will be considered as inert. Lability is defined as the input of metal species into an overall flux toward a binding surface/interface (i.e. a chemical sensor or an organism) which is the sum of the diffusive mass transport and the metal complex dissociation fluxes.[686]

The dynamic approach was explored in the field by ClE-AdSV by Xue and Sigg at low concentrations found in natural waters.[687] Tercier-Waeber *et al.* showed that GIME-voltammetry is efficient for simultaneous measurement of the dynamic fraction of Cu(II), Pb(II), Cd(II) and Zn(II), at the part per trillion level using square wave anodic stripping voltammetry, and of Mn(II) and Fe(II), at the part per billion level using square wave cathodic sweep voltammetry.[688] Several dynamic DMT procedures were tested to shorten the analysis time.[689,690] It is possible to assess the evolution of the isotopically exchangeable concentration as a function of time by performing isotopic exchange kinetic experiments, which allow the identification of the different types of exchangeable compartments.[685,691]

The simultaneous use of dynamic and equilibrium techniques to study trace element binding with heterogeneous ligands like OM can result in a great advantage as they provide complementary data.[692] The reactive transport softwares (ECOSAT and ORCHESTRA) incorporate a state of the art description of thermodynamic reactions, particularly for OM and sorption at the mineral colloid surfaces, combined to a mass transport code. However, thermodynamic modelling alone is a simplified limiting case of the general dynamic situation.[693] New approaches including a dynamic description of the equilibrium were developed to evaluate metal bioavailability. User-friendly codes for such computations, FLUXY[694] and MHEDYN[695,696] are available and are described below. The parameter values required for such calculations are given in Buffle *et al.*[694] and Zhang *et al.*[697] The latter modeled the values of the chemical rate constants and diffusion coefficients needed for metal ion flux calculations with the most appropriate models for the three major types of complexing agent considered in aquatic systems: the small and simple ligands, the fulvic and humic compounds, and the colloidal "particles" or aggregates.

The simplest available code for the computation of steady state metal fluxes is FLUXY[694] (http://www.unige.ch/cabe/dynamic/). The code includes two major assumptions, (i) the existence of excess of ligand (L) compared to the metal ion concentration (M), and (ii) in a series of successive ML complexes, the reaction is the rate-limiting step for the flux calculation. The code was applied for Pb, Cd, Zn, Cu and Ni to typical case studies of aquatic systems: culture medium with simple ligands, solutions with heterogeneous FA and suspension of aggregates with a broad size distribution.[698] The calculations, for an interface attracting trace elements (i.e. adsorbing or absorbing) with planar geometry, show that in the culture medium, Pb and Cd are strongly complexed by simple ligands and form labile complexes which all contribute to the flux, whereas Zn forms labile but not very stable complexes, i.e., the contribution of the zinc plays a major role on the total flux.

A more complex approach, MHEDYN, computes the contribution to metal flux, the degree of lability and the concentration profile of any species, in the transient and steady state regimes, without requirement of ligand excess compared to metal, like in FLUXY. The MHEDYN code was extensively tested to simulate the metal flux and dynamic speciation at bio-interfaces for copper.[695,696] These results suggest that the contribution of Cu-complexes with particles/aggregates is usually small, even though it depends on the diffusion layer thickness (i.e. the region in the vicinity of an electrode where the concentrations are different from their value in the bulk solution, DLT), whereas the contribution of Cu-fulvic complexes is highly significant or even predominant. In a similar study focussing on Pb uptake, Zhang and Buffle showed that changes of the DLT greatly influence the overall lability of the Pb complexes in aquatic systems.[699] The size of the DLT can be found either from the simulated or measured concentration profiles (yielding the "true" diffusion layer thickness) or, in the case of the Nernst ("linear") diffusion layer by extrapolating the concentration gradient at the electrode surface to the distance at which the concentration takes its bulk value. At large DLT values, typically > 1 mm, almost all complexes of Pb are labile, while at very small DLT values, only the contribution of free metal ion to the total flux is relevant. However, under realistic conditions where DLT > 5–10 μm, the Pb-complexes formed with carbonate and part of the FA are the most labile. The role of FA on the total flux may vary from very important, at sufficiently large metal/FA ratio to negligible at high pH and low metal/FA ratios, since the lability of metal–FA complexes drastically depends on those factors.[695]

In 2013, Duval proposed a new theory for the dynamics of metal uptake by a spherical microorganism whose peripheral structure consists of a charged bioactive surface surrounded by a soft (ion-permeable) charged layer.[700] This formalism is developed to explicitly consider the concomitant steady-state conductive diffusion transport of metals from bulk medium to the bioactive surface and the kinetics of intracellular metal internalisation described by a Michaelis–Menten mechanism.[700] The spatial distribution of metals at the microorganism/solution interface results from an explicit solution of the Nernst–Planck equation with differentiated metal diffusion coefficients inside and outside the microorganism soft surface layer. This new model offers a better definition of the complex relationships existing between toxicity of metals and extent of metal bioaccumulation and physicochemical properties of microorganisms.

Despite the many efforts that were devoted to determine the labile metal species, different labile species are determined by different methods, depending on the detection window, both in thermodynamic and kinetic sense. For this reason, the comparison between dynamic methods is very difficult, complicating the efforts necessary to integrate the experimental kinetic data on computer codes and models. In addition, the natural OM often aggregates with various types of solids or polymeric compounds. This clearly affects their diffusive transport, having a critical impact on the contribution of metal–aggregate complexes to the total flux. In addition to diffusion, the effective chemical kinetics of complex association/dissociation also depends on the particle/aggregate size (metal diffusion inside the aggregate may also affect the overall kinetics of metal–aggregate interaction due to the inhomogeneous site distribution inside the aggregate) and the electric double layer field.[701] All these aspects are crucial and challenging to incorporate in models of metal dynamic speciation.

V. Conclusion

Despite its great importance in the global biogeochemical cycles and potential role in the transportation of trace elements, natural OM is often considered as a "black box," especially in the models of ecosystem functioning. Thanks to numerous analytical developments, the chemist nowadays has a large array of techniques at his disposal to investigate the chemical structure of the natural OM. They led to conspicuous advances in the knowledge of the composition, reactivity and fate of this complex material, which in addition exhibits a large spatial and temporal heterogeneity. Despite this continuous progress, some analytical limitations still exist. They are mainly related to the difficulties in simultaneously analyzing the mineral and organic phases and in deriving quantitative data. Modeling the organo-mineral interactions, although it is still sometimes semi-empirical rather than mechanistic, can help in understanding behaviors that cannot be probed directly and thus can overcome some analytical limitations. However, an effort should be pursued to include the detailed composition of the OM in models describing the interactions between OM and the other components of soil and sediments. An improved quantification of DOM reactivity and variability in soil solutions can only be achieved through a precise knowledge of overall DOM composition. It would result in an accurate prediction of trace element speciation in soils and soil solutions. More integrative analysis of the various DOM fractions is required to further characterize the quality of DOM to improve the model performances. Moreover, it is still important to obtain more accurate measurement by improving the detection limit of the trace element speciation analytical techniques in the solutions. Meanwhile, the role of the competitive binding of Fe^{3+} and Al^{3+} to OM on the trace element speciation deserves more attention.

It is experimentally shown that certain fractions of the OM are preferentially adsorbed onto the mineral surfaces in various conditions of pH, ionic strength and coverage ratio and result in differences in reactivity between the sorbed and the dissolved OM fractions. These variations in reactivity should be taken into account to model the fate of the pollutant/OM/mineral ternary systems.

Currently, the major problem is our limited level of understanding of the metal dynamic fate and the associated lack of efficient modelling tools. The efforts devoted to determine the labile metal species contributing to the bioavailable fraction are unfortunately depending on each method detection window, from a thermodynamic and kinetic point of view.

As described in the present review, most studies on natural OM presently focus on carbon despite the increasing interest in the biogeochemical cycles of other nutrients such as nitrogen, sulfur and phosphorus. However, the characterization of the chemical functions comprising these elements still constitutes an analytical challenge and further developments are needed in this context. Another limitation in the study of the natural OM is that most of this OM occurs in an insoluble and even often non-hydrolysable form. As a result, the available analytical methods are more restricted and there were several attempts to recover soluble fractions that could be representative of the whole OM but no satisfactory solution was reached so far. Similarly, methods which are presently developed on model compounds should now be applied to natural samples. Moreover, as detailed in this chapter, physical and chemical fractionations commonly yield OM which is not representative of that naturally occurring in the environment. A significant breakthrough was achieved through the development of the *in situ* measurements, especially thanks to the evolvement of synchrotron-based techniques. However, these techniques probe a tiny fraction of the sample, which also leads to questions about its representativeness and to difficulties in changing the scale. Data retrieved at this microscale should be confronted to information derived from larger scale analyses to test their general character. More generally, combining several analytical approaches is necessary to avoid any bias associated with the use of a single technique. This is especially true due to the complexity and heterogeneity of the natural OM.

VI. Acknowledgments

The authors thank Marie Alexis for providing NMR spectra, Elodie Salmon and François Baudin for their help and expertise in ESI-FTICR-MS and in Rock-Eval pyrolysis, respectively. We appreciate the constructive suggestions and the fine tuning of the layout supplied by Joel Dyon whose professionalism conjured the figures from our drawings.

VII. References

1. Hunt, J. M. *Petroleum Geochemistry and Geology*, 2nd ed., Freeman, New York, **1996**.
2. Wollast, R. In *The Sea*. Brink, K. H.; Robinson, A. R. (eds.). University of Brussels Press, **1998**, pp. 213–252.

3. Hedges, J. I.; Keil, R. *Mar. Chem.* **1995**, *4*, 81–115.

4. Meybeck, M. In *Interactions of C, N, P and S*. Wollast, R.; Mackenzie, F. T.; Chou, L. (eds.). Springer, Berlin, Heidelberg, New York, **1993**, pp. 163–193.

5. Hedges, J. I.; Oades, J. M. *Org. Geochem.* **1997**, *27*(7), 319–361.

6. Tranvik, L. J.; Downing, J. A.; Cotner, J. B.; Loiselle, S. A.; Striegl, R. G.; Ballatore, T. J.; Dillon, P.; Finlay, K.; Fortino, K.; Knoll, L. B.; Kortelainen, P. L.; Kutser, T.; Larsen, S.; Laurion, I.; Leech, D. M.; Mccallister, S. L.; Mcknight, D. M.; Melack, J. M.; Overholt, E.; Porter, J. A.; Prairie, Y.; Renwick, W. H.; Roland, F.; Sherman, B. S.; Schindler, D. W.; Sobek, S.; Tremblay, A.; Vanni, M. J.; Verschoor, A. M.; Von Wachenfeldt, E.; Gesa, A.; Weyhenmeyer, G. A. *Limnol. Oceanogr.* **2009**, 54, 2298–2314.

7. Harrison, J. A.; Caraco, N.; Seitzinger, S. P. *Global Biogeochem. Cycl.* **2005**, *19*(4).

8. Hansell, D. A.; Carlson, C. A.; Repeta, D. J.; Schlitzer, R. *Oceanography* **2009**, *22*(4), 202–211.

9. Dittmar, T.; Stubbins, A. Dissolved organic matter in aquatic systems. In *Treatise on Geochemistry,* 2nd ed., Elsevier, Oxford, **2014**, 125–156.

10. Jobbagy.; Jackson. *Ecol. Appl.* **2000**, *10*, 423–436.

11. Lal, R. *Science* **2004**, *304*, 1623.

12. Hedges, J. I.; Eglinton, G.; Hatcher, P. G.; Kirchman, D. L.; Arnosti, C.; Derenne, S.; Evershed, R. P.; Kögel-Knabner, I.; De Leeuw, J. W.; Littke, R.; Michaelis W.; Rullkötter, J. *Strbl* **2000**, *31*, 945–958.

13. Tissot, B.; Welte, D. H. *Petroleum Formation and Occurrence*. Springer, Berlin, **1984**.

14. Vandenbroucke, M.; Largeau, C. *Org. Geochem.* **2007**, *38*, 719–833.

15. Van Krevelen, D. W. *Coal*. Elsevier Publishing Co., New York, **1961**.

16. Kögel-Knabner, I. *Soil Biol. Biochem.* **2002**, *34*, 139–162.

17. Lal, R., *Critical Reviews in Plant Sciences*, Taylor & Francis, **2003**.

18. Sollins, P.; Kramer, M. G.; Swanston, C.; Latjha, K.; Filley, T.; Aufdenkampe, A. K.; Wagai, R.; Bowden, R. D. *Biogeochemistry* **2009**, *96*, 209–231.

19. Miltner, A.; Bombach, P.; Schmidt-Brücken, B.; Kästner, M. *Biogeochemistry* **2012**, *111*, 41–45.

20. Simpson, A. J.; Simpson, M. J.; Smith, E.; Kelleher, B. P. *Environ. Sci. Technol.* **2007**, *41*, 8070–8076.

21. Grandy, S.; Neff, J. C. *Sci. Total Environ.* **2008**, *404*, 297–307.

22. Mambelli, S.; Bird, J. A.; Gleixner, G.; Dawson, T. E.; Torn, M. S. *Org. Geochem.* **2011**, *42*(9), 1099–1108.

23. Liang, C.; Balser, T. C. *Nat. Rev. Microbiol.* **2011**, *9*(1), 75.

24. Cotrufo, M. F.; Wallenstein, M. D.; Boot, C. M.; Denef, K.; Paul, E. A. *Global Change Biol.* **2013**, *19*, 988–995.

25. Kögel-Knabner, I.; Amelung, W. *Treatise in Geochemistry*. Elsevier, **2014**.

26. Kögel-Knabner, I. *Soil Biol. Biochem.* **2016**.

27. Derenne, S.; Nguyen Tu, T. T. *Comptes Rendus Geoscience* **2014**, *346*(3), 53–63.

28. Durand, B.; Nicaise, G. In *Kerogen, Insoluble Organic Matter from Sedimentary Rocks*. Durand, B. (ed.). Editions Technip, Paris, **1980**, pp. 13–34.

29. Huizinga, B. J.; Tannenbaum, E.; Kaplan, I. R. *Geochim. Cosmochim. Acta* **1987**, *51*, 1083–1097.

30. Nierop, K. G. J.; Van Bergen, P. F. *J. Anal. Appl. Pyrolysis* **2002**, *63*, 197–208.

31. Faure, P.; Schlepp, L.; Mansuy-Huault, L.; Elie, M.; Jardé, E.; Pelletier, M. *J. Anal. Appl. Pyrolysis* **2006**, *75*, 1–10.

32. Wu, Z.; Steel, K. M. *Fuel* **2007**, *86*, 2194–2200.

33. Strydom, C. A.; Bunt, J. R.; Schobert, H. H.; Raghoo, M. *Fuel Process. Technol.* **2011**, *92*, 764–770.

34. Robl, T. L.; Davis, B. H. *Org. Geochem.* **1993**, *20*, 249–255.

35. Meshram, P.; Purohit, B. K.; Sinha, M. K.; Sahu, S. K.; Pandey, B. D. *Renew. Sustain. Energ. Rev.* **2015**, *41*, 745–761.

36. Suleimenova, A.; Bake, K. D.; Ozkan, A.; Valenza Ii, J. J.; Kleinberg, R. L.; Burnham, A. K.; Ferralis, N.; Pomerantz, A. E. *Fuel* **2014**, *135*, 492–497.

37. Gelinas, Y.; Prentice, K.; Baldock, J. A.; Hedges, J. *Est.* **2001**, *35*, 3519–3525.

38. Ruppenthal, M.; Oelmann, Y.; Wilcke, W. *Environ. Sci. Technol.* **2013**, *47*, 949–957.

39. Dai, K.; Johnson, C. E. *Geoderma* **1999**, *93*, 289–310.

40. Smernik, R. J.; Oades, J. M. *Geoderma* **2000**, *96*, 159–171.

41. Gonçalves, C. N.; Dalmolin, R. S. D.; Dick, D. P.; Knicker, H.; Klamt, E.; Kögel-Knabner, I. *Geoderma* **2003**, *116*, 373– 392.

42. Schmidt, M.; Gleixner, G. *Eur. J. Soil Sci.* **2005**, *56*, 407–416.

43. Rumpel, C.; Rabia, N.; Derenne, S.; Quénéa, K.; Eusterhues, K.; Kögel-Knabner, I.; Mariotti, A. *Org. Geochem.* **2006**, *37*, 1437–1451.

44. Eusterhues, K.; Rumpel, C.; Kögel-Knabner, I. *Org. Geochem.* **2007**, *38*, 1356–1372.

45. Zegouagh, Y.; Derenne, S.; Dignac, M.-F.; Bariuso, E.; Mariotti, A.; Largeau, C. *J. Anal. Appl. Pyrolysis* **2004**, *71*, 119–135.

46. Tambach, T. J.; Veld, H.; Griffioen, *J. Appl. Geochem.* **2009**, *24*, 2144–2151.

47. Lou, L.; Luo, L.; Cheng, G.; Wei, Y.; Mei, R.; Xun, B. *et al. Bioresour. Technol.* **2012**, *112*, 61– 66.

48. Sandron, S.; Rojas, A.; Wilson, R.; Davies, N. W.; Haddad', P. R.; Shellie, R. A.; Nesterenko, P. N.; Kelleher, B. P.; Paull, B. *Environ. Sci.: Processes Impacts* **2015**, *17*, 1531.

49. Sun, L; Perdue, E. M.; Mccarthy, J. F. *Water Res.* **1995**, *29*, 1471–1477.

50. Zhang, Y.; Huang, W.; Ran, Y.; Mao, J. *Marine Poll. Bull.* **2014**, *85*, 60–66.

51. Ouellet *et al.*, *Environ. Sci. Technol.* **2008**, *42*, 2490–2495.

52. Koprivnjak, J. F.; Perdue, E. M.; Pfromm, P. H. *Water Res.* **2006**, *40*, 3385–3392.

53. Koprivnjak, J. F.; Pfromm, P. H.; Ingall, E.; Vetter, T. A.; Schmitt-Kopplin, P.; Hertkorn, N.; Frommberger, M.; Knicker, H.; Perdue, E. M. *Geochim. Cosmochim. Acta* **2009**, *73*, 4215–4231.

54. Vetter, T. A.; Perdue, E. M.; Ingall, E.; Koprivnjak, J.-F.; Pfromm, P. H. *Sep. Purif. Technol.* **2007**, *56*, 383–387.

55. Gurtler, B. K.; Vetter, T. A.; Perdue, E. M.; Ingall, E.; Koprivnjak, J. F.; Pfromm, P. H. *Membr J. Sci.* **2008**, *323*, 328–336.

56. Chen, C.; Dynes, J. J.; Wang, J.; Karunakaran, C.; Sparks, D. L. *Environ. Sci. Technol.* **2014**, *48*, 6678–6686.

57. Helms, J. R.; Mao, J.; Chen, H.; Perdue, E. M.; Green, N. W.; Hatcher, P. G.; Mopper, K.; Stubbins, A. *Mar. Chem.* **2015**, *177*, 278–287.

58. Green, N. W.; Perdue, E. M.; Aiken, G. R.; Butler, K. D.; Chen, H.; Dittmar, T.; Niggemann, J.; Stubbins, A. *Mar. Chem.* **2014**, *161*, 14–19.

59. Chon, K.; Lee, Y.; Traber, J.; Von Gunten, U. *Water Res,* **2013**, 47, 5381–5391.

60. Chen, M.; Hur, J. *Water Res.* **2015**, *79*, 10–25.

61. Von Lützow, M.; Kögel-Knabner, I.; Ekschmitt, K.; Flessa, H.; Guggenberger, G.; Matzner, E.; Marschner, B. *Soil Biol. Biochem.* **2007**, *39*, 2183–2207.

62. Christensen, B. T. *Adv. Soil Sci.* **1992**, *20*, 1–90.

63. Golchin, A.; Oades, J. M.; Skjemstad, J. O.; Clarke, P. *Aust. J. Soil Res.* **1994**, *32*, 285–309.

64. Cerli, C.; Celi, L.; Kalbitz, K.; Guggenberger, G.; Kaiser, K. *Geoderma* **2012**, *170*, 403–416.

65. Griepentrog, M.; Schmidt, M. W. I. *J. Plant Nutr. Soil Sci.* **2013**, *176*, 500–504.

66. Crow, S. E.; Swanston, C. W.; Lajtha, K.; Brooks, J. R.; Keirstead, H. *Biogeochemistry* **2007**, *85*, 69–90.

67. Basile-Doelsch, I.; Amundson, R.; Stone, W. E. E.; Borschneck, D.; Bottero, J. Y.; Moustier, S.; Masin, F.; Colin, F. *Geoderma* **2007**, *137*, 477–489.

68. Basile-Doelsch, I.; Brun, T.; Borschneck, D.; Masion, A.; Marol, C.; Balesdent, J. *Geoderma* **2009**, *151*, 77–86.

69. De Junet, A.; Basile-Doelsch, I.; Borschneck, D.; Masion, A.; Legros, S.; Marol, C.; Balesdent, J.; Templier, J.; Derenne, S. *J. Anal. Appl. Pyrolysis* **2013**, *99*, 92–100.

70. Wagai, R.; Kajiura, M.; Asano, M.; Hiradate, S. *Geoderma* **2015**, *241*, 295–305.

71. Wagai, R.; Mayer, L. M.; Kitayama, K. *J. Plant Nutr. Soil Sci.* **2009**, *55*, 13–25.

72. Sohi, S. P.; Mahieu, N.; Arah, J. R. M.; Powlson, D. S.; Madari, B.; Gaunt, J. L. *Soil Sci. Soc. Am. J.* **2001**, *65*, 1121–1128.

73. Poeplau, C.; Don, A. *J. Plant Nutr. Soil Sci.* **2014**, *177*, 137–140.

74. Castanha, C.; Trumbore, S.; Amundson, R. *Radiocarbon* **2008**, *50*, 83–97.

75. Heckman, K.; Grandy, A. S.; Gao, X.; Keiluweit, M.; Wickings, K.; Carpenter, K.; Chorover, J.; Rasmussen, C. *Geochim. Cosmochim. Acta* **2013**, *121*, 667–683.

76. Poirier, N.; Sohi, S. P.; Gaunt, J. L.; Mahieu, N.; Randall, E. W.; Powlson, D. S.; Evershed, R. P. *Org. Geochem.* **2005**, *36*, 1174–1189.

77. Heckman, K.; Throckmorton, H.; Clingensmith, C.; Gonzalez Vila, F. J.; Horwath, W. R.; Knicker, H.; Rasmussen, C. *Soil Biol. Biochem.* **2014**, *77*, 1–11.

78. Courtier-Murias, D.; Simpson, A. J.; Marzadori, C.; Baldoni, G.; Ciavatta, C.; Fernández, J. M.; López-De-Sá, E. G.; Plaza, C. *Agric. Ecosyst. Environ.* **2013**, *171*, 9–18.

79. Boeni, M.; Bayer, C.; Dieckow, J.; Conceição, P. C.; Dick, D. P.; Knicker, H.; Salton, J. C.; Motta Macedo, M. C. *Agric. Ecosyst. Environ.* **2014**, *190*, 80–86.

80. Rovira, P.; Vallejo, V. R. *Soil Biol. Biochem.* **2003**, *35*, 245–261.

81. Sollins, P.; Swanston, C.; Kleber, M.; Filley, T.; Kramer, M.; Crow, S.; Caldwell, B. A; Lajtha, K.; Bowden, R. *Soil Biol. Biochem.* **2006**, *38*, 3313–3324.

82. Mikutta, R.; Kaiser, K.; Dörr, N.; Vollmer, A.; Chadwick, O. A.; Chorover, J.; Kramer, M. G.; Guggenberger, G. *Geochim. Cosmochim. Acta* **2010**, *74*, 2142–2164.

83. Bock, M. J.; Mayer, L. M. *Mar. Geol.* **2000**, *163*, 65–75.

84. Sampere, T. P.; Bianchi, T. S.; Wakeham, S. G.; Allison, M. A. *Cont. Shelf Res.*, **2008**, *28*, 2472–2487.

85. Wakeham, S. G.; Canuel, E. A.; Lerberg, E. J.; Mason, P.; Sampere, T. P.; Bianchi, T. S. *Mar. Chem.* **2009**, *115*, 211–225.

86. Schreiner, K. M.; Bianchi, T. S.; Eglinton, T. I.; Allison, M. A.; Hanna, A. J. M. *J. Geophys. Res. Biogeosci.* **2013**, *118*, 808–824.

87. Wang, J.; Yao, P.; Bianchi, T. S.; Li, D.; Zhao, B.; Cui, X.; Pana, H.; Zhang, T.; Yu, Z. *Chem. Geol.* **2015**, *402*, 52–67.

88. Chenu, C.; Plante, A. F. *Eur. J. Soil Sci.* **2006**, 57, 569–607.

89. Christensen, B. T. *Eur. J. Soil Sci.* **2001**, *52*, 345–353.

90. Breulmann, M.; Masyutenko, N. P.; Kogut, B. M.; Schroll, R.; Dörfler, U.; Buscot, F.; Schulz, E. *Sci. Total Environ.* **2014**, *497*, 29–37.

91. Breulmann, M.; Boettger, T.; Buscot, F.; Gruendling, R.; Schulz, E. *Sci. Total Environ.* **2016**, *545*, 30–39.

92. Maillard, É.; Angers, D. A.; Chantigny, M.; Bittman, S.; Rochette, P.; Lévesque, G.; Hunt, D.; Parent, L.-É. *Agric. Ecosyst. Environ.* **2015**, 202, 108–119.

93. Toriyama, J.; Hak, M.; Imaya, A.; Hirai, K.; Kiyono, Y. *For. Ecol. Manag.* **2015**, 335, 147–155.

94. Dou, X.; Xu, X.; Shu, X.; Zhang, Q.; Cheng, X. *Ecol. Engin.* **2016**, 87, 263–270.

95. Tisdall, J. M.; Oades, J. M. *J. Soil Sci.* **1982**, *33*, 141–163.

96. Asano, M.; Wagai, R. *Geoderma* **2014**, *216*, 62–74.

97. Bimüller, C.; Kreyling, O.; Kölbl, A.; Von Lützow, M.; Kögel-Knabner, I. *Soil Tillage Res.* **2016**, *160*, 23–33.

98. Puget, P; Chenu, C; Balesdent, J. *Eur. J. Soil Sci.* **2000**, *51*, 595–605.

99. Six, J.; Elliott, E. T.; Paustian, K. *Soil Biol. Biochem.* **2000**, *32*, 2099–2103.

100. Six, J.; Bossuyt, H.; Degryze, S.; Denef, K. *Soil Till. Res.* **2004**, *79*, 7–31.

101. Six, J.; Paustian, K. *Soil Biol. Biochem.* **2014**, *68*, A4–A9.

102. Guo, L. D.; Santschi, P. H. *Marine Chem.* **1996**, *55*, 113–127.

103. Benner, R. In *Biogeochemistry of Marine Dissolved Organic Matter*. Hansell, D. A.; Carlson, C. A. (eds.). Academic Press, New York, **2002**, pp. 59–85.

104. Kruger, B. R.; Dalzell, B. J.; Minor, E. C. *Aquat. Sci.* **2011**, *73*, 405–417.

105. Mopper, K.; Stubbins, A.; Ritchie, J. D.; Bialk, H. M.; Hatcher, P. G. *Chem. Rev.* **2007**, *107*, 419–442.

106. Roubeuf, V.; Mounier, S.; Benaim, J. Y. *Org. Geochem.* **2000**, *31*, 127–131.

107. Matilainen, A.; Gjessing, E.; Lahtinen, T.; Hed, L.; Bhatnagar, A.; Sillanpää, M. *Chemosphere* **2011**, *83*(11), 1431–1442.

108. Alasonati, E.; Slaveykova, V. I.; Gallard, H.; Croué, J.-P.; Benedetti, M. F. *Water Res.* **2010**, *44*, 223–231.

109. Bhatia, D.; Bourven, I.; Simon, S.; Bordas, F.; Van Hullebusch, E. D.; Rossano, S.; Lens, P. N. L.; Guibaud, G. *Bioresour. Technol.* **2013**, *131*, 159–165.

110. Everett, C. R.; Chin, Y.-P.; Aiken, G. A. *Limnol. Oceanogr.* **1999**, *44*, 1316–1322.

111. Pelekani, C.; Newcombe, G.; Snoeyink, V. L.; Hepplewhite, C.; Assemi, S.; Beckett, R. *Environ. Sci. Technol.* **1999**, *33*, 2807–2813.

112. Pérez, M. A. P. P.; Moreira-Turcq, P.; Gallard, H.; Allard, T.; Benedetti, M. F. *Chem. Geol.* **2011**, *286*, 158–168.

113. Qin, X.; Liu, F.; Wang, G.; Weng, L. *J. Sep. Sci.* **2012**, *35*, 3455–3460.

114. Alberts, J. J. J.; Tak, C. S. M.; Egeberg, P. K.; Takacs, M. *Org. Geochem.* **2002**, *33*, 817–828.

115. Woods, G. C.; Simpson, M. J.; Kelleher, B. P.; Mccaul, M.; Kingery, W. L.; Simpson, A. J. *Environ. Sci. Technol.* **2009**, *44*, 624–630.

116. Espinoza, L. A. T.; Ter Haseborg, E.; Weber, M.; Frimmel, F. H. *Appl. Catal. B Environ.* **2009**, *87*, 56–62.

117. Kawasaki, N.; Matsushige, K.; Komatsu, K.; Kohzu, A.; Nara, F. W.; Ogishi, F.; Yahata, M.; Mikami, H.; Goto, T.; Imai, A. *Water Res.* **2011**, *45*, 6240–6248.

118. Lepane, V.; Leeben, A.; Malashenko, O. *Aquat. Sci. - Res. Across Boundaries* **2004**, *66*, 185–194.

119. Minor, E. C.; Simjouw, J. P.; Boon, J. J.; Kerkhoff, A. E.; Van Der Horst, J. *Mar. Chem.* **2002**, *78*, 75–102.

120. Giddings, J. C.; Myers, M. N.; Caldwell, K. D. *Sep. Sci. Technol.* **1981**, *16*, 549–575.

121. Beckett, R.; Jue, Z.; Giddings, J. C. C. *Environ. Sci. Technol.* **1987**, *21*, 289–295.

122. Guéguen, C.; Cuss, C. W. *J. Chromatogr. A* **2011**, *1218*, 4188–4198.

123. Stolpe, B.; Guo, L.; Shiller, A. M.; Hassell, V. M. *Mar. Chem.* **2010**, *118*, 119–128.

124. Wells, M. L. *Mar. Chem.* **2004**, *89*, 89–102.

125. Yohannes, G.; Wiedmer, S. K. K.; Jussila, M.; Riekkola, M. L. M.-L. *Chromatographia* **2005**, *61*, 359–364.

126. Zanardi-Lamardo, E.; Clark, C. D.; Zika, R. G. *Anal. Chim. Acta* **2001**, *443*, 171–181.

127. Moon, J.; Kimb, S. H.; Cho, J. (2006). *Colloids Surf. A Physicochem. Eng. Asp.* **2006**, *287*, 232–236.

128. Pifer, A. D.; Miskin, D. R.; Cousins, S. L.; Fairey, J. L. *J. Chromatogr. A* **2011**, *1218*, 4167–4178.

129. Pifer, A. D.; Fairey, J. L. *Water Res.*, **2012**, *46*, 2927– 2936.

130. Zanardi-Lamardo, E.; Clark, C. D.; Moore, C. A.; Zika, R. G. *Environ. Sci. Technol.* **2002**, *36*, 2806–2814.

131. Balch, J.; Guéguen, C. *Chemosphere* **2015**, *119*, 498–503.

132. Boehme, J.; Wells, M. *Mar. Chem.* **2006**, 101, 95–103.

133. Eusterhues, K.; Rumpel, C.; Kleber, M.; Kögel-Knabner, I. *Org. Geochem.* **2003**, *34*, 1591–1600.

134. Eusterhues, K.; Rumpel, C.; Kögel-Knabner, I. *Org. Geochem.* **2005**, 1567–1575.

135. Kleber, M.; Mikutta, R.; Torn, M. S.; Jahn, R. *Eur. J. Soil Sci.* **2005**, *56*, 717–725.

136. Mikutta, R.; Kleber, M.; Torn, M. S.; Jahn, R. *Biogeochemistry* **2006**, 77, 25–56.

137. Leavitt, S. W.; Follett, R. F.; Paul, E. A. *Radiocarbon* **1996**, *38*, 231–239.

138. Paul, E. A.; Collins, H. P.; Leavitt, S. W. *Geoderma* **2001**, *104*, 239–256.

139. Zimmermann, M.; Leifeld, J.; Abiven, S.; Schmidt, M. W. I.; Fuhrer, J. *Geoderma* **2007**, *139*,171–179.

140. Sleutel, S.; Leinweber, P.; Begum, S. A.; Kader, M. A.; De Neve, S. *Geoderma* **2009**, *149*, 257–266.

141. Plante, A. F.; Conant, R. T.; Paul, E. A.; Paustian, K.; Six, J. *Eur. J. Soil Sci.* **2006**, *57*, 456–467.

142. Jagadamma, S.; Lal, R. *Biol. Fertil. Soils* **2010**, *46*(6), 543–554.

143. Thuriès, L.; Pansu, M.; Larré-Larrouy, M. C.; Feller, C. *Soil Biol. Biochem.* **2002**, *34*, 239–250.

144. Abiven, S.; Recous, S.; Reyes, V.; Oliver, R. *Biol. Fertil. Soils* **2005**, *42*(2), 119–128.

145. Jensen, L. S.; Salo, T.; Palmason, F.; Breland, T. A.; Henriksen, T. M.; Stenberg, B.; Esala, M. *Plant Soil* **2005**, *273*(1-2), 307–326.

146. Lashermes, G.; Nicolardot, B.; Parnaudeau, V.; Thuries, L.; Chaussod, R.; Guillotin, M. L.; Tricaud, A. *Eur. J. Soil Sci.* **2009**, *60*(2), 297–310.

147. Van Soest, P. J. *J. Assoc. Off. Agric.Chem.* **1963**, *46*, 829–835.

148. Van Soest, P. V.; Robertson, J. B.; Lewis, B. A. *J. Dairy Sci.* **1991**, *74*(10), 3583–3597.

149. Peltre, C.; Dignac, M. F.; Derenne, S.; Houot, S. *Waste Manag.* **2010**, *30*(12), 2448–2460.

150. Saviozzi, A.; Levi-Minzi, R.; Riffaldi, R.; Vanni, G. *Biol. Fertil. Soils* **1997**, 25(4), 401–406.

151. Parnaudeau, V.; Nicolardot, B. *J. Environ. Qual.* **2004**, *33*(5), 1885–1894.

152. Parnaudeau, V.; Dignac, M. F. *J. Anal. Appl. Pyrolysis* **2007**, *78*(1), 140–152.

153. Leavitt, S. W.; Danzer, S. R. *Anal. Chem.* **1993**, *65*, 87–89.

154. Cranwell, P. A. *Geochim. Cosmochim. Acta* **1978**, *42*, 1523–1532.

155. Goosens, H.; Düren, R. R.; De Leeuw, J. W.; Schenk, P. A. *Org. Geochem.* **1989**, 14, 27–41.

156. Rieley, G.; Collier, R. J.; Jones, D. M.; Eglinton, G. *Org. Geochem.* **1991**, 17, 901–912.

158. Jaffé, R.; Elismé, T.; Cabrera, A. C. *Org. Geochem.* **1996**, 25, 9–17.

159. Bull, I. D.; Van Bergen, P. F.; Nott, C. J.; Poulton, P. R.; Evershed, R. P. *Org. Geochem.* **2000**, 31, 389–408.

160. Naafs, D. F. W.; Van Bergen, P. F.; Boogert, S. J.; De Leeuw, J. W. *Soil Biol. Biochem.* **2004**, *36*, 297–308.

161. Quénéa, K.; Derenne, S.; Largeau, C.; Rumpel, C.; Mariotti, A. *Geoderma* **2006**, *136*(1), 136–151.

162. Kirk, T. K.; Obst, J. R. *Meth. Enzymol.* **1988**, 161, 89–101.

163. Dignac, M.-F.; Pechot, N.; Thevenot, M.; Lapierre, C.; Bahri, H.; Bardoux, G.; Rumpel, C. *J. Anal. Appl. Pyrolysis* **2009**, *85*, 426–430.

164. Björkman, A. *Nature* **1954**, 1057.

165. Stevenson, F. J. *Humus Chemistry*, Wiley, New York, **1982**.

166. Baldock, J. A.; Nelson, P. N. *Soil organic matter.* In *Sumner.* Malcolm, E. (ed.). Handbook of Soil Science. CRC Press, Boca Raton, Florida, USA, **2000**, pp. B25–B84.

167. Kleber, M.; Johnson, M. G. *Advances in Agronomy*, Volume 106, Chapter 3, **2010**, pp. 77–142.

168. Lehmann, J.; Kleber, M. *Nature* **2015**, *528*, 60–68.

169. Piccolo, A.; Conte, P.; Cozzolino, A. *Eur. J. Soil Sci.* **1999**, *50*, 687–694.

170. Simpson, A. J. *Magn. Reson. Chem.* **2002**, *40*, S72–S82.

171. Thurman, E. M. **1985**. Martinus Nijhoff / W. Junk Publishers.

172. Benner, R.; Pakulski, D.; Mccarthy, M.; Hedges, J. I.; Hatcher, P. G. *Science, New Series* **1992**, *255*, 1561–1564.

173. Simjouw, J.-P.; Minor, E. C.; Mopper, K. *Mar. Chem*, **2005**, *96*, 219–235.

174. Zherebker, A. Y.; Perminova, I. V.; Konstantinov, A. I.; Volikov, A. B.; Kostyukevich, Y. I.; Kononikhin, A. S.; Nikolaev, E. N. *J. Anal. Chem.* **2016**, *71*, 372–378.

175. Aiken, G. R. Humic substances. In *Soil, Sediment, and Water*. Aiken, G. R.; Mcknight, D. M.; Wershaw, R. L.; Maccarthy, P. (eds.). Wiley, New York, **1985**, pp. 363–385.

176. Schwede-Thomas, S. B.; Chin, Y.-P.; Dria, K. J.; Hatcher, P.; Kaiser, E.; Sulzberger, B. *Aquat. Sci.* **2005**, *67*, 61–71.

177. Dittmar, T.; Koch, B.; Hertkorn, N.; Kattner, G. *Limnol. Oceanogr. Meth.* **2008**, *6*: 230–235.

178. Peuravuori, J.; Monteiro, A.; Eglite, L.; Pihlaja, K. *Talanta* **2005**, *65*, 408–422.

179. Li, H.; Minor, E. C. *Environ. Sci. Process. Impact.* **2015**, *17*, 1829–1840.

180. Liška, I. *J. Chromatogr. A* **2000**, *885*, 3–16.

181. Santos, P. S. M.; Otero, M.; Filipe, O. M. S.; Santos, E. B. H.; Duarte, A. C. *Talanta* **2010**, *83*, 505–512.

182. Minor, E. C.; Swenson, M. M.; Mattson, B. M.; Oyler, A. R. *Environ. Sci. Process. Impact.* **2014**, *16* (9), 2064–2079.

183. Andrew, A. A.; Del Vecchio, R.; Zhang, Y.; Subramaniam, A.; Blough, N. V. *Front. Chem.* **2016**, 4(4), 1–12.

184. Kim, S.; Kramer, R. W.; Hatcher, P. G. *Anal. Chem.* **2003**, *75*(20), 5336–5344.

185. Ratpukdi, T.; Rice, J. A.; Chilom, G.; Bezbaruah, A.; Khan, E. *Water Environ. Res.* **2009**, *81*, 2299–2308.

186. Swenson, M. M.; Oyler, A. R.; Minor, E. C. *Limnol. Oceanogr. Meth.* **2014**, *12*, 713–728.

187. Rosario-Ortiz, F. L.; Snyder, S.; Suffet, I. H. *Environ. Sci. Technol.*, **2007**, *41*, 4895–4900.

188. Philibert, M.; Rosario-Ortiz, F.; Suffet, M. *Water Sci. Technol.* **2012**, 66(11), 2418–2424.

189. von Lützow, M.; Kögel-Knabner, I.; Ludwig, B.; Matzner, E.; Flessa, H.; Ekschmitt, K.; Guggenberger, G.; Marschner, B.; Kalbitz, K. *J. Plant Nutr. Soil Sci.* **2008**, *171*, 111–124.

190. Meentemeyer, V.; Box, E. O.; Thompson, R. *Bioscience* **1982**, *32*(2), 125–128.

191. Jandl, R.; Rodeghiero, M.; Martinez, C.; Cotrufo, M. F.; Bampa, F.; Van Wesemal, B.; Harrison, R. B.; Guerrini, I. A.; Dess Richter, D. J.; Rustad, L.; Lorenz, K.; Chabbi, A.; Miglietta, F. *Sci. Total Environ.* **2014**, 468–469.

192. Ekschmitt, K.; Liu, M.; Vetter, S.; Fox, O.; Wolters, V. *Geoderma* **2005**, *128*, 167–176.

193. Marschner, B.; Brodowski, S.; Dreves, A.; Gleixner, G.; Gude, A.; Grootes, P. M.; Hamer, U.; Heim, A.; Jandl, G.; Ji, R.; Kaiser, K.; Kramer, C.; Leinweber, P.; Rethemeyer, J.; Schäffer, A.; Schmidt, M. W. I.; Schwark, L.; Wiesenberg, G. L. B. *J. Plant Nutr. Soil Sci.* **2008**, *171*, 91–110.

194. Von Lützow, M.; Kögel-Knabner, I.; Ekschmitt, K.; Matzner, E. Guggenberger, G.; Marschner, B.; Flessa, H. *Eur. J. Soil Sci.* **2006**, *57*, 426–445.

195. Tegelaar, E. W.; De Leeuw, J. W.; Derenne, S.; Largeau, C. *Geochim. Cosmochim. Acta* **1989**, *53*, 3103–3106.

196. Lichtfouse, E.; Chenu, C.; Baudin, F.; Leblond, C.; Da Silva, M.; Behar, F.; Derenne, S.; Largeau, C.; Wehrung, P.; Albrecht, P. *Org. Geochem.* **1998**, *28*, 411–415.

197. Kellerman *et al. Nat. Geosci.* **2015**.

198. Torn, M. S.; Trumbore, S. E.; Chadwick, O. A.; Vitousek, P. M.; Hendricks, D. M. *Nature* **1997**, *389*, 170–173.

199. Rumpel, C.; Kögel-Knabner, I.; Bruhn, F. *Org. Geochem.* **2002**, *33*, 1131–1142.

200. Barré, P.; Plante, A. F.; Cecillon, L.; Lutfalla, S.; Baudin, F.; Bernard, S.; Christensen, B. T.; Eglin, T.; Fernandez, J. M.; Houot, S.; Kätterer, T.; Le Guillou, C.; Macdonald, A.; Van Oort, F.; Chenu, C. *Biogeochemistry* **2016**, *130*, 1–12.

201. Zegouagh, Y.; Derenne, S.; Largeau, C.; Bertrand, P.; Sicre, M. A.; Saliot, A.; Rousseau, B. *Org. Geochem.* **1999**, *30*(2), 101–117.

202. Garcette-Lepecq, A.; Derenne, S.; Largeau, C.; Bouloubassi, I.; Saliot, A. *Org. Geochem.* **2000**, *31*, 1663–1683.

203. Flaig, W.; Beutelspacher, H.; Rietz, E. In *Soil Components, Organic Components*, Vol. 1. Gieseking, J. E. (ed.). Springer Verlag, New York, **1975**, pp. 1–211.

204. Sutton, R.; Sposito, G. *Environ. Sci. Technol.* **2005**, *39*, 9009–9015.

205. Haumaier, L.; Zech, W. *Org. Geochem.* **1995**, *23*(3), 191–196.

206. Hempfling, R.; Ziegler, F.; Zech, W.; Schulten, H. R. *J. Plant Nutr. Soil Sci.* **1987**, *150*(3), 179–186.

207. Kelleher, B. P.; Simpson, M. J.; Simpson, A. J. *Geochim. Cosmochim. Acta* **2006**, *70*(16), 4080–4094.

208. Espitalié, J.; Deroo, G.; Marquis, F. *Rev. Inst. Fr. Pétrol.* **1985**, *40*(5), 563–579.

209. Lafargue, E.; Marquis, F.; Pillot, D. *Revue de l'institut français du pétrole* **1998**, *53*(4), 421–437.

210. Disnar, J. R.; Guillet, B.; Keravis, D.; Di-Giovanni, C.; Sebag, D. *Org. Geochem.* **2003**, *34*, 327–343.

211. Sebag, D.; Disnar, J.-R.; Guillet, B.; Di Giovanni, C.; Verrechia, E. P.; Durand, A. *Eur. J. Soil Sci.* **2006**, *57*, 344–355.

212. Sebag, D.; Verrecchia, E. P.; Cécillon, L.; Adatte, T.; Albrecht, R.; Aubert, M.; Bureau, F.; Cailleau, G.; Copard, Y.; Decaens, T.; Disnar, J.-R.; Hetényi, M.; Nyilas, T.; Trombino, L. *Geoderma* **2016**, *284*, 185–203.

213. Saenger, A.; Cécillon, L.; Poulenard, J.; Bureau, F.; De Daniéli, S.; Gonzalez, J.-M.; Brun, J. J. *Geoderma* **2015**, *241*, 279–288.

214. Saenger, A.; Cécillon, L.; Sebag, D.; Brun, J. J. *Org. Geochem.* **2013**, *54*, 101–114.

215. Delarue, F.; Disnar, J.-R.; Copard, Y.; Gogo, S.; Jacob, J.; Laggoun-Défarge, F. *Org. Geochem.* **2013**, *61*, 66–72.

216. Plante, A. F.; Fernandez, J. M.; Leifeld, J. *Geoderma* **2009**, *153*, 1–10.

217. Leifeld, J. *Org. Geochem.* **2007**, *38*, 112–127.

218. Lopez-Capel, E.; Bol, R.; Manning, D. A. C. *Rapid Comm. Mass Spec.* **2005**, *19*(22), 3192–3198.

219. Kleber, M.; *Environ. Chem.* **2011,** *7*, 320–332.

220. De La Rosa, J. M.; González-Perez, J. A.; González-Vázquez, R.; Knicker, H.; López-Capel, E.; Manning, D. A. C.; González-Vila, F. J. *Catena* **2008**, *74*, 296–303.

221. Manning, D. A. C.; Lopez-Capel, E.; Barker, S. *Mineral. Mag.* **2005**, *69*, 425–435.

222. Lopez-Capel, E.; Krull, E. S.; Bol, R.; Manning, D. A. C. *Rapid Comm. Mass Spec.* **2008**, *22*, 1751–1758.

223. Marschner, B.; Kalbitz, K. *Geoderm*a **2013**, *113*, 232–235.

224. Kalbitz, K.; Schwesig, D.; Sch merwitz, J.; Kaiser, K.; Haumaier, L.; Glaser, B.; Ellerbrock, R.; Leinweber, P. *Soil Biol. Biochem.* **2003**, *35*, 1129–1142.

225. Strauss, E. A.; Lamberti, G. A. *Freshwater Biol.* **2002**, *47*, 65–74.

226. Killops; Stephen, D.; Killops Vanessa, J.; *Introduction to Organic Geochemistry*, 2nd ed.

227. Edwards, A.; Cresser, M. *Acidification Of Freshwaters*, Cambridge Environmental Chemistry Series, **1987**.

228. Brandstetter, A.; Sletten, R. S.; Mentler, A.; Wenzel, W. W.; *J. Soil Sci. Plant Nutr.* **1996**, *159*, 605–607.

229. Matsunaga, T.; Nagao, S.; Ueno, T.; Takeda, S.; Amano, H.; Tkachenko, Y. *Appl. Geochem.* **2004**, *19*, 1581–1599,

230. Dilling, J.; Kaiser, K. *Water Res.* **2002**, *36*, 5037–5044.

231. Twardowski, M. S. E.; Boss, J. M.; Sullivan, P. L.; Donaghay. *Mar. Chem.* **2004**, *89*, 69–88.

232. Stedmon, C. A.; Markager, S.; Kaas, H. *Estuar. Coast. Shelf Sci.* **2000**, *51*, 267–278.

233. Helms, J.; Stubbins, A.; Ritchie, J.; Minor, E.; Kieber, D.; Mopper, K. *Limnol. Oceanograph.* **2008**, *53*(3), 955–969.

234. Helms, J. R.; Mao, J.; Schmidt-Rohr, K.; Abdulla, H.; Mopper, K. *Geochim. Cosmochim. Acta* **2013**, *121*, 398–413.

235. Coble, P. G. *Mar. Chem.* **1996**, *51*, 325–346.

236. Thacker, S.; Tipping, E.; Baker, A.; Gondar, D. *Water Res.* **2005**, *39*, 4559–4573.

238. Hudson, N.; Baker, A.; Reynolds, D. *River Res. Appl.* **2007**, *23*, 631–649.

238. Hudson, N. J.; Baker, A.; Ward, D.; Brunsdon, C.; Reynolds, D.; Carliell-Marquet, C.; Browning S. *Sci. Total Environ.* **2008**, *391*, 149–158.

239. Parlanti *et al.* **2000**.

240. Murphy Kathleen, R.; Butler Kenna, D.; Spencer Robert, G. M; Stedmon Colin, A.; Boehme Jennifer, R.; Aiken George, R. *Environ. Sci. Technol.* **2010**, *44(24)*, 9405–9412.

241. Zsolnay, A.; Baigar, E.; Jimene, Z. B.; Steinberg, B.; Saccomandi, F. *Chemosphere* **1999**, *38*, 45–50.

242. Kalbitz, K.; Geyer, W.; Geyer, S. *Biogeochem.* **1999**, *47*, 219–238.

243. Huguet, A.; Vacher, L.; Relexans, S.; Saubusse, S.; Froidefond, J. M.; Parlanti, E. *Org. Geochem.* **2009**, *40*(6), 706–719.

244. Huguet, A.; Vacher, L.; Saubusse, S.; Etcheber, H.; Abril, G.; Relexans, S.; Ibalot, F.; Parlanti, E. *Org. Geochem.* **2010**, *41*(6), 595–610.

245. Stedmon, C. A.; Markager, S.; Bro, R. *Mar. Chem.* **2003**, *82*, 239–254.

246. Wickland, K. P.; Neff, J. C.; Aiken, G. R. *Ecosystems* **2007**, *10*, 9101–9104.

247. Fellman, J. B.; D'amore, D. V.; Hood, E.; Boone, R. D. *Biogeochemistry* **2008**, *88*, 169–

248. Cuss, C. W.; Guéguen, C. *Anal. Chim. Acta* **2012**, *733*, 98–102.

249. Cuss, C. W.; Guéguen, C. *Front. Microbiol.* **2012**, *3*, 166.

250. Yamashita, Y.; Jaffé, R. *Environ. Sci. Technol.* **2008**, *42*, 7374–7379.

251. Stedmon, C. A.; Markager, S. *Limnol. Oceanogr.* **2005**, *50*, 686–697.

252. Cory, R. M.; Mcknight, D. M. *Environ. Sci Technol.* **2005**, *39*(21), 8142–8149.

253. Maie, N.; Yamashita, Y.; Cory, R. M.; Boyer, J. N.; Jaffé, R. *Appl. Geochem.* **2012**, *27*, 917–929.

254. Cuss, C. W.; Guéguen, C. *Chemosphere* **2013**, *92*, 1483–1489.

255. Cuss, C. W.; Guéguen, C. *Water Res.* **2015**, *68*, 487–497.

256. Helton, A. M.; Wright, M. S.; Bernhardt, E. S.; Poole, G. C.; Cory, R. M.; Stanford, J. A. *J. Geophys. Res. Biogeosci.* **2015**.

257. Milori, D. M. B. P.; Galeti, H. V. A.; Martin-Neto, L.; Dieckow, J.; González-Pérez, M.; Bayer, C.; Salton, J. *Soil Sci. Soc. Am. J.* **2006**, *70*, 57–63.

258. Tadini, A. M.; Nicolodelli, G.; Mounier, S.; Montes, C. R.; Milori, D. M. B. P. *Sci. Total Environ.* **2015**, *537*, 152–158.

259. Haberhauer, G.; Rafferty, B.; Strebl, F.; Gerzabek, M. H. *Geoderma* **1998**, *83*, 331–342.

260. Piccolo, A.; Zaccheo, P.; Genevini, P. G.; *Biores. Technol.* **1992**, *40*, 275–282.

261. Poirier, N.; Derenne, S.; Rouzaud, J.; Largeau, C.; Mariotti, A.; Balesdent, J.; Maquet, J. *Org. Geochem.* **2000**, *31*, 813–827.

262. Quénéa, K.; Derenne, S.; Largeau, C.; Rumpel, C.; Mariotti, A. *Org. Geochem.* **2005**, *36*, 349–362.

263. Ellerbrock, R. H.; Gerke, H. H. *Adv. Agro.* **2013**, *121*, 117–177.

264. Leifeld, J. *Eur. J. Soil Sci.*, **2006**, *57*, 846–857.

265. Ludwig, B.; Nitschke, R.; Terhoeven-Urselmans, T.; Michel, K.; Flessa, H. *J. Plant Nutr. Soil Sci.* **2008**, *171*(3), 384–391.

266. Tremblay, L.; Alaoui, G.; Léger, M. N. *Environ. Sci. Technol.* **2011**, *45*(22), 9671–9679.

267. Dalal, R. C.; Henry, R. J. *Soil Sci. Soc. Am. J.* **1986**, *50*(1), 120–123.

268. Cécillon, L.; Barthès, B. G.; Gomez, C.; Ertlen, D.; Génot, V.; Hedde, M.; Brun, J. J. *Eur. J. Soil Sci.* **2009**, *60*(5), 770–784.

269. Golchin, A.; Oades, J. M.; Skjemstad, J.; Clarke, *Aust. J. Soil Res.* **1995**, *33*, 59–76.

270. Kögel-Knabner, I. *Geoderma* **1997**, 80, 243–270.

271. Preston, C. M.; Nault, J. R.; Trofymow, J. A. *Ecosystems* **2009**, 12, 1078–1102.

272. Nelson, P. N.; Baldock, J. A. *Biogeochemistry* **2005**, *72*(1), 1–34.

273. Simpson, M. J.; Simpson, A. J. *J. Chem. Ecol.* **2012**, *38*, 768–784.

274. Cao, X.; Yang, J.; Mao, J. *Int. J. Coal Geol.* **2013**, *108*, 83–90.

275. Templier, J.; Derenne, S.; Croué, J.-P.; Largeau, C. *Org. Geochem.,* **2005**, *36*, 1418–1442.

276. Lankes, U.; Lüdemann, H. D.; Frimmel, F. H. *Water Res.* **2008**, *42*(4), 1051–1060.

277. Leenheer, J. A.; Croué, J. P. *Environ Sci. Technol.* **2003**, 37, 18–26.

278. Hockaday, W. C.; Purcell, J. M.; Marshall, A. G.; Baldock, J. A.; Hatcher, P. G. *Limnol. Oceano. Meth.* **2009**, *7*(1), 81–95.

279. Leenheer, J. A.; Nanny, M. A.; Mcintyre, C. *Environ. Sci. Technol.* **2002**, *37*(11), 2323–2331.

280. Lam *et al. Environ. Sci. Technol.* **2007**, *41*, 8240.

281. Hertkorn *et al. Biogeosciences* **2016**, *13*, 2257–2277.

282. Kögel-Knabner, I.; Hatcher, P. G.; Tegelaar, E. W.; De Leeuw, J. W. *Sci. Total Environ.* **1992**, *113*, 89–106.

283. Baldock, J. A.; Oades, J. M.; Nelson, P. N.; Skene, T. M.; Golchin, A.; Clarke, P. *Aust. J. Soil Res.* **1997**, *35*(5), 1061–1083.

284. Simpson, A. J.; Kingery, W. L.; Shaw, D. R.; Spraul, M.; Humpfer, E.; Dvortsak, P. *Environ. Sci. Technol.* **2001**, 35, 3321–3325.

285. Salmon, E.; Behar, F.; Hatcher, P. G. *Org. Geochem.* **2011**, *42*(4), 301–315.

286. Kramer, M. G.; Sollins, P.; Sletten, R. S.; Swart, P. K. *Ecology* **2003**, *84*, 2021–2025.

287. Glaser, B. *J. Plant Nutr. Soil Sci.* **2005**, *168*, 633–648.

288. Balesdent, J.; Mariotti, A.; Guillet, B. *Soil Biol. Biochem.* **1987**, *19*, 25–30.

289. Balesdent, J.; Balabane, M. *Soil Biol. Biochem.* **1992**, *24*, 97–101.

290. Amelung, W.; Brodowski, S.; Sandhage-Hofmann, A.; Bol, R. *Adv. Agron.* **2008**.

291. Smith, B. N.; Epstein, S. *Plant Physiol.* **1971**, *47*, 380–384.

292. Bender, M. *Phytochemistry* **1971**, *10*, 1239–1244.

293. O'leary, M. *Phytochemistry* **1981**, *20*, 553–567.

294. Balesdent, J. *Eur. J. Soil Sci.* **1996**, *47*, 485–493.

295. Cerri, C.; Feller, C.; Balesdent, J.; Victoria, R.; Plenecasssagne, A. *Compte Rendu De L'académie Des Sciences De Paris* **1985**, *300*, 423–428.

296. Martin, A.; Mariotti, A.; Balesdent, J.; Lavelle, P.; Vuattoux, R. *Soil Biol. Biochem.* **1990**, *22*, 517–523.

297. Balesdent, J.; Besnard, E.; Arrouays, D.; Chenu, C. *Plant Soil* **1998**, *201*, 49–57.

298. Gregorich, E. G.; Monreal, C. M.; Ellert, B. H. *Can. J. Soil Sci.* **1995**, *75*(2), 161–167.

299. Balesdent, J.; Mariotti, A. In *Mass Spectrometry of Soils.* S.; B. T.; Y. (ed.). Marcel Dekker, New York, **1996**, pp. 83–111.

300. Balesdent, J.; Girardin, C.; Mariotti, A. *Ecology* **1993**, *74*, 1713–1721.

301. Agren, G. I.; Bosatta, E.; Balesdent, J. *Soil Sci. Soc. Am. J.* **1996**, *60*, 1121–1126.

302. Benner, R.; Fogel, M. L.; Sprague, E. K.; Hodson, R. E. *Nature* **1987**, *329*, 708–710.

303. Boström, B.; Comstedt, D.; Ekblad, A. *Oecologia* **2007**, *153*, 89–98.

304. Amato, M.; Ladd, J. N.; Ellington, A.; Ford, G.; Mahoney, J. E.; Taylor, A. C.; Walsgott, D. *Soil Res.* **1987**, *25*(1), 95–105.

305. Angers, D. A.; Recous, S.; Aita, C. *Eur. J. Soil Sci.* **1997**, *48*, 295–300.

306. Kuzyakov, Y.; Domanski, G. *J. Plant Nutr. Soil Sci.* **2000**, *163*(4), 421–431.

307. Zeller, B.; Colin-Belgrand, M.; Dambrine, E.; Martin, F.; Bottner, P. *Oecologia* **2000**, *123*(4), 550–559.

308. D'annunzio, R.; Zeller, B.; Nicolas, M.; Dhôte, J. F.; Saint-André, L. *Soil Biol. Biochem.* **2008**, *40*(2), 322–333.

309. Girardin, C.; Rasse, D.; Biron, P.; Ghashghaie, J.; Chenu, C. *Rapid Comm. Mass Spec.* **2009**, *23*, 1792–1800.

310. Hatton, P. J.; Kleber, M.; Zeller, B.; Moni, C.; Plante, A. F.; Townsend, K.; Derrien, D. *Org. Geochem.* **2012**, *42*(12), 1489–1501.

311. Vidal, A.; Quenea, K.; Alexis, M.; Tu, T. N.; Mathieu, J.; Vaury, V.; Derenne, S. *Geoderma* **2017**, *285*, 9–18.

312. Hertkorn, N.; Frommberger, M.; Witt, M.; Koch, B. P.; Schmitt-Kopplin, P.; Perdue, E. M. *Anal. Chem.* **2008**, *80*(23), 8908–8919.

313. Reemtsma, T. *J. Chromatogr. A* **2009**, *1216*(18), 3687–3701.

314. Sleighter, R. L.; Hatcher, P. G. *J. Mass Spec.* **2007**, *42*(5), 559–574.

315. Fievre, A.; Solouki, T.; Marshall, A. G.; Cooper, W. T. *Energy Fuels,* **1997**, *11*, 554–560.

316. Kujawinski, E. B.; Freitas, M. A.; Zang, X.; Hatcher, P. G.; Green-Church, K. B.; Jones, R. B. *Org. Geochem.* **2002**, *33*(3), 171–180.

317. Koch, B. P.; Witt, M.; Engbrodt, R.; Dittmar, T.; Kattner, G. *Geochim. Cosmochim. Acta* **2005**, *69*, 3299 –3308.

318. Minor, E. C.; Steinbring, C. J.; Longnecker, K.; Kujawinski, E. B. *Org. Geochem.* **2012**, *43*, 1–11.

319. Tfaily, M. M.; Hamdan, R.; Corbett, J. E.; Chanton, J. P.; Glaser, P. H.; Cooper, W. T. *Geochim. Cosmochim. Acta* **2013**, *112*, 116–129.

320. Schmidt, F.; Koch, B. P.; Witt, M.; Hinrichs, K.-U. *Geochim. Cosmochim. Acta* **2014**, *141*, 83–96.

321. Valle, G. M. J.; Schmitt-Kopplin, P.; Hertkorn, N.; Bastviken, D.; Luek, J.; Harir, M.; Bastos, W.; Enrich-Prast, A. *Biogeosciences* **2016**, *13*, 4279–4290.

322. Ohno, T.; He, Z.; Sleighter, R. L.; Honeycutt, C. W.; Hatcher, P. G. *Environ. Sci. Technol.* **2010**, *44*, 8594–8600.

323. Ikeya, K.; Sleighter, R. L.; Hatcher, P. G.; Watanabe, A. *Geochim. Cosmochim. Acta* **2015**, *153*, 169–182.

324. Di Donato, N.; Chen, H.; Waggoner. D.; Hatcher, P. G. *Geochim. Cosmochim. Acta* **2016**, *178*, 210–222.

325. Hartman, B. E.; Chen, H.; Hatcher, P. G. *Int. J. Coal Geol.* **2015**, *144*, 15–22.

326. D'andrilli, J.; Cooper, W. T.; Foreman, C. M.; Marshall, A. G. *Rapid Comm. Mass Spec.* **2015**, *29*, 2385–2401.

327. Dungait, J. A. J.; Hopkins, D. W.; Gregory, A. S.; Whitmore, A. P. *Global Change Biol.* **2012**, *18*, 1781–1796.

328. Lichtfouse, E. *Tetrahedron Lett.* **1995**, *36*, 529–530.

329. O'malley, V. P.; Burke, R. A.; Schlotzhauer, W. S. *Org. Geochem.* **1997**, *27*, 567–581.

330. Meir-Augenstein, W. *J. Chromatograph. A*, **1999**, *842*, 351–371.

331. Cayet, C.; Lichtfouse, E. *Org. Geochem.* **2001**, *32*, 253–258.

332. Derrien, D.; Balesdent, J.; Marol, C.; Santaella, C. *Rapid Comm. Mass Spec.* **2003**, *17*, 2626–2631.

333. Dignac, M.-F.; Bahri, H.; Rumpel, C.; Rasse, D. P.; Bardoux, G.; Balesdent, J.; Girardin, C.; Chenu, C.; Mariotti, A. *Geoderma* **2005**, *128*, 3.

334. Maillard. *Comptes Rendus Hebdomadaires Des Seances De L Academie Des Sciences* **1912**, *154*, 66–68.

335. Rubinsztain, Y.; Ioselis, P.; Ikan, R.; Aizenshat, Z. *Org. Geochem.* **1984**, *6*, 791–804.

336. Yamamoto.; Ishiwatari. *Org. Geochem.* **1989**, *14*, 479–489.

337. Poirier, N.; Derenne, S.; Balesdent, J.; Chenu, C.; Bardoux, G.; Mariotti, A.; Largeau, C. *Eur. J. Soil Sci.* **2006**, *57*(5), 719–730.

338. Henrichs.; Sugai. *Geochim. Cosmochim. Acta* **1993**, *57*, 823–835.

339. Cosovic, B.; Vojvodić, V.; Bošković, N.; Plavšić, M.; Lee, C. *Org. Geochem.* **2010**, *41*, 200–205.

340. Goldberg, D. *Black Carbon in The Environment*, John Wiley, New York, **1985**.

341. Schmidt, M. W.; Noack, A. G. *Global Biogeochem. Cycl.* **2000**, *14*(3), 777–793.

342. Seiler, W.; Crutzen, P. J. *Clim. Change* **1980**, *2*, 207–247.

343. Jaffé, R.; Ding, Y.; Niggemann, J.; Vähätalo, A. V.; Stubbins, A.; Spencer, R. G.; Campbell, J.; Dittmar, T. *Science* **2013**, *340*(6130), 345–347.

344. Qu, X.; Fu, H.; Mao, J.; Ran, Y.; Zhang, D.; Zhu, D. *Carbon* **2016**, *96*, 759–767.

345. Koelmans, A.; Jonker, M.; Cornelissen, G.; Bucheli, T.; Noort, P. V.; Gustafsson Ö. *Chemosphere* **2006**, *63*(3), 365–377.

346. Kaal, J.; Cortizas, A. M.; Eckmeier, E.; Casais, M. C.; Estévez, M. S.; Boado, F. C. *J. Archaeol. Sci.* **2008**, *35*(8), 2133–2143.

347. Beesley, L.; Moreno-Jiménez, E.; Gomez-Eyles, J. L.; Harris, E.; Robinson, B.; Sizmur, T. *Environ. Poll.* **2011**, *159*, 3269–3282.

348. Sohi, S. P.; Krull, E.; Lopez-Capel, E.; Bol, R. *Adv. Agro.* **2010**, *105*, 47–82.

349. Gonzalez-Pérez, J. A.; González-Vila, F. J.; Almendros, G.; Knicker, H. *Environ. Int.* **2004**, *30*, 855–870.

350. Spokas, K. A.; Cantrell, K. B.; Novak, J. M.; Archer, D. W.; Ippolito, J. A.; Collins, H. P.; Boateng, A. A.; Lima, I. M.; Lamb, M. C.; Mcaloon, A. J.; Lentz, R. D.; Nichols, K. A. *J. Environ. Qual.* **2012**, *41*, 973–989.

351. Gustafsson, O.; Bucheli, T.; Kukulska, Z.; Andersson, M.; Largeau, C.; Rouzaud, J.; Reddy, C.; Eglinton, T. *Global Biogeochem. Cycl.* **2001**, *15*, 881–890.

352. Gustafsson, O.; Haghseta, F.; Chan, C.; Macfarlane, J.; Gschwend, P. M. *Environ. Sci. Technol.* **1997**, *31*, 203–209.

353. Lim, B.; Cachier, H. *Chem. Geol.* **1996**, *131*, 143–154.

354. Simpson, M. J.; Hatcher, P. G. *Org. Geochem.* **2004**, *35*, 923–935.

355. Skjemstad, J.; Clarke, P.; Taylor, J. A.; Oades, J. M.; Mcclure, S. G. *Aust. J. Soil Res.* **1996**, *34*, 251–271.

356. Hammes, K.; Schmidt, M. W. I.; Smernik, R. J. *et al. Global Biogeochem. Cycles* **2007**, *21*, 3016.

357. Knicker, H. *Biogeochemistry,* **2007**, *85*, 91–118.

358. Meredith, W.; Ascough, P. L.; Bird, M. I.; Large, D. J.; Snape, C. E.; Sun, Y.; Tilston, E. L. *Geochim. Cosmochim. Acta* **2012**, *97*, 131–147.

359. Schmidt, M.; Skjemstad, J.; Czimeczik, C.; Glaser, B.; Prentice, K. M.; Gelinas, Y.; Kuhlbusch, T. A. J. *Global Biogeochem. Cycles* **2001**, *15*, 163–167.

360. Glaser, B.; Haumaier, L.; Guggenberger, G.; Zech, W. *Org. Geochem.* **1998**, *29*, 811–819.

361. Brodowski, S.; Rodionov. A.; Haumaier, L.; Glaser, B. *Org. Geochem.* **2005**.

362. Schneider, M. P.; Smittenberg, R. H.; Dittmar, T.; Schmidt, M. W. *Org. Geochem,* **2011**, *42*(3), 275–282.

363. Hatten.; Goñi. *PloS One* **2016**, *11*(3), E0151957.

364. Kappenberg, A.; Bläsing, M.; Lehndorff, E.; Amelung, W. *Org. Geochem.* **2016**, *94*, 47–51.

365. Kuo, L. J.; Herbert, B. E.; Louchouarn, P. *Org. Geochem.* **2008**, *39*(10), 1466–1478.

366. Kaal, J.; Brodowski, S.; Baldock, J. A.; Nierop, K. *Org. Geochem.* **2008**.

367. Liang, B.; Lehmann, J.; Solomon, D.; Sohi, S. P.; Thies, J. E.; Skjemstad, J. O.; Luizao, F. J.; Engelhard, M. H.; Neves, E. G.; Wirick, S. U. *Geochim. Cosmochim. Acta* **2008**, *72*, 6069–6078.

368. Smernik, R. J.; Baldock, J. A.; Oades, J. M.; Whittaker, A. K. *Solid State Nucl. Magn. Res.* **2002**, *22*(1), 50–70.

369. Heymann, K.; Lehmann, J.; Solomon, D.; Schmidt, M. W.; Regier, T. *Org. Geochem.* **2011**, *42*(9), 1055–1064.

370. Glaser, B.; Balashov, E.; Haumaier, L.; Guggenberger, G.; Zech, W. *Org. Geochem.* **2000**, *31*, 669–678.

371. Knicker, H.; Müller, P.; Hilscher, A. *Geoderma* **2007**, *142*, 178–196.

372. Masiello, C. A.; Druffel, E. R. M. *Science* **1998**, *280*(5371), 1911–1913.

373. Kuzyakov, Y; Subbotina, I.; Chen, H.; Bogomolova, I.; Xu, X. *Soil Biol. Biochem.* **2009**, *41*, 210–219.

374. Spokas, K. *Carbon Management* **2010**, *1*, 289–303.

375. Knicker. *Org. Geochem.* **2010**, *41*, 947–950.

376. Ding, Y.; Watanabe, A.; Jaffé, R. *Org. Geochem.* **2014**, *68*, 1–4.

377. Sinninghe-Damsté; De Leeuw. *Org. Geochem.* **1989**, *16*, 1077–1101.

378. Schaeffer, P.; Reiss, C.; Albrecht, P. *Org. Geochem.* **1995**, *23*, 567–581.

379. Schaeffer-Reiss, C.; Schaeffer, P.; Putschew, A.; Maxwell, J. R. *Org. Geochem.* **1998**, *29*, 1857–1873.

380. Van Kaam-Peters, H. M. E.; Schouten, S.; Köster, J.; Sinninghe Damsté, J. S. *Geochim. Cosmochim. Acta.* **1998**, *62*, 3259–3283.

381. Mongenot. *Org. Geochem.* **1999**, *30*, 39–56.

382. Barakat, A. O.; Scholz-Böttcher, B. M.; Rullkötter, J. *Fuel* **2012**, *96*, 176–184.

383. Barakat, A. O.; Scholz-Böttcher, B. M.; Rullkötter, J. *Fuel* **2013**, *104*, 788–797.

384. Orr. *Org. Geochem.* **1986**, *10*, 499–516.

385. Peng, P. A.; Morales-Izquierdo, A.; Lown, E. M.; Strausz, O. P. *Energy Fuels* **1999**, *13*(2), 248–265.

386. Damsté, J. S. S.; de las Heras, F. X. C.; van Bergen, P. F.; de Leeuw, J. W. *Geochim. Cosmochim. Acta* **1993**, *57*(2), 389–415.

387. Herzsprung, P.; Hertkorn, N.; Tümpling, W. V.; Harir, M.; Friese, K.; Schmitt-Kopplin, P. *Anal. Bioanal. Chem.* **2016**, *408*(10), 2461–2469.

388. Schmidt, F.; Elvert, M.; Koch, B. P.; Witt, M.; Hinrichs, K.-U. *Geochim. Cosmochim. Acta 73*, 3337–3358.

389. Riederer, M.; Matzke, K.; Ziegler, F.; Kögel-Knabner, I. *Org. Geochem.* **1993**, *20*, 1063–1076.

390. Derenne, S.; Largeau, C. *Soil Sci.* **2001**, *166*, 833–847.

391. Hatcher, P. G.; Spiker, E. C.; Szeverenyi, N. M.; Maciel, G. E. *Nature* **1983**, *305*, 498–501.

392. Poirier, N.; Derenne, S.; Balesdent, J.; Mariotti, A.; Massiot, D.; Largeau, C. *Eur. J. Soil Sci.* **2003**, *54*, 243–255.

393. Berkaloff, C.; Casadevall, E.; Largeau, C.; Metzger, P.; Peracca, S.; Virlet, J. *Phytochemistry* **1983**, *22*, 389–397.

394. Derenne, S.; Largeau, C.; Casadevall, E.; Berkaloff, C. *Phytochemistry* **1989**, *28*, 1137–1142.

395. Derenne, S.; Largeau, C.; Berkaloff, C.; Rousseau, B.; Wilhelm, C.; Hatcher, P. G. *Phytochemistry* **1992**, *31*, 1923–1929.

396. Gelin, F.; Volkman, J. K.; Largeau, C.; Derenne, S.; Sinninghe Damste´, J. S.; De Leeuw, J. W. *Org. Geochem.* **1999**, *30*, 147–159.

397. Versteegh, G. J. M.; Blokker, P. *Phycol. Res.* **2004**, *52*, 325–339.

398. Kodner, R. B.; Summons, R. E.; Knoll, A. H. *Org. Geochem.* **2009**, *40*(8), 854–862.

399. Allard, B.; Templier, J.; Largeau, C. *Org. Geochem.* **1998**, *28*, 543–548.

400. Blokker, P.; Schouten, S.; Van Den Ende, H.; De Leeuw, J. W.; Hatcher, P. G.; Sinninghe Damsté, J. S. *Org. Geochem.* **1998**, *29*, 1453–1468.

401. Zelibor Jr. J. L.; Romankiw, L.; Hatcher, P. G.; Colwell, R. R. *Appl. Environ. Microbiol.* **1988**, *54*, 1051–1060.

402. Obeid, W.; Salmon, E.; Hatcher, P. G. *Org. Geochem.* **2014**, *76*, 259–269.

403. Scholz, M. J.; Weiss, T. L.; Jinkerson, R. E.; Jing, J.; Roth, R.; Goodenough, U.; Posewitz. M. C.; Gerken, H. G. *Eukary. Cell* **2014**, 13(11), 1450–1464.

404. Derenne, S.; Largeau, C.; Taulelle, F. *Geochim. Cosmochim. Acta* **1993**, *57*, 851–857.

405. Gelin, F.; Boogers, I.; Noordeloos, A. A. M.; Sinninghe Damsté, J. S.; Riegman, R.; De Leeuw, J. W. *Org. Geochem.* **1997**, *26*, 659–675.

406. Largeau, C.; Derenne, S.; Casadevall. E.; Kadouri, A.; Sellier, N. In *Advances in Organic Geochemistry.* Leythaeuser, D.; Rullkötter, J. (eds.). Pergamon Press, Oxford, **1986**, 1023–1032.

407. Derenne, S.; Largeau, C.; Casadevall, E.; Berkaloff, C.; Rousseau, B. *Geochim. Cosmochim. Acta* **1991**, *55*, 1041–1050.

408. Derenne, S.; Largeau, C.; Hatcher, P. G. *Org. Geochem.* **1992**, *18*, 417–422.

409. Gillaizeau, B.; Derenne, S.; Largeau, C.; Berkaloff, C.; Rousseau, B. *Org. Geochem.* **1996**, *24*, 671–679.

410. Zhang, Z.; Volkman, J. K.; Xie, X.; Snowdon, L. R. *Org. Geochem.* **2016**, *91*, 89–99.

411. Goth, K.; De Leeuw, J. W.; Püttmann, W.; Tegelaar, E. W. *Nature* **1998**, *336*(6201), 759–761.

412. Derenne, S.; Largeau, C.; Hetényi, M.; Brukner-Wein, A.; Connan, J.; Lugardon, B. *Geochim. Cosmochim. Acta* **1997**, *61*, 1879–1889.

413. Biller, P.; Ross, A. B.; Skil, I. S. C. *Org. Geochem.* **2015**, *81*, 64–69.

414. Chalansonnet, S.; Largeau, C.; Casadevall, E.; Berkaloff, C.; Peniguel, G.; Couderc, R. *Org. Geochem.* **1988**, *13*, 1003–1010.

415. Allard, B.; Templier, J.; Largeau, C. S. *Org. Geochem.* **1997**, *26*, 691–703.

416. Kolattukudy, P. E. *Can. J. Bot.* **1984**, *62*(12), 2918–2933.

417. Heredia, A. *Biochim. Biophys. Acta* **2003**, *1620*, 1–7.

418. Li-Beisson; Yonghua; Verdier; Gaëtan; Xu; Lin; Beisson, F. In: Els. John Wiley & Sons Ltd, Chichester, **2016**.

419. Hunneman, D. H.; Eglinton, G. *Phytochemistry* **1972**, *11*, 1989–2001.

420. Goñi, M. A.; Hedges, J. I. *Geochim. Cosmochim. Acta* **1990**, *54*, 3065–3072.

421. Espelié, K. E.; Kolattukudy, P. E. *Plant Physiol.* **1979**, *63*, 433–435.

422. Espelié, K. E. *Planta* **1982**, *155*, 166–175.

423. Zeier, J.; Schreiber, L. *Plant Physiol.* **1997**, *113*, 1223–1231.

424. Zeier, J.; Schreiber, L. *Planta* **1998**, *206*, 349–361.

425. Bernards, M. A.; Lewis, N. G. *Phytochemistry* **1998**, *47*, 915–933.

426. Franke, R.; Briesen, I.; Wojciechowski, T.; Faust, A.; Yephremov, A.; Nawrath, C.; Schreiber, L. *Phytochemistry* **2005**, *66*, 2643–2658.

427. Graça, J. *Front. Chem.* **2015**, *3*.

428. Kolattukudy, P. T. *Ann. Rev. Plant Physiol.* **1981**, *32*(1), 539–567.

429. Bernards, M. A. *Revue Canadienne De Botanique* **2002**, *80*(3), 227–240.

430. Mendez-Millan, M.; Dignac, M. F.; Rumpel, C.; Derenne, S. *Org. Geochem.* **2010**, *41*(2), 187–191.

431. Augris, N.; Balesdent, J.; Mariotti, A.; Derenne, S.; Largeau, C. *Org. Geochem.* **1998**, *28*, 119–124.

432. Nierop, K. *Org. Geochem.* **1998**, *29*, 1009–1016.

433. Lopes, M. H.; Pascoal Neto, C.; Barros, A. S.; Rutledge, D.; Delgadillo, I.; Gil, A. M. *Biopolymers* **2000**, *57*, 344–351.

434. Deshmukh, A. P.; Simpson, A.; Hatcher, P. G. *Phytochemistry* **2003**, *64*, 1163–1170.

435. Lopes, M. H.; Sarychev, A.; Pascoal Neto, C.; Gil, A. M. *Solid State Nucl. Magn. Reson.* **2000**, *16*, 109–121.

436. Järvinen, R.; Silvestre, A. J.; Gil, A. M.; Kallio, H. *J. Food Compos. Anal.* **2011**, *24*(3), 334–345.

437. Baker. *New Phytol.* **1970**, *69*, 1053–1058.

438. Graça; Pereira. *Phytochem. Anal.* **2000**, *11*, 45–51.

439. Riederer, M.; Schönherr, J. *Planta* **1998**, *174*, 127–138.

440. Del Rio, J.; Hatcher, P. G. *Org. Geochem.* **1998**, *29*, 1441–1451.

441. Nierop, K. G. J. *J. Anal. Appl. Pyrolysis* **2001**, *61*, 111–132.

442. Nip, M.; Tegelaar, E. W.; De Leeuw, J. W.; Schenck, P. A.; Et Holloway, P. J. *Naturwissenschaften* **1986**, *73*, 579–585.

443. Tegelaar, E. W.; Hollman, G.; Van Der Vegt, P.; De Leeuw, J. W.; Holloway, P. J. *Org. Geochem.* **1995**, *23*, 239–250.

444. Villena, J. F.; Domínguez, E.; Stewart, D.; Heredia, A. *Planta* **1999**, *208*(2), 181–187.

445. Boom, A.; Sinninge Damste, J. S.; De Leeuw, J. W. *Org. Geochem.* **2005**, *36*, 595–601.

446. Turner, J. W.; Hartman, B. E.; Hatcher, P. G. *Org. Geochem.* **2013**, *57*, 41–53.

447. Guzmán-Delgado *et al. Physiol. Plant.* **2016**, *157*, 205–220.

448. Deshmukh, A. P.; Simpson, A.; Hadad, C. M.; Hatcher, P. G. *Org. Geochem.* **2005**, *36*, 1072–1085.

449. Tsubaki, S.; Sakumoto, S.; Uemura, N.; Azuma, J. *Food Chem.* **2013**, *138*, 286–290.

450. De Leeuw, J. W.; Largeau, C. In *Organic Geochemistry: Principles and Applications.* Engel, M. H.; Macko, S. A. (eds.). Plenum Press, New York, **1993**, pp. 23–72.

451. Nip, M.; De Leeuw, J. W.; Schenck, P. A. *Geochim. Cosmochim. Acta* **1989**, *53*, 671–683.

452. Tegelaar, E. W.; Wattendorff, J.; Deleeuw, J. W. *Rev. Paleobot. Palynol.* **1993**, *77*, 149–170.

453. Van Bergen, P. F.; Scott, A. C.; Barrie, P. J.; De Leeuw, J. W.; Collinson, M. E. *Org. Geochem.* **1994**, *21*, 107–112.

454. Van Bergen, P. F.; Goñi, M.; Collinson, M. E.; Barrie, P. J.; Sinninghe Damsté, J. S.; De Leeuw, J. W. *Geochim. Cosmochim. Acta* **1994**, *58*, 3823–3844.

455. Mösle, B.; Finch, P.; Collinson, M. E.; Scott, A. C. *J. Anal. Appl. Pyrolysis*, **1997**, *40*, 585–597.

456. Gupta, N. S.; Michels, R.; Briggs, D. E. G.; Collinson, M. E.; Evershed, R. P.; Pancost, R. D. *Org. Geochem.* **2007**, *38*, 28–36.

457. Aucour, A.-M.; Faure, P.; Gomez, B.; Hautevelle, Y.; Michels, R.; Thévenard, F. *Org. Geochem.* **2009**, *40*, 784–794.

458. Baldock, J. A.; Oades, J. M.; Waters, A. G.; Peng, X.; Vassallo, A. M.; Wilson, M. A. *Biogeochemistry* **1992**, *16*, 1–42.

459. Mikutta, R.; Mikutta, C.; Kalbitz, K.; Scheel, T.; Kaiser, K.; Jahn, R. *Geochim. Cosmochim. Acta* **2007**, *71*, 2569–2590.

460. Bull, I. D.; Nott, C. J.; Van Bergen, P. F.; Poulton, P. R.; Evershed, R. P. *Soil Biol Biochem.* **2000**, *32*, 1367–1376.

461. Nierop, K. G. J.; Naafs, D.; Verstraten, J. M. *Org. Geochem.* **2003**, *34*, 719–729.

462. Nierop, K. G.; Verstraten, J. M. *Org. Geochem.* **2003**, *34*(4), 499–513.

463. Feng, X.; Simpson, M. *Org. Geochem.* **2007**, *38*, 1558–1570.

464. Wiesenberg, G.; Schwarzbauer, J.; Schmidt, M. W. I.; Scwark, L. *Org. Geochem.* **2004**, *35*, 1371–1393.

465. Feng, X.; Simpson, M. J. *J. Environ. Monit.* **2011**, *13*, 1246–1254.

466. Bachmann, J.; Guggenberger, G.; Baumgartl, T.; Ellerbrock, R. H.; Urbanek, E.; Goebel, M. O.; Andfischer, W. R. *J. Plant Nutr. Soil Sci.* **2008**, *171*(1), 14–26.

467. Rasse, D. P.; Rumpel, C.; Dignac, M.-F. *Plant Soil* **2005**, *269*, 341–356.

468. Kirk, T. K.; Farrell. *Ann. Rev. Microbiol.* **1987**, *41*, 465–505.

469. Filley, T. R.; Hatcher, P. G.; Shortle, W. *Org. Geochem,* **2000**, *31*, 181–198.

470. González, J. A.; González-Vila, F. J.; Almendros, G.; Zancada, M. C.; Polvillo, O.; Martín, F. *J. Anal. Appl. Pyrolysis* **2003**, *68*, 287–298.

471. Hedges, J. I.; Mann, D. C. *Geochim. Cosmochim. Acta* **1979**, *43*(11), 1803–1807.

472. Hedges; Hertel. *Anal. Chem.* **1982**, *54*, 174–178.

473. Saiz-Jimenez, C.; De Leeuw, J. W. *Adv. Org. Geochem.* **1986**, *10*, 869–876.

474. Martin, F.; Del Rio, J. C.; Gonzalez-Vila, F. J.; Verdejo, T. *J. Anal. Appl. Pyrolysis* **1995**, *35*, 1–13.

475. Clifford, D. J.; Carson, D. M.; Mckinney, D. E.; Bortiatynski, J. M.; Hatcher, P. G. *Org. Geochem.* **1995**, *23*, 169–175.

476. Hatcher, P. G.; Minard, R. D. *Org. Geochem.* **1995**, *23*, 991–994.

477. Challinor, J. M. *J. Anal. Appl. Pyrolysis* **2001**, *61*, 3–34.

478. Opsahl, S.; Benner, R. *Geochim. Cosmochim. Acta* **1995**, *59*, 4889–4904.

479. Kögel, I. *Soil Biol. Biochem.* **1986**, *18*(6), 589–594.

480. Hedges, J. I.; Blanchette, R. A.; Weliky, K.; Devol, A. H. *Geochim. Cosmochim. Acta* **1988**, *52*, 2717–2726.

481. Benner, R.; Welily, K.; Hedges, J. I. *Geochim. Cosmochim. Acta* **1990**, *54*, 1991–2001.

482. Otto, A.; Simpson, M. J.; *Biogeochemistry* **2006**, *80*(2), 121–142.

483. Guggenberger, G.; Christensen, B. T.; Zech, W. *Eur. J. Soil Sci.* **1994**, *45*, 449–458.

484. Guggenberger, G.; Zech, W.; Thomas, R. J. *Biol. Biochem.* **1995**, *27*, 1629–1638.

485. Heim, A.; Schmidt, M. W. I. *Eur. J. Soil Sci.* **2007**, *58*, 599–608.

486. Kiem, R.; Kögel-Knabner, I. *Soil Biol Biochem.* **2003**, *35*(1), 101–118.

487. Rasse, D.; Mulder, J.; Moni, C.; Chenu, C. *Soil Sci. Soc. Am. J.* **2006**, *70*, 2097–2105.

488. Thevenot, M.; Dignac, M.-F.; Rumpel, C. *Soil Biol Biochem.* **2010**, *42*, 1200–1211.

489. Klotzbücher *et al. Og* **2011**, *42*, 1271–1278.

490. Derrien, D.; Marol, C.; Balabane, M.; Balesdent, J. *Eur. J. Soil Sci.* **2006**, *57*, 547–557.

491. Gleixner, G.; Bol, R.; Balesdent, J. *Rapid Comm. Mass Spec.* **1999**, *13*, 1278–1283.

492. Gleixner, G.; Poirier, N.; Bol, R.; Balesdent, J. *Org. Geochem.* **2002**, *33*, 357–366.

493. Glaser, B.; Turrión, M.-B.; Alef, K. *Soil Biol. Biochem.* **2004**, *36*, 399–407.

494. Amelung, W. In *Assessment Methods for Soil Carbon*. Lal, R.; Kimble, J. M.; Follett, R. F.; Stewart, B. A. (eds.). Lewis Publishers, Boca Raton, **2001**, pp. 233–272.

495. Tremblay; Benner. *Limnol. Oceanogr.* **2009**, *54*(3), 681–691.

496. Allard, B.; Derenne, S. *Org. Geochem.* **2013**, *65*, 103–117.

497. Zeglin, L. H.; Kluber, L. A.; Myrold, D. D. *Biogeochemistry* **2013**, *112*(1–3), 679–693.

498. Kaiser, K.; Zech, W. *Eur. J. Soil Sci.* **2000**, *51*, 403–411.

499. Decock, C.; Denef, K.; Bodé, S.; Six, J.; Boeckx, P. *Rapid Comm. Mass Spec.* **2009**, *23*, 1201–1211.

500. Bodé, S.; Denef, K.; Boeckx, P. R. *Rapid Comm. Mass Spec.* **2009**, *23*, 2519–2526.

501. Dipplod, M. A.; Boesel, S.; Gunina, A.; Kuzyakov, Y.; Glaser, B. *Rapid Comm. Mass Spec.* **2014**, *28*, 569–576.

502. Baas, M.; Briggs, D. E. G.; Van Heemst, J. D. H.; Kear, A. J.; De Leeuw, J. W. *Geochim. Cosmochim. Acta* **1995**, *59*, 945–951.

503. Stankiewicz, B. A.; Hutchins, J. C.; Thomson, R.; Briggs, D. E. G.; Evershed, R. P. *Science* **1997**, *276*, 1541–1543.

504. Flannery, M. B.; Stott, A. W.; Briggs, D. E. G.; Evershed, R. P. *Org. Geochem.* **2001**, *32*, 745–754.

505. Wysokowski, M.; Zatoń, M.; Bazhenov, V. V.; Behm, T.; Ehrlich, A.; Stelling, A.; Ehrlich, A. L.; Stelling, M.; Hog, L.; Ehrlich, H. *Paleobiology* **2014**, *40*(4), 529–540.

506. Rillig, M. C.; Caldwell, B. A.; Wösten, H. A. B.; Sollins, P. *Biogeochemistry* **2007**, *85*, 25–44.

507. Knicker. *Soil Biol. Biochem.* **2011**, *43*, 1118–1129.

508. Miltner, A.; Kindler, R.; Knicker, H.; Richnow, H.-H.; Kästner, M. *Org. Geochem.* **2009**, *40*, 978–985.

509. Nannipieri, P.; Eldor, P. *Soil Biol. Biochem.* **2009**, *41*, 2357–2369.

510. Amelung, W.; Zhang, X.; Flach, K. W. *Geoderma,* **2006**, *130*, 207–217.

511. Knicker, H.; Hatcher, P. G. *Naturwissenschaften* **1997**, *84*, 231–234.

512. Zang, X.; Van Heemst, J. D. H.; Dria, K. J.; Hatcher, P. *Org. Geochem.* **2000**, *31*, 679–695.

513. Riboulleau, A.; Mongenot, T.; Baudin, F.; Derenne, S.; Largeau, C. *Org. Geochem.* **2002**, *33*, 1127–1130.

514. Zang, X.; Hatcher, P. *Org. Geochem.* **2002**, *33*, 201–211.

515. Clemente, J. S.; Simpson, A. J.; Simpson, M. J. *Org. Geochem.* **2011**, *42*, 1169–1180.

516. Tomaszewski, J.; Schwarzenbach, R.; Sander, M. *Environ. Sci. Technol.* **2011**, *45*(14), 6003–6010.

517. Hsu; Hatcher. *Geochim. Cosmochim. Acta* **2005**, *69*, 4521–4533.

518. Smernik, R. J.; Baldock, J. A. *Biogeochemistry* **2005**, *75*, 507–528.

519. Vairavamurthy; Wang. *Environ. Sci. Technol.* **2002**, *36*, 3050–3056.

520. Knicker, H.; González-Vila, F. J.; Polvillo, O.; González, J. A.; Almendros, G. *Soil Biol. Biochem.* **2005**, *37*, 701–718.

521. Knicker, H.; Hilscher, A.; Gonzàlez-Vila, F.; Almendros, G. *Org. Geochem.* **2008**, *39*, 935–939.

522. Keil, R. G.; Tsamakis, E.; Fuh, C. B.; Giddings, J. C.; Hedges, J. I. *Geochim. Cosmochim. Acta* **1994**, *58*(2), 879–893.

523. Mayer, L. M. *Chem. Geol.* **1994**, *114*(3–4), 347–363.

524. Mayer, L. M. *Geochim. Cosmochim. Acta* **1994**, *58*, 1271–1284.

525. Kleber, M.; Eusterhues, K.; Keiluweit, M.; Mikutta, C.; Mikutta, R.; Nico, P. S. *Adv. Agro.* **2015**, *130*, 1–140.

526. Sollins, P.; Homann, P.; Caldwell, B. A. *Geoderma* **1996**, *74*, 65–105.

527. Salmon, V.; Derenne, S.; Lallier-Verges, E.; Largeau, C.; Beaudoin, B. *Org. Geochem.* **2000**, *31*(5), 463–474.

528. Kaiser, K.; Guggenberger, G. *Eur. J. Soil Sci.* **2003**, *54*, 219–236.

529. Baldock, J. A.; Skjemstad, J. *Org. Geochem.* **2000**, *31*, 697–710.

540. Kögel-Knabner, I.; Guggenberger, G.; Kleber, M.; Kandeler, E.; Kalbitz, K.; Scheu, S.; Eusterhues, K.; Leinweber, P. *J. Plant Nutr. Soil Sci.* **2008**, *171*, 61–82.

541. Ladd, J.; Amato, M.; Oades, J. M. *Soil Res.* **1985**.

542. Skjemstad, J.; Janik, L. J.; Head, M. J.; Mcclure, S. G. *J. Soil Sci.* **1993**, *44*, 485–499.

543. Puget, P.; Chenu, C.; Balesdent, J. *Eur. J. Soil Sci.* **1995**, *46*, 449–459.

544. Monreal, C. M.; Schulten, H.-R.; Kodama, H. *Can. J. Soil Sci.* **1997**, *77*, 379–388.

545. Moni, C.; Chenu, C.; Rumpel, C.; Virto, I.; Chabbi, A. *Eur. J. Soil Sci.* **2010**, *61*, 958–969.

546. Schulten, H.-R.; Leinweber, P. *Eur. J. Soil Sci.* **1999**, *50*, 237–248.

547. Feng, W.; Plante, A. F.; Aufdenkampe, A. K.; Six, J. *Soil Biol. Biochem.* **2014**, *69*, 398–405.

547. Feng, W.; Klaminder, J.; Boily, J.-F. *J. Phys. Chem. A* **2015**, *119*, 12790–12796.

548. Hassink, J.; Whitmore, A. P.; Kubát, J. *Eur. J. Agro.* **1997**, *7*, 189–199.

548. Swanston, C.; Torn, M. S.; Hanson, P. J.; Southon, J. R.; Garten, C. T.; Hanlon, E. M.; Ganio, L. *Geoderma* **2005**, *128*, 52–62.

549. Keil, R. G.; Mayer, L. M. *Mineral Matrices and Organic Matter, Treatise On Geochemistry*, 2nd. ed., Elsevier, **2014**, pp. 338–359.

549. Wiesmeier, M.; Hübner, R.; Spörlein, P.; Geuß, U.; Hangen, E.; Reischl, A.; Kögel-Knabner, I. *Global Change Biol.* **2014**, *20*(2), 653–665.

550. Kalbitz, K.; Kaiser, K. *J. Plant Nutr. Soil Sci.* **2008**, *171*(1), 52–60.

550. Plante, A. F.; Pernes, M.; Chenu, C. *Geoderma* **2005**, *129*, 186–199.

551. Kaiser, K. *Org. Geochem.* **2003**, *34*(11), 1569–1579.

552. Ransom, B.; Kim, D.; Kastner, M.; Wainwright, S. *Geochim. Cosmochim. Acta* **1998**.

553. Shang, C.; Tiessen, H. *Soil Sci.* **1997**, *162*, 795–807.

554. Igarashi, S.; Nomura, N.; Mishima, F.; Akiyama, Y.; Nishijima, S. *Physica C* **2014**, *504*, 144–147.

555. Kaiser, K.; Guggenberger, G. *Org. Geochem.* **2000**, *31*, 711–725.

556. Kiem, R.; Kögel-Knabner, I. *Org. Geochem.* **2002**, *33*(12), 1699–1713.

557. Barré, P.; Fernandez-Ugalde, O.; Virto, I.; Velde, B.; Chenu, C. *Geoderma*, **2014**, *235*, 382–395.

558. Six, J.; Conant, R. T.; Paul, E. A.; Paustian, K. *Plant Soil* **2002**, *241*, 155–176.

559. Vogel, C.; Babin, D.; Pronk, G. J.; Heister, K.; Smalla, K.; Kögel-Knabner, I. *Soil Biol. Biochem.* **2014**, *79*, 57–67.

560. Feng, X.; Simpson, A. J.; Simpson, M. J. *Org. Geochem.* **2005**, *36*(11), 1553–1566.

561. Kalbitz, K.; Schwesig, D.; Rethemeyer, J.; Matzner, E. *Soil Biol. Biochem.* **2005**, *37*, 1319–1331.

562. Galindo, C.; Del Nero, M. *Environ. Sci. Technol.* **2014**, *48*(13), 7401–7408.

563. Galindo, C.; Del Nero, M. *Rsc. Adv.* **2015**, *5*(89), 73058–73067.

564. Lv, J.; Zhang, S.; Wang, S.; Luo, L.; Cao, D.; Christie, P. *Environ. Sci. Technol.* **2016**, *5*(5), 2328–2336.

565. Basile-Doelsch, I.; Amundson, R.; Stone, W. E. E.; Masiello, C. A.; Bottero, J. Y.; Colin, F.; Masin' F.; Borschneck, D.; Meunier, D. *Eur. J. Soil Sci.* **2005**.

566. Wattel-Koekkoek, E. J. W.; Van Genuchten, P. P. L.; Buurman, P.; Van Lagen, B. *Geoderma* **2001**, *99*, 27–49.

567. Kaiser, K.; Zech, W. *Soil Sci. Soc. Am. J.* **1997**, *61*(1), 64–69.

568. Kleber, M.; Sollins, P.; Sutton, R. *Biogeochemistry* **2007**, *85*, 9–24.

569. Davis, J. A.; Gloor, R. *Environ. Sci. Technol.* **1981**, *15*(10), 1223–1229.

570. Day, G. M.; Hart, B. T.; Mckelvie, I. D.; Beckett, R. *Colloids Surf. A* **1994**, *89*(1), 1–13.

571. Kaiser, K.; Guggenberger, G.; Haumaier, L.; Zech, W. *Eur. J. Soil Sci.* **1997**, *48*(2), 301–310.

572. Gu, B.; Schmitt, J.; Chem, Z.; Liang, L.; Mccarthy, J. F. *Environ. Sci. Technol.* **1994**, *28*, 38– 46.

573. Hur, J.; Schlautman, M. A. J. *Coll. Inter. Sci.* **2003**, *264*(2), 313–321.

574. Hur, J.; Schlautman, M. A. J. *Coll. Inter. Sci.* **2004**, *277*(2), 264–270.

575. Reiller, P.; Amekraz, B.; Moulin, C. *Environ. Sci. Technol.* **2006**, *40*(7), 2235– 2241.

576. Claret, F.; Scha`Fer, T.; Brevet, J.; Reiller, P. E. *Environ. Sci. Technol.* **2008**, *42*(23), 8809–8815.

577. Saito, T.; Koopal, L. K.; Vanriemsdijk, W. H.; Nagasaki, S.; Tanaka, S. *Langmuir* **2004**, *20*(3), 689– 700.

578. Janot, N.; Reiller, P. E.; Korshin, G. V.; Benedetti, M. F. *Environ. Sci. Technol.* **2010**, *44*(17), 6782–6788.

579. Janot, N.; Reiller, P. E.; Zheng, X.; Croué, J.-P.; Benedetti, M. F. *Water Res.* **2012**, *46*(3), 731–740.

580. Dümig, A.; Häusler, W.; Steffens, M.; Kögel-Knabner, I. *Geochim. Cosmochim. Acta* **2012**, *85*, 1–18.

581. Yu, W. H.; Li, N.; Tong, D. S.; Zhou, C. H.; Lin, C. X. C.; Xu, C. Y. *Appl. Clay Sci.* **2013**, *80*, 443–452.

582. Bingham, A. H.; Cotrufo, M. F. *Stoten* **2016**, *551*, 116–126.

583. Yeasmin, S.; Singh, B.; Kookana, R. S.; Farrell, M.; Sparks, D.; Johnston, C. J. *Coll. Inter. Sci.* **2014**, *432*(15), 246–257.

584. Keiluweit *et al. Geochim. Cosmochim. Acta* **2012**, *95*, 213–226.

585. Totsche, K. U.; Rennert, T.; Gerzabek, M. H.; Kögel-Knabner, I.; Samalla, K.; Spiteller, M.; Vogel, H.-J. *J. Plant Nutr. Soil Sci.* **2010**, *173*, 88–99.

586. Mi, B.; Elimelech, M. *J. Memb. Sci.* **2010**, *348*(1), 337–345.

587. Yamamura, H.; Kimura, K.; Okajima, T.; Tokumoto, H.; Watanabe, Y. *Environ. Sci. Technol.* **2008**, *42*(14), 5310–5315.

588. Boks, N. P.; Norde, W.; Van Der Mei, H. C.; Busscher, H. J. *Microbiology* **2008**, *154*, 3122–3133.

589. Aubry, C.; Gutierre, L.; Croue, J.-P. *Water Res.* **2013**.

590. Chassé, A. W.; Ohno, T.; Higgins, S. R.; Ariaa, A.; Yildirim, N.; Parr, T. B. *Environ. Sci. Technol.* **2015**, *49*(16), 9733–9741.

591. Chi, F. H.; Amy, G. L. *J. Coll. Inter. Sci.* **2004**, *274*, 380–391.

592. Kalinichev, A. G.; Kirkpatrick, R. J. *Eur. J. Soil Sci.* **2007**, *58*, 909–917.

593. Sposito, G. *Soil Sci. Soc. Am. J.* **1984**, *48*, 531–536.

594. Nierop, K. G.; Jansen, B.; Verstraten, J. M. *Sci. Total Environ.* **2002**, *300*(1), 201–211.

595. Hughey, C. A.; Hendrickson, C. L.; Rodgers, R. P.; Marshall, A. G.; Qian, K. *Anal. Chem.* **2001**, *73*, 4676–4681.

596. Schumacher, M.; Christl, I.; Scheinost, A. C.; Jacobsen, C.; Kretzschmar, R. *Environ, Sci. Technol.* **2005**, *39*(23), 9094–9100.

597. Lehmann, J.; Kinyangi, J.; Solomon, D. *Biogeochemistry* **2007**, *85*(1), 45–57.

598. Lehmann, J.; Solomon, D.; Kinyang, J.; Dathe, L.; Wirick, S.; Jacobsen, S. *Nature Geosci.* **2008**, *1*, 238–242.

599. Wan, J. M.; Tyliszczak, T.; Tokunaga, T. K. *Geochim. Cosmochim. Acta* **2007**, *71*, 5439–5449.

600. Vogel, C.; Mueller, C. W.; Höschen, C.; Buegger, F.; Heister, K.; Schulz, S.; Schloter, M.; Kögel-Knabner, I. *Nature Comm.* **2014**, *5*, 2947.

601. Furukawa, Y. *Org. Geochem,* **2000**, *31*, 735–744.

602. Arnarson, Keil. *Org. Geochem,* **2001**, *32*, 1401–1415.

603. Herrmann, A. M.; Ritz, K.; Nunan, N.; Clode, P. L.; Pett-Ridge, J.; Kilburn, M. R.; Murphy, D. V.; O'Donnell, A. G.; Stockdale, E. A. *Soil Biol. Biochem.* **2007**, *39*, 1835–1850.

604. Mueller, K. K.; Lofts, S.; Fortin, C.; Campbell, P. G. *Environ. Chem.* **2012**, *9*(4), 356–368.

605. Remusat, L.; Hatton, P.-J.; Nico, P. S.; Zeller, B.; Kleber, M.; Derrien, D. *Environ. Sci. Technol.* **2012**, *46*, 3943–3949.

606. Kaiser, K.; Mikutta, R.; Guggenberger, G. *Soil Sci. Soc. Am. J.* **2007**, *71*(3), 711–719.

607. Guggenberger, G.; Kaiser, K. *Geoderma* **2003**, *13*, 293–310.

608. Stewart, C. E.; Paustian, K.; Conant, R. T.; Plante, A. F.; Six, J. *Biogeochemistry* **2007**, *86*, 19–31.

609. Kool, D. M.; Chung, H.; Tate, K. R.; Ross; Des, J.; Newton, P. C. D.; Six, J. *Global Change Biol.* **2007**, *13*, 1282–1293.

610. Beare, M. H.; Cabrera, M. L.; Hendrix, P. F.; Coleman, D. C. *Soil Sci. Soc. Am. J.* **1994**, *58*(3), 787–795.

611. Chenu, C.; Stotzky, G. In *Interactions between Soil Particles and Microrganisms*. Huang, P.; Bollag, J.-M.; Senesi, N. (eds.). John Wiley, New York, **2002**, pp. 3–40.

612. Sexstone, A. J.; Revsbech, N. P.; Parkin, T. B.; Tiedje, J. M. *Soil Sci. Soc. Am. J.* **1985**, *49*(3), 645–651.

613. Rabbi, S. M. F.; Daniel, H.; Lockwood, P. V. *et al. Scientific Rep.* **2016**, *6*.

614. Remigio, P.; Van Oort, F.; Barré, P.; Billiou, D.; Chenu, C. *Geoderma* **2016**, *275*, 48–54.

615. Oades, J. M. *Geoderma* **1993**, *56*, 377–400.

616. Balesdent, J.; Chenu, C.; Balabane, M. *Soil Till. Res.* **2000**, *53*(3), 215–230.

617. Denef, K.; Six, J.; Paustian, K.; Merckx, R. *Soil Biol. Biochem.* **2001**, *33*, 2145–2153.

618. Plante, A. F.; Mcgill, W. B. *Soil Till. Res.* **2002**, *66*(1), 79–92.

619. Jastrow, J. D.; Miller, R. M.; Boutton, T. W. *Soil Sci. Soc. Am. J.* **1995**, *60*, 801–807.

620. Denef, K.; Zotarelli, L.; Boddey, R.; Six, J. *Soil Biol. Biochem.* **2007**, *39*, 1165–1172.

621. Arrouays, D.; Pelissier, P. *Soil Sci.* **1994**, *157*, 185–191.

622. Besnard, E.; Chenu, C.; Balesdent, J.; Puget, P.; Arrouays, D. *Eur. J. Soil Sci.* **1996**, *47*, 495–503.

623. Virto, I.; Barré, P.; Chenu, C. *Geoderma* **2008**, *146*, 326–335.

624. Feller, C.; Chenu, C. *Etude Et Gestion Des. Sols.* **2012**, *19*, 235–248.

625. Elliott, E. T.; Coleman, D. C. *Ecol. Bull.* **1988**, *39*, 23–32.

626. Strong, D. T.; Wever, H. D.; Merckx, R.; Recous, S. *Eur. J. Soil Sci.* **2004**, *55*(4), 739–750.

627. Kinyangi, J.; Solomon, D.; Liang, B.; Lerotic, M.; Wirick, S.; Lehmann, J. *Soil Sci. Soc. Am. J.* **2006**, *70*, 1708–1718.

628. Pesavento, M.; Alberti, G.; Biesuz, R. *Anal. Chim. Acta* **2009**, *631*, 129–141.

629. Oelkers, E. H.; Schott, J. *Rev. Mineral. Geochem.* **2009**, *70*.

630. Golubev, S. V.; Bauer, A.; Pokrovsky, O. S. *Geochim. Cosmochim. Acta* **2006**, *70*, 4436–4451.

631. Ganor, J.; Reznik, I. J.; Rosenberg, Y. O. *Rev. Mineral.* **2009**, *70*, 250–369.

632. Golubev, S. V.; Pokrovsky, O. S. *Experimental Chem. Geol.* **2006**, *235*, 377–389.

633. Ganor, J.; Cama, J.; Metz, V. *J. Coll. Interf. Sci.* **2003**, *264*(1), 67–75.

634. Bowser, C. J.; Jones, B. F. *Am. J. Sci.* **2002**, *302*, 582–662.

635. Taylor, L. L.; Leake, J. R.; Quirk, J.; Hardy, K.; Banwart, S. A.; Beerling, D. J. *Geobiology* **2009**, *7*(2), 171–191.

636. Bray, A. W.; Oelkers, E. H.; Bonneville, S.; Wolff-Boenisch, D.; Potts, N. J.; Fones, G.; Benning, L. G. *Geochim. Cosmochim. Acta* **2015**, *164*, 127–145.

637. Batley, G. E.; Apte, S. C; Stauber, J. L. *Aust. J. Chem.* **2004**, *57*, 903–919.

638. Benedetti, M. F.; Milne, C. J.; Kinniburgh, D. J.; Riemsdijk, W. H. V.; Koopal, L. K. *Environ. Sci. Technol.* **1995**, *29*, 446–457.

639. Pretsch, E. *Trends Anal. Chem.* **2007**, *26*, 47–51.

640. Ren, Z.-L.; Tella, M.; Bravin, M. N.; Comans, R. N. J.; Dai, J.; Garnier, J.-M. *et al.* *Chem. Geol.* **2015**, *398*, 61–69.

641. Weng, L.; Riemsdijk, W. H. V.; Temminghoff, E. J. M. *Anal. Chem.* **2005**, *77*, 2852–2861.

642. Zhang, H.; Davison, W. *Anal. Chem.* **1995**, *72*, 3391–3400.

643. Zhang, H, Davison, W. *Anal. Chem.* **2000**, *72*, 4447–4457.

644. Galceran, J; Puy, J.; Salvador, J.; Cecılia, J.; Leeuwen, H. P. V. *J. Electroanal. Chem.* **2001**, *505*, 85–94.

645. Galceran, J.; Companys, E.; Puy, J.; Cecilia, J.; Garces, J. L. *J. Electroanal. Chem.* **2004**, *566*, 95–109.

646. Town, R. M.; Leeuwen, H. P. V. *J. Electroanal. Chem.* **2003**, *541*, 51–65.

647. Town, R. M.; Leeuwen, H. P. V. *J. Electroanal. Chem.* **2002**, 535, 11–25.

648. Town, R. M.; Leeuwen, H. P. V. *Aust. J. Chem.* **2004**, *57*, 983–992.

649. Croot, P. L.; Johansson, M. *Electroanalysis* **2000**, *12*(8), 565–576.

650. Laglera, L.; Filella, M. *Mar. Chem.* **2015**, *173*, 100–113.

651. Pei, J.; Tercier-Waeber, M. L.; Buffle, J.; Fiaccabrino, G. C.; Koudelka-Hep, M. *Anal. Chem.* **2001**, *73*(10), 2273–2281.

652. Belmont-Hébert; Tercier, M.-L.; Buffle, J.; Fiaccabrino, G. C.; M. Koudelka-Hep, M. *Anal. Chem.* **1998**, *70*, 2949–2956.

653. Lofts, S.; Tipping, E. *Environ. Chem.* **2011**, *8*(5), 501–516.

654. Koopal, L.; VanRiemsdijk, W.; Dewit, J.; Benedetti, M. *J. Coll. Interf. Sci.* **1994**, *166*, 51–60.

655. Sips, R. *J. Chem. Phys.* **1948**, *16*(5), 490–495.

656. Domingos, R. F.; Franco, C.; Pinheiro, J. P. *Environ. Sci. Pollut. Res. Int.* **2015**, *22*(4), 2900–2906.

657. Mota, A. M.; Pinheiro, J. P.; Goncalves, M. L. S. *J. Phys. Chem. A* **2012**, *116*, 6433–6442.

658. Van Leeuwen, H. P.; Town, R. M.; Buffle, J.; Cleven, R. F.; Davison, W.; Puy, J.; Van Riemsdijk, W. H.; Sigg, L. *Environ. Sci. Technol.* **2005**, *39*(22), 8545–8556.

659. Hamilton-Taylor, J.; Ahmed, I. A. M.; Davison, W.; Zhang, H. *Environ. Chem.* **2011**, *8*, 461–465.

660. Dudal, Y.; Gérard, F. *Earth Sci. Rev.* **2004**, *66*, 199–216.

661. Filella, M. *J. Mol. Liquids* **2008**, *143*, 42–51.

662. Kinniburgh, D. G.; Riemsdijk, W. H. V; Koopal, L. K.; Borkovec, M.; Benedetti, M. F; Avena, M. J. *Coll. Surf. A Physicochem. Eng. Asp.* **1999**, *151*, 147–166.

663. Tipping, E.; Lofts, S.; Sonke, J. E. *Environ. Chem.* **2011**, *8*, 225–235.

664. Keizer, M. G.; Riemsdijk, W. H. V. Manual Program. Agricultural University of Wageningen, Wageningen, **1994**.

665. Zhang, H. *Environ. Sci. Technol.* **2004**, *38*, 1421–1427.

666. Gustafsson, J. P. *J. Coll. Interf. Sci.* **2001**, *244*, 102–112.

667. Xiong, J.; Koopal, L. K.; Tan, W.; Fang, L.; Wang, M.; Zhao, W.; Fan, L.; Jing, Z.; Weng, L. *Environ. Sci. Technol.* **2013**, *47*(20), 11634–11642.

668. Bryan, S. E.; Tipping, E.; Hamilton-Taylor, J. *Toxicol. Pharmacol.* **2002**, *133*, 37–49.

669. Ge, Y.; Macdonald, D.; Sauvé, S.; Hendershot, W. *Environ. Model Soft.* **2005**, *20*, 353–359.

670. Meylan, S.; Odzak, N.; Behra, R.; Sigg, L. *Anal. Chim. Acta* **2004**, *510*, 91–100.

671. Groenenberg, J. E.; Koopmans, G. F.; Comans, R. N. J. *Environ. Sci. Technol.* **44**, 1340–1346.

672. Unsworth, E. R.; Warnken, K. W.; Zhang, H.; Davison, W.; Black, F.; Buffle, J.; Cao, J.; Cleven, R.; Galceran, J.; Gunkel, P.; Kalis, E.; Kistler, D.; Van Leeuwen, H. P.; Martin, M.; Noël, S.; Nur, Y.; Odzak, N.; Puy, J.; Van Riemsdijk, W.; Sigg, L.; Temminghoff, E.; Tercier-Waeber, M.-L.; Toepperwien, S.; Town, R. M.; Weng, L.; Xue, H. *Environ. Sci. Technol.* **2006**, *40*, 1942–1949.

673. Han, S.; Naito, W.; Hanai, Y.; Masunaga, S. *Water Res.* **2013**, *47*(14), 4880–4892.

674. Vega, F. A.; Weng, L. *Water Res.* **2013**, *47*(1), 363–372.

675. Powell, P. E.; Cline, G. R.; Reid, C. P. P.; Szaniszlo, P. J. *Nature* **1980**, *287*, 833–834.

676. Mawji, E.; Gledhill, M.; Milton, J. A.; Tarran, G. A.; Ussher, S.; Thompson, A.; Wolff, G. A.; Worsfold, P. J.; Achterberg, E. P. *Environ. Sci. Technol.* **2008**, *42*(23), 8675–8680.

677. Schalk, I. J.; Hannauer, M.; Braud, A. *Environ. Microbiol.* **2011**, *13*, 2844–2854.

678. Saha, R.; Saha, N.; Donofrio, R. S.; Bestervelt, L. L. *J. Basic Microbiol.* **2013**, *53*, 303–317.

679. Buss, H. L; Lüttge, A.; Brantley, S. L. *Chem. Geol.* **2007**, *240*, 326–342.

680. Jorand, F.; Boué-Bigne, F.; Block, J. C.; Urbain, V. *Water Sci. Technol.* **1998**, *37*, 307–315.

681. Lamelas, C.; Benedetti, M.; Wilkinson, K. J.; Slaveykova, V. I. *Chemosphere* **2006**, *65*, 1362–1370.

682. Domingos, R. F.; Benedetti, M. F.; Pinheiro, J. P. *Anal. Chim. Acta* **2007**, *589*, 261–268.

683. Gelabert, A.; Sivry, Y.; Ferrari, R.; Akrout, A.; Cordier, L.; Nowak, S.; Benedetti, M. F. *Environ. Toxicol. Chem.* **2014**, *33*(2), 341–349.

684. Sivry, Y.; Gelabert, A.; Cordier, L.; Ferrari, R.; Lazar, H.; Juillot, F.; Menguy, M.; Benedetti, M. F. *Chemosphere* **2014**, *95*, 519–526.

685. Zelano, I.; Sivry, Y.; Quantin, C.; Gélabert, A.; Tharaud, M.; Jouvin, D. E.; Montarges-Pelletier, J.; Garnier, R.; Pichon, S.; Nowak, S.; Miska, O.; Abollino; Miska, S. *Coll. Surf A Physicochem. Engin. Asp.* **2013**, *435*, 36–47.

686. Van Leeuwen, H. P. *Electroanalysis* **2001**, *13*, 826–830.

687. Xue, B.-B.; Sigg, L. In *Analyses of Trace Element Biogeochemistry*. Tailefert, M.; Rozan, T. F. (eds.). American Chemical Society, Washington DC, 2002, p. 336.

688. Tercier-Waeber, M.-L. *et al. Electroanalysis* **2000**, *12*, 27–34.

689. Kalis, E. J. J.; Weng, L.; Temminghoff, E. J. M.; Riemsdijk, W. H. V. *Anal. Chem.* **2007**, *79*, 1555–1563.

690. Marang, L.; Reiller, P.; Pepe, M.; Benedetti, M. F. *Environ. Sci. Technol* **2006**, *40*, 5496–5501.

691. Zelano, I.; Sivry, Y.; Quantin, C.; Gélabert, A.; Tharaud, M.; Nowak, S.; Benedetti, M. F. *Appl. Geochem.* **2016**, *64*, 146–156.

692. Domingos, R. F.; Lopez, R.; Pinheiro, J. P. T. *Environ. Chem.* **2008**, *5*, 24–32.

693. Sigg, L. *et al. Environ. Sci. Technol.* **2006**, *40*, 1934–1941.

694. Buffle, J.; Zhang, Z.; Startchev, K. *Environ. Sci. Technol.* **2007**, *41*, 7609–7620.

695. Alemani, D.; Buffle, J.; Zhang, Z.; Galceran, J.; Chopard, B. *Environ. Sci. Technol.* **2008**, 2021–2027.

696. Alemani, D.; Buffle, J.; Zhang, Z.; Galceran, J.; Chopard, B. *Environ. Sci. Technol.* **2008**, *42*, 2028–2033.

697. Zhang, Z; Buffle, J.; Alemani, D. *Environ. Sci. Technol.* **2007**, *41*, 7621–7631.

698. Zhang, Z; Buffle, J.; Startchev, K.; Alemani, D. *Environ. Chem.* **2008**, *5*, 204–217.

699. Zhang, Z; Buffle, J. *Geochim. Cosmochim. Acta* **2009**, *73*, 1236–1243.

700. Duval, J. F. L. *Phys. Chem. Chem. Phys.* **2013**, *15*, 7873–7888.

701. Duval, J. F. L. *J. Phys. Chem. A* **2009**, *113*, 2275–2293.

Index